MONOGRAPHS OF THE PHYSIOLOGICAL SOCIETY

*Editors: H. Davson, A. D. M. Greenfield,
R. Whittam, G. S. Brindley*

Number 24 WATER TRANSPORT IN CELLS AND
TISSUES

1. Sympathetic Control of Human Blood Vessels*
   by H. Barcroft and H. J. C. Swann
2. Man in a Cold Environment*
   by A. C. Burton and O. G. Edholm
3. Neural Control of the Pituitary Gland*
   by G. W. Harris
4. Physiology of Gastric Digestion*
   by A. H. James
5. The Mammalian Cerebral Cortex*
   by B. Delisle Burns
6. Physiology of the Retina and Visual Pathway, 2nd Edition
   by G. S. Brindley
7. Blood Flow in Arteries*
   by D. A. McDonald
8. Physiology of the Salivary Glands
   by A. S. V. Burgen and N. G. Emmelin
9. Biological Effects of Freezing and Supercooling
   by Audrey U. Smith
10. Renal Function
    by W. J. O'Connor
11. Secretory Mechanisms of the Gastro-Intestinal Tract*
    by R. A. Gregory
12. Substances Producing Pain and Itch
    by C. A. Keele and Desiree Armstrong
13. Transport and Diffusion in Red Blood Cells
    by R. Whittam
14. Physiology of the Splanchnic Circulation
    by J. Grayson and D. Mendel
15. Physiology of Puberty
    by B. T. Donovan and J. J. van der Werff ten Bosch
16. Pulmonary and Bronchial Vascular Systems
    by I. de Burgh Daly and Catherine Hebb
17. The Auditory Pathway
    by I. C. Whitfield
18. The Climatic Physiology of the Pig
    by L. E. Mount
19. Electrophysiological Analysis of Synaptic Transmission
    by J. I. Hubbard, R. Llinás, and D. Quastel
20. Mechanisms of Urine Concentration and Dilution in Mammals
    by S. E. Dicker
21. Biogenesis and Physiology of Histamine
    by G. Kahlson and Elsa Rosengren
22. The Physiology of Lactation
    by A. T. Cowie and J. S. Tindal
23. Mammalian Muscle Receptors and Their Central Actions
    by Peter B. C. Matthews

Volumes marked * are now out of print.

WATER TRANSPORT IN CELLS AND TISSUES

C. R. HOUSE
B.Sc., M.Sc., Ph.D.
Lecturer in Veterinary Physiology
University of Edinburgh

LONDON
EDWARD ARNOLD (PUBLISHERS) LTD.

© C. R. House, 1974

First published 1974
by Edward Arnold (Publishers) Ltd.
25 Hill Street, London W1X 8LL

ISBN 0 7131 4215 4

All Rights Reserved. No part of this publication may be reproduced, stored in a retrieval system, or transmitted in any form or by any means, electronic, mechanical, photo-copying, recording or otherwise, without the prior permission of Edward Arnold (Publishers) Ltd.

Printed in Great Britain by
The Camelot Press Ltd, London and Southampton

PREFACE

The flux equations of irreversible thermodynamics that govern the movements of water and uncharged solutes across membranes require the determination of three membrane coefficients—the hydraulic conductivity, L_p, the solute permeability, ω_s, and the reflexion coefficient, σ_s. Most of the experimental work discussed in this monograph is concerned with the significance of L_p, ω_s and σ_s in certain artificial and biological membranes. To this end I decided, perhaps wrongly, to give some theoretical background to the use of this phenomenological approach. That decision posed unforeseen difficulties, because in writing the opening chapters I became increasingly aware of the danger of producing a theoretical treatment not only unintelligible to the novice but also useless to the specialist. In order to resolve the dilemma I have taken a dual course and so Chapter 2 contains on the one hand, the gist of the theory with its experimental significance (see *Practical preamble*) and, on the other, a more prolonged treatment of the phenomenological equations for those who wish to learn about their development.

In preparing this monograph I have incurred inevitably a huge debt of gratitude.

By far the largest is that owed to my wife, Olive, who not only typed the drafts and final copy so meticulously but also gave me the necessary encouragement and peace to finish the task.

Its a great pleasure to thank Drs J. M. Diamond, B. L. Ginsborg, P. H. Tuft and G. A. P. Wyllie for their comments on parts of this book and especially Professor J. Dainty for his valuable advice on the whole manuscript.

I am also grateful to many colleagues who were consulted and gave me unpublished information generally cited in the text as personal communications.

The figures were prepared by Mr Colin Warwick of the Department of Veterinary Physiology, University of Edinburgh. His assistance and the way he gave it were greatly appreciated.

Finally, for their permission to reproduce figures I am indebted to numerous authors and also to certain journals and publishers. The source of these illustrations is given in the legends and the bibliography.

C. R. H.

Edinburgh, 1973

CONTENTS

Chapter		Page
1	The anatomy and physiology of water	1
2	Phenomenological description of transport processes	36
3	Theoretical aspects of transport in porous membranes	77
4	Some experiments on artificial membranes	103
5	Water permeabilities of animal and plant cells	152
6	Water relations of cells	192
7	Fluid dynamics in the embryo	263
8	Transport across the capillary wall	287
9	Permeability characteristics of epithelia	318
10	Active salt and water transport	390
References		471
Author Index		527
Subject Index		541

1
THE ANATOMY AND PHYSIOLOGY OF WATER

Introduction	1
The water molecule	4
Structure of ice	6
Ordinary ice	6
Other forms of ice	7
Structure of water	8
Nets and rings	10
Cages and guests	10
'Flickering clusters'	11
Conclusion	14
'Anomalous' water	14
Transport processes in ice and water	15
Electrical properties	16
Self-diffusion	19
Final comment	21
Interaction of substances with water	22
Electrolytes	22
Anaesthetics	24
Hydrocarbons	25
Macromolecules	26
The state of water in the cell	28
Nuclear magnetic resonance studies	30
'Non-solvent' behaviour of cellular water	32
X-ray diffraction studies	34
Conclusion	35

Introduction

WATER is an intriguing substance. It is the most common example of the liquid state and yet the most anomalous liquid. Our familiarity with water almost obstructs our recognition of its peculiar properties. For example, when we compare the melting and boiling points of hydride molecules, such as water and ammonia, we find that water has the highest temperature limits for its liquid state (Fig. 1.1). Even its neighbours, NH₃ and HF, are similarly but less markedly anomalous when they are compared to hydrides, such as HCl and H₂S, derived from other parts of the periodic table. The fact that water is not gaseous at room temperature is a sign of its relatively strong intermolecular forces. Not

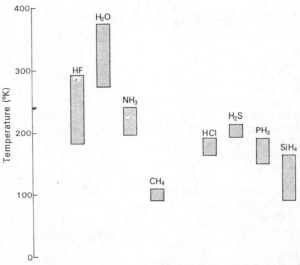

FIG. 1.1. Temperature ranges for the liquid state of various hydrides.

only does water stand out as an unusual hydride but also it emerges as an unusual liquid in comparison with other liquids. To carry out such a comparison effectively one needs to define a normal liquid and to characterize its behaviour, and as a prelude to that one requires a satisfactory theory of the liquid state.

For some time now we have had good theories for both gases and solids; by comparison, the theory of the liquid state is poorly developed. The condensed state of liquids lacks regular structure and, although the packing of atoms in liquids is as tight as that in solids, liquids have no shear strength because of the absence of structural regularity. It is ironic that the first important piece of work on the liquid state was done on water (Bernal & Fowler, 1933). Bernal (1964, 1965) has reviewed his contributions to the structure of ideal liquids and, in particular, he has discussed a model of the liquid state which he constructed with the aid of a computer. The condition that the distance between any pair of atoms must be greater than an arbitrary minimum distance was the only specification which Bernal enforced on the random choice of atomic positions. The theoretical solution of this problem gave a composite picture of interatomic distance which agreed with measurements made on real liquids with X-ray techniques. Moreover, Bernal's model of the liquid state exhibits a striking feature, namely the presence of rows of atoms. Indeed, the random structure of this model of the liquid state has paradoxically a sort of poor one-dimensional regularity since there are atomic rows which are approximately equidistant. Alignment of the atoms in the rows is, of course, not perfect, and the rows are oriented in all directions. Because of the existence of the channels surrounding each row, the atomic rows would tend to move along their axes in preference to moving obliquely to their axes. Thus, the application of a shear stress to liquids will cause flow because the movement of atoms is then such as to relieve the stress.

One can appreciate the anomalous nature of water by comparing the packing of its molecules with that of typical or normal liquids; probably the best examples of the normal liquid are the noble gases, notably argon (Rowlinson, 1959). In the typical liquid the molecules can be considered naïvely to be hard spheres interacting on contact; for this case, molecular packing will be highly stable when the number of contacts is maximal and this occurs when each sphere has 12 immediate neighbours. This array is the face-centred cubic lattice (Barlow, 1883). Now, the density of a liquid with its molecules packed in this economical manner can be computed after the intermolecular distance has been obtained from X-ray scattering measurements. The distance between water molecules is 2·9 Å and, therefore, the density of liquid water should be $1·7$ g cm^{-3} if the molecules are in closest packing. The conclusion must be that liquid water is a relatively open structure and, in fact, the X-ray data of Morgan & Warren (1938) reveal that each water molecule has only about four or five neighbours.

The inference from the discovery that water has a relatively open structure is that the forces between water molecules must be strongly directional. Actually these forces are a consequence of the uneven distribution of electrical charge on the water molecule itself.

The water molecule

The water molecule has a permanent dipole moment which indicates that it possesses an asymmetrical distribution of electric charge. Thus, the possible linear and symmetrical H–O–H structure is ruled out immediately. In fact, the molecule has a triangular shape (Fig. 1.2a) where the H–O–H angle is about 104·5° and the

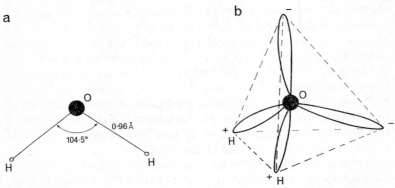

FIG. 1.2. Diagrammatic representations of the water molecule. The bond angle and length are shown in **a** and the tetrahedral arrangement of electric charge in **b**.

O–H bonds are about 0·96 Å long. Measurements of the dipole moment have given values close to $1·83 \times 10^{-18}$ e.s.u. cm (Moelwyn-Hughes, 1964). In order to get this dipole moment into quantitative perspective one should remember that the dipole moment of two equal charges, say of an electron, but of opposite sign, which are separated by 1 Å, will yield a dipole moment of $4·8 \times 10^{-18}$ e.s.u. cm.

How does this uneven distribution of charge arise in the water molecule? When the hydrogen atoms combine with an oxygen atom they contribute two electrons to the oxygen's outer shell of six electrons. In the water molecule, therefore, the eight electrons in the outer shell of the oxygen atom can be grouped into pairs where two of the pairs participate in the covalent bonds and the remaining two pairs are referred to as lone pairs. The pairs of electrons occupy four orbits extending tetrahedrally from the oxygen nucleus (Fig. 1.2b). However, the tetrahedral geometry of the orbits is not perfect, otherwise the orbits extending along the O–H axes would be separated by an angle of about 109° (the tetrahedral angle) whereas the H–O–H angle is about 105°. The oxygen nucleus exerts a pull on the two pairs of electrons, which lie in orbits tetrahedrally oriented towards the protons, and this leaves the hydrogen atoms with a net positive charge. In the other two tetrahedral orbits pointing away from the O–H bonds there are two pairs of unshared electrons. Therefore, each water molecule has two regions of high electron density at two corners of a tetrahedron, described about the oxygen atom, and two positive regions in the vicinity of the hydrogen atoms at the other two corners of the tetrahedron (Fig. 1.2b).

The water molecule, therefore, has an electrical polarity. Consequently each water molecule tends to have four neighbours by attaching itself by so-called hydrogen bonds to two other molecules at the positively charged sites and to another pair of molecules at the negatively charged sites.

The hydrogen bonds play an important role in the nature and geometry of the bonding between each water molecule and its neighbours. The attractive force constituting the hydrogen bond must act over a distance of about 1·75 Å since in ordinary ice the oxy-hydrogen bonding length is 1·01 Å and the distance between oxygen atoms in ice is 2·76 Å. In general this bond exists between strongly electronegative atoms, such as fluorine or oxygen, and hydrogen. It is a relatively weaker bond than the covalent oxy-hydrogen bond within a water molecule and yet stronger than the van der Waals type of binding which is generated by the mutual polarization of the water molecules. The energy of the covalent bond in water can be obtained from the energy of formation of a water molecule at 0°K, i.e. 220 kcal mole^{-1}. Thus, the O–H bond energy, is quoted as 110 kcal mole^{-1} or half of the energy of formation. A wide range of values exists for the energy of the hydrogen bond probably because of the numerous operational definitions of this parameter which have been employed. For example, the hydrogen bond energy can be obtained from the lattice energy of ice; the latter is defined as the difference in energy between one mole of immobile water molecules isolated from one another at 0°K and one mole of immobile water molecules in an ice lattice at 0°K. The lattice energy is 13·4 kcal mole^{-1} and since each mole of water has two moles of hydrogen atoms and hence two moles of hydrogen bonds, the hydrogen bond energy is 6·7 kcal per mole of hydrogen bonds. On the other hand, there have been many theoretical attempts to compute the hydrogen bond energy; for example, Campbell, Gelernter, Heinen & Moorti (1967) found values in the range 4–4·5 kcal per mole of hydrogen bonds from calculations of the electrostatic energy. Although the energy of the hydrogen bond is only about 5% of the O–H bond energy in water, it significantly determines the interactions and orientations of water molecules.

Structure of ice

The regular order of the ice lattice is the structural limit to which liquid water is attaining and, therefore, one might expect to be able to extrapolate in some way from the structure of ice to the structure of liquid water. In fact, this approach has been tried with some success. When it is realized, however, that there are at the latest count (Kamb, 1972) thirteen different forms of ice, the hope that the structure of liquid water may be solved satisfactorily by analogy seems vain.

Ordinary ice

Ice I, so-called ordinary ice, has a structure (Fig. 1.3) similar to that of the hexagonal form of silica known as tridymite. The structure can be visualized as a system of layers containing oxygen atoms. Each layer, which is constituted by a network of puckered six-membered rings, is the mirror image of its adjacent layer. This structure is one which satisfies the condition that each water molecule has only four neighbours, to which it is hydrogen-bonded. Thus, each oxygen atom is encompassed tetrahedrally by four others. The distance between neighbouring oxygen atoms is 2·76 Å. This co-ordinated system of tetrahedra produces an open lattice but one in which inter-molecular attraction is strong. There are cavities in the lattice which are large enough to accommodate independent water molecules and this fact led Forslind (1952) to propose a so-called interstitial model of liquid water which will be discussed later.

The resemblance between many of the physical properties of water vapour and ordinary ice led some workers, notably Bernal & Fowler (1933) and Pauling (1935), to conclude that the water molecules in ice are not ionized but remain intact. This view has

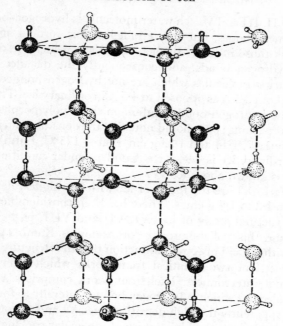

FIG. 1.3. The lattice structure of ordinary ice. The covalent bonds between oxygen atoms and the smaller hydrogen atoms are represented as solid rods whereas the hydrogen bonds are shown as interrupted lines. Each oxygen atom has two covalent and two hydrogen bonds which are tetrahedrally orientated. From Pauling: *The Nature of the Chemical Bond*, 3rd Ed. Fig. 12.6. Copyright (1960, Cornell University Press). Used by permission of Cornell University Press.

been repeatedly confirmed and it is now known that the dimensions of water molecules in ice I are similar to those of isolated molecules. The H–O–H angle probably just exceeds that cited for the isolated molecule, i.e. 104·5° and the O–H bond is 1·01 Å, slightly longer than that in the isolated water molecule.

Other forms of ice

In addition to ordinary ice (ice I) there are ten high-pressure polymorphs of ice (ices II to XI, cubic ice, Ic) and, finally, an ice of possibly amorphous structure called vitreous ice. Despite the numerous types of ice, each form apparently retains a tetrahedral arrangement of water molecules even though appreciable distortion of hydrogen bonds occurs in some cases.

In ices II, III and V each water molecule is hydrogen-bonded to its four neighbours, but the tetrahedra thus formed are not so regular as those in ice I. In fact, there is considerable distortion of the hydrogen bonds, and consequently the distance between neighbouring molecules which are not hydrogen-bonded to each other is about 3·5 Å as opposed to 4·5 Å in ordinary ice. The distortion of the hydrogen bonds confers on these polymorphs a closer molecular packing; one should note that the densities of ices II, III and V are 1·17, 1·14 and 1·23 g cm^{-3} (at $-175°$C, 1 atm) whereas that of ordinary ice is 0·94 g cm^{-3} under similar conditions.

Little is known about ice IV.

Ices VI to XI are the forms of ice with the highest densities in the range 1·2 to 1·6 g cm^{-3}. There has been considerable progress in the structural study of ices VI, VII and VIII, particularly by Kamb and his collaborators. For example, Kamb (1965) has shown with the aid of X-ray diffraction that the structure of ice VI is composed of two identical frameworks which interpenetrate but do not interconnect. Each framework comprises a chain of water molecules which is hydrogen bonded laterally to four similar chains. Apparently the high densities of ice VI and the other poly morphs result from the high degree of molecular packing achieved by such interpenetrating structures. Such packing is attained when each molecule in one lattice occupies a cavity formed by the alternative lattice.

When water vapour condenses at a temperature below $-160°$C vitreous ice is formed. This type of ice exhibits diffuse X-ray and electron diffraction patterns so it must have a glassy structure. Cubic ice is formed irreversibly as vitreous ice is warmed. The arrangement of oxygen atoms in cubic ice is isomorphous with that of carbon atoms in diamond and once again each water molecule is hydrogen-bonded tetrahedrally to its four nearest neighbours. As in ordinary ice at the same temperature the distance between neighbouring oxygen atoms is about 2·75 Å. The density of cubic ice is, as one would expect, quite close to that of ordinary ice.

Structure of water

There is no widely accepted theory of the structure of water. Some of the earliest views, that it was a mixture of ice and steam which varied proportionately with temperature (Röntgen, 1892), are surprisingly similar to some modern theories. In 1946 Eucken suggested that there is an equilibrium in water between single molecules and small polymers, particularly $(H_2O)_8$, and he produced with the aid of three (adjustable) equilibrium constants a theoretical description of many properties of water. It is now clear that a satisfactory theory of water structure must do more than predict values for various macroscopic properties. Actually the chief requirement in this field is not for more theoretical work, by itself, but for the design of experiments which will reveal some structural aspects of water.

An essential component of any molecular description of the structure of liquid water is a quantitative estimate of the number of hydrogen bonds. The maximal number of hydrogen bonds in liquid water is double the number of water molecules present since each molecule is capable of forming four hydrogen bonds and these bonds are shared with its neighbours. In ordinary ice each water molecule achieves its full tetrahedral co-ordination, and when ice melts some of the hydrogen bonds are broken. The latent heat of vaporization of ordinary ice is about 12 kcal mole^{-1} or 6 kcal per mole of hydrogen bonds. On the other hand, the latent heat of fusion of ice is 1·4 kcal mole^{-1} or 0·7 kcal per mole of hydrogen bonds. Therefore, according to the view that the energy of vaporization can be attributed entirely to the rupturing of hydrogen bonds, only about 12% of the total number of hydrogen bonds in ice are broken when ice melts. Other independent arguments suggest that liquid water at 0°C possesses about 85% of the original maximal number of hydrogen bonds present in the ice lattice, but there is no shortage of alternative estimates; in a review of such estimates Falk & Ford (1966) cited eighteen values in the range 28–98%. Although liquid water is regarded as a quasi-crystalline structure, there are no permanent crystalline aggregates. The proportion of quasi-crystalline liquid in relation to the unbonded water molecules decreases as the temperature increases, although some quasi-crystalline structure still exists in water at 100°C.

Although there is now agreement that water is a quasi-crystalline liquid arising from extensive hydrogen bonding between molecules, there is still controversy about the nature of the structural array. These disparate views on water structure reflect separate ideas on the forms that extensive hydrogen bonding may take.

Nets and rings

The first structural theory of water was conceived by Bernal & Fowler (1933) and their theory involved three types of water structure. At temperatures below 4°C the first form of water, water I, took an ice-like form similar to tridymite while at ordinary room temperature the second water structure, water II, resembled quartz; below the boiling point the third type of water structure, water III, emerged as a close-packed ideal liquid. At any point on the temperature scale Bernal and Fowler considered that the liquid was homogeneous. Bernal (1964) has generalized this model of water structure and he concluded that the structure is generated by the conditions that each molecule has four neighbours and that there are chains of successive neighbours forming closed rings. The structure of liquid water is visualized as a network of molecular rings arranged randomly. The rings may have four, five, six or even more members, but five-membered rings are preferred according to Bernal. The predominance of puckered rings of five molecules in Bernal's theory is congruent with the discovery of a new form of silica called keatite containing five-membered rings. In retrospect it seems that the structure of keatite was analogous to that of water II in Bernal and Fowler's earlier theory of water structure and incidentally to that of ice III.

Bernal's work on the theory of liquids invites the speculation that the instantaneous array of water molecules is a random framework of hydrogen bonds (Grant, 1957; Frank & Quist, 1961). Each hydrogen bond is expected to be oriented almost tetrahedrally with its neighbours and this suggests that the resulting structure ought to be composed of pentagonal and hexagonal rings of water molecules. This random net model of water, however, also contains meshes larger than those described by pentagonal and hexagonal rings of molecules. Moreover, according to this model a fraction of the total possible number of hydrogen bonds would not be formed. Consequently, the network might be considerably more labile than a relatively more crystalline type of water structure.

Cages and guests

X-ray scattering measurements on liquid water at 1·5°C (Morgan & Warren, 1938) formed the basis of an interstitial model of water first proposed by Samoilov (1946). According to this model liquid water is composed of ice-like lattices probably similar to that of ice I, containing large internal cavities which are occupied by interstitial water molecules. In other words, guest water molecules are trapped in cages within an ice-like lattice. Later, Forslind (1952) devised a similar model after his study of the progressive growth of lattice defects in ice during an increase in the ambient temperature.

The X-ray studies of water by Danford & Levy (1962) are also compatible not only with the existence of an expanded lattice of 'tridymite-ice' but also with the occupation of the cavities in this structure by interstitial water molecules. It is relevant, perhaps, that the structures of many hydrates of noble gases and hydrocarbons have been described as clathrates with the 'foreign' molecule occupying the cavities of hydrogen-bonded cages of water molecules. In 1960 Pauling suggested that water is a clathrate hydrate of itself; however, Pauling's hydrate model of water has been shown recently to be in discord with certain X-ray scattering measurements (Narten, Danford & Levy, 1967).

'Flickering clusters'

The cluster model of water has been proposed by Frank & Wen (1957) on the basis of 'co-operative' hydrogen bonding (Frank, 1958). In particular, Frank argued that the formation of a hydrogen bond to a water molecule perturbs the molecule so that it is easier for it to form additional hydrogen bonds with other molecules. As a result of this 'co-operative' phenomenon hydrogen bonds are not made and broken singly; this produces short-lived clusters of hydrogen-bonded molecules within a condensed phase of non-hydrogen-bonded water molecules (see Fig. 1.4).

Like the cage or interstitial models of water the cluster models are basically mixture models which assume that water molecules play two alternative roles. In both sets of model there is undoubtedly a rapid exchange of water molecules between the unbonded or interstitial position and the lattice position. This dual state of water molecules is supported by the evidence of certain workers, notably Litowitz & Carnevale (1955), who studied the absorption of ultrasound in water. They found that the shrinkage of water follows the increment in pressure with a time lag which reflects the finite time required for the transfer of molecules between two different states. At $0°C$ this relaxation time is 4×10^{-12} sec. In water the particular molecular process is the transfer of molecules from the cage to the interstitial position or alternatively from the cluster to the condensed phase of unbonded molecules.

It must be emphasized that clusters are not polymers in equilibrium with each other and with unbonded molecules since water

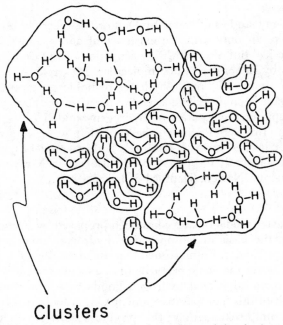

FIG. 1.4. Schematic representation of water containing hydrogen-bonded clusters and unbonded water molecules. In the clusters the molecules are co-ordinated tetrahedrally although not depicted as such in the diagram (Nemethy & Scheraga, 1962a: Fig. 5).

has a single dielectric relaxation time of about 10^{-11} sec (the time lag between the onset of an electric field and the induced polarization of water); moreover, there is a common energy of activation, 4·6 kcal mole^{-1}, associated with dielectric relaxation, self-diffusion, shear viscosity and bulk viscosity. We may conclude that one fundamental process underlies all of these phenomena. This process may be the formation and decay of clusters or alternatively of a lattice network of water molecules. This implies that clusters have a half-life of the order of 10^{-11} sec which is several hundred times the period of molecular vibration. Thus, the cluster model leads one to think of a dynamic, rather than a static, structural order in water. Frank & Wen (1957) coined the phrase 'flickering clusters' to describe these transient structures.

The 'flickering cluster' concept has been adopted by Nemethy & Scheraga (1962a) who analysed this scheme theoretically and compared their results with several macroscopic properties of water. They estimated that at room temperature each cluster contains about 50 water molecules (see Table 1.1). Moreover, their estimates of the number of unbonded water molecules are in good agreement with the experimental measurements of Buijs & Choppin (1963) also cited in the table. According to Nemethy and Scheraga the clusters are separated by water molecules having no hydrogen bonds (Fig. 1.4). It presents serious difficulties, however, to think of the molecules in the dense phase surrounding the clusters as unbonded molecules since they must interact strongly in order to account for the strong cohesion of water. Nemethy and Scheraga suggest, for example, that the dipole interaction between the pairs of molecules *may* be strong enough to partially account for the cohesion of these 'free mononers'.

TABLE 1.1. Temperature dependence of cluster size and the mole fraction of unbonded molecules in liquid water

Temperature °C	Number of water molecules in cluster	Mole fraction of unbonded water molecules
0	91	0·24
6	–	0·27*
10	72	0·27
20	57	0·29
21	–	0·31*
30	47	0·32
34	–	0·33*
40	38	0·34
47	–	0·36*
50	32	0·36
60	28	0·38
70	25	0·39
80	23	0·41
90	22	0·42
100	21	0·44

* Values obtained from Buijs & Choppin (1963). (Modified from Nemethy & Scheraga, 1962a.)

Conclusion

Although all of the experimental evidence indicates water is a quasi-crystalline liquid, the actual structure of liquid water is not known. In brief, there are two divergent views about the structural order of water. First, the structure may take the form of discrete molecular clusters, or secondly, the molecular array may exist as a random network of hydrogen bonds extending throughout the liquid. The interesting suggestion has been made by Franks & Good (1966) that both types of structure exist in liquid water; in particular, they argued that clusters predominate above 35°C whereas the network persists at lower temperatures. Perhaps that is so, but like so many speculations about the structure of liquid water it remains to be thoroughly substantiated by experimental evidence.

'Anomalous' water

Certain experimental evidence has been obtained, which indicates that the interface between a solid surface and a liquid may influence the structure and properties of the liquid. In his review of the literature on this topic Henniker (1949) concluded that several sets of experimental data strongly suggest the existence of a region, adjacent to the interface, where there is a structural orientation of the liquid and that this zone may be as large as several hundred angstroms for liquid water. Bangham & Bangham (1968) have also discussed the state of liquids, and in particular water, at solid–liquid interfaces. In the last decade or so Derjaguin and his collaborators have made an extensive study of the physical properties of polar liquids, including water, at solid surfaces (see Derjaguin, 1965). Since Derjaguin (1966) has concluded that the usual state of water is thermodynamically metastable, he has suggested that ordinary water should be called 'metawater' and that 'anomalous' water be called 'orthowater'. However, it is hard to understand why ordinary water—the less stable of the two forms—should be the more abundant! In a recent report Derjaguin concluded that 'anomalous' water can exist as a stable entity in the bulk liquid state (Derjaguin, Churaev, Fedyakin, Talaev & Ershova, 1967). According to Derjaguin and his collaborators the solid surface, which is apparently essential for the formation of 'anomalous' water, is not required for its continued existence.

Most of Derjaguin's work has been performed on columns of 'anomalous' water formed by condensation of water vapour in narrow quartz capillaries with diameters of about 10 μm. The unusual nature of the 'anomalous' water is characterized by the following properties: (a) more viscous than normal water; (b) larger density (1·2–1·3 g cm^{-3}); (c) lower vapour pressure than that of normal water; (d) phase separation at temperatures below 0°C; and (e) different behaviour during thermal expansion. Recent studies by Willis, Rennie, Smart & Pethica (1969) have confirmed some of the peculiar properties of 'anomalous' water; however, they expressed the doubt that the anomalous behaviour of water might have been caused by minute quantities of impurities such as silicon (see also Pethica, Thompson & Pike, 1971). The question of impurities in 'anomalous' water has been taken up by several workers, notably Bascom, Brooks & Worthington (1970) who showed that the residues obtained by condensing 'anomalous' water contained silicon and sodium atoms. On the other hand, Barnes, Cherry, Finney & Petersen (1971) have demonstrated that organic impurities from certain parts of the experimental apparatus involved in the production and handling of 'anomalous' water are probably the main sources of its peculiar properties. Thus, it seems exceedingly unlikely that a pure form of 'anomalous' water does exist. It is surprising, nevertheless, that such minute levels of impurities can modify the bulk properties of water so dramatically.

Despite the failure to demonstrate that pure 'anomalous' water can exist as an entity, there is no doubt that the behaviour of water in the vicinity of solid surfaces is different from that of bulk water. What role that anomalous behaviour may play in membrane transport and more generally in the water relations of cells is difficult to predict. The need to answer that question is an urgent one, especially when we consider how much physiological work on the hydrostatic and hydrodynamic behaviour of water in animals and plants relies heavily on the accepted properties of ordinary water.

Transport processes in ice and water

The rates of molecular displacements in ice and liquid water determine the outstanding differences in some of their properties. Viscosity, for example, obviously depends upon the rates of molecular translation and reorientation, and there is a huge difference between the viscosities of ice and water; the viscosity of water at 0°C is about 2×10^{-2} poise whereas that of ice is about 10^{14} poise. A great deal has been learned about the physiology of water by examining transport processes, such as viscosity, self-diffusion and electrical conduction, and some of this work will now be briefly discussed.

Electrical properties

Ordinary ice. The static dielectric constant, ϵ_0, of ice I has been determined by Auty & Cole (1952). It is about 90 at 0°C and rises to 130 at -70°C. Such large values indicate that water molecules are able to change their orientations in the lattice and the rate of reorientation has been investigated by analysis of the frequency dependence of the dielectric constant, ϵ. The reorientation of water molecules accounts for the major fraction of ϵ at relatively low frequencies of the applied electrical field, but when the frequency is increased the dielectric constant decreases because the molecules cannot reorient themselves at a fast enough rate. From these experiments it is possible to estimate the dielectric relaxation time, τ_d; for ice I at 0°C τ_d is about 2×10^{-5} sec. The reorientation of water molecules is, of course, achieved by thermal agitation and from the value of τ_d it is clear that typical water molecules undergo about 50,000 reorientations per second.

How do these frequent reorientations of water molecules come about? According to Bjerrum (1951) the reorientations can be explained by postulating the existence of a small number of defects in the ice lattice. In this model thermal movements cause a water molecule to rotate through 120° about one its hydrogen bond axes, O–H $\cdots$ O. This specific rotation generates vacant positions between pairs of oxygen atoms, O $\cdots$ O, (L-defect) and also between one doubly occupied bond, O–H H–O, (D-defect). These L- and D- defects become separated by rotation of adjacent water molecules and subsequently they migrate through the ice lattice by this rotational mechanism (Fig. 1.5). The Bjerrum defects have equal charges with the D-defect having a positive charge due to one excess proton and the L-defect have a corresponding negative charge. Although the presence of these defects accounts for dielectric relaxation it does not do so for the direct-current (d.c.) conductivity. The movement of a D-defect, for instance, transfers the defect state across the lattice but not a proton whereas electrolysis experiments have demonstrated that ions, i.e. protons, are the carriers of charge (Granicher, Jaccard, Scherrer & Steinemann, 1957). In fact, the conductivity of ordinary ice relies heavily upon the ionic dissociation of water molecules into H_3O^+ and OH^- ions. In other words, the dissociation of a water molecule means that

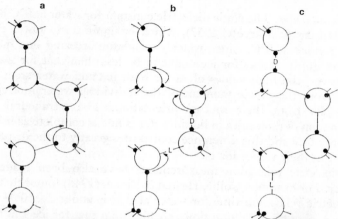

Fig. 1.5. Formation and diffusion of Bjerrum L- and D-defects in the ordinary ice lattice. The unperturbed lattice is shown in **a** while the formation and separation of a pair of L- and D-defects is depicted in **b** and **c** respectively (Jaccard, 1972: Fig. 4).

one of its protons jumps on to an adjacent water molecule to give an H_3O^+. Thus, the dissociation of a water molecule creates an ionic defect in the ice lattice and subsequent proton jumps cause the migration of the ionic defect through the lattice. Under the influence of an applied electrical field, therefore, the current will be carried by proton jumps which will, of course, perturb the ice lattice. All of the water molecules whose orientations have been disturbed by this process will need to undergo reorientation before another proton can use the same path. This requirement, however, does not interfere with conduction since the number of jumping protons that may arrive at a given molecule in one second is about 30, whereas the same molecule probably experiences about 30,000 reorientations in that interval. Although the number of jumping protons is small the rate of jumping is exceedingly rapid because the residence time of a jumping proton with a water molecule is only about 10^{-13} sec. The net result of this rapid jumping process in ice is that the mobility of protons is about 10 to 100 times greater than that in liquid water and, in fact, it is about 2% of the electron mobility in semi-conductors and some metals. For this reason Kavanau (1964) has suggested that ice might be considered a 'proton semi-conductor'.

Liquid water. The static dielectric constant for water at 0°C is 88 (Malmberg & Maryott, 1957), which is significantly lower than that for ice at 0°C. Since water expands on freezing one might expect glibly that ϵ_0 for ice ought to be less than that for water. However, the high values of ϵ_0 for both ice and water stem not only from the dipole moment of the individual water molecules but also from their mutual orientation. The tetrahedral co-ordination of molecules in liquid water is not as complete as it is in ordinary ice and this difference contributes greatly to the disparity between their values for ϵ_0.

Dielectric relaxation measurements have also been made on water. For example, Collie, Hasted & Ritson (1948) found that the dielectric relaxation time for water at 0°C is about 2×10^{-11} sec, that is, about one million times smaller than that for ice I at 0°C. Again, the dielectric relaxation time for liquid water can be interpreted as a consequence of reorientation of water molecules although there is no widely accepted mechanism of reorientation. A possible explanation is that the external electric field induces water molecules to form 'flickering clusters' with their dipoles oriented in the direction of the field. Indeed, Frank & Wen (1957) believe that τ_d is practically identical to the half-life of a cluster. Eisenberg & Kauzmann (1969), however, have argued from the standpoint of energetics that such clusters are likely to consist of only one or two molecules and this is at variance with the proposed size of the 'flickering clusters'. Another possible explanation for the time course of dielectric relaxation involves molecular reorientations arising from orientational defects in the randomly arranged network of hydrogen bonds in the water lattice.

As in ordinary ice the large d.c. conductivity of liquid water can be explained by taking account of proton transfers. Hydrogen ions are formed by the dissociation of water molecules and they are strongly bound to water molecules to form H_3O^+ ions or possibly larger complexes. In fact, Eigen & De Maeyer (1958) suggest that

$H_9O_4^+$ is probably the commonest hydrated form of H^+ ions. The mean lifetime of a proton–water molecule complex is about 10^{-12} sec (Eigen, 1964) while the mean interval between successive associations of a proton with a given molecule is about 5×10^{-4} sec (Eisenberg & Kauzmann, 1969). The latter is considerably longer than the time (10^{-11} sec) required for, say, reorientation of an individual water molecule.

The proton mobility in liquid water at 25°C is 3.6×10^{-3} cm² volt⁻¹ sec⁻¹ whereas that of a sodium ion is only 5×10^{-4} cm² volt⁻¹ sec⁻¹. In contrast, the proton mobility in ordinary ice is about 8×10^{-2} cm² volt⁻¹ sec⁻¹, even larger than that for liquid water. The reason for the high mobility of hydrogen ions was given originally by Grotthuss (1806) who suggested that there are fast proton transfers between H_3O^+ ions and other water molecules. Eigen & De Maeyer (1958), in particular, have also suggested that the high mobilities of H^+ and OH^- ions in both water and ice result from rapid movement of protons which is facilitated by the hydrogen bonds between water molecules. For instance, consider the series of proton jumps depicted in Fig. 1.6a. One of the protons in the H_3O^+ ion jumps along a hydrogen bond on to a neighbouring molecule which subsequently loses a proton to its neighbour and so on. In this way net charge is rapidly transported without any individual proton or H_3O^+ ion actually crossing the lattice. Similarly the transfer of OH^- ions can be achieved by proton jumps of that sort (Fig. 1.6**b**). The mobility of H^+ ions in liquid water is less than that in ordinary ice because the network of hydrogen bonds is irregular in the liquid according to Eigen & De Maeyer (1958).

Self-diffusion

Ordinary ice. The diffusion of intact water molecules through the lattice of ice I is characterized by a self-diffusion coefficient, D_w, of about 10^{-10} cm² sec⁻¹ at $-2°C$ (Kuhn & Thurkauf, 1958) and an activation energy of 13·5 kcal mole⁻¹ (Dengel & Riehl, 1963). Self-diffusion of water molecules can occur only if there are defects in the ice lattice, and Haas (1962) suggested that self-diffusion is accomplished by migrating interstitial water molecules. According to Haas these molecules are associated with the Bjerrum L- and D-defects in the ice lattice and, moreover, the rate of diffusion must be related to the rate of migration of these

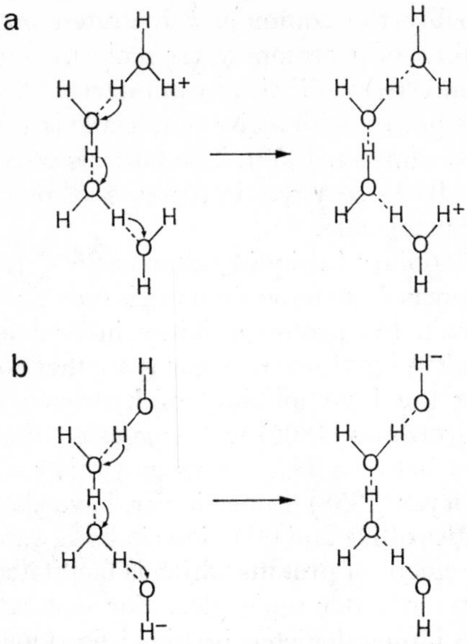

FIG. 1.6. Representation of how electric charge may be transferred across several water molecules by a series of proton jumps. The direction of the proton jumps from one molecule to its neighbour is indicated by small arrows. In **a** positive charge is transferred from the water molecule at the top (left) to that at the bottom (right) by a consecutive series of proton jumps. The transfer of negative charge by a similar process is shown in **b**.

orientational defects. Haas marshalled some arguments in favour of this model but Onsager & Runnels (1963) challenged the theory on the grounds that an individual water molecule probably makes a diffusional jump of about three lattice positions rather than one lattice position per jump as Haas suggested.

Liquid water. Values for D_w in water have been obtained by several investigators (see Kohn, 1965). In particular, Wang, Robinson & Edelman (1953) employed three different forms of labelled water, namely HDO, HTO and $H_2^{18}O$, and found no significant difference between the magnitudes of D_w. Since the deuterated and tritiated water molecules do not diffuse more rapidly than the $H_2^{18}O$ molecules, it has been concluded (Wang et al., 1953) that the rapid transfer of protons, which plays such an important role in electrical conductivity, does not influence the self-diffusion of water molecules.

The temperature dependence of D_w has been studied by Wang et al (1953) and their data are shown in Table 1.2. Over that temperature range (0–55°C) the activation energy for self-diffusion is 4·6 kcal mole^{-1}.

TABLE 1.2. Temperature dependence of self-diffusion coefficient for liquid water

Temperature °C	$D_w \times 10^5$ (cm^2 sec^{-1})		
	HDO	HTO	H$_2$^{18}O
1·1			1·44
4·9			1·55
5·0		1·39	
10·0	1·57		1·90
15·0		1·83	
18·0	2·06		2·35
25·0	2·34	2·44	2·66
35·0		3·04	3·49
45·0	3·87	3·83	4·38
55·0	4·95		5·45

Modified from Wang et al., 1953.

It is possible to account for the size of D_w by assuming that self-diffusion occurs as a consequence of molecular jumping. According to this view each jump of a water molecule into a new position is preceded by a relatively long wait in an equilibrium position. The time between jumps has been estimated as 4×10^{-12} sec (Singwi & Sjölander, 1960) which is about forty times the period of molecular vibration. Wang (1965) has estimated that the average distance between such equilibrium positions is about 3·7 Å which is about fifty times the amplitude of molecular vibrations. There seems to be no strong support for the view that self-diffusion involves a series of small jumps as opposed to the 'jump and wait' mechanism mentioned above.

Final comment

Although there is no adequate picture of how molecular movements proceed in water several features do stand out. For example, no matter what kinetic phenomenon is considered, with the possible exception of thermal conduction, the rate at which translations and reorientations of water molecules occur is quite uniform. In water at 0°C this rate is about 10^{11} per second and in ordinary ice it is 10^5 per second. Indeed, the processes of translation and reorientation of water molecules seem intimately related to each other since not only the kinetics of processes such as self-diffusion, viscosity and mechanical and electrical relaxation but also their temperature dependences are similar. It could be argued, in fact, that the basic element of all these kinetic effects is molecular rotation as it probably rate-limits the translational movement of water molecules.

Interaction of substances with water

Electrolytes

According to the accepted view of ionic hydration (Frank & Wen, 1957) there are concentric domains surrounding an ion in aqueous solution (Fig. 1.7). First, there is a region of immobilized

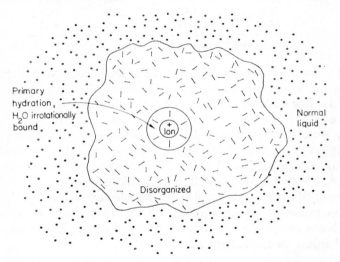

FIG. 1.7. Model of structure of water in the neighbourhood of an ion (Klotz, 1970: Fig. 11).

water molecules immediately adjacent to the ion where the polarizing effect of the ionic electric field radially orients the water molecules. Beyond this region there is one in which the organization of water is more random than that of ordinary water; in the secondary zone the electric field of the central ion is evidently strong enough to interfere with the normal structural order of water. Finally, the water recaptures its usual structure in the outermost region.

There still remains some difficulty in explaining the significant differences in the interactions of water molecules with cations as opposed to their interactions with anions. The degree of hydration of anions is considerably less than that of cations as a general rule. It is visualized that water molecules in the immediate vicinity of anions are freer to rotate and, consequently, to engage in hydrogen bonds with other water molecules than the so-called irrotationally bound water molecules of cations. Although this concept of the differences between the nature of water's interaction with anions and cations has been supported by measurements of dielectric constants (Haggis, Hasted & Buchanan, 1952; Harris & O'Konski, 1957), the actual basis of the different interactions is not entirely understood (see Kavanau, 1964).

From the observations that relatively small ions and multivalent ions, such as Li^+, Na^+, F^-, Ca^{++}, Mg^{++} and Al^{+++}, increase the viscosity of water, it has been concluded that these ions promote structure in water. The structural order, which such ions generate by virtue of their strong electric fields, extends beyond the first layer of water into the intermediate domain of disrupted structure. Several different pieces of experimental evidence support that view.

As a partial consequence of the relatively weak electric fields surrounding large monovalent ions, such as NH_4^+, Rb^+, Cs^+, Br^-, I^- and NO_3^-, there is a tendency for an enhanced disruption of water structure in the intermediate domain of the water surrounding them. This structure-breaking effect is consistent with the increase in fluidity of water which these ions produce. Moreover, the ability of large monovalent cations and of most anions to break water structure exceeds their weak tendencies to order the structure of their immediate neighbours. The changes in the dielectric relaxation time (Haggis *et al.*, 1952) and in the self-diffusion coefficient of water in electrolyte solutions (Wang, 1954), are particular examples of data which substantiate the claim that these ions disrupt water structure.

The foregoing argument about the structure-breaking role of large monovalent cations partially rests on their effects on viscosity of water. Perhaps it is somewhat naïve, however, to expect that an increase in the fluidity of water stems solely from structure-breaking.

As a footnote to the description of ionic hydration it is pertinent to indicate what sort of life span an irrotationally bound water molecule enjoys. The range of lifetimes is enormous. For univalent cations, such as Li^+, Na^+, K^+, Rb^+ and Cs^+, there is a uniform lifetime of about 10^{-9} sec whereas the lifetime of a hydrated proton (H_3O^+) is 10^{-12} sec. For multivalent cations, on the other hand, the lifetimes are widely divergent; for example, the values in seconds are 10^{-9} (Ca^{++}, Sr^{++}, Ba^{++}), 10^{-8} (Mn^{++}), 10^{-7} (Cu^{++}), 10^{-6} (Mg^{++}), 10^{-5} (Ni^{++}, Fe^{+++}), 10^{-2} (Be^{++}), 7·5 (Al^{+++}) and $1·5 \times 10^5$ (Cr^{+++}). These lifetimes can be found in the papers of Connick & Poulson (1959), Baldwin & Taube (1960), Eigen (1963), Wicke (1966) and Fiat & Connick (1968). Even the relatively short lifetime (10^{-9} sec) of water in the hydration crust of some univalent cations is significantly longer than the half-life (10^{-11} sec) of molecules in the 'flickering clusters' (Frank, 1958) and much longer than the period (10^{-13} sec) of molecular vibration or rotation.

Anaesthetics

Inspection of the physical properties of anaesthetics has suggested two possible mechanisms of anaesthesia. On the one hand, Meyer (1899) suggested that anaesthetic potency depends upon the partition coefficient of the anaesthetic between the water and lipids of the brain. The relatively large solubility of anaesthetics in lipids and in protein solutions, which had been observed by Meyer (1899) and by Moore & Roaf (1905) and re-examined for fluorinated compounds by Miller, Paton and Smith (1965), does suggest that anaesthetics act on membrane lipids. On the other hand, both Pauling (1961) and Miller (1961) argued that anaesthetics act in the aqueous phases of the central nervous system and cited the strong correlation between the stability of hydrates of anaesthetics and their potency. Consider, for example, one of the most unusual anaesthetics—xenon. This noble gas is incapable of forming ordinary chemical compounds by covalent or ionic bonding; however, it has the striking ability to form clathrates with water. X-ray diffraction studies have revealed that crystals of xenon hydrate have an identical structure to that of hydrates of other molecules such as methane and chlorine (Clausen, 1951*a,b,c*). This structure consists of cages formed by 20 water molecules joined tetrahedrally by hydrogen bonds so as to form twelve-sided polygons with each face being pentagonal. These pentagonal dodecahedra occupy space in configurations which encompass volumes large enough to accommodate atoms, such as xenon, or small molecules such as chlorine. The hydrate of chloroform is constituted by a slightly different arrangement of dodecahedral clusters of water molecules.

Anaesthesia cannot be simply due to the formation of hydrates of anaesthetics in the central nervous system since huge pressures would be required to stabilize the hydrates at body temperature. Pauling suggested that the side chains of membrane proteins and solutes in the body fluids contribute to the stability of these clathrates and he argued that the anaesthetics themselves act as stabilizers for hydrates occurring in the vicinity of membrane proteins. This mechanism of anaesthesia can be envisaged, therefore, as an enhancement of structural water in axonal and synaptic membranes which interferes with the normal processes of transmission. Both Miller and Pauling reasoned that hydrate formation in the absence of gas molecules would result in anaesthesia too, and they surmised that anaesthesia induced by hypothermia might be explained by this mechanism.

At present it is inappropriate to favour either of these theories of anaesthesia because there is no compelling evidence for one or the other.

Hydrocarbons

Many investigators have studied the thermodynamic properties of aqueous solutions of hydrocarbons in order to improve our structural picture of liquid water. Examination of the solubility of hydrocarbons in water leads to the conclusion that the mixing of water and hydrocarbons is not ideal since, if this were so, we would expect no change in volume to occur. Actually, a decrease in volume occurs and also there is a large negative change in entropy which Frank & Evans (1945) have attributed to the ordering of water by the hydrocarbons. The current model, accounting for the departure from ideal behaviour by hydrocarbon solutions, is that the water lattice forms a partial cage around the hydrocarbon. This concept originated from the observation that hydrates of hydrocarbons and rare gases had crystalline structures (Stackelberg & Müller, 1954). In fact, the hydrocarbon affects the energies of its neighbouring water molecules and consequently there is a greater degree of hydrogen bonding in this water (Nemethy & Scheraga, 1962b). The reduction in the volume of hydrocarbon solutions occurs because the hydrocarbon occupies space previously empty in the pure water lattice.

In analysing the interactions of hydrocarbons with each other in aqueous solution Nemethy & Scheraga (1962b) partially refuted an earlier hypothesis that the source of attraction was due solely to van der Waals forces. They showed that the van der Waals forces contributed only about 50% of the total free energy of the formation of the interaction. This type of interaction has been called the hydrophobic bond. When the non-polar side chains of a protein in aqueous solution approach each other, the total number of water molecules involved in the partial cages will decrease and this change in the structure of the adjacent water cages generates to a certain extent the formation of the hydrophobic bond. Obviously this type of interaction plays quite an important role in the conformation of polypeptide chains. Hydrophobic bonding has been invoked (Kauzmann, 1959) to explain the fact that non-polar amino acids can exist in the interior of the proteins without any contact with the aqueous medium.

Macromolecules

In dealing with the types of hydration which may occur in biologically important macromolecules, it is possible to dismiss the charged groups from this part of the discussion since they will hydrate basically like ions in solution. This leaves two distinguishable types of surfaces on macromolecules where hydration may take place. First, there is the non-polar surface, such as the protein backbone, and secondly, there is the surface with hydrogen-bonding sites, such as the $C=O$ or $N-H$ groups on proteins.

There has been an exceedingly extensive discussion of the hydration of non-polar groups (Frank & Evans, 1945; Nemethy & Scheraga, 1962a,b) and it seems very probable that the water structures around such groups are clathrates composed of pentagonal rings. There is no reason to believe, however, that the water structure near a flat non-polar surface would consist entirely of clathrate-type structure. Nevertheless, since the local concentration of non-polar groups in a macromolecule may be high, one might expect also co-operative effects between adjacent groups to induce a stabilization of the neighbouring water molecules.

The hydration of hydrophilic surfaces on macromolecules, where there are arrays of hydrogen-bonding sites, has a different character from that of hydrophobic surfaces. In particular, if the array of hydrogen-bonding sites fits a regular water lattice, such as ice I, the structure may become stabilized in the liquid. There have been numerous speculative studies of the way that macromolecules forming either helical structures or sheets may fit into water lattices. It must be emphasized that the proposed water structures in the vicinity of these macromolecular surfaces are not rigidly ice-like; however, these surfaces probably impose a restriction on the rotation of adjacent water molecules. Nuclear magnetic resonance† (NMR) experiments on hydrated collagen indicated that some of the water molecules in the hydration crust rotate anistropically about an axis parallel to the fibre axis (Berendsen, 1962; Berendsen & Migchelsen, 1965). Berendsen has suggested that the collagen macromolecules stabilize chains and possibly three-dimensional arrays of water molecules in the hydration coat. In fact, the proposed chains of water molecules must lie parallel to the fibre axis and between the collagen molecules. Thus, hydrated collagen definitely possesses oriented water molecules and Berendsen & Migchelsen (1966) have concluded that the water molecules exist in this state for about 10^{-4}–10^{-3} sec.

In this system there must be an extensive number of hydrogen

† The fundamental principle underlying this technique is that any atom possessing a nuclear magnetic moment will interact with an imposed magnetic field and consequently the atom will absorb electromagnetic radiation of a particular frequency. Since hydrogen has a magnetic moment, any molecule, such as water, containing hydrogen will interact with the magnetic field. The important feature of this tool is that the nature of the absorption signal is highly dependent on the ability of the hydrogen atoms to rotate. For example, if the hydrogen atom is attached to a mobile molecule, such as a water molecule in aqueous solution, then the absorption signal is very sharp. If the molecule containing hydrogen is somewhat immobilized, the absorption signal is broad.

Recently Walter & Hope (1971) have discussed the use of NMR in the study of the state of water in cells, and the reader is referred to their review for a full account of this topic.

bonds between the chains of water molecules and the collagen molecules. Since the fracture and re-formation of hydrogen bonds between the chains and the collagen molecules will occur relatively frequently on the time scale of molecular vibration, the continual re-shuffling of linkages between the water molecules and collagen offers a mechanism for chain rotation. Corresponding studies (Berendsen & Migchelsen, 1965) on hydrated silk fibroin have shown that the anisotropic direction for rotation of associated water molecules is perpendicular to the fibre axis. The nature of the restriction on rotation is dependent, therefore, on the character of the macromolecular surface and, consequently upon the type of water lattice which will fit itself to that surface.

The proposed structures of small polypeptides and proteins, such as gramicidin S and tobacco mosaic virus (TMV) protein (Warner, 1961, 1964), have characteristics which are likely to provide ordered arrays of hydration. In particular, these structures have a hydrophobic and a hydrophilic side. The latter type of surface fits the water lattice in a hexagonal array, and it would be expected to promote long-range order (up to, perhaps, 30 Å from the surface) of an expanded ice I lattice containing numbers of interstitial molecules. Of course, it is exceedingly difficult to estimate the dimensional extent to which water structure is influenced. Jacobson (1953) promulgated the view that a hydrophilic macromolecular surface may induce a water lattice with less defects than that of bulk water to extend outwards to distances of 500–1000 Å. Now it seems generally accepted, however, that water will be present in a regular array over a distance of 10–20 Å from such a hydrophilic surface. Within this domain the water structure will be somewhat different from the usual ice lattice. Beyond that region extends a further zone where the water molecules are less rigidly oriented. This secondary layer of partially organized water structure ends about 100 Å from the surface. The arguments supporting this view of macromolecular hydration have been discussed by Bernal (1965).

The state of water in the cell

Recent research has shown that the state of cellular water is not identical to that of pure water or, indeed, of aqueous solutions of proteins. The cell may be envisaged as a complex system of membranes and heterogeneous aqueous regions, which probably differ markedly from one cell type to another. In each type of aqueous domain within the typical cell, there exist different molecular environments for water molecules. For example, when we use the blanket term cellular water we are actually referring to water within membranes, water in contact with macromolecular surfaces, water inside organelles and, finally, water in the cytoplasm. The list is by no means complete. A closer examination of the individual components of cellular water reveals that a large proportion of this water is intimately associated with membranes.

Both Finean (1957) and Fernandez-Moran (1959) have stressed that water is an important component of membrane systems and the implication from their work is that water has a structural role in the organization of membranes. In particular, Fernandez-Moran (1962) has suggested that water must exist in some type of lattice in the neighbourhood of the polar groups of the membrane proteins and lipids. The proposition that this ordered water in membranes may be functionally important as well as structurally significant probably originates from the views of Szent-Györgyi (1957). There are numerous realms in which ordered water might exert its influence on cell membrane behaviour; for instance, alterations in the water structure at distinctive regions in excitable membranes might be responsible for the transient perturbations in ionic permeabilities which occur. However, our ignorance of membrane structure is so great that such speculations are largely pointless. In addition to the possibility that there exist water lattices within the membrane phase, there remains the question of ordered water at the surfaces of macromolecules either at membrane surfaces or in structural filaments.

The study of the state of cellular water should be set against the backcloth of controversy. On one side, we have the almost classical picture of the cell as a membranous sac containing various organelles suspended in an aqueous solution of ions and proteins. The other view of the cell belongs to the minority. This faction regards the cell as an ordered phase, which is constituted by a matrix of a quasi-crystalline water with embedded solutes. The books and review of Ling (1962, 1965) and of Troshin (1961, 1966), who are among the chief protagonists of this view, should be consulted for full accounts. In order to resolve this dichotomy, unambiguous evidence about the nature of cellular water is required. The experimental data gathered for this purpose have been reviewed from time to time by several authors (Gortner, 1932; Höber, 1945; Szent-Györgyi, 1957; Tait & Franks, 1971).

It should be stressed at the beginning that the experimental evidence about the state of water in cells and tissues is not very extensive but, despite this, the work which has been published evidently demonstrates that there is some immobilized water within cells. For example, Hearst & Vinograd (1961) concluded that some of the water closely associated with DNA excluded ions; a rather similar observation on hydrated proteins was made earlier by Perutz (1946). It is only in the last decade or so that these studies on macromolecular hydration have been parallelled by examinations of the state of cellular water.

Briefly, one might describe the experimental attack on the problem of cellular water by saying that it has three facets. First, there have been investigations of the non-solvent behaviour of intracellular water; secondly, there is some X-ray diffraction work and, finally, we have the NMR experiments on living cells and tissues. Because the last approach is the most powerful tool of the trade, it is given prime place in the following discussion.

Nuclear magnetic resonance studies

Apparently the use of this technique in biological system was pioneered by Odeblad and his collaborators. In particular Odeblad examined the state of water in red blood cells (Odeblad, Bhar & Lindström, 1956), in human cervical mucus (Odeblad & Bryhn, 1957), in eye tissues (Huggert & Odeblad, 1959a,b) and in vaginal cell sediments (Odeblad, 1959). The general conclusion from all of these studies was that some intracellular water was relatively immobilized although neither the nature of its structural array nor its location could be determined. Invariably Odeblad and his colleagues observed a broadening of the proton absorption signal in the NMR spectrum of cellular water relative to that in pure water. In the erythrocyte, for example, it was found that approximately 75% of the protons had a fourfold broadening of the NMR spectrum.

Similar broadening of the NMR spectrum of water was observed in fish muscle (Sussman & Chin, 1966) and in the sciatic nerve of the frog (Swift & Fritz, 1969). The preliminary work of

Chapman & McLauchlan (1967) on the NMR spectrum of water in the sciatic nerve of the rabbit revealed a partial splitting of the adsorption signal into two peaks; the resolution of the peaks was dependent on the orientation of the nerve axis in the apparatus. These workers concluded that the bulk of tissue water was partially ordered in some manner.

Especially interesting is the work of Hopkins (1960) and Bratton, Hopkins & Weinberg (1965) on the state of water in cardiac muscle and in skeletal muscle respectively. The NMR studies of Bratton and his colleagues showed that a certain fraction of the intracellular water in skeletal muscle was restrained severely from rotational motion. They estimated that the immobilized fraction of the total intracellular water was likely to be about 0·2%. Furthermore, they observed that about 20% of this immobilized fraction was released during isometric contraction. This finding confirmed the earlier study by Hopkins on cardiac muscle, which also apparently liberates a similar proportion of its immobilized water during contraction.

Despite the contention that the interpretation of the NMR studies described above is complicated, for example, by the possible presence of magnetic inhomogeneties in the biological sample and of several different fractions of tissue water, all of these investigations strongly suggest that some proportion of cellular water has a higher degree of crystallinity than ordinary water. The troublesome ambiguity in the interpretation of NMR spectra of cellular water has been removed recently by Cope (1969) who studied the structure of water in rat muscle and brain. The tissue water in these experiments was partially replaced by heavy water which serves as a NMR probe for microscopic electric fields in the sample in contrast to water which serves as a NMR probe for microscopic magnetic fields. In liquid water, for example, one expects that electric fields will tend to cancel each other because the electric dipoles (water molecules) are randomly oriented. If there is some order in the structure of the tissue water, an electrical gradient will exist and this will broaden the NMR spectrum of heavy water. Aside from the steady-state NMR spectrum, pulsed NMR experiments can be performed in order to estimate the relaxation time for the return of the probe molecules (D_2O) to their equilibrium positions. It is expected that the presence of structural order in the liquid sample will reduce the relaxation time. Armed with these techniques, Cope examined the extent of quasi-crystallinity in the tissue water of rat muscle and brain. Cope's NMR analyses confirmed earlier conclusions that tissue water exhibited more organization than pure water. Indeed, Cope's work produced an added bonus, for it revealed the presence of distinct fractions of tissue water in both muscle and brain. Within each type of tissue some restriction on exchange of water molecules between the fractions was evident and Cope suggested that there must be some anatomical compartmentalization of tissue water. Quantitative estimates of the size of these fractions were obtained; approximately 27% of tissue water in muscle and 13% in brain exist as separate components with an even greater degree of quasi-crystallinity than that of remaining fractions. Cope raised the possibility that each of the two components of tissue water may not be homogeneous itself and he suggested that the highly ordered fraction of water might consist of protein hydration (amounting to to less than 1% of total tissue water) plus a relatively large liquid fraction. The latter proposal is similar to the conclusions of Bratton *et al.* (1965). Hazlewood, Nichols & Chamberlain (1969) have also concluded from their NMR analysis of water in skeletal muscle that there are at least two components of tissue water and each fraction has a higher structural order than ordinary water. Once again the minor fraction, which is about 8% of total tissue water, is evidently more quasi-crystalline than the larger fraction.

Cope asserts that his data are compatible with a more or less continuous distribution of water structures within the cell. That the nature of cellular water is complex has been underlined not only by the studies of Cope and others but also by different evidence. This is not an entirely unexpected conclusion since our present knowledge about cells indicates that there is a diversity of macromolecular surfaces within them.

'Non-solvent' behaviour of cellular water

In osmotic swelling and shrinkage experiments on many types of cells it has been observed almost invariably that these cells do not behave *apparently* as perfect osmometers. Ponder (1948) expressed this spurious behaviour in terms of a factor 'R' equivalent to the apparent fraction of the total cellular water taking part in the osmotic response. The discrepancy between estimated values of 'R' and unity has been ascribed to the fraction of cellular water which is not free to act as a solvent. In the case of erythrocyte, which has been commonly cited in this connexion, the difference of 'R' from unity has been attributed to the water associated directly with haemoglobin (Savitz, Sidel & Solomon, 1964). Dick (1959a, 1966) has discussed in great detail the interpretations of 'R' values in erythrocytes and in other cells, such as muscle. With regard to the erythrocyte, Dick concluded that the osmotic behaviour of this cell in a swollen condition can be described adequately by taking into account the physico-chemical behaviour of haemoglobin in aqueous solution. The corresponding interpretation of the osmotic shrinkage observations cannot be so easily made because of the complex behaviour of haemoglobin under these conditions. In principle, a similar analysis of the familiar osmotic behaviour of other cells, such as ovarian eggs and muscle fibres, could be made. The proteins in these cells and their behaviour in aqueous solutions are not so well described as that of haemoglobin and, therefore, the task of predicting their behaviour in swelling and shrinkage experiments would be formidable. These experimental and theoretical difficulties illustrate the ambiguity generated by the use of apparent 'non-solvent' behaviour of water as an index of structural order in cellular water. For example, LeFevre (1964) and Miller (1964) have obtained estimates of 0·79 and 1·00 for the fraction of cellular water in the erythrocyte which is available for the solution of glucose. The notable disparity between those values remains unexplained.

A test for 'non-solvent' behaviour of cellular water, which is considerably less doubtful than the use of 'R' values, is the NMR spectrum of sodium ions in biological tissues. Numerous investigators (e.g. Troshin, 1961; Shaw & Simon, 1955; Dick & Lea, 1964; Ling & Ochsenfeld, 1966) have shown by various techniques that a large proportion of total intracellular sodium is 'sequestered'. The term 'sequestered' is chosen to describe an immobilized sodium fraction in a manner which is free from any mechanistic implications. Cope (1965, 1967a), however, has found that approximately 70% of the total sodium content of muscle gave no observable NMR signal and he inferred that this fraction is 'complexed' with macromolecules such as actomyosin. Similar studies on kidney and brain tissues revealed in each case that about 60% of the total sodium content is 'complexed'. The term 'complexed' is taken to mean 'chemically bound' as distinct from the non-committal use of 'sequestered' sodium, and the NMR analysis seems to bear out the validity of Cope's description. *In vitro* experiments confirmed that actomyosin 'complexed' sodium ions and, further, these NMR studies demonstrated that the solubility of sodium ions in actomyosin gels is considerably lower than that in water. According to Cope the reduced solubility of sodium in the vicinity of actomyosin implied a high degree of structural order in the neighbouring water. The general inference from Cope's work is that the sequestering of intracellular sodium in several types of cells is associated with the highly organized nature of water in the neighbourhood of macromolecules.

Finally, it should be emphasized that any estimate of the nonsolvent fraction of total cellular water is necessarily the lower limit of that proportion of cellular water which has a higher structural order than pure water. This must be so, since there is no *a priori* reason to believe that the quasi-crystalline water, which has been detected in living tissues by NMR spectroscopy, can completely exclude small solutes, such as sodium ions and glucose molecules.

X-ray diffraction studies

The interpretation of low-angle X-ray diffraction patterns from preparations of isolated membranes has offered an estimate of the degree of hydration which is required obligatorily for the combination of lipid with protein in cell membranes. On this basis, Finean, Coleman, Green & Limbrick (1966) found that about 30% of the dried weight of myelin could be attributed to water intimately associated with its structure. Finean *et al.* (1966) also found that about 20% of the dry weight of rat erythrocyte was constituted by water. This water is essential to the structural integrity of the plasma membrane. Using an entirely different technique Ladbrooke, Jenkinson, Kamat & Chapman (1968) have confirmed those estimates. These workers found that the thermal transition points, which are characteristic of free lipid phases, are absent from membrane lipids, but they noted that, when the level of hydration of myelin was reduced to about 20% of the dry weight, the thermal transitions appeared. Thus, both these methods demonstrated that the lipid phase was liberated from the structure of the membrane when the water content fell below a level of 20%. Consequently we may conclude that the maintenance of the lipoprotein structure in cell membranes depends upon that fraction of membrane hydration which is relatively immobilized. The work of Worthington & Blaurock (1968) on the structure of myelin seems to indicate that there are discrete layers of water about 15 Å thick between the membranes. Although such layers are compatible with the size of the fraction of immobilized water in myelin, Finean (1969) has cited evidence against this view. Finean stressed particularly the observation (Finean & Millington, 1957) that the periodicity of myelin increases when a nerve is placed in hypertonic solutions and he contended that this finding is not consistent with the proposal that there are discrete layers of water within the myelin.

Conclusion

Although the preceding review of the state of water in cells has been far from extensive, it still illustrates the important point that some small fraction of the water in cells exists in a more organized form than that of pure water. Does this conclusion challenge the 'membrane theory' of transport, where the cytoplasm is assumed to behave almost like an ideal electrolyte solution, and does it favour the 'sorption theory' where the exchange and transport of substances including water are rate-controlled by the characteristics of adsorption on to fixed-charge sites in the cytoplasm? In my opinion the evidence is not in conflict with the membrane theory. On the contrary possibly the most important site of organized or immobilized water is the cell membrane itself. This may be of crucial significance because of the strategic position that the membrane occupies both in the structure and function of the cell. Only further experiments will reveal exactly what influence such organized water molecules exert on transport phenomena in cells.

2
PHENOMENOLOGICAL DESCRIPTION OF TRANSPORT PROCESSES

Practical preamble	36
The phenomenological equations of irreversible thermodynamics	40
Transport of non-electrolytes and water	44
Membrane coefficients	49
Transport of ions and water	53
Membrane coefficients	57
Transport of heat and water	60
Phenomenological description of active transport	63
Transport of non-electrolytes and water across heterogeneous membranes	66
Membranes in parallel	66
Membranes in series	70

Practical preamble

THIS introductory section contains the bare theoretical bones that are necessary to understand the experimental work described in later chapters. Essentially we are concerned with the flows of water and uncharged solutes across a membrane.

Consider a homogeneous membrane separating outer and inner (ideal) dilute solutions designated by superscripts o and i. Net fluxes of solute and water from outside to inside are taken as positive. The difference in solute concentration across the membrane is given by the outside concentration minus the inside one. Differences of hydrostatic and osmotic pressure are defined similarly.

The subscripts v, s and i refer to volume, permeant solute and impermeant solute respectively.

The following symbols are used.

J_v	$cm^3\ cm^{-2}\ sec^{-1}$	Net volume flux
J_s	$mole\ cm^{-2}\ sec^{-1}$	Net solute (s) flux

R	atm cm³ mole⁻¹ deg⁻¹	Gas constant
T	deg	Absolute temperature
Δp	atm	Hydrostatic pressure difference
$\Delta \pi_i$	atm	Osmotic pressure difference due to i (equals $RT\Delta c_i$ where Δc_i (mole cm⁻³) is the concentration difference of i).
$\Delta \pi_s$	atm	Osmotic pressure difference due to s (equals $RT\Delta c_s$ where Δc_s is the concentration difference of s).
$\bar{c}_s$	mole cm⁻³	Average concentration of s across membrane.
L_p	cm sec⁻¹ atm⁻¹	Hydraulic conductivity of membrane.
ω_s	mole cm⁻² atm⁻¹ sec⁻¹	'Solute permeability' of membrane, related to the conventional permeability P_s (cm sec⁻¹) by $\omega_s RT = P_s$.
σ_s		Reflexion coefficient of membrane.

The equations derived from irreversible thermodynamics to describe volume and solute flows are (see page 53)

$$J_v = L_p(\Delta p - \Delta \pi_i - \sigma_s \Delta \pi_s)$$
$$J_s = \omega_s \Delta \pi_s + J_v \bar{c}_s (1 - \sigma_s)$$

In order to use these equations L_p, ω_s and σ_s must be determined.

Hydraulic conductivity. The volume flow equation shows that when $\Delta \pi_i$ and $\Delta \pi_s$ are both zero

$$J_v = L_p \Delta p$$

L_p may be determined, therefore, by measuring J_v for a given Δp. This method has been used for single cells (e.g. squid axon, Vargas, 1968*a*; see page 166), capillaries (e.g. frog mesentery, Landis, 1927; see page 294), and epithelia (e.g. canine gastric mucosa, Moody & Durbin, 1969; see page 336).

Alternatively when Δp and $\Delta \pi_s$ are both zero the volume flow is given by

$$J_v = -L_p \Delta \pi_i$$

where the negative sign indicates that volume moves in the

opposite direction to that of $\Delta\pi_i$. This relation also yields L_p and it has been used for single cells (e.g. sea urchin eggs, Lucké, 1940; see page 163), capillaries (e.g. frog mesentery, Landis & Sage, 1971; see page 297) and epithelia (e.g. rat proximal tubule, Ullrich, Rumrich & Fuchs, 1964; see page 323).

Solute permeability. The solute flux equation shows that when J_v is zero

$$J_s = \omega_s \Delta\pi_s$$

or, more conveniently,

$$J_s = P_s \Delta c_s$$

In current practice P_s is determined by measuring the unidirectional flux of labelled s for a given concentration difference of labelled s when there is no volume flow. Solute permeabilities are not considered in detail in this book but some work, for example, on epithelia is described (e.g. rabbit gall bladder, Smulders & Wright, 1971; see page 348).

This procedure can also be used to obtain the diffusional permeability to labelled water which is conventionally called P_d to distinguish it from L_p. Attempts have been made to measure P_d for single cells (e.g. human erythrocytes, Paganelli & Solomon, 1957; see page 153), capillaries (e.g. *vasa recta* of rat kidney, Morgan & Berliner, 1968; see page 300) and epithelia (e.g. frog skin, Dainty & House, 1966b; see page 321).

Reflexion coefficient. This membrane coefficient can be determined in a number of ways that are indicated by inspection of the volume flow equation above.

First, consider that $\Delta\pi_i$ is zero. Then the volume flow will be halted, i.e. $J_v = 0$, when

$$\Delta p = \sigma_s \Delta\pi_s$$

Knowing $\Delta\pi_s$ and the appropriate value of Δp necessary to make J_v equal to zero one can obtain $\sigma_s = \Delta p/\Delta\pi_s$. This method has been used for artificial membranes but not for biological ones with the notable exception of Kedem & Katchalsky's (1958) estimates of σ_s from the data of Pappenheimer, Renkin & Borrero (1951) on the capillaries of the dog's hind limb (see page 306).

Secondly, consider that $\Delta p = 0$. Then the volume flow will be

halted when
$$\Delta\pi_i = -\sigma_s \Delta\pi_s$$

In practice the appropriate value of $\Delta\pi_s$ for a known $\Delta\pi_i$ can be found by interpolation after a number of volume flow experiments have been performed at different values of Δc_s. This method has been used, for example, for single cells (e.g. human erythrocytes, Goldstein & Solomon, 1960; see page 253).

Thirdly, consider again that $\Delta p = 0$ and also that $\Delta\pi_i = 0$. Then the volume flow for a known $\Delta\pi_s$ is given by

$$J_v = -L_p \sigma_s \Delta\pi_s$$

By measuring J_v and L_p (in a previous experiment) one can obtain σ_s. This method has been used, for example, for capillaries (e.g. rabbit heart capillaries, Vargas & Johnson, 1964; see page 306).

Finally, consider again that $\Delta p = 0$. Suppose that two different volume fluxes, J_v' and J_v'', are obtained for experiments where first $\Delta\pi_i$ only is set up across the membrane and subsequently an identical value of $\Delta\pi_s$ only is established.

Then the separate experiments are described by

$$J_v' = -L_p \Delta\pi_i$$
$$J_v'' = -L_p \sigma_s \Delta\pi_s \text{ (where } \Delta\pi_s = \Delta\pi_i\text{)}$$

Thus, the reflexion coefficient is obtained from

$$\sigma_s = \frac{J_v''}{J_v'}$$

This method has been used, for example, for single cells (e.g. *Nitella translucens*, Dainty & Ginzburg, 1964d; see page 253) and epithelia (e.g. rabbit gall bladder, Wright & Diamond, 1969a,b; see page 351).

The preceding description outlines briefly how L_p, ω_s and σ_s may be determined practically. The significance of both L_p and ω_s is quite plain and that of the new membrane parameter—the reflexion coefficient—is discussed later (see pages 52 and 83).

The rest of this chapter deals more fully with the equations described above and with others between the flows of solutes and water across homogeneous membranes and the following operational 'forces': the difference of pressure, Δp, the difference of osmotic pressure, $\Delta\pi$, the difference of electric potential, $\Delta\psi$, and the difference of temperature, $\varDelta T$, across the membrane. The basic aim of the following account of the phenomenological approach to solute and water transport across membranes is to offer simply an introduction to this theoretical field and thereby set the scene for the discussion of transport phenomena in artificial and biological membranes. It is certainly not my intention to present a rigorous treatment of the theory of irreversible thermodynamics as applied to membrane transport. Those readers who wish to study such an account should consult the textbook of Katchalsky & Curran (1965).

The phenomenological equations of irreversible thermodynamics

Until recently a set of self-consistent flux equations describing solute and water flows across membranes did not exist. During the course of experimental work on membrane transport certain flux equations became accepted somewhat reluctantly. Several workers (Frey-Wyssling, 1946; Ussing, 1952; Pappenheimer, 1953) attempted to modify these conventional flux equations to account for the contribution which solute flow makes to volume flow across the membrane, and also for the effect of 'solvent drag' on solute flow. The advent of irreversible thermodynamics offered a solution to these theoretical difficulties. In 1958 Kedem & Katchalsky published a treatment of the solute and volume flow equations, derived from irreversible thermodynamics, which had direct application to permeability studies on biological membranes.

Although certain basic elements of irreversible thermodynamics were discussed in the last century, this theoretical approach did not gain a sound foothold until 1931 when Onsager published his work on the relations between forces and flows. Thereafter numerous workers (Prigogine, 1961; De Groot, 1952) extended this operational approach to the treatment of several phenomena. The applications of irreversible thermodynamics to transport across cell membranes are illustrated, for example, in the work of Kedem & Katchalsky (1958, 1961, 1963a,b,c) and Katchalsky & Kedem (1962). Irreversible or non-equilibrium thermodynamics is founded on classical thermodynamics, and it deals with the time course of irreversible processes. The particular irreversible processes, in which we are interested, are the movements of solutes and water across membranes and we can use irreversible thermodynamics to describe the rates of these processes. Whereas classical thermodynamics indicates that the entropy of an isolated system, undergoing irreversible change, increases towards a maximum at equilibrium, irreversible thermodynamics considers the rate of change of entropy, dS/dt, where time is denoted by t. An important aspect of this theoretical approach is that the total entropy change, dS, of a system can be split into two components. First, there is a transfer of entropy, d_eS, between the system and its surroundings and, secondly, there is the increment, d_iS, in the internal entropy which is generated by the irreversible process occurring within the system. Therefore, we have

$$dS = d_eS + d_iS \qquad 2.1$$

For a closed system experiencing reversible change, the change in entropy is given by the heat gained, dQ, divided by the absolute temperature, that is

$$d_eS = \frac{dQ}{T} \qquad 2.2$$

If the same closed system now undergoes an irreversible change, the entropy change is greater than that due to the heat absorbed. Thus

$$dS = \frac{dQ}{T} + d_iS \qquad 2.3$$

While d_eS may be positive or negative, d_iS is invariably greater than, or equal to, zero. When dealing with irreversible processes in an isothermal system it is frequently convenient to consider the function Φ given by

$$\Phi = T\frac{d_iS}{dt} \qquad 2.4$$

which Lord Rayleigh called the *dissipation function*. This function is a measure of the dissipation of free energy by the irreversible processes. Provided that the irreversible processes are *slow* and the system is *not too far from equilibrium* then it can be shown that Φ is the sum of the products of fluxes and their appropriate driving forces (see Chapter 7, Katchalsky & Curran, 1965). In the language of irreversible thermodynamics one refers to the flux of substance i by the symbol J_i and the conjugate (appropriate) force on i by X_i.

In nature there are many examples of linear relations between fluxes and their conjugate forces. For example, Fick demonstrated that the rate of diffusion of a substance is proportional to the negative gradient of its concentration. Extensive studies of the relations between flows and their conjugate forces revealed, however, that there could exist coupling between a flow and a nonconjugate force. Seebeck, for example, showed that a temperature gradient in a bimetallic system produced a current flow and Peltier demonstrated subsequently that current passage through that system generated heat transfer between the dissimilar metals. Thus, any theoretical treatment of the relation between flows and forces must include not only the coupling between fluxes and their conjugated forces but also possible coupling between flows and nonconjugate forces. First, any flow $\mathcal{J}_i$ may be expressed as a power series in X_i and if the system is relatively close to equilibrium (X_i small) then all powers of X_i larger than the first may be ignored. This approximation is certainly invalid when the range of values of X_i is extended beyond certain limits. Secondly, the relation between $\mathcal{J}_i$ and the non-conjugate force X_j has been assumed to be linear too and in 1931 Onsager embodied both of these relations between flows and forces in a set of equations known as the *phenomenological equations*. These equations can be expressed concisely in the form

$$\mathcal{J}_i = \sum_{i=1}^{n} L_{ij} X_j \quad (i = 1, 2, 3, \ldots n) \qquad 2.5$$

or in the extended form

$$\begin{aligned}
\mathcal{J}_1 &= L_{11}X_1 + L_{12}X_2 + L_{13}X_3 + \cdots + L_{1n}X_n \\
\mathcal{J}_2 &= L_{21}X_1 + L_{22}X_2 + L_{23}X_3 + \cdots + L_{2n}X_n \\
\mathcal{J}_3 &= L_{31}X_1 + L_{32}X_2 + L_{33}X_3 + \cdots + L_{3n}X_n \\
&\cdots \\
\mathcal{J}_n &= L_{n1}X_1 + L_{n2}X_2 + L_{n3}X_3 + \cdots + L_{nn}X_n
\end{aligned} \qquad 2.6$$

This system of equations describes n different flows in terms of n forces. $L_{11}, L_{22} \ldots L_{nn}$ are the classical coefficients signifying the relations between the fluxes and their corresponding conjugate forces. On the other hand, L_{ij} ($i \neq j$) are the cross coefficients signifying the dependence of fluxes on non-conjugate forces. Provided that the fluxes and conjugate forces are chosen so that the sum of their products, i.e. $\sum_{i=1}^{n} J_i X_i$, is equal to the rate of entropy production in the system, then the cross coefficients must satisfy the Onsager reciprocal relations, namely

$$L_{ij} = L_{ji} \quad (i \neq j) \qquad 2.7$$

In an isothermal system this pre-condition for the Onsager reciprocal relations corresponds to equating $\sum_{i=1}^{n} J_i X_i$ with $T d_i S/dt$, or in other words

$$\Phi = \sum_{i=1}^{n} J_i X_i \qquad 2.8$$

Thus, in isothermal systems the sum of the products of the fluxes and their conjugate forces must equal the dissipation function, and we can employ equation 2.8 to select appropriate forces and flows for the sake of experimental convenience. This selection procedure will be illustrated later.

An important consequence of the expression 2.7 is that instead of the n^2 coefficients apparently required to describe n flows in equation 2.6 only $n(n+1)/2$ independent coefficients are required. Hence, if we wish to describe the flows of, say, water and glucose across a membrane then three independent transport coefficients are required.

Since the phenomenological equations have a linear form, they can be expressed in an alternative way. The transformation renders corresponding relations for the forces, expressed as linear functions of the flows, namely

$$\begin{aligned} X_1 &= R_{11} J_1 + R_{12} J_2 + R_{13} J_3 + \cdots + R_{1n} J_n \\ X_2 &= R_{21} J_1 + R_{22} J_2 + R_{23} J_3 + \cdots + R_{2n} J_n \\ X_3 &= R_{31} J_1 + R_{32} J_2 + R_{33} J_3 + \cdots + R_{3n} J_n \\ &\cdots \\ X_n &= R_{n1} J_1 + R_{n2} J_2 + R_{n3} J_3 + \cdots + R_{nn} J_n \end{aligned} \qquad 2.9$$

Whereas the coefficients L_{ij} have the characteristics of generalized conductances, the alternative set of coefficients R_{ij} represent generalized resistances.

The phenomenological equations of irreversible thermodynamics describe the relations between fluxes and forces in continuous systems. For example, these equations can be applied to the analysis of isothermal diffusion of solute in a volume of solvent. Schlögl (1969) has contended that the phenomenological equations should not be used to describe transport of material across biological cell membranes since these thin layers cannot be considered to be continuous systems. In principle, this objection is valid. In practice, there is no adequate alternative theory of membrane transport which expresses itself in measurable parameters and is entirely free from some theoretical objection or other. The formal framework of irreversible thermodynamics, as it is applied to membrane transport, is probably the most useful one we have at present. Consequently it is developed in the following discussion.

Transport of non-electrolytes and water

Consider the membrane system represented in Fig. 2.1. The system consists of two compartments separated by a *homogeneous* membrane of thickness Δx (cm) and area A (cm²), and the compartments are filled with different aqueous solutions of the same non-electrolyte s. The outer and inner compartments are designated by the superscripts o and i respectively and the convention adopted here and elsewhere in this book is that 'outside' is on the left and 'inside' is on the right. Let us assume that there is no difference of temperature across the membrane but that there is a difference of hydrostatic pressure, Δp, and of solute concentration, Δc_s, across the membrane. The permeation of solute and water which results from these gradients involves discontinuous transitions; that is, both species have to cross phase boundaries between the membrane and the solutions. In addition, the driving forces on solute and water are established within the membrane. In order to facilitate the adaptation of the thermodynamic formalism, developed for continuous systems, to this discontinuous system several conditions must be satisfied. It must be assumed that each compartment is so well agitated that no local concentration gradients of s exist at the surfaces of the membrane. Moreover, it must be assumed that our discontinuous system is in a state of stationary flow; that is, the net fluxes of solute and water across the membrane are independent of time. The latter condition implies, for example, that the flux of s has the same magnitude and direction at all points in the membrane. Under these circumstances certain 'forces' cause flows of water and solute across the membrane. Before embarking on the simple mathematics of this case let us guess what the final result might look like. First, one would expect that Δp will produce an hydraulic flow of water and Δc_s might establish an osmotic flow.

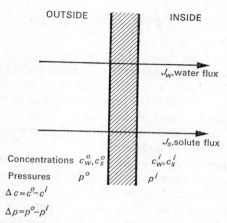

FIG. 2.1. Basic transport system showing flows of solute and water across a homogeneous membrane. The symbols and sign conventions used in the text are also shown in the diagram.

Secondly, Δc_s will produce a diffusional flux of s and Δp might generate an ultrafiltration of s across the membrane. Putting these predictions into the framework of the phenomenological equations 2.6 and expressing them as simply as possible, we expect to find that

water flow = **hydraulic flow** + osmotic flow

and

solute flow = ultrafiltration + **diffusion**

where the bold type signifies the flows driven by their conjugate forces. Let us now return to the theoretical treatment of this problem.

Let us denote the net fluxes of solute and water across the membrane from outside to inside by J_s and J_w respectively. These fluxes represent the number of moles passing across unit area of membrane in unit time and they are expressed in units of mole cm^{-2} sec^{-1}. The theory of irreversible thermodynamics shows that the appropriate conjugate forces X_s and X_w in this case are the negative gradients of the chemical potentials for solute and water. The gradient of the chemical potential, μ_s, for the non-electrolyte s at any point x in the membrane is given by

$$\frac{d\mu_s}{dx} = \bar{V}_s\frac{dp}{dx} + RT\frac{d}{dx}(\ln \gamma_s c_s) \qquad 2.10$$

where the sign of the chemical potential gradient, $d\mu_s/dx$, is taken as positive when μ_s increases with x, $\bar{V}_s$ is the partial molar volume in cm^3 mole^{-1}, γ_s is the activity coefficient, c_s is the concentration of s in mole cm^{-3}, dp/dx is the pressure gradient in atm cm^{-1}, R is the gas constant in atm cm^3 deg^{-1} mole^{-1} and T is the absolute temperature. In this scheme the units of μ_s are atm cm^3 mole^{-1} which are equivalent to joule mole^{-1}.

It is customary in treating biological systems, where the activity, $\gamma_s c_s$, of the solute cannot be determined accurately, to replace activity in equation 2.10 by the mole fraction, n_s, for s. This procedure is also adopted for water and n_w denotes its mole fraction. Thus, the conjugate forces for water and solute flows are

$$X_w = -\frac{d\mu_w}{dx} = -\bar{V}_w\frac{dp}{dx} - RT\frac{d}{dx}(\ln n_w) \qquad 2.11$$

$$X_s = -\frac{d\mu_s}{dx} = -\bar{V}_s\frac{dp}{dx} - RT\frac{d}{dx}(\ln n_s) \qquad 2.12$$

We can now employ the dissipation function to obtain new relations between the flows and forces. The treatment outlined below is essentially that of Kedem & Katchalsky (1958) and Katchalsky (1961).

Consider a volume element of unit area and thickness dx within the membrane. If this volume is homogeneous, and provided there is no electrical gradient across the membrane, the dissipation function, Φ_e, for the element is

$$\Phi_e = J_w\left(-\frac{d\mu_w}{dx}\right) + J_s\left(-\frac{d\mu_s}{dx}\right) \qquad 2.13$$

Equation 2.13 is based on the general relation 2.8. Since J_w and J_s are independent of x it is possible to integrate equation 2.13 across the membrane from $x = 0$ to $x = \Delta x$ to find the dissipation function per unit area, Φ_m, for the entire membrane. Thus,

$$\Phi_m = J_w \int_0^{\Delta x} \left(-\frac{d\mu_w}{dx}\right) dx + J_s \int_0^{\Delta x} \left(-\frac{d\mu_s}{dx}\right) dx \qquad 2.14$$

Provided that the chemical potentials at the membrane surfaces are identical to the corresponding chemical potentials in the adjacent solutions (see Katchalsky & Curran, 1965), Φ_m may be expressed as

$$\Phi_m = J_w(\mu_w^0 - \mu_w^i) + J_s(\mu_s^0 - \mu_s^i) \qquad 2.15$$

Thus, Φ_m is the sum of the products of each flow with its corresponding 'force' namely $\Delta\mu_w$ and $\Delta\mu_s$ which are the differences of the chemical potentials between the bathing solutions. It is more convenient to choose the differences of hydrostatic pressure, Δp, and of osmotic pressure, $\Delta\pi$ (or, $RT\Delta c_s$), as the generalized 'forces' rather than $\Delta\mu_w$ and $\Delta\mu_s$ since the latter cannot be determined easily. Although this is a convenient theoretical transformation it raises a very serious difficulty. Mauro (1960, 1965) has produced not only strong theoretical arguments but also experimental evidence demonstrating that during the osmotic flow across porous semi-permeable membranes there is a discontinuity in the hydrostatic pressure at the mouth of the pores which are excluding solute molecules. In this case it is wrong to equate the hydrostatic pressure in the solution with the pressure at the surface of the porous membrane. Since the theory of transport described here treats the case of permeation in a *homogeneous* membrane, that particular difficulty will be ignored. However, it does cast very serious doubt on the validity of applying this continuous analysis of membrane transport to *heterogeneous* membranes, especially porous ones.

For ideal solutions in the outer and inner compartments we have

$$\Delta\mu_w = \bar{V}_w \Delta p + RT\Delta (\ln n_w) \qquad 2.16$$

and

$$\Delta\mu_s = \bar{V}_s \Delta p + RT\Delta (\ln n_s) \qquad 2.17$$

Provided both solutions are dilute then equations 2.16 and 2.17 become

$$\Delta\mu_w = \bar{V}_w \Delta p - \frac{RT}{\bar{c}_w} \Delta c_s \qquad 2.18$$

$$\Delta\mu_s = \bar{V}_s \Delta p + \frac{RT}{\bar{c}_s} \Delta c_s \qquad 2.19$$

where $\Delta c_s = c_s^o - c_s^i$, and $\bar{c}_w$ and $\bar{c}_s$ are the mean concentrations of water and solute across the membrane. When dilute solutions bathe both sides of the membrane there is a negligible difference of the water concentration and $\bar{c}_w$ may be taken equal to that of pure water. Under these conditions, where the membrane separates ideal solutions having approximately similar solute concentrations, $\bar{c}_s$ is the numerical average of c_s^o and c_s^i. Introducing equations 2.18 and 2.19 into equation 2.15 for Φ_m yields

$$\Phi_m = \mathcal{J}_w \left(\bar{V}_w \Delta p - \frac{RT}{\bar{c}_w} \Delta c_s \right) + \mathcal{J}_s \left(\bar{V}_s \Delta p + \frac{RT}{\bar{c}_s} \Delta c_s \right) \qquad 2.20$$

Rearranging equation 2.20 gives

$$\Phi_m = (\mathcal{J}_w \bar{V}_w + \mathcal{J}_s \bar{V}_s) \Delta p + \left(\frac{\mathcal{J}_s}{\bar{c}_s} - \frac{\mathcal{J}_w}{\bar{c}_w} \right) RT \Delta c_s \qquad 2.21$$

Thus Φ_m can be represented by a different set of flows and forces, and equation 2.21 reveals that $X_v = \Delta p$ and $X_D = RT\Delta c_s$ while the conjugate flows are the total *volume flow*, $\mathcal{J}_v$, per unit area, i.e.

$$\mathcal{J}_v = \mathcal{J}_w \bar{V}_w + \mathcal{J}_s \bar{V}_s \qquad 2.22$$

and the so-called *exchange flow*, $\mathcal{J}_D$, per unit area. In fact, $\mathcal{J}_D$ is the relative velocity of s with respect to water, described by

$$\mathcal{J}_D = \frac{\mathcal{J}_s}{\bar{c}_s} - \frac{\mathcal{J}_w}{\bar{c}_w} \qquad 2.23$$

The units of the new flows are different from that of the old flows, $\mathcal{J}_s$ and $\mathcal{J}_w$; in fact, both $\mathcal{J}_v$ and $\mathcal{J}_D$ have the units of velocity, cm sec^{-1}, but the units of $\mathcal{J}_v$ are commonly quoted as cm^3 cm^{-2} sec^{-1} to indicate that it is the volume passing across unit area in unit time. The new set of flows and forces may be written in the form of the phenomenological equations to yield

$$\mathcal{J}_v = L_p \Delta p + L_{pD} \Delta \pi \qquad 2.24$$
$$\mathcal{J}_D = L_{Dp} \Delta p + L_D \Delta \pi \qquad 2.25$$

where the osmotic pressure difference, $\Delta \pi$, has replaced $RT\Delta c_s$. Equations 2.24 and 2.25 indicate that three independent coefficients are required to describe the relations between forces and flows in this system since $L_{pD} = L_{Dp}$ by the Onsager reciprocal relations.

Before going on to discuss the nature of the transport coefficients specified by the phenomenological equations 2.24 and 2.25 it is worth noting that these flow equations confirm our original guess about the outcome of the preceding analysis. In other words the volume flux contains both a hydraulic term ($L_p \Delta p$) and an osmotic term ($L_{pD} \Delta \pi$) and the exchange flux contains both a diffusional term ($L_D \Delta \pi$) and an ultrafiltration term ($L_{Dp} \Delta p$).

Membrane coefficients

In general we need three independent coefficients to describe a membrane separating two aqueous solutions of a non-electrolyte. However, if the particular membrane under study is impermeable to the solute then only one coefficient is necessary for its characterization. In this case the exchange flow is given by

$$J_D = -\frac{J_w}{\bar{c}_w} \qquad 2.26$$

since $(J_s/\bar{c}_s) = 0$. Equation 2.26 may be re-written as

$$J_D = -\frac{J_w \bar{V}_w}{\bar{c}_w \bar{V}_w} \qquad 2.27$$

and since $\bar{V}_w \bar{c}_w$ is unity for dilute solutions (i.e. $\bar{c}_w = 55 \cdot 5 \times 10^{-3}$ mole cm^{-3} and $\bar{V}_w = 18$ cm^3 mole^{-1}) we obtain

$$J_D = -J_v \qquad 2.28$$

If we add equations 2·24 and 2·25 together and substitute $J_D = -J_v$ we find

$$(L_p + L_{Dp})\Delta p + (L_{pD} + L_D)\Delta \pi = 0 \qquad 2.29$$

which is satisfied only if the $(L_p + L_{Dp})$ and $(L_{pD} + L_D)$ are both zero. These necessary conditions can be expressed as

$$L_p = -L_{Dp} = L_D \qquad 2.30$$

since $L_{Dp} = L_{pD}$. Thus, the ideal semipermeable membrane which excludes solute transport is completely described by one coefficient, L_p.

Let us return to the case of the 'leaky' membrane permeable to both water and solute. In practice we must determine three coefficients which are related to L_p, L_{pD} and L_D in order to characterize the membrane. These practical coefficients are the *hydraulic conductivity*, the *solute permeability* and the *reflexion coefficient*. The first of these is identical to L_p but the remaining two practical coefficients are not identical to L_{pD} and L_D but are, in fact, parameters involving L_p, L_{pD} and L_D. In other words, the nature of the independent membrane coefficients is dictated by experimental convenience and not by the formal nature of the phenomenological equations for volume and exchange flows.

Hydraulic conductivity. Let us assume that the membrane separates two identical solutions and that a difference of hydrostatic pressure, Δp, is applied. According to equation 2.24 there will be a volume flow given by

$$J_v = L_p \Delta p \qquad 2.31$$

In this type of experiment J_v will be measured in units of cm³ cm⁻² sec⁻¹ and Δp in atm. These measurements determine L_p, known as the hydraulic conductivity or the mechanical filtration coefficient. L_p is our first practical coefficient and it is expressed in units of cm sec⁻¹ atm⁻¹. Of course, L_p may be expressed in different units, such as cm³ dyne⁻¹ sec⁻¹ if Δp is measured in dyne cm⁻². There is a small advantage in quoting L_p in cm sec⁻¹ atm⁻¹ because it is possible to measure L_p in a different type of experiment, where the driving force is not Δp but an osmotic pressure difference which is also conveniently expressed in atm. We shall discuss this type of L_p determination later.

In the hydraulic flow experiment described above there will also be a flow of solute if L_{pD} is not zero. The exchange flow is given by

$$J_D = L_{pD} \Delta p \qquad 2.32$$

and it expresses the rate of ultrafiltration of the solute in response to the hydrostatic pressure gradient.

Solute permeability. Instead of determining the exchange flow across the membrane it is more convenient to measure the solute flux, J_s. Indeed, it is possible to obtain an expression for J_s from the equations for J_v and J_D. By rearranging and adding equations 2.22 and 2.23 we find

$$\mathcal{J}_s = \frac{(\mathcal{J}_v + \mathcal{J}_D)\bar{c}_s}{\bar{V}_s \bar{c}_s + 1} \qquad 2.33$$

In the derivation of equation 2.33 use is made of the approximation $\bar{V}_w \bar{c}_w = 1$. Moreover, for dilute solutions we can assume that $\bar{V}_s \bar{c}_s \ll 1$ and equation 2.33 becomes

$$\mathcal{J}_s = (\mathcal{J}_v + \mathcal{J}_D)\bar{c}_s \qquad 2.34$$

Let us assume that there is a difference of solute concentration, Δc_s, across the membrane. Determinations of the solute permeability should be performed under conditions of zero volume flow. This condition is achieved by the application of a pressure gradient given by

$$\Delta p = -\frac{L_{pD}}{L_p} RT \Delta c_s \qquad 2.35$$

obtained by setting $\mathcal{J}_v = 0$ in equation 2.24. In this type of experiment, therefore, the solute flux is measured in the absence of volume flow. The conventional solute permeability P_s is determined by

$$\mathcal{J}_s = P_s \Delta c_s \qquad 2.36$$

where $\mathcal{J}_s$ is expressed in mole cm^{-2} sec^{-1} and Δc_s in mole cm^{-3}; hence P_s is expressed in cm sec^{-1}. How does P_s relate to L_p, L_{pD} and L_D? This is revealed by substituting equations 2.24, 2.25 and 2.35 into equation 2.34 to yield

$$\mathcal{J}_s = \left[\frac{L_p L_D - L_{pD}^2}{L_p}\right] \bar{c}_s RT \Delta c_s \qquad 2.37$$

In equation 2.37 the coefficient $[(L_p L_D - L_{pD}^2)/L_p]\bar{c}_s$ is the second practical coefficient defined by Kedem & Katchalsky (1958) and referred to as the solute permeability; it is denoted by ω_s. Furthermore, ω_s is related to P_s by

$$\omega_s RT = P_s \qquad 2.38$$

The units of ω_s are mole cm^{-2} atm^{-1} sec^{-1}.

Thus, the first two practical coefficients, L_p and ω_s, are rather similar to, but more rigorously defined than, the conventional water and solute permeabilities used in membrane transport studies.

Reflexion coefficient. The remaining practical coefficient is the reflexion coefficient, σ_s, introduced originally by Staverman (1951) and employed by Kedem & Katchalsky (1958) in their treatment of the phenomenological equations. This coefficient is defined by

$$\sigma_s = -\frac{L_{pD}}{L_p} \qquad 2.39$$

or,

$$\sigma_s = -\left(\frac{\mathcal{J}_D}{\mathcal{J}_v}\right)_{\Delta\pi = 0} \qquad 2.40$$

If we perform an experiment in which a suitable pressure gradient Δp is applied across a membrane to bring the volume flux to zero, then according to equation 2.24

$$\Delta p = -\frac{L_{pD}}{L_p} \Delta\pi = \sigma_s RT \Delta c_s \qquad 2.41$$

or in other words

$$\sigma_s = \frac{\Delta p}{RT \Delta c_s} \qquad 2.42$$

Thus the reflexion coefficient is determined by the applied pressure divided by the theoretical van't Hoff osmotic pressure, namely $RT\Delta c_s$. For an ideal semipermeable membrane one will recall that $L_p = -L_{pD}$, i.e. $\sigma_s = 1$; under those conditions the solute concentration gradient will develop its full van't Hoff osmotic pressure. Thus, if we establish an osmotic gradient $RT\Delta c_i$, across a membrane with an impermeant solute i the volume flux will be given by

$$\mathcal{J}_v = L_p RT \Delta c_i \qquad 2.43$$

when $\Delta p = 0$. This kind of experiment has been used frequently to determine the hydraulic conductivity of the membrane. For the semipermeable membrane one can see the physical meaning of Staverman's reflexion coefficient because the solute does not permeate but is 'reflected' back. For the permeable membrane the full reflexion is not observed, i.e. $\sigma_s < 1$, and the pressure gradient required to make $\mathcal{J}_v = 0$ is less than the van't Hoff osmotic pressure.

The value of σ_s depends not only on the characteristics of the membrane but also on the nature of the solute. In this sense it is similar to the solute permeability ω_s but it differs from ω_s insofar as its magnitude expresses the degree of interaction between the solute and water within the membrane. This point will be discussed fully in Chapter 3.

Since the reflexion coefficient is defined as the ratio of two pressures it has no units. Its magnitude normally lies between 0 and 1. When $\sigma_s = 0$, the membrane does not discriminate between the permeant solute and water. It has generally been assumed that the reflexion coefficients for non-electrolytes never attain negative values as is the case for electrolytes (Kedem & Katchalsky, 1961, 1963b). The existence of negative reflexion coefficients means that the membrane exhibits anomalous osmosis where volume flow occurs in the opposite direction to that indicated by the apparent osmotic gradient. Nevertheless, negative reflexion coefficients for a non-electrolyte (polyethylene glycol) have been recorded in Vycor glass membranes by Elias (1961a,b) and Talen & Staverman (1965a,b).

With the practical coefficients L_p, ω_s and σ_s it is now possible to express the equations 2.24 and 2.25 in the relatively more useful forms of equations for $\mathcal{J}_v$ and $\mathcal{J}_s$. The equation for $\mathcal{J}_v$ is

$$\mathcal{J}_v = L_p(\Delta p - \sigma_s RT \Delta c_s) \qquad 2.44$$

If there exists in addition to $RT\Delta c_s$ an osmotic pressure difference, $\Delta \pi_i$, due to certain impermeant solutes, then it is sufficiently accurate to re-write the expression for $\mathcal{J}_v$ as

$$\mathcal{J}_v = L_p[(\Delta p - \Delta \pi_i) - \sigma_s RT \Delta c_s] \qquad 2.45$$

The appropriate equation for $\mathcal{J}_s$ is

$$\mathcal{J}_s = \omega_s RT \Delta c_s + \mathcal{J}_v(1 - \sigma_s)\bar{c}_s \qquad 2.46$$

Equations 2.45 and 2.46 are, therefore, our basic equations relating the observable flows of volume and solute to the driving 'forces', Δp and $\Delta \pi$.

Transport of ions and water

The basic treatment of solute and water transport discussed above deals with the relatively simple case of permeation of uncharged molecules through a homogeneous membrane. When the non-electrolyte solutions in the outer and inner compartments of our system are replaced by different aqueous solutions of the same salt, an additional flow—electric current—is introduced into the phenomenological equations. Its conjugate force is the gradient of electric potential, E, across the membrane. This problem has been investigated theoretically by several authors (Staverman, 1952; Lorimer, Boterenbrood & Hermans, 1956; Kedem & Katchalsky, 1963a).

Each of the flows in this case will have three components:

water flow   = **hydraulic flow**   + osmotic flow
                                    + electro-osmotic flow
solute flow = ultrafiltration   + **diffusion**
                                    + electrophoresis
current flow = streaming current + diffusion current
                                    + **electric current**

where the bold type signifies the flows driven by their conjugate forces. We expect, therefore, that the phenomenological equations governing the flows and forces will take the following form

$$J_v = L_p \Delta p + L_{pD} \Delta \pi + L_{pE} \Delta E \qquad 2.47$$

$$J_D = L_{Dp} \Delta p + L_D \Delta \pi + L_{DE} \Delta E \qquad 2.48$$

$$J_E = L_{Ep} \Delta p + L_{ED} \Delta \pi + L_E \Delta E \qquad 2.49$$

In equation 2.47 for the volume flux there is an additional term $L_{pE}\Delta E$ expressing the coupling between the volume flow and the electrical gradient. This coupling is observed as the phenomenon of *electro-osmosis*. In equation 2.48 for the solute flow there is again a new term $L_{DE}\Delta E$ expressing another type of coupling which is observed in the form of *electrophoresis*. Finally, we come to the new flow equation 2.49 for the electric current. Obviously $L_E\Delta E$ denotes the expected electric current driven by ΔE but the other terms in the equation signify components of the observed current flow which arise due to the presence of Δp and $\Delta \pi$. The first term $L_{Ep}\Delta p$ is a *streaming current* owing to the pressure-driven flow of ions whereas the second term $L_{ED}\Delta \pi$ is a *diffusion current* owing to the ionic diffusion driven by $\Delta \pi$.

In order to obtain the appropriate membrane coefficients and the practical flux equations describing the flows of ions and water across the membrane we must again turn to the dissipation function.

For the sake of simplicity we shall assume that the membrane separates aqueous solutions of the same uni-univalent electrolyte, say NaCl. The treatment which follows is that of Kedem & Katchalsky (1963a) and Katchalsky & Curran (1965) except that these authors considered the general case where the electrolyte is not necessarily uni-univalent.

Let the cation and anion be denoted by c and a respectively. The net fluxes, J_c and J_a, of cation and anion are defined in the same way as J_s was defined in the preceding section. The theory of irreversible thermodynamics shows that the appropriate conjugate force on an ion j is the negative gradient of the electrochemical potential, $\tilde{\mu}_j$, for j. The gradient of $\tilde{\mu}_j$ at any point x in the membrane is given by

$$\frac{d\tilde{\mu}_j}{dx} = \bar{V}_j \frac{dp}{dx} + RT \frac{d}{dx} (\ln \gamma_j c_j) + z_j F \frac{d\psi}{dx} \qquad 2.50$$

which is distinguished from the chemical potential gradient for a non-electrolyte (cf. equation 2.10) by the presence of the additional term $z_j F d\psi/dx$; z_j is the algebraic valency of j, F is the Faraday and $d\psi/dx$ is the electric potential gradient. Using the same theoretical procedures as those employed for the permeation of nonelectrolytes and water, it is possible to show that the dissipation function for the flows of ions and water across a homogeneous membrane is

$$\Phi_m = J_c \Delta\tilde{\mu}_c + J_a \Delta\tilde{\mu}_a + J_w \Delta\mu_w \qquad 2.51$$

where $\Delta\tilde{\mu}_j = \tilde{\mu}_j{}^o - \tilde{\mu}_j{}^i$. Φ_m can be changed into a more convenient form than that of equation 2.51. The transformation is achieved by defining the electromotive force, ΔE, across the membrane as measured between electrodes reversible to the anion. In this case

$$\Delta E = \frac{\Delta\tilde{\mu}_a}{z_a F} = -\frac{\Delta\tilde{\mu}_a}{F} \qquad 2.52$$

since $z_a = -1$. Of course the electromotive force measured in this way differs from $\Delta\psi$ measured by electrodes such as calomel electrodes with salt bridges. ΔE is chosen in preference to $\Delta\psi$ because relatively simple expressions for Φ_m can be derived. Since the electrodes in the outer and inner compartments are reversible to the anion, the cation is neither removed nor emitted at the electrodes and hence J_c is equivalent to the net flux, J_s, of neutral salt s. Instead of the diffusional fluxes of ions, $\mathcal{J}_c$ and $\mathcal{J}_a$, we employ $\mathcal{J}_s$ and the net flow of current, I, where current flow is given by

$$I = F(\mathcal{J}_c - \mathcal{J}_a) \qquad 2.53$$

With those transformations of flows are associated transformations of the forces. Instead of $\Delta\tilde{\mu}_c$ and $\Delta\tilde{\mu}_a$ we employ ΔE and the difference of the chemical potential, $\Delta\mu_s$, for the salt, where $\Delta\mu_s$ is given by

$$\Delta\mu_s = \Delta\tilde{\mu}_c + \Delta\tilde{\mu}_a \qquad 2.54$$

$\Delta\mu_s$ is equal to $\bar{V}_s \Delta p + (RT\Delta c_s/\bar{c}_s)$, where $\bar{V}_s$ is the partial molar volume of s, Δc_s is the difference of salt concentration, c_s, across the membrane and $\bar{c}_s$ is the mean concentration of s across the membrane. After these changes in flows and forces have been made we find that they are compatible with

$$\Phi_m = \mathcal{J}_s \Delta\mu_s + \mathcal{J}_w \Delta\mu_w + IE \qquad 2.55$$

Although this form of the dissipation function is a more useful one than the original form, it would be preferable if it contained the observable flow $\mathcal{J}_v$ rather than $\mathcal{J}_w$. In fact, it is possible to achieve a further transformation of Φ_m so as to include $\mathcal{J}_v$. This gives

$$\Phi_m = \mathcal{J}_v(\Delta p - \Delta\pi) + \mathcal{J}_s \frac{\Delta\pi_s}{\bar{c}_s} + IE \qquad 2.56$$

where $\Delta\pi_s = RT\Delta c_s$ and $\Delta\pi$ is the sum of $\Delta\pi_s$ and any additional difference in osmotic pressure, $\Delta\pi_i$, due to impermeant solutes i. This expression for Φ_m indicates that we can write the following equations for the flows

$$\mathcal{J}_v = L_{11}(\Delta p - \Delta\pi) + L_{12}\frac{\Delta\pi_s}{\bar{c}_s} + L_{13}\Delta E \qquad 2.57$$

$$\mathcal{J}_s = L_{21}(\Delta p - \Delta\pi) + L_{22}\frac{\Delta\pi_s}{\bar{c}_s} + L_{23}\Delta E \qquad 2.58$$

$$I = L_{31}(\Delta p - \Delta\pi) + L_{32}\frac{\Delta\pi_s}{\bar{c}_s} + L_{33}\Delta E \qquad 2.59$$

These equations form a set of relations describing the fluxes of salt, volume and electric current across the membrane. Before they can be expressed in a practical form we need to specify the appropriate membrane coefficients which must be determined.

Membrane coefficients

Because of the Onsager reciprocal relations only six, rather than nine, independent practical coefficients need to be measured to specify the transport properties of the membrane. In order to measure the coefficients certain forces and flows must be held at zero and these are written in subscript positions on the following coefficients.

The first three coefficients are the so-called straight coefficients defined below.

Hydraulic conductivity, L_p. This is determined by estimating the volume flux driven by a given pressure gradient, while $\Delta \pi_s$ and I are held at zero. It is defined by

$$L_p = [\mathcal{J}_v/(\Delta p - \Delta \pi)]_{\Delta \pi_s, I} \qquad 2.60$$

Salt permeability, ω_s'. This is determined by estimating the solute flux driven by a given concentration gradient across the membrane while $(\Delta p - \Delta \pi)$ and I are held at zero. It is defined by

$$\omega_s' = [\mathcal{J}_s/\Delta \pi_s]_{(\Delta p - \Delta \pi), I} \qquad 2.61$$

and clearly ω_s' is related to the conventional salt permeability, P_s (equal to $\mathcal{J}_s/\Delta c_s$), by $\omega_s' RT = P_s$.

Specific or electric conductance, κ'. This is determined by recording the current flow in response to an applied electric field, while $(\Delta p - \Delta \pi)$ and $\Delta \pi_s$ are held at zero. It is defined by

$$\kappa' = [I/E]_{(\Delta p - \Delta \pi), \Delta \pi_s} \qquad 2.62$$

Apart from these straight coefficients we need to determine three so-called coupling coefficients defined below.

Reflexion coefficient, σ_s. This is determined by estimating the solute flux associated with a given volume flux driven, say, by a hydrostatic pressure gradient. During this measurement both $\Delta \pi_s$ and I must remain zero. It is defined by

$$\bar{c}_s(1 - \sigma_s) = [\mathcal{J}_s/\mathcal{J}_v]_{\Delta \pi_s, I} \qquad 2.63$$

Electro-osmotic permeability, β. This is determined by estimating the volume flux associated with current flow across the membrane when $(\Delta p - \Delta \pi)$ and $\Delta \pi_s$ are held at zero. It is defined by

$$\beta = [\mathcal{J}_v/I]_{(\Delta p - \Delta \pi), \Delta \pi_s} \qquad 2.64$$

Transport number, τ_c'. This is determined by estimating the salt flux associated with a given current while $(\Delta p - \Delta \pi)$ and $\Delta \pi_s$ are held at zero. It is defined by

$$\frac{\tau_c'}{F} = [\mathcal{J}_s/I]_{(\Delta p - \Delta \pi), \Delta \pi_s} \qquad 2.65$$

Experimental procedures for determining these practical coefficients have been discussed by Staverman (1952), Kedem & Katchalsky (1963a) and Katchalsky & Curran (1965).

An alternative set of practical coefficients may be employed to describe this system (see set I, Table 2.1). With the practical coefficients outlined above, the phenomenological equations can be transformed into

$$\mathcal{J}_v = L_p(\Delta p - \Delta \pi_i) - \sigma_s L_p \Delta \pi_s + \beta I \qquad 2.66$$

$$\mathcal{J}_s = \bar{c}_s L_p (1 - \sigma_s)(\Delta p - \Delta \pi) + \omega_s' \Delta \pi_s + \frac{\tau_c'}{F} I \qquad 2.67$$

$$I = \kappa' \beta (\Delta p - \Delta \pi) + \kappa' \frac{\tau_c'}{F} \frac{\Delta \pi_s}{\bar{c}_s} + \kappa' \Delta E \qquad 2.68$$

Employing the alternative transport coefficients (set I, Table 2.1) one obtains another equivalent set of phenomenological equations

$$\mathcal{J}_v = L_p(\Delta p - \Delta \pi) + L_p(1 - \sigma_s)\Delta \pi_s - \frac{P_E L_p I}{\kappa} \qquad 2.69$$

$$\mathcal{J}_s = \omega_s \Delta \pi_s + \frac{\tau_c}{F} I + \bar{c}_s (1 - \sigma_s) \mathcal{J}_v \qquad 2.70$$

$$I = \kappa \Delta E - P_E \mathcal{J}_v + \frac{\kappa \tau_c}{F} \frac{\Delta \pi_s}{\bar{c}_s} \qquad 2.71$$

These alternative sets of equations constitute the basic phenomenological description of passive transport of salt, volume and current across a *homogeneous* membrane and each set requires the determination of six practical coefficients. Again the flows are expressed in terms of readily measured driving 'forces' Δp, $\Delta \pi$ and ΔE.

TABLE 2.1. *The two sets of practical coefficients for a membrane permitting the flows of volume, salt and electric current*

		SET I	SET II
straight coefficients	hydraulic conductivity	$L_p = \left[\dfrac{\mathcal{J}_v}{\Delta p - \Delta\pi}\right]_{\Delta\pi_s, I}$	$L_p = \left[\dfrac{\mathcal{J}_v}{\Delta p - \Delta\pi}\right]_{\Delta\pi_s, I}$
	salt permeability	$\omega_s = \left[\dfrac{\mathcal{J}_s}{\Delta\pi_s}\right]_{\mathcal{J}_v, I}$	$\omega_s' = \left[\dfrac{\mathcal{J}_s}{\Delta\pi_s}\right]_{(\Delta p - \Delta\pi), I}$
	specific conductance	$\kappa = \left[\dfrac{I}{E}\right]_{\mathcal{J}_v, \Delta\pi_s}$	$\kappa' = \left[\dfrac{I}{E}\right]_{(\Delta p - \Delta\pi), \Delta\pi_s}$
coupling coefficients	reflexion coefficient	$\bar{c}_s(1-\sigma_s) = -\left[\dfrac{\Delta p - \Delta\pi}{\Delta\pi_s/\bar{c}_s}\right]_{\mathcal{J}_v, I}$	$\bar{c}_s(1-\sigma_s) = \left[\dfrac{\mathcal{J}_s}{\mathcal{J}_v}\right]_{\Delta\pi_s, I}$
	electro-osmotic pressure / permeability	$P_E = \left[\dfrac{\Delta p - \Delta\pi}{E}\right]_{\mathcal{J}_v, \Delta\pi_s}$	$\beta = \left[\dfrac{\mathcal{J}_v}{I}\right]_{(\Delta p - \Delta\pi), \Delta\pi_s}$
	transport number	$\tau_c = F\left[\dfrac{\mathcal{J}_s}{I}\right]_{\mathcal{J}_v, \Delta\pi_s}$	$\tau_c' = F\left[\dfrac{\mathcal{J}_s}{I}\right]_{(\Delta p - \Delta\pi), \Delta\pi_s}$

The subscripts indicate the flows or forces which are held at zero. (After Kedem & Katchalsky, 1963a.)

Transport of heat and water

Up to this point the theoretical treatment of membrane transport has considered only isothermal systems. Several workers have demonstrated that there is coupling between the flow of gases (Denbigh & Raumann, 1950) or of water (Rastogi, Blokhra & Aggarwala, 1964) and temperature gradients established across artificial membranes. It is not an easy matter, however, to predict the direction of the flow of matter since this depends on certain characteristics of the membrane.

In order to discuss the coupling between the flows of heat and water we shall consider a simple system which consists of a *homogeneous* permeable membrane separating two well-mixed compartments containing pure water at different temperatures and pressures; the entire system is insulated thermally from its surroundings.

Once again we can use the phenomenological equations to describe the relations between the flows and the forces. For example, we can write

$$J_v = L_p \Delta p + L_{pQ} \Delta T \qquad 2.72$$
$$J_Q = L_{Qp} \Delta p + L_Q \Delta T \qquad 2.73$$

where J_Q is the flow of heat and ΔT is the difference of temperature across the membrane. The term $L_{pQ}\Delta T$ in equation 2·72 expresses the coupling between the volume flow and the temperature gradient. This coupling is observed as the phenomenon of *thermo-osmosis*. The equation 2.73 for heat flow shows the straightforward thermal-conduction term $L_Q \Delta T$ plus an additional component of heat flow, $L_{Qp}\Delta p$, due to the pressure gradient. The latter type of heat flow has been called *pressure-pyresis*.

According to Katchalsky & Curran (1965) the most suitable form of the dissipation function for this case is

$$\Phi_m = J_w \Delta \mu_w + J_s \Delta T \qquad 2.74$$

where J_S is the flow of entropy. Thus the phenomenological equations can be re-written as

$$J_w = L_{11}\Delta\mu_w + L_{12}\Delta T \qquad 2.75$$
$$J_S = L_{21}\Delta\mu_w + L_{22}\Delta T \qquad 2.76$$

with $L_{12} = L_{21}$.

It may seem that some benefit has been lost by considering $\mathcal{J}_S$ rather than $\mathcal{J}_Q$ since the former is unobservable; however, we refer back to heat transfer at a later stage of the treatment.

For a system consisting of water alone $\Delta\mu_w$ can be written as

$$\Delta\mu_w = -\bar{S}_w \Delta T + \bar{V}_w \Delta p \qquad 2.77$$

where $\bar{S}_w$ is the partial molar entropy of water. Substituting for $\Delta\mu_w$ in equations 2.75 and 2.76 yields

$$\mathcal{J}_w = L_{11}\bar{V}_w \Delta p + (L_{12} - L_{11}\bar{S}_w)\Delta T \qquad 2.78$$
$$\mathcal{J}_S = L_{21}\bar{V}_w \Delta p + (L_{22} - L_{21}\bar{S}_w)\Delta T \qquad 2.79$$

When a constant temperature difference, ΔT, is maintained across the membrane, water flow occurs and consequently a pressure difference is generated across the membrane. Thus, Δp builds up to a value, which finally makes $\mathcal{J}_w = 0$, and this difference in pressure is obtained from equation 2.78 as

$$\left(\frac{\Delta p}{\Delta T}\right)_{J_w=0} = -\frac{L_{12}}{L_{11}} \cdot \frac{1}{\bar{V}_w} + \frac{\bar{S}_w}{\bar{V}_w} \qquad 2.80$$

The coupling of the flows of water and entropy is represented by the term (L_{12}/L_{11}); in fact, the coupling term may be considered in terms of a new quantity, the *entropy of transfer* S^*, which is defined as the entropy transported by a unit flow of matter under isothermal conditions. S^* may be obtained from equations 2.78 and 2.79 by setting $\Delta T = 0$ to yield

$$S^* = \left(\frac{\mathcal{J}_S}{\mathcal{J}_w}\right)_{\Delta T=0} = \frac{L_{21}}{L_{11}} \qquad 2.81$$

and, therefore, we can rewrite equation 2.80 as

$$\left(\frac{\Delta p}{\Delta T}\right)_{J_w=0} = -\frac{(S^* - \bar{S}_w)}{\bar{V}_w} \qquad 2.82$$

It is found that the difference between S^* and $\bar{S}_w$ obeys the relation

$$Q^* = T(S^* - \bar{S}_w) \qquad 2.83$$

where Q^* is called the *heat of transfer* (see for example Katchalsky & Curran, 1965).

Substituting equation 2.83 into equation 2.82 gives

$$\left(\frac{\Delta p}{\Delta T}\right)_{J_w=0} = -\frac{Q^*}{\bar{V}_w T} \qquad 2.84$$

2. DESCRIPTION OF TRANSPORT PROCESSES

In this system the result of the isothermal transfer of a unit quantity of water is the flow of Q^* units of heat. The latter quantity can be visualized by considering the example where the membrane separating the compartments is replaced by a vapour phase. Since the transfer of one mole of water across the vapour barrier requires both the vaporization of one mole on one side and the condensation of that mole on the other side, one compartment will lose a quantity of heat identical to the heat of vaporization whereas the other compartment will gain that quantity. Thus, the heat of vaporization will be transferred across the barrier to maintain the uniform temperature. In this hypothetical example Q^* equals the heat of vaporization of water. Of course, experimental studies must be performed to establish the magnitude of Q^* for the particular membrane under study.

It must be stressed that the characteristics of the membrane determine, to a large extent, the magnitude of Q^*. Moreover, it is possible that Q^* may be zero, positive or negative (see equation 2.83).

Spanner (1954) has shown, with the aid of equation 2.84, that the hydraulic conductivity of a membrane is related to Q^* by

$$\frac{\partial}{\partial T} \ln \left\{ \frac{L_p T}{\bar{V}_w} \right\} = \frac{Q^*}{RT^2} \qquad 2.85$$

and, since $\bar{V}_w$ is practically independent of T, we have

$$\frac{1}{L_p T} \frac{\partial}{\partial T} (L_p T) = \frac{Q^*}{RT^2} \qquad 2.86$$

On the assumption that Q^* is not strongly dependent on T Spanner performed the integration of equation 2.86 between $T = T_1$, and $T = T_2$ to obtain

$$\ln \left\{ \frac{^2L_p T_2}{^1L_p T_1} \right\} = \frac{Q^*}{R} \left[\frac{1}{T_1} - \frac{1}{T_2} \right] \qquad 2.87$$

where ^2L_p and ^1L_p are the hydraulic conductivities at T_2 and T_1. For $T_2 - T_1 = 10°K$ at ordinary temperatures the ratio $(^2L_p/^1L_p)$ is called the Q_{10} and equation 2.87 reduces to the approximate expression

$$\ln \{1 \cdot 034 \, Q_{10}\} = \frac{10 Q^*}{R T_1^2} \qquad 2.88$$

This relation yields a simple method of estimating Q^* from the temperature dependence of L_p.

The Q_{10} values for hydraulic or osmotic water flow across artificial or biological membranes are likely to be equal to or larger than unity. That means that the heat of transfer will probably be positive and consequently thermo-osmosis will move water from the hot to the cold side of the membrane. Spanner (1954) has discussed the significance of this phenomenon for water movement across biological membranes. His calculations show that quite small temperature gradients of, say, 0·01 centigrade degree could, in some instances, generate counter-pressures of about 1 atm. However, both Spanner (1954) and Katchalsky & Curran (1965) have concluded particularly for biological membranes that such temperature gradients could not be sustained in the face of the rapid thermal conduction which is likely to occur across thin membranes.

The significance of thermo-osmosis for water absorption and secretion in animal epithelial membranes will be discussed in Chapter 10.

Phenomenological description of active transport

In principle, it is possible to include in the phenomenological description of transport processes across a membrane certain coupling terms between the flux (J_v, I, J_s) and the metabolic reactions occurring within the membrane. That particular approach leads to a definition of active transport. Kedem (1961) pioneered the phenomenological description of active transport by expressing the phenomenological equations in the alternative form

$$\Delta \tilde{\mu}_i = \sum_{k=1}^{n} R_{ik} J_k + R_{ir} J_r \qquad 2.89$$

where $\Delta \tilde{\mu}_i$ is the difference of the electrochemical potential for a species i across the membrane and J_r is the rate of the metabolic reaction which is coupled to the transport of i. Equation 2.89 can be rearranged to give an expression for the flux of i, namely

$$J_i = \frac{\Delta \tilde{\mu}_i}{R_{ii}} - \sum_{\substack{k=1 \\ k \neq i}}^{n} \frac{R_{ik}}{R_{ii}} J_k - \frac{R_{ir}}{R_{ii}} J_r \qquad 2.90$$

According to Kedem we may conclude that i is actively transported if the value of R_{ir} is not zero. This definition of active transport rests on the assumption that the chemical reaction, which is dependent upon metabolism, *occurs within the membrane*. Moreover, the incorporation of chemical reactions occurring within the membranes into the phenomenological treatment of forces and flows rests upon certain other assumptions. For example, at first glance it seems that any phenomenological equation which describes the coupling between a chemical reaction and the flux of a substance across a membrane, violates Curie's principle. According to Curie's principle a diffusional flux, or any vectorial quantity, cannot be coupled to chemical reactions because the latter are scalars. Curie's principle is applicable, however, only to homogeneous or isotropic phases (Moszynski, Hoshiko & Lindley, 1963). Hence, we must presume that certain metabolic reactions occur in biological membranes, that such membranes are anisotropic and that such reactions may be coupled to flows of ions or water. None of these suppositions is difficult to accept. Of course, as soon as we do accept them we must acknowledge that we are no longer dealing with a homogeneous membrane and consequently that the strength of the phenomenological approach is probably undermined. Another assumption in the phenomenological approach is that the active transport system is linear in its behaviour; no experimental support for this view exists. Finally, it might be contended that the linear relations of irreversible thermodynamics should not be applied to chemical reactions. For sufficiently slow rates of chemical reaction it is known that the rate of the reaction is linearly related to the affinity. The affinity has been defined by DeDonder (see Prigogine, 1961) as $\left(\sum_i \nu_i \mu_i\right)$ where ν_i and μ_i denote the stoichiometric coefficient and the chemical potential of the reactant i.

In their phenomenological treatment of active transport Hoshiko & Lindley (1967) postulated that the flows across a membrane undergoing metabolic reactions can be related linearly to the rate of the chemical reaction even although such a reaction may be related non-linearly to the affinity. According to Hoshiko & Lindley (1967) the dissipation function for a membrane, separating two different aqueous solutions of the same permeant salt is

$$\Phi_m = J_v(\Delta p - \Delta \pi) + J_s \frac{\Delta \pi_s}{\bar{c}_s} + IE + J_r A_r \qquad 2.91$$

where A_r is the affinity of the metabolic reaction possibly coupled to the flow of material across the membrane. In addition to the six practical coefficients (set I, Table 2.1) this system requires three additional coefficients specifying the effect of active transport and a final metabolic rate coefficient. Hoshiko and Lindley designated these active coefficients as follows:

Electrogenicity coefficient, ϵ, given by

$$\epsilon = [I/\mathcal{J}_r]_{\Delta\mu_s, \Delta E, J_v} \qquad 2.92$$

Active salt-pumping coefficient, U, given by

$$U = [\mathcal{J}_s/\mathcal{J}_r]_{\Delta\mu_s, I, J_v} \qquad 2.93$$

Volume pump coefficient, V, given by

$$V = [\mathcal{J}_v/\mathcal{J}_r]_{\Delta\mu_s, I, (\Delta p - \Delta\pi)} \qquad 2.94$$

Metabolic rate coefficient, k, given by

$$k = [\mathcal{J}_r/A_r]_{\Delta\mu_s, I, J_v} \qquad 2.95$$

Hoshiko and Lindley described straightforward techniques for measuring the new coefficients and their paper should be consulted for practical details. Under their scheme the practical flux equations for $\mathcal{J}_v$, $\mathcal{J}_s$, I and $\mathcal{J}_r$ are

$$\mathcal{J}_v = L_p(\Delta p - \Delta\pi) + L_p(1-\sigma_s)\Delta\pi_s - \frac{P_E L_p}{\kappa}I + V\mathcal{J}_r \qquad 2.96$$

$$\mathcal{J}_s = \omega_s \Delta\pi_s + \bar{c}_s(1-\sigma_s)\mathcal{J}_v + \frac{\tau_c}{F}I + U\mathcal{J}_r \qquad 2.97$$

$$I = \frac{\kappa\tau_c}{F}\frac{\Delta\pi_s}{\bar{c}_s} + \kappa\Delta E - P_E\mathcal{J}_v + \epsilon\mathcal{J}_r \qquad 2.98$$

$$\mathcal{J}_r = kU\frac{\Delta\pi_s}{\bar{c}_s} + \frac{k\epsilon}{\kappa}I + \frac{kV}{L_p}\mathcal{J}_v + kA_r \qquad 2.99$$

The phenomenological description of active transport, illustrated in the work of Kedem (1961) and Hoshiko & Lindley (1967), is exceedingly useful. As a starting point these workers have adopted a rigorous definition of active transport, namely: '*primary active transport of a substance is defined as the ability to generate a gradient of that substance with no fluxes except that due to the driving metabolic reaction*'.

Before the expressions governing the coupling between forces and flows can be employed, however, the specific and important metabolic reactions in active transport must be identified. This crucial problem has not yet been solved. In this respect, at least, the influence of irreversible thermodynamics on the study of active transport is in its infancy.

Transport of non-electrolytes and water across heterogeneous membranes

So far the discussion on passive transport of solutes and water has been concerned with homogeneous membranes but now we shall turn our attention to heterogeneous barriers. Such heterogeneity can be divided into two classes: in the first, there is an array of different membranes in parallel whereas in the second different membranes are arranged in series with one another. A phenomenological description of transport in composite membranes, both parallel and series arrays, has been given by Kedem & Katchalsky (1963b,c). Their analysis shows that such heterogeneous membranes have certain transport characteristics which distinguish them from their homogeneous components. In their treatment of composite membranes Kedem and Katchalsky considered the flows of volume, salt and current and the 'forces' of hydrostatic pressure, salt concentration and electric potential. For the sake of simplicity, however, we shall consider below only the flows of volume and an uncharged solute and the 'forces' of hydrostatic pressure and solute concentration.

Membranes in parallel

Consider a membrane composed of two different regions a and b arranged in parallel (Fig. 2.2). This membrane separates two aqueous solutions of a non-electrolyte s and, as in previous treatments, the flows of solute and volume are independent of time and position and are perpendicular to the membrane surfaces. Let γ^a and γ^b denote the fractions of the membrane's area occupied by the regions a and b. Then, the flows of solute and water across each region are governed by the phenomenological equations 2.24 and 2.25. For example, for region a we have

$$J_v{}^a = L_p{}^a \Delta p + L_{pD}{}^a \Delta \pi \qquad 2.100$$
$$J_D{}^a = L_{Dp}{}^a \Delta p + L_D{}^a \Delta \pi \qquad 2.101$$

Moreover, the flows of solute and volume across the membrane can be expressed in terms of the component flows across a and b. For example, the volume flux $\mathcal{J}_v$ across the membrane is given by

$$\mathcal{J}_v = \gamma^a \mathcal{J}_v{}^a + \gamma^b \mathcal{J}_v{}^b \qquad 2.102$$

A consequence of the fact that the phenomenological equations describe the behaviour of a and b and that the total flows are

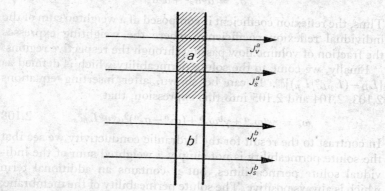

FIG. 2.2. Composite membrane system containing two dissimilar membranes a and b in parallel.

linear functions of the component flows is that the Onsager coefficients for the membrane are given by

$$L_p = \gamma^a L_p{}^a + \gamma^b L_p{}^b \qquad 2.103$$
$$L_{pD} = \gamma^a L_{pD}{}^a + \gamma^b L_{pD}{}^b \qquad 2.104$$
$$L_D = \gamma^a L_D{}^a + \gamma^b L_D{}^b \qquad 2.105$$

This argument can be applied to any number of flows across a membrane composed of any number of parallel elements (see Kedem & Katchalsky 1963*b*) and it will still hold.

Given the above relations for the Onsager coefficients it is possible to express the practical coefficients L_p, σ_s and ω_s in terms of the corresponding coefficients for a and b. In fact, we have already done this for the hydraulic conductivity, which in this case is identical to the Onsager coefficient L_p and is described by equation 2.103. Thus, L_p is a suitably weighted sum of the individual hydraulic conductivities of a and b, where the weighting involves the relative areas of a and b. It is convenient in this connexion to assess what fractions of $\mathcal{J}_v$ pass through a and b. These fractions are denoted by α^a and α^b, given by

$$\alpha^a = \frac{\gamma^a L_p{}^a}{L_p} \qquad \alpha^b = \frac{\gamma^b L_p{}^b}{L_p} \qquad 2.106$$

Recalling that the reflexion coefficient is given by $(-L_{pD}/L_p)$ it can be shown that

$$\sigma_s = \alpha^a \sigma_s{}^a + \alpha^b \sigma_s{}^b \qquad 2.107$$

Thus, the reflexion coefficient is composed of a weighted sum of the individual reflexion coefficients, where the weighting expresses the fraction of volume flow passing through the respective regions.

Finally, we come to the solute permeability which is defined as $[L_D - (L_{pD}{}^2/L_p)]\bar{c}_s$. It can be shown, after inserting equations 2.103, 2.104 and 2.105 into this expression, that

$$\omega_s = \gamma^a \omega_s{}^a + \gamma^b \omega_s{}^b + (\sigma_s{}^b - \sigma_s{}^a)^2 \alpha^a \alpha^b L_p \bar{c}_s \qquad 2.108$$

In contrast to the result for the hydraulic conductivity we see that the solute permeability is not simply a weighted sum of the individual solute permeabilities, but it contains an additional term which is always positive. The solute permeability of the membrane, therefore, will be larger than the appropriately weighted sum of $\omega_s{}^a$ and $\omega_s{}^b$ except when $\sigma_s{}^b = \sigma_s{}^a$.

The physical significance of the deviation in the solute permeability of the composite membrane can be grasped when one considers the conditions for its measurement. During the measurement of ω_s there must be no volume across the membrane (see page 51). In the membrane this condition is satisfied when

$$\gamma^a \mathcal{J}_v{}^a + \gamma^b \mathcal{J}_v{}^b = 0 \qquad 2.109$$

In practice a hydrostatic pressure difference, Δp, is applied across the membrane to bring $\mathcal{J}_v$ to zero while the flow of solute is driven across the membrane by $\Delta \pi_s$. It can be shown that

$$\Delta p = (\alpha^a \sigma_s{}^a + \alpha^b \sigma_s{}^b) \Delta \pi_s \qquad 2.110$$

and this relation permits one to determine the separate volume flows $\gamma^a \mathcal{J}_v{}^a$ and $\gamma^b \mathcal{J}_v{}^b$ through a and b. It turns out that

$$\gamma^a \mathcal{J}_v{}^a = \alpha^b L_p (\sigma_s{}^b - \sigma_s{}^a) \Delta \pi_s \qquad 2.111$$
$$\gamma^b \mathcal{J}_v{}^b = -\alpha^b L_p (\sigma_s{}^b - \sigma_s{}^a) \Delta \pi_s \qquad 2.112$$

Thus, although there is no net volume flow through the membrane there are equal and opposite flows through a and b, provided $\sigma_s{}^b \neq \sigma_s{}^a$. For example, let us assume that a is more permeable to solute than b, i.e. $\sigma_s{}^b > \sigma_s{}^a$; $\gamma^a \mathcal{J}_v{}^a$ will occur in the same direction as solute diffusion and it will exert more solvent-drag than that by $\gamma^b \mathcal{J}_v{}^b$ in the opposite direction. Therefore, solute flow will be augmented by preferential solvent drag through the permeable region as opposed to that in the less permeable region. This argument holds too when b is designated the permeable region in preference to a. Thus, when a membrane is composed of two dissimilar elements the circulation of volume flow in the membrane gives rise to net solute flow provided the reflexion coefficients are different for a and b.

The volume flows $\gamma^a \mathcal{J}_v{}^a$ and $\gamma^b \mathcal{J}_v{}^b$ are accompanied by solute flows $\gamma^a \mathcal{J}_v{}^a (1-\sigma_s{}^a) \bar{c}_s$ and $\gamma^b \mathcal{J}_v{}^b (1-\sigma_s{}^b) \bar{c}_s$ respectively. Therefore, the total solute flow generated by circulation of volume flows is

$$[\gamma^a \mathcal{J}_v{}^a (1-\sigma_s{}^a) + \gamma^b \mathcal{J}_v{}^b (1-\sigma_s{}^b)] \bar{c}_s \qquad 2.113$$

Recalling that $\gamma^a \mathcal{J}_v{}^a + \gamma^b \mathcal{J}_v{}^b = 0$ and that $\gamma^a \mathcal{J}_v{}^a$ is given by equation 2.111, the expression 2.113 for solute flow can be rewritten as

$$(\sigma_s{}^b - \sigma_s{}^a)^2 a^a a^b L_p \bar{c}_s \Delta \pi_s \qquad 2.114$$

When there is a gradient of solute across the membrane the solute flow will consist of two terms. The first component will be the diffusional flux, $(\gamma^a \omega_s{}^a + \gamma^b \omega_s{}^b) \Delta \pi_s$, while the second will be a consequence of volume circulation and is given by 2.114. Thus, circulation within the membrane means that ω_s is not simply $(\gamma^a \omega_s{}^a + \gamma^b \omega_s{}^b)$ but that it contains an additional term $(\sigma_s{}^b - \sigma_s{}^a)^2 a^a a^b L_p \bar{c}_s$ in agreement with equation 2.108 for ω_s.

The paper by Kedem & Katchalsky (1963b) shows how the phenomenon of circulation influences the practical coefficients, L_p, σ_s, $\omega_s{}'$, κ', β and τ'_c for a membrane permitting the flows of volume, salt and electric current. In this case only κ', β and τ'_c are not affected by the circulation of flows between the parallel components of the membrane. In particular, Kedem and Katchalsky showed that a membrane composed of positively and negatively charged parallel components could have a salt permeability much larger than that computed purely on the basis of its components. Indeed, the salt permeability may be increased to such an extent that anomalous osmosis occurs.

Membranes in series

Consider a membrane composed of two different membranes a and b arranged in series (Fig. 2.3). The membranes separate two aqueous solutions of a non-electrolyte s and again we consider that the flows are perpendicular to the membrane surfaces and independent of time and position. Let us assume that inserted between a and b is an infinitely thin layer of aqueous solution of s whose concentration is denoted by $c_s{}^{ab}$; the superscript ab signifies a property of the intermediate layer. Furthermore, it is assumed that the chemical potentials for s and water in this layer are equal to the corresponding chemical potentials at the adjacent surfaces of a and b. It is also considered that the intermediate layer offers no resistance to the flows.

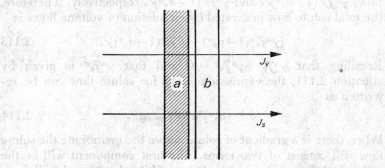

FIG. 2.3. Composite membrane system containing two dissimilar membranes a and b in series. An infinitely thin layer of solution separates a and b.

Kedem & Katchalsky (1963c) have analysed the general properties of membranes in series by expressing the forces as linear functions of the flows (cf. equation 2.9). At face value it appears that the overall resistance of the membrane to, say, volume flow ought to be the sum of the individual resistances of a and b. In other words we would expect that the hydraulic conductivity of the membrane would be given by

$$\frac{1}{L_p} = \frac{1}{L_p{}^a} + \frac{1}{L_p{}^b} \qquad 2.115$$

However, Kedem and Katchalsky argued that this conclusion does not invariably hold because the resistances ought to be independent of flows and forces whereas it is quite probable that they are not. This failure to meet the requirements for summation of the resistances arises because the solute concentration of the intermediate layer will be determined by the flows. In the case of volume flow the resistance of the membrane may be dependent on the magnitude and direction of the flow due to osmotic gradients being established between the intermediate layer and the bathing solutions. Let us, therefore, estimate the forces developed in the intermediate layer in relation to the flows of solute and volume which determine its composition.

For such a double-layered membrane bathed by ideal solutions it can be shown that the difference of osmotic pressure, $\Delta\pi_s$, between the outer and inner solutions is given by

$$\Delta\pi_s = \Delta\pi_s{}^a + \Delta\pi_s{}^b \qquad 2.116$$

where $\Delta\pi_s{}^a$ and $\Delta\pi_s{}^b$ are the osmotic pressure differences across a and b and are given by $\Delta\pi_s{}^a = \pi_s{}^o - \pi_s{}^{ab}$ and $\Delta\pi_s{}^b = \pi_s{}^{ab} - \pi_s{}^i$ respectively. Similarly the hydrostatic pressure difference, Δp, is given by

$$\Delta p = \Delta p^a + \Delta p^b \qquad 2.117$$

where $\Delta p^a = p^o - p^{ab}$ and $\Delta p^b = p^{ab} - p^i$.

Consider first the hydraulic conductivity of the membrane. It is determined by measuring the volume flow, J_v, driven by Δp when the membrane is bathed by identical solutions at a concentration c_s. For membrane a we have

$$J_v = L_p{}^a \Delta p^a - \sigma_s{}^a L_p{}^a \Delta \pi_s{}^a \qquad 2.118$$
$$J_s = \omega_s{}^a \Delta \pi_s{}^a + (1 - \sigma_s{}^a) J_v c_s{}^* \qquad 2.119$$

where $c_s{}^*$ is the average concentration of s across a; it is probably safe to assume that $c_s{}^* = (c_s + c_s{}^{ab})/2$ although Kedem & Katchalsky (1963c) defined $c_s{}^*$ in a different way to take account of large deviations between c_s and $c_s{}^{ab}$. The corresponding flux equations for membrane b are

$$J_v = L_p{}^b \Delta p^b - \sigma_s{}^b L_p{}^b \Delta \pi_s{}^b \qquad 2.120$$
$$J_s = \omega_s{}^b \Delta \pi_s{}^b + (1 - \sigma_s{}^b) J_v c_s{}^* \qquad 2.121$$

where the average solute concentration across b is again $c_s{}^*$. In this experiment $\Delta\pi_s = 0$ and, hence, equation 2.116 shows that

$$\Delta\pi_s{}^a = -\Delta\pi_s{}^b \qquad 2.122$$

From equations 2.117, 2.118, 2.120 and 2.122 it can be deduced that

$$\mathcal{J}_v \left(\frac{1}{L_p{}^a}+\frac{1}{L_p{}^b}\right) = \Delta p - \Delta\pi_s{}^a(\sigma_s{}^a - \sigma_s{}^b) \qquad 2.123$$

This equation shows that the presence of the intermediate layer renders invalid the notion that the overall resistance to volume flow is simply the sum of the individual resistances, namely $(1/L_p{}^a + 1/L_p{}^b)$. Furthermore, we can estimate the size of the osmotic pressure gradient $\Delta\pi_s{}^a$ which is established between the intermediate layer and the outer solution. From equations 2.119, 2.121 and 2.122 we obtain

$$\Delta\pi_s{}^a = c_s{}^* \frac{(\sigma_s{}^a - \sigma_s{}^b)}{\omega_s{}^a + \omega_s{}^b} \mathcal{J}_v \qquad 2.124$$

This relation allows us to assess whether volume flow causes solute accumulation or depletion in the intermediate layer. For example, let us assume volume flow occurs from outside to inside under the action of Δp. If the outer membrane is less permeable than the inner (i.e. $\sigma_s{}^a > \sigma_s{}^b$) then $\Delta\pi_s{}^a$ will be positive meaning that there is depletion of solute in the intermediate layer. On the other hand, if a is more permeable than b (i.e. $\sigma_s{}^a < \sigma_s{}^b$) solute will be accumulated in the layer as a result of volume flow. These changes in $c_s{}^{ab}$ exert an influence on the volume flow and by substituting equation 2.124 into 2.123 we obtain an expression for the overall hydraulic conductivity,

$$\frac{1}{L_p} = \frac{1}{L_p{}^a}+\frac{1}{L_p{}^b}+c_s{}^* \frac{(\sigma_s{}^a - \sigma_s{}^b)^2}{\omega_s{}^a + \omega_s{}^b} \qquad 2.125$$

Thus, the resistance to volume flow will always be larger than $(1/L_p{}^a + 1/L_p{}^b)$ provided the components are dissimilar, i.e. $\sigma_s{}^a \neq \sigma_s{}^b$, and the positive term $c_s{}^* (\sigma_s{}^a - \sigma_s{}^b)^2/(\omega_s{}^a + \omega_s{}^b)$ expresses the effect of osmotic gradients established between the intermediate layer and the bathing solutions.

Kedem & Katchalsky (1963c) drew attention to the fact that $c_s{}^*$ is dependent on volume flow and they obtained the following expression

$$c_s^* = c_s \left[1 - \frac{(\sigma_s{}^a - \sigma_s{}^b)J_v}{2RT(\omega_s{}^a + \omega_s{}^b)} \right] \qquad 2.126$$

which holds for slow rates of volume flow. Therefore, it is now possible to express the functional relationship between J_v and Δp by introducing equation 2.126 into $J_v = L_p \Delta p$ to yield

$$\Delta p = \left[\frac{1}{L_p{}^a} + \frac{1}{L_p{}^b} + \frac{c_s(\sigma_s{}^a - \sigma_s{}^b)^2}{(\omega_s{}^a + \omega_s{}^b)} \right] J_v - \frac{c_s(\sigma_s{}^a - \sigma_s{}^b)^3 J_v{}^2}{2RT(\omega_s{}^a + \omega_s{}^b)^2} \qquad 2.127$$

which shows that J_v is not linearly dependent on the driving force, Δp. The same conclusion applies when the driving force is an osmotic pressure gradient due to the presence of impermeant solutes. Inspection of the above relation between J_v and Δp reveals that the character of the non-linear behaviour will depend on which membrane is the more permeable one. For example, let us consider that membrane b is more permeable than a, i.e. $\sigma_s{}^a > \sigma_s{}^b$. Then, a pressure-driven inflow of volume will cause depletion of solute in the intermediate layer. The effect of solute depletion is to reduce the overall resistance to volume flow (see equation 2.125). On the other hand, a pressure-driven efflux of volume will cause accumulation of solute within the intermediate layer and consequently the effect of the osmotic gradients established between the layer and the bathing solutions will be to increase the resistance. This argument is summarized schematically in Fig. 2.4a which shows the shape of the relation between J_v and Δp predicted by equation 2.127.

For the case where membrane a is more permeable than b volume inflow will cause solute accumulation whereas volume efflux will produce solute depletion in the intermediate layer. Thus, the changes in resistance to volume flow accompanying inflows and effluxes will be opposite to those described above (Fig. 2.4b).

In a series-membrane system, therefore, one may observe rectification of volume flow if the individual membranes have different reflexion coefficients. This rectification will also occur in osmotic experiments where the gradients are established by the presence of impermeant solutes.

Thus, the resistance to volume flow is not simply the sum of the resistances of a and b. Does this apply also to solute transport? The solute permeability ω_s must be determined when volume flow

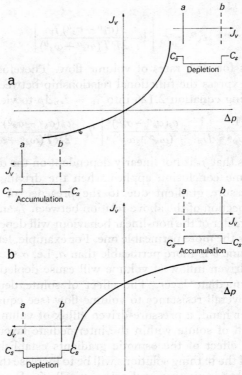

FIG. 2.4. Rectification of volume flow across a double-membrane system in which the different membranes a and b are separated by a layer of solution. It is assumed that membrane a is less permeable to solute than b. In part **a** of the figure an influx of volume causes solute depletion in the central compartment while a volume efflux creates solute accumulation. The rectification of volume flow arises from these alterations in solute concentration within the system. In part **b** of the figure the same argument applies but because the membranes are arranged in a different order the double-membrane rectifies volume flow in the reverse direction (see text).

is abolished. When this is satisfied the solute flows across membranes a and b are identical and are given by

$$J_s = \omega_s^a \Delta \pi_s^a \quad \quad 2.128$$
$$J_s = \omega_s^b \Delta \pi_s^b \quad \quad 2.129$$

respectively. Under those conditions $\Delta \pi_s^a + \Delta \pi_s^b = \Delta \pi_s$ and

$\mathcal{J}_s = \omega_s \Delta \pi_s$ for the entire membrane. It can be shown that

$$\mathcal{J}_s \left(\frac{1}{\omega_s{}^a} + \frac{1}{\omega_s{}^b} \right) = \Delta \pi_s \qquad 2.130$$

and therefore

$$\frac{1}{\omega_s} = \frac{1}{\omega_s{}^a} + \frac{1}{\omega_s{}^b} \qquad 2.131$$

This expression shows that the overall resistance $(1/\omega_s)$ to solute transport is equal to the sum of the separate resistances in a and b. It is possible that such a double-membrane system may rectify solute transport. For example, let us consider that $(1/\omega_s{}^a)$ is much larger than $(1/\omega_s{}^b)$ and that $(1/\omega_s{}^a)$ is dependent on the concentration of s. In this situation the solute permeability of the entire membrane determined with a given gradient will not be equal to that determined when the gradient is reversed. Thus, the permeability to solute flow in one direction will differ from that in the other. Of course this will not occur if both $\omega_s{}^a$ and $\omega_s{}^b$ are independent of solute concentration.

Finally, we consider the reflexion coefficient of the double-membrane system. σ_s is measured when the volume flow is abolished by the application of a suitable pressure gradient, Δp. Under those conditions the solute flows across membranes a and b are both equal to $\mathcal{J}_s$. Thus, we have

$$\omega_s{}^a \Delta \pi_s{}^a = \omega_s \Delta \pi_s \qquad 2.132$$
$$\omega_s{}^b \Delta \pi_s{}^b = \omega_s \Delta \pi_s \qquad 2.133$$

Since there is no volume flow across either a or b we know what

$$\Delta p^a = \sigma_s{}^a \Delta \pi_s{}^a \qquad 2.134$$
$$\Delta p^b = \sigma_s{}^b \Delta \pi_s{}^b \qquad 2.135$$

Because σ_s is defined as $(\Delta p / \Delta \pi_s)$ when $\mathcal{J}_v = 0$ it can be expressed as

$$\sigma_s = \frac{\Delta p^a + \Delta p^b}{\Delta \pi_s{}^a + \Delta \pi_s{}^b} \qquad 2.136$$

Introducing equations 2.132, 2.133, 2.134 and 2.135 into 2.136 gives

$$\sigma_s = \sigma_s{}^a \frac{\omega_s}{\omega_s{}^a} + \sigma_s{}^b \frac{\omega_s}{\omega_s{}^b} \qquad 2.137$$

Thus, the reflexion coefficient for the membrane is a suitably weighted sum of the individual reflexion coefficients. σ_s, like ω_s, will exhibit no asymmetry provided that the individual coefficients are not dependent on solute concentration.

Thus, a double-membrane system bathed by non-electrolyte solutions may exhibit asymmetrical resistances to volume flow and possibly also to solute flow. Kedem & Katchalsky (1963c) extended their arguments to the case where the system was bathed by ionic solutions and they concluded that similar asymmetries were likely to occur in the flows of electric current, volume and possibly of salt too.

3
THEORETICAL ASPECTS OF TRANSPORT IN POROUS MEMBRANES

Restricted motion of solutes	78
Significance of the reflexion coefficient	83
Diffusional and viscous flows of water	86
Solvent drag	93
Flux ratio for solutes	93
Flux ratio for labelled water	95
Electro-osmosis	97
Summary	102

THE phenomenological description of membrane transport, as it was developed in the preceding chapter, concerned itself almost exclusively with the fluxes of solute and water across homogeneous membranes. However, the behaviour of certain artificial and biological membranes suggests they are heterogeneous. This behaviour has several facets; for example, water flow may exert a drag on solute movement. A common way of accounting for this and other phenomena involving coupling between solute and water flows is to postulate that the membrane contains water-filled pores. Many workers have tackled the question of membrane porosity with particular reference to biological cell membranes. The result is that we now have a variety of experimental tools each of which has been fashioned from its own theoretical base. In this connexion the phenomenological approach can also play a useful role, especially when its practical coefficients are expressed in terms of frictional coefficients which describe the frictional forces experienced by solute and water molecules.

In the following discussion we will examine five distinct characteristics of membrane behaviour which indicate porosity. To a certain extent all of these express the degree of interaction that can occur between individual molecules and between molecules and the membrane.

Restricted motion of solutes

Let us consider that we have a membrane of area A and thickness Δx and that it possesses uniform cylindrical pores passing through it perpendicular to its surfaces. On the assumption that solutes pass solely through the aqueous pores we would expect to find that the membrane offered some restriction to solute diffusion and, moreover, that the restriction was larger for bigger molecules. In other words, solute permeability should be related to the size of the solute. There have been numerous theoretical attempts to describe membrane permeability in terms of a 'membrane–pore' model. In this respect the most notable success was achieved by Pappenheimer, Renkin & Borrero (1951) who described the net solute flux $\mathcal{J}_s$ by the following modification of Fick's law

$$\mathcal{J}_s = D_s A_s \frac{\Delta c_s}{\Delta x} \qquad 3.1$$

where D_s ($cm^2\ sec^{-1}$) is the solute diffusion coefficient in free solution, Δc_s is the difference in concentration of solute across the membrane and Δx is the total path length which in an ideal membrane without tortuous aqueous channels is equal to membrane thickness. The units of $\mathcal{J}_s$ and Δc_s are those employed in Chapter 2. In equation 3.1, A_s is a numerical factor representing certain restrictions imposed by the membrane on solute diffusion. Of course, the most obvious source of restriction is the fact that the total area of the pores is less than the membrane area; however, there are other sources of restriction (see later). In this type of experiment we measure the conventional solute permeability P_s given by $\mathcal{J}_s/\Delta c_s$ or

$$P_s = \frac{D_s A_s}{\Delta x} \qquad 3.2$$

and since D_s is known, the determination of P_s actually yields an estimate of $(A_s/\Delta x)$. Only in certain artificial membranes do we know the thickness of the membrane and even in those cases we still need to take account of the tortuosity of the channels. Thus, according to this view of solute permeability, determinations of P_s give us knowledge of $(A_s/\Delta x)$ which we must try to interpret in terms of solute diffusion in aqueous channels.

An alternative way of expressing the solute flux equation is in the form

$$J_s = D_s' \frac{A_p}{A} \frac{\Delta c_s}{\Delta x} \qquad 3.3$$

where D_s' is called the *restricted diffusion coefficient* of the solute and A_p is the total area of the pores; for n pores of radius r, $A_p = n\pi r^2$. This form has no advantage over the first since A_p is difficult to estimate in some artificial membranes and impossible to determine directly in biological membranes. However, both descriptions are equivalent and thus

$$D_s A_s = \frac{D_s' A_p}{A} \qquad 3.4$$

In studies of solute diffusion across artificial porous membranes (e.g. Renkin, 1954) it has been found that $(A_s/\Delta x)$ decreases with increasing molecular weight of the solute. It seems that this decrease in $(A_s/\Delta x)$ can be attributed to two effects: first, a frictional force experienced by solute as it diffuses along the pore and, secondly, a steric hindrance to solute diffusion at the entrance to the pore.

Faxen (1923) has analysed the hydrodynamical problems associated with the movement of spheres in fluid capillaries. His theory demands that the fluid in the capillary be treated as a continuum, and, further, that the spheres should be considerably larger than the fluid molecules. It seems unlikely that both of these theoretical conditions are satisfied in the case of solute diffusion through porous membranes, whether they be biological or artificial in character. Faxen's equation is

$$\frac{g_s^o}{g_s} = 1 - 2\cdot104 \left(\frac{a_s}{r}\right) + 2\cdot09 \left(\frac{a_s}{r}\right)^3 - 0\cdot95 \left(\frac{a_s}{r}\right)^5 \qquad 3.5$$

where g_s^o and g_s are the frictions exerted on the solute molecule in free solution and at the wall of the capillary respectively and a_s is the radius of the solute molecule. Renkin (1954) used equation 3.5 to express the frictional restriction to solute diffusion through porous membranes.

An additional restriction to solute diffusion is a geometrical one stemming from the fact that the size of the solute molecule may not be inconsequential in relation to the diameter of the pore. Such a restriction will impede the entry of the solute molecule at the mouth of the pore. Pappenheimer et al. (1951) suggested that the probability of the solute molecule entering the pore is equal to the central area of the pore, which can accommodate the molecule, divided by the total area of the pore; this ratio is $\pi(r-a_s)^2/\pi r^2$. Thus, the chance of no collision occurring between the solute molecule and the rim of the pore is $(1-a_s/r)^2$.

Both frictional and steric hindrances to solute diffusion in porous membranes are reflected in the fact that the total apparent area, A_{sd}, for solute diffusion, which is given by AA_s, is less than A_p. Actually A_{sd} is the virtual area which would be required to explain the observed solute flux as a consequence of free solute diffusion. Thus, (A_{sd}/A_p) embodies the correction for steric hindrance at the pore's orifice and frictional interactions within the pore itself, and according to the derivation of Renkin (1954) described above it is given by

$$\frac{A_{sd}}{A_p} = \left(1-\frac{a_s}{r}\right)^2 \left[1-2\cdot 104\left(\frac{a_s}{r}\right)+2\cdot 09\left(\frac{a_s}{r}\right)^3-0\cdot 95\left(\frac{a_s}{r}\right)^5\right] \quad 3.6$$

Solomon (1968) has emphasized the point that the restraints to solute diffusion in a porous membrane are included in the factor A_s. The restrictions to diffusion stem from three sources. First, the pore area is less than the membrane area. Secondly, there is friction within the pore and this is given by (g_s^o/g_s) and, finally, there is steric hindrance at its mouth expressed by $(1-a_s/r)^2$. A summary of these considerations may be expressed algebraically as

$$A_s = \left(\frac{A_p}{A}\right)\left(\frac{A_{sd}}{A_p}\right) \quad 3.7$$

Hence, we can obtain an expression for $(A_s/\Delta x)$ in terms of a_s and r by substituting equation 3.6 into equation 3.7 to give

$$\frac{A_s}{\Delta x} = \frac{A_p}{A\Delta x}\left(1-\frac{a_s}{r}\right)^2$$
$$\times \left[1-2\cdot 109\left(\frac{a_s}{r}\right)+2\cdot 09\left(\frac{a_s}{r}\right)^3-0\cdot 95\left(\frac{a_s}{r}\right)^5\right] \quad 3.8$$

Since $(A_p/A\Delta x)$ can be estimated in some artificial membranes, equation 3.8 provides a description of the dependence of solute permeability or $(A_s/\Delta x)$ upon molecular radius in terms of a 'pore-model'.

Renkin (1954) also described the movement of solute (ultrafiltration) which occurs when there is a hydrostatic pressure gradient across a porous membrane. He recognized that the steric hindrance term $(1-a_s/r)^2$ had to be altered to take account of the radial velocity profile in the pore and to that end he used the steric hindrance expression of Ferry (1936), namely

$$2\left(1-\frac{a_s}{r}\right)^2 - \left(1-\frac{a_s}{r}\right)^4 \qquad 3.9$$

Thus, when the expression 3.9 is substituted for $(1-a_s/r)^2$ the overall restriction to solute movement through the pores becomes

$$\frac{A_{sf}}{A_p} = \left[2\left(1-\frac{a_s}{r}\right)^2 - \left(1-\frac{a_s}{r}\right)^4\right]$$
$$\times \left[1 - 2\cdot 104\left(\frac{a_s}{r}\right) + 2\cdot 09\left(\frac{a_s}{r}\right)^3 - 0\cdot 95\left(\frac{a_s}{r}\right)^5\right] \qquad 3.10$$

where A_{sf} is the apparent area for solute transport in this case. Renkin (1954) has compared the different degrees of restriction which solute molecules experience during diffusion and ultrafiltration (Fig. 3.1). The former restriction in the pores is expressed, of course, by (A_{sd}/A_p) and the latter by (A_{sf}/A_p). It is evident from Fig. 3.1 that a given solute molecule ought to experience less hindrance during ultrafiltration than during diffusion through an aqueous pore.

Many of the important assumptions in this hydrodynamical treatment of solute movement through small pores cannot be strongly upheld; for instance, the properties ascribed to the solute and water molecules within the pores are physically unrealistic. This approach with its inherent limitations, however, does present two independent tests for membrane porosity by seeking to explain the variation in both (A_{sd}/A_p) and (A_{sf}/A_p) for a variety of test solutes in terms of (a_s/r). The success of this approach to solute transport through porous membranes depends ultimately on its ability to produce a picture of membrane porosity which is self-consistent and compatible with the results of other methods.

Since it is not possible to measure A_p for biological membranes it is advantageous to study water flow as well as solute flow. Using Renkin's (1954) filtration analysis one can describe the

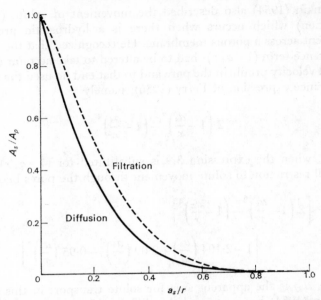

FIG. 3.1. The ratio of the apparent area (A_s) for solute transport to the total pore area (A_p) of unit area of membrane as a function of the ratio of solute molecular radius, a_s, to the equivalent pore radius, r. The apparent area A_s assumes the different values A_{sd} and A_{sf} depending on whether solute transport arises from diffusion or filtration. The corresponding ratios A_{sd}/A_p and A_{sf}/A_p have been obtained as functions of a_s/r from equations (3.6) and (3.10) respectively and are shown in the diagram as solid and interrupted lines (Renkin, 1954: Fig. 2).

apparent area, A_{wf}, for filtration of water by

$$\frac{A_{wf}}{A_p} = \left[2\left(1-\frac{a_w}{r}\right)^2 - \left(1-\frac{a_w}{r}\right)^4\right]\left(\frac{g_w^o}{g_w}\right) \quad 3.11$$

where a_w is the radius of the water molecule and (g_w^o/g_w) represents the friction exerted on a water molecule in solution relative to that exerted by the pore wall. (g_w^o/g_w) is given by

$$\frac{g_w^o}{g_w} = 1 - 2 \cdot 104\left(\frac{a_w}{r}\right) + 2 \cdot 09\left(\frac{a_w}{r}\right)^3 - 0 \cdot 95\left(\frac{a_w}{r}\right)^5 \quad 3.12$$

which has an identical form to equation 3.5 for solute molecules. With the aid of equations 3.10 and 3.11 it is possible to obtain the operational quantity, (A_{sf}/A_{wf}), given by

$$\frac{A_{sf}}{A_{wf}} = \frac{\left[2\left(1-\frac{a_s}{r}\right)^2 - \left(1-\frac{a_s}{r}\right)^4\right]}{\left[2\left(1-\frac{a_w}{r}\right)^2 - \left(1-\frac{a_w}{r}\right)^4\right]}$$

$$\times \frac{\left[1 - 2\cdot104\left(\frac{a_s}{r}\right) + 2\cdot09\left(\frac{a_s}{r}\right)^3 - 0\cdot95\left(\frac{a_s}{r}\right)^5\right]}{\left[1 - 2\cdot104\left(\frac{a_w}{r}\right) + 2\cdot09\left(\frac{a_w}{r}\right)^3 - 0\cdot95\left(\frac{a_w}{r}\right)^5\right]} \quad 3.13$$

Durbin, Frank & Solomon (1956) suggested that (A_{sf}/A_{wf}) was identical to $(1-\sigma_s)$ and later Durbin (1960) employed measurements of the reflexion coefficient for various solutes as a test for porosity in three different artificial membranes. Durbin's method of describing σ_s as a function of (A_{sf}/A_{wf}) and hence of (a_s/r) has been used for the red cell membrane by Goldstein & Solomon (1960).

The role of the reflexion coefficient as a test for membrane porosity will now be discussed.

Significance of the reflexion coefficient

Consider a permeable membrane separating pure water from a dilute solution of some non-electrolyte s. The reflexion coefficient is the ratio of the apparent osmotic pressure (Δp) to the theoretical van't Hoff osmotic pressure $(RT\Delta c_s)$ which would be observed if s were not permeant, that is

$$\sigma_s = \frac{\Delta p}{RT\Delta c_s}$$

which is determined in the absence of volume flow.

Durbin *et al.* (1956) and Durbin (1960), as has already been mentioned, attempted to relate the reflexion coefficients for certain solutes to the steric and frictional effects encountered by the solutes during diffusion and filtration across the membrane. They devised the equation

$$\sigma_s = 1 - \frac{A_{sf}}{A_{wf}} \quad 3.14$$

in order to establish the magnitude of (A_{sf}/A_{wf}) from measurements of σ_s. This approach has been criticized on the grounds that σ_s may be less than unity even when there are no pores in the membrane (Dainty, 1963a). The reason for this is given below.

Dainty (1963a) employed the following arguments to show how σ_s is related to the nature of solute and water movements through the membrane. If the solute s is entirely impermeant then water will flow across the membrane from the pure water to the solution and this water flow can be stopped by the application of a suitable hydrostatic pressure Δp across the membrane. In this case $\Delta p = RT\Delta c_s$, and $\sigma_s = 1$. Consider now that s is a permeant species and that it moves across the membrane through separate pathways from that of the water molecules. The application of Δp again makes the chemical potential for water in the solution equal to that of pure water and no water flow occurs. Nevertheless, there is a volume flow across the membrane due to the flux of s down its concentration gradient; in this case, the volume flow is $\bar{V}_s\omega_sRT\Delta c_s$. By decreasing the difference in hydrostatic pressure to a certain value, $\Delta p'$, it is possible to cancel the volume flow of solute by an opposing volume flow of water. The water flow will be $L_p(RT\Delta c_s - \Delta p')$; hence, it can be shown that the apparent osmotic pressure, $\Delta p'$, under these conditions is obtained when $\bar{V}_s\omega_sRT\Delta c_s = L_p(RT\Delta c_s - \Delta p')$, i.e. when

$$\Delta p' = RT\Delta c_s \left[1 - \frac{\omega_s\bar{V}_s}{L_p}\right] \qquad 3.15$$

Thus, for the case where solute and water move through the membrane by separate pathways, the reflexion coefficient $(\Delta p'/RT\Delta c_s)$ is described by

$$\sigma_s = 1 - \frac{\omega_s\bar{V}_s}{L_p} \qquad 3.16$$

which is identical to the relation derived by Kedem & Katchalsky (1958). In the interesting case where solute and water may cross the membrane by the same route, such as a water-filled pore, there will be some frictional interaction between them. Because the solute flowing across the membrane may drag some water along with it, the volume flow across the membrane from the solution to the water will be larger than $\bar{V}_s\omega_sRT\Delta c_s$. In order to maintain the net volume flow at zero, it is necessary to reduce the difference in hydrostatic pressure below $\Delta p'$ in this case. Thus, if there is an interaction between solute and water as they cross the membrane the reflexion coefficient for s must be less than $(\Delta p'/RT\Delta c_s)$, i.e.

$$\sigma_s < 1 - \frac{\omega_s \bar{V}_s}{L_p} \qquad 3.17$$

This inequality constitutes an experimental test for the existence of pores in membrane or, more correctly, for interactions between solute and water molecules within the membrane.

Explicit expressions for σ_s have been calculated by Kedem & Katchalsky (1961) and Dainty & Ginzburg (1963). Choosing the 'frictional pore-model' of Spiegler (1958), Kedem & Katchalsky (1961) expressed various parameters of the membrane in terms of the distribution and frictional coefficients for solutes. In particular, they found that the reflexion coefficient is given by

$$\sigma_s = 1 - \frac{\omega_s \bar{V}_s}{L_p} - \frac{K f_{sw}}{\varphi_w(f_{sw}+f_{sm})} \qquad 3.18$$

where

$$\frac{K f_{sw}}{\varphi_w(f_{sw}+f_{sm})} = \frac{A_{sd}}{A_{wd}} \qquad 3.19$$

where A_{wd} is the apparent area available for water diffusion. K is the average distribution (or partition) coefficient (defined by equation 2–7, Kedem & Katchalsky, 1961) of the solute between the membrane and solution, f_{sw} and f_{sm} are the coefficients of friction between one mole of solute and an infinite quantity of water and the membrane respectively, and φ_w is the overall cross-sectional area of the pores per unit area of membrane. On the other hand, Dainty & Ginzburg (1963) analysed the transport of solute and water through a conventional 'lipid-pore model' for the cell membrane. They assumed that both solute and water moved through the lipid component and through pores. Frictional interactions between solute and water were considered to occur only in the pores. Dainty and Ginzburg deduced from this model that

$$\sigma_s = 1 - \frac{\omega_s \bar{V}_s}{L_p} - \frac{K_s{}^c f_{sw}{}^c}{f_{sw}{}^c + f_{sm}{}^c} \qquad 3.20$$

where

$$\frac{K_s{}^c f_{sw}{}^c}{f_{sw}{}^c + f_{sm}{}^c} = \frac{A_{sf}}{A_{wf}} \qquad 3.21$$

$K_s{}^c$ is the distribution coefficient defined by the ratio of the mean concentration of solute in the pore to the mean solute concentration $\bar{c}_s$ and $f_{sw}{}^c$ and $f_{sm}{}^c$ are the frictional coefficients in the pores between one mole of solute and water and between one mole of solute and the membrane respectively. Both of these treatments of frictional interactions between solute and water within porous membranes yield the convergent conclusion that $\sigma_s < 1 - \omega_s \bar{V}_s/L_p$ in accord with the inequality 3.17.

If the magnitude of σ_s is to be used to estimate the pore radius, provided that this is warranted, then there is a dichotomy between the use of (A_{sd}/A_{wd}) and (A_{sf}/A_{wf}) in these expressions for σ_s. Since the model of Dainty & Ginzburg (1963) is more realistic than that of Kedem & Katchalsky (1961) it appears that the use of (A_{sf}/A_{wf}) is to be preferred. Solomon (1968) has come to a similar conclusion on the grounds that the theory of Kedem & Katchalsky (1961) relies on the use of a tortuosity factor which is related to the effective path length through the membrane. Moreover, Solomon noted that, when the tortuosity factor was omitted from the derivation of σ_s, the equality between $(1 - \omega_s \bar{V}_s/L_p - \sigma_s)$ and (A_{sd}/A_{wd}) was abolished.

The approach outlined above indicates how measurements of σ_s, ω_s and L_p for a given membrane can determine whether or not aqueous channels exist in the membrane. If the observed reflexion coefficient is less than $(1 - \omega_s \bar{V}_s/L_p)$ then one can infer that solute and water flows interact with each other probably in aqueous channels. In this case (A_{sf}/A_{wf}) is given by $(1 - \omega_s \bar{V}_s/L_p - \sigma_s)$.

Diffusional and viscous flows of water

Some information about the porosity of membranes can be obtained from studies of the rates of diffusion and viscous flow of water across them. Let us assume that the porous membrane already described separates two compartments, each being filled with water containing different concentrations of labelled water, w^*. According to Mauro (1957) the rate at which labelled water diffuses across the membrane is described by a similar expression to that for solute diffusion, namely

$$J_{w^*} = D_w A_w \frac{\Delta c_{w^*}}{\Delta x} \qquad 3.22$$

DIFFUSIONAL AND VISCOUS FLOWS OF WATER 87

where $\mathcal{J}_{w*}$ is the net flux of labelled water, A_w is a factor analogous to A_s in equation (3.1) and Δc_{w*} is the difference in concentration of labelled water across the membrane. D_w and Δx have their usual meanings. Measurements of $\mathcal{J}_{w*}$ and Δc_{w*} yield an estimate of the diffusional permeability, P_d (cm sec^{-1}), of the membrane to water, since P_d is defined by

$$P_d = \frac{D_w A_w}{\Delta x} \qquad 3.23$$

Mauro also considered the nature of the water flux which occurs when a difference in hydrostatic pressure is applied across the membrane. He suggested that the diffusional component, $\mathcal{J}_w^d$, of this water flux at any point x in the membrane is given by

$$\mathcal{J}_w^d = -\frac{D_w A_w c_w}{RT}\frac{d\mu_w}{dx} \qquad 3.24$$

where c_w is the concentration of water at x and $(d\mu_w/dx)$ is the gradient of the chemical potential for water at x. In this case $d\mu_w = \bar{V}_w dp$ and since $c_w \bar{V}_w = 1$, equation 3.24 becomes

$$\mathcal{J}_w^d = -\frac{D_w A_w}{RT}\frac{dp}{dx} \qquad 3.25$$

If we assume that the membrane is 'uniform' and that the applied Δp gives rise consequently to a linear gradient, dp/dx, which is identical to $(\Delta p/\Delta x)$, then $\mathcal{J}_w^d$ can be written as

$$\mathcal{J}_w^d = \frac{D_w A_w}{RT\Delta x}\Delta p \qquad 3.26$$

where Δp has been measured in the opposite sense to that of dp/dx. If the diffusional flux is the sole component of the total flux which is observed under these circumstances, then we can identify equation (3.26) with $\mathcal{J}_v = L_p \Delta p$. Hence, we obtain

$$L_p = \frac{D_w A_w \bar{V}_w}{RT\Delta x} \qquad 3.27$$

after equating $\mathcal{J}_v$ with $\mathcal{J}_w^d \bar{V}_w$. Substituting equation 3.23 into equation 3.27 gives

$$\frac{L_p RT}{\bar{V}_w} = P_d \qquad 3.28$$

This relation permits the hydraulic conductivity, L_p (cm sec^{-1}

atm^{-1}), to be expressed in the same units as P_d, i.e. cm sec^{-1}. It must be stressed that equation 3.28 will hold if, and only if, the total flux of water driven by a hydrostatic pressure gradient is solely diffusional in character. The expression $(L_p RT/\bar{V}_w)$ is often called the *filtration permeability*, P_f, or the *osmotic permeability*, P_{os}, for water, and both P_f and P_{os} are usually expressed in the same units as those of P_d.

It has been found in a large number of studies on both artificial and biological membranes that $(L_p RT/\bar{V}_w) > P_d$. This discrepancy suggests quite strongly that the flow of water due to a difference in its chemical potential is composed of both diffusional and quasi-viscous components. Since the existence of viscous flow across a membrane indicates the presence of aqueous channels or 'pores' in the membrane, there have been numerous comparisons of P_d and $(L_p RT/\bar{V}_w)$ for biological membranes.

Let us assume that there is a difference in pressure, Δp, across the membrane and that it contains n aqueous 'pores' of radius r. Further, let us assume that both diffusional and viscous flows of water pass through A_p only; thus A_w becomes identical to A_p. According to equations 3.23 and 3.26 the diffusional flux of water, $\mathcal{J}_w^d$, is given by

$$\mathcal{J}_w^d = \frac{D_w A_p \Delta p}{\Delta x RT} = \frac{P_d}{RT} \Delta p \qquad 3.29$$

From equation 3.29 it is evident that P_d is equal to $n\pi r^2 D_w/\Delta x$. In addition to the diffusional flux there will be a viscous flux, $\mathcal{J}_w^v$, through the pores. According to Poiseuille's law the volume flow, $\mathcal{J}_w^v \bar{V}_w$, driven by Δp is given by

$$\mathcal{J}_w^v = \frac{n\pi r^4}{8\eta \bar{V}_w} \frac{\Delta p}{\Delta x} \qquad 3.30$$

where η is the viscosity of fluid in the pores. Equation 3.30 can be expressed in the language of irreversible thermodynamics as

$$\mathcal{J}_w^v = \frac{L_p' \Delta p}{\bar{V}_w} \qquad 3.31$$

where L_p' is a parameter, analogous to L_p, which refers to the filtration permeability of the pores. By dividing equation 3.31 by equation 3.30 and introducing $P_d = n\pi r^2 D_w/\Delta x$ it can be shown

that
$$r^2 = 8\eta D_w \frac{L_p'}{P_d} \qquad 3.32$$

Since L_p' cannot always be obtained directly it is necessary to show how it can be eliminated from equation 3.32.

Consider the total water flux across the membrane. In this case it is equal to $(\mathcal{J}_w{}^d + \mathcal{J}_w{}^v)$ and it can also be described by

$$\frac{\mathcal{J}_v}{\bar{V}_w} = \frac{L_p \Delta p}{\bar{V}_w} \qquad 3.33$$

where $\mathcal{J}_v$ is the observed volume flux and L_p is the hydraulic conductivity of the membrane. Thus, we can equate $(L_p \Delta p / \bar{V}_w)$ with the sum of equations 3.29 and 3.31 to yield

$$\frac{L_p \Delta p}{\bar{V}_w} = \frac{P_d \Delta p}{RT} + \frac{L_p' \Delta p}{\bar{V}_w} \qquad 3.34$$

The relation 3.32 for the pore radius can be rewritten with the aid of equation 3.34 as

$$r = \sqrt{\frac{8 D_w \eta_w \bar{V}_w}{RT}\left[\frac{L_p RT}{\bar{V}_w P_d} - 1\right]} \qquad 3.35$$

where η_w, the viscosity of water, has been substituted for η, the viscosity of the pore fluid. Kedem & Katchalsky (1961) have found an identical expression for r on the basis of a frictional interpretation of the flux equations for labelled water diffusion and hydraulic volume flow; see also Nevis (1958).

When the appropriate values for D_w, η_w, $\bar{V}_w$ and RT at 25°C are substituted into equation 3.35 the relation simplifies to

$$r = 3\cdot 8 \sqrt{\frac{L_p RT}{\bar{V}_w P_d} - 1} \qquad \text{angströms} \qquad 3.36$$

This method of establishing whether or not the membrane is porous depends upon reliable estimates of $(L_p RT / \bar{V}_w)$ and P_d.

It might be contended that it is invalid to assign the values for D_w and η_w to the corresponding coefficients for the pore-fluid. Unfortunately there is no definite answer to this source of dubiety. Fedyakin (1962), for example, studied fluid flow in narrow glass capillaries (100–1000 Å) and found that the viscosity of the water in the capillaries was larger than η_w and linearly dependent on the capillary radius. From that observation and other experiments, Derjaguin (1965) has concluded that the water in such narrow capillaries is 'anomalous' (see Chapter 1). However, the interpretation of these experiments is a controversial matter and it is not safe to conclude from them that the water in porous membranes is 'anomalous'. This caution is supported by the data of Madras, McIntosh & Mason (1949) who concluded that viscous flow through a swollen cellophane membrane was characterized by the bulk viscosity of water. In a sense it is somewhat irrelevant to question the validity of Poiseuille's law and the assignment of particular values for η_w and D_w since the ideal geometry of uniform pores traversing the membrane is not realized either in artificial or biological membranes.

When this theoretical treatment is applied to barriers, such as biological membranes, where the pore radius is likely to be small, say $r = 2a_w$, then some account must be taken of the effective areas for diffusion and filtration of water in the membrane. To obtain a suitable correction for these restrictions we can apply the analysis of the previous section on solute diffusion by employing A_{w*d} and A_{w*f} as the equivalents of A_{sd} and A_{sf}. Solomon and his collaborators (e.g. Paganelli & Solomon, 1957), have applied such a correction to the analysis by rewriting equation 3.35 for the pore radius as

$$r^2 = \lambda \left(\frac{A_{w*d}}{A_{w*f}} \right) \qquad 3.37$$

where

$$\lambda = \frac{8 D_w \eta_w \bar{V}_w}{RT} \left[\frac{L_p RT}{\bar{V}_w P_d} - 1 \right] \qquad 3.38$$

Equation 3.37 for r^2 has been obtained from the original derivation by replacing A_p in equation 3.29 by A_{w*d} and $n\pi r^2$ in equation 3.30 by A_{w*f}. By dividing equation 3.6 by equation 3.10 and putting $a_s = a_w$, it follows that

$$\frac{A_{w*d}}{A_{w*f}} = \frac{1}{2 - \left(1 - \frac{a_w}{r}\right)^2} \qquad 3.39$$

By substituting for (A_{w*d}/A_{w*f}) in equation 3.37 the expression for the pore radius becomes

$$r = -a_w + \sqrt{(2a_w^2 + \lambda)} \qquad 3.40$$

This relation has been used by Paganelli & Solomon (1957) to estimate the pore radius in biological membranes. However, they stressed that the magnitude of r obtained from equation 3.40 was an *equivalent pore radius*. Thus, r is basically an operational parameter which describes the biological membrane in terms of an ideal membrane containing right cylindrical pores in which bulk flow and diffusion of water are governed by the relations of Poiseuille and Fick.

Inspection of equation 3.35 suggests that a pressure-driven flow across a membrane will have a purely diffusional character only when $(L_p RT/\bar{V}_w)$ and P_d are identical, i.e. at $r = 0$. It seems physically realistic to expect, however, that at some critical non-zero value of the pore-radius, r_c, the transport will be obligatorily diffusional because each molecule will not be able to 'overtake' its neighbour. Longuet-Higgins & Austin (1966) have pursued that kind of argument in their analysis of the mechanism of osmotic flow. They asserted that the water permeabilities, expressed separately by hydrodynamical and diffusional models, are identical only when $r = r_c$ and that r_c is given by $\sqrt{(D_w \eta_w \bar{V}_w/RT)}$. Accordingly r_c is about 4 Å at 25°C whereas the radius of the water molecule is about 1·5 Å. Longuet-Higgins and Austin considered that osmotic flow must be diffusional when the pore radius is less than r_c and quasi-laminar through pores larger than r_c. Their analysis assumes that the diffusional component of the water flux is absent when $r > r_c$ and consequently their value for r_c must be an overestimate.

A description of both the hydrodynamical and diffusional aspects of water flow across porous membranes can be based on the laws of Poiseuille and Fick when the pore radius is much larger than the radius of the water molecule. The application of these laws to pores of molecular width is extremely dubious. In such narrow pores water molecules may move in single file, and this mechanism of transport (Hodgkin & Keynes, 1955) was originally proposed to account for anomalous flux ratios of potassium in the squid axon. Calculations based on the 'single-file' model for water transport were initially performed by Harris (1960) and later by Lea (1963) and Dick (1966) in order to explain the discrepancy between (L_pRT/V_w) and P_d for cell membranes. This view of water flow in long narrow pores can account for the different values for the osmotic and diffusional water permeabilities provided that there are several water molecules in each pore. The numbers of water molecules that are demanded for cell membranes lie in the range 1–69 (Dick, 1966). Hirsch (1967) has questioned the relevance of this model for water flow through porous cell membranes on the grounds that pores of the requisite narrow dimensions ought to contain no water molecules for 98% of the time and only one molecule for approximately 2% of the time.

Finally, the discrepancy between the osmotic and tracer permeabilities of the membrane to water has been discussed by Essig (1966) with the aid of a frictional approach to water transport. In particular, Essig considered the degree of frictional coupling between the flows of water and labelled water across the membrane. Such coupling can be described by the phenomenological coefficients of irreversible thermodynamics (Kedem & Essig, 1965). From his frictional treatment of the flows of w and w^* across a *homogeneous* membrane Essig concluded that the ratio of the exchange resistance, R_{w*}, for isotope flow to the resistance, R_w, for total water flow was given by

$$\frac{R_{w*}}{R_w} = 1 + \frac{f_{w*w}}{f_{wm}} \qquad 3.41$$

where f_{w*w} and f_{wm} are the frictional coefficients between w^* and w and between w and the membrane. In Essig's treatment (R_{w*}/R_w) is equivalent to $(L_pRT/\bar{V}_wP_d)$. The extent to which (R_{w*}/R_w) deviates from unity depends on f_{w*w} and f_{wm}. For example, the degree of isotope interaction is large $(f_{w*w} \gg f_{wm})$ for water flow in capillaries of large radius, whereas for narrow capillaries (R_{w*}/R_w) tends to unity because $f_{wm} \gg f_{w*w}$. In principle, a significant degree of isotope interaction may occur in membranes without aqueous channels. For example, Sidel & Hoffman (1961) and Thau, Bloch & Kedem (1966) found values of 2·4 and 2·1 for (R_{w*}/R_w) in studies of water transport through liquid membranes without aqueous pores. Consequently the size of $(L_pRT/\bar{V}_wP_d)$ cannot be regarded as a valid criterion for pores in membranes except when $(L_pRT/\bar{V}_wP_d) \gg 1$. Moreover, the alternative 'long narrow pore' explanation for the discrepancy between $(L_pRT/\bar{V}_w)$ and P_d is considered to be weak on physical and theoretical grounds (Hirsch, 1967; Kedem & Essig, 1965), especially since certain liquid membranes possess no structural pathways compatible with viscous flow or single-file transport of water molecules and yet have values of $(L_p RT/\bar{V}_w P_d)$ exceeding unity.

The theoretical relations between the viscous and diffusional components of water flow through narrow pores (≤ 10 Å) has been discussed by Mikulecky (1967). He has drawn attention to the arbitrariness of the assumptions underlying the approach used here and elsewhere (Mauro, 1957; Longuet-Higgins & Austin, 1966; Lakshminarayanaiah, 1967) to distinguish between the diffusional and viscous components of pressure-driven flow across a porous membrane. According to Onsager (1945) these components are inseparable. Although expression 3.35 for the equivalent pore radius rests on a questionable theoretical approach it still provides a useful experimental test which, of course, must be substantiated by other independent tests.

Solvent drag

Viscous flow of water through a porous membrane will disturb the unidirectional fluxes of small solutes passing through aqueous pores. Hence, the unidirectional flux of the solute in the same direction as bulk flow will be enhanced by solvent drag while that in the opposite direction will be hindered. Solvent drag, of course, can also alter the unidirectional fluxes of labelled water across the membrane and from this we can estimate the size of the equivalent pores.

Ussing (1949) and Koefoed-Johnson & Ussing (1953) obtained expressions for the ratio of unidirectional fluxes and one of the terms in their equation explicitly described the drag that water might exert on solute transfer across the membrane. Let us now look at the flux-ratio equation first for solutes and then for labelled water movement.

Flux ratio for solutes

The relation between solvent drag and the observed flux ratio for solute transport has been derived by Hoshiko & Lindley (1964) and Kedem & Essig (1965) within the framework of irreversible thermodynamics.

Consider a membrane separating two dilute solutions of an uncharged solute s. Let us assume that there are two isotopes, a and b, of the solute and that the outer compartment contains a solution of a whereas the inner compartment contains b alone. Provided that a and b have the same chemical properties, the unidirectional fluxes of s can be written (Hoshiko & Lindley, 1964) as

$$\mathcal{J}_a = \frac{c_a^o}{2}[2\omega_s RT + (1-\sigma_s)\mathcal{J}_v] \qquad 3.42$$

$$\mathcal{J}_b = -\frac{c_b^i}{2}[2\omega_s RT - (1-\sigma_s)\mathcal{J}_v] \qquad 3.43$$

where $\mathcal{J}_a$ and $-\mathcal{J}_b$ denote the influx, $\mathcal{J}_s^{in}$, and efflux, $\mathcal{J}_s^{out}$ of s respectively. Equations 3.42 and 3.43 have been obtained from the solute flux equation 2.46 with the conditions that $c_a^i = 0$ and $c_b^o = 0$; in this case $\bar{c}_s$, becomes $c_a^o/2$ and $c_b^i/2$ respectively. To simplify matters let us take $c_a^o = c_b^i$; this is a convenient condition for the experimental determination of the flux ratio. Thus, the flux ratio is

$$\frac{\mathcal{J}_s^{in}}{\mathcal{J}_s^{out}} = \frac{\left[\dfrac{2\omega_s RT}{(1-\sigma_s)\mathcal{J}_v}+1\right]}{\left[\dfrac{2\omega_s RT}{(1-\sigma_s)\mathcal{J}_v}-1\right]} \qquad 3.44$$

Since

$$\ln\left\{\frac{x+1}{x-1}\right\} = \frac{2}{x}$$

when x is large, the relation for ($\mathcal{J}_s^{in}/\mathcal{J}_s^{out}$) can be reduced (by taking the logarithms of both sides) to

$$\ln\left\{\frac{\mathcal{J}_s^{in}}{\mathcal{J}_s^{out}}\right\} = \frac{(1-\sigma_s)\mathcal{J}_v}{\omega_s RT} \qquad 3.45$$

provided that $2\omega_s RT \gg (1-\sigma_s)\mathcal{J}_v$. This expression for the flux ratio shows that its logarithm is directly proportional to the volume flow across the membrane. Hence, a plot of $\ln(\mathcal{J}_s^{in}/\mathcal{J}_s^{out})$ against $\mathcal{J}_v$ should be linear if solvent drag exerts an influence on the movement of solute. Moreover, the slope of such a plot should be given by $(1-\sigma_s)/\omega_s RT$. The flux ratio test, therefore, has two facets; first the plot should be linear and secondly its gradient should be compatible with independent estimates of σ_s and $\omega_s RT$.

Both Hoshiko & Lindley (1964) and Kedem & Essig (1965)

obtained a general form for the flux-ratio equation, which included not only the solvent-drag term but also the usual terms arising from the electrochemical potentials. Their expression was identical to that of Ussing (1949). One point that does emerge from these treatments is that the flux ratio for a solute will depend on whether the volume flow is driven by a hydrostatic pressure or an osmotic pressure gradient due to additional impermeant solutes. The flux-ratio equation for the former driving force contains an additional term, ($V_s \Delta p/RT$), not present when osmotic flow occurs. The additional contribution to the flux ratio for s is due to the fact that Δp increases the chemical potential gradient of s over and above its influence on the transport of s via solvent drag. Thus, each plot of the flux ratio for s against the volume flow will be linear provided solvent drag occurs, but in the case of pressure-driven flow a component of the solute flux ratio will be independent of entrainment of solute and water flows. The latter component is likely to be relatively small, if not entirely insignificant (see Hoshiko & Lindley, 1964).

Flux ratio for labelled water

Let us consider the experiment where the unidirectional fluxes, $\mathcal{J}_{w*}^{in}$ and $\mathcal{J}_{w*}^{out}$, are determined for a given membrane. If there is a volume flow $\mathcal{J}_v$ induced, say, by an osmotic gradient $-RT\Delta c_i$ due to impermeant solutes then the flux ratio for labelled water will be given by equation 3.45. In this case, however, the test solute is labelled water and the membrane cannot discriminate between this and water; consequently the reflexion coefficient σ_{w*} is zero and equation 3.45 becomes

$$\ln\left\{\frac{\mathcal{J}_{w*}^{in}}{\mathcal{J}_{w*}^{out}}\right\} = \frac{\mathcal{J}_v}{\omega_{w*}RT} \qquad 3.46$$

where $\omega_{w*}RT$ is the conventional permeability to labelled water or P_d. In this experiment $\mathcal{J}_v$ is given by $-L_p RT \Delta c_i$ which can be rewritten as $\bar{c}_w RT L_p \ln(n_w^o/n_w^i)$ where n_w^o and n_w^i are the mole fractions of water in the outer and inner solutions (see equations 2.17 and 2.18). Thus, we can obtain equation 3.46 in the form

$$\ln\left\{\frac{\mathcal{J}_{w*}^{in}}{\mathcal{J}_{w*}^{out}}\right\} = \frac{L_p RT}{\bar{V}_w P_d} \ln\left\{\frac{n_w^o}{n_w^i}\right\} \qquad 3.47$$

since $\bar{V}_w \bar{c}_w = 1$ for dilute solutions.

This relation is identical to that derived by Koefoed-Johnsen & Ussing (1953) who expressed the relation in the form

$$\ln\left\{\frac{\mathcal{J}_{w*}^{\text{in}}}{\mathcal{J}_{w*}^{\text{out}}}\right\} = \left(1 + \frac{G_w}{g_w'}\right) \ln\left\{\frac{a_w^o}{a_w^i}\right\} \qquad 3.48$$

where a_w^o and a_w^i are the activities of the water in the outer and inner solutions, G_w is the frictional coefficient for self-diffusion of water and g_w' is the friction exerted upon one mole of water at unit velocity. In fact, G_w is given by (RT/D_w) and g_w' is defined for viscous flow across a uniform membrane by

$$\mathcal{J}_w = A_p \frac{1}{g_w'} \frac{\Delta p}{\Delta x} \qquad 3.49$$

Thus, we can rearrange equation 3.49 to give

$$g_w' = \frac{\Delta p A_p}{\mathcal{J}_w \Delta x} = \frac{\bar{V}_w P_d}{L_p' D_w} \qquad 3.50$$

where L_p' has been described previously (equation 3.31) and P_d has been substituted for $(D_w A_p/\Delta x)$.

Replacing G_w by (RT/D_w) and g_w' by $(\bar{V}_w P_d/L_p' D_w)$ in equation 3.48 we get

$$\ln\left\{\frac{\mathcal{J}_{w*}^{\text{in}}}{\mathcal{J}_{w*}^{\text{out}}}\right\} = \left(1 + \frac{RT L_p'}{\bar{V}_w P_d}\right) \ln\left\{\frac{a_w^o}{a_w^i}\right\} \qquad 3.51$$

Introducing equation 3.34 for L_p' into the above relation gives an equation identical to 3.47.

The equation of Koefoed-Johnsen & Ussing (1953) for the unidirectional water fluxes relates the flux ratio to $(L_p RT/\bar{V}_w P_d)$ and by using equation 3.35 for the equivalent pore radius we can transform equation 3.47 into

$$\ln\left\{\frac{\mathcal{J}_{w*}^{\text{in}}}{\mathcal{J}_{w*}^{\text{out}}}\right\} = \left(1 + \frac{r^2 RT}{8 D_w \eta_w \bar{V}_w}\right) \ln\left\{\frac{n_w^o}{n_w^i}\right\} \qquad 3.52$$

Evidently measurements of the unidirectional fluxes of labelled water can be used to estimate the equivalent pore radius with this equation. Essig (1966), however, has shown that the frictional model of water transport replaces the term $(1 + G_w/g_w')$ in equation 3.48 with the corresponding term $(1 + f_{w*w}/f_{wm})$. As we have mentioned before, f_{w*w} may exceed f_{wm} in artificial membranes without aqueous pores. Therefore, the flux-ratio test for pores, just like the value of $(L_p RT/\bar{V}_w P_d)$ is not an unambiguous criterion, although it is still a useful one.

Electro-osmosis

The existence of electro-osmosis in a membrane can be considered as good evidence for the presence of water-filled channels. Electro-osmosis, as we have indicated before (page 54), is a volume flow induced by current flow across the membrane. If electro-osmosis is present then so too should be the converse electrokinetic phenomenon–streaming potential. The streaming potential is the potential difference which is generated by pressure-driven volume flow across the membrane. Both of these phenomena are intimately interrelated and they signify a coupling between the flows of volume and current which generally is a consequence of water-filled channels carrying an electrical charge.

Electro-osmosis, and other electrokinetic phenomena for that matter, have been described successfully by the phenomenological equations. Let us assume that the membrane is bathed by identical salt solutions and that the 'forces' Δp and ΔE operate. The appropriate phenomenological equations become

$$\mathcal{J}_v = L_{11}\Delta p + L_{12}\Delta E \qquad 3.53$$
$$I = L_{21}\Delta p + L_{22}\Delta E \qquad 3.54$$

with $L_{12} = L_{21}$. Once again we can transform these equations into practical equations for $\mathcal{J}_v$ and I by using the membrane coefficients L_p, β and κ' described (Set II, Table 2.1) in Chapter 2. These coefficients are related to the Onsager coefficients by

$$L_p = \left[\frac{\mathcal{J}_v}{\Delta p}\right]_I = L_{11} - \frac{L_{12}^2}{L_{22}} \qquad 3.55$$

$$\beta = \left[\frac{\mathcal{J}_v}{I}\right]_{\Delta p} = \frac{L_{12}}{L_{22}} \qquad 3.56$$

$$\kappa' = \left[\frac{I}{\Delta E}\right]_{\Delta p} = L_{22} \qquad 3.57$$

(see Katchalsky & Curran, 1965). For this set of coefficients the hydraulic conductivity, L_p, must be determined for zero current flow, the electro-osmotic permeability, β, must be determined at zero pressure difference and, finally, the specific conductance, κ', must also be determined at zero pressure difference. Consequently a set of practical equations can be written as

$$\mathcal{J}_v = L_p \Delta p + \beta I \qquad 3.58$$
$$I = \kappa'\beta\Delta p + \kappa'\Delta E \qquad 3.59$$

Electro-osmosis is a very clear example of a flow, $\mathcal{J}_v$, being driven by its nonconjugate force, ΔE, and as we have seen this can be expressed quantitatively as the ratio of the electro-osmotic flow to the current. According to equation 3.59 the application of a pressure gradient, Δp, will generate a streaming potential, ΔE, when there is zero current flow. The streaming potential per unit pressure difference, $(\Delta E/\Delta p)$, at $I = 0$ is equal to $-\beta$ according to equation 3.59. In other words, a membrane which permits electro-osmosis should also produce a streaming potential when pressure-driven volume flow occurs. Further, the observations of these converse electrokinetic phenomena should satisfy the relation

$$\left[\frac{\mathcal{J}_v}{I}\right]_{\Delta p} = -\left[\frac{\Delta E}{\Delta p}\right]_I \qquad 3.60$$

since the magnitude of each ratio is equal to β. The negative sign appears in the above equation because the solution in the pore carries a net charge of opposite sign to that on the pore wall. For example, if the charge on the wall is negative then the application of Δp will cause a flow of volume, and hence of positive charge, inwards thus giving rise to a potential difference which is negative ($\Delta E = E^o - E^i$). On the other hand, a positive current flow will generate a positive volume flow. Thus $(\Delta E/\Delta p)$ will be equal, but of opposite sign, to $(\mathcal{J}_v/I)$.

The important point about the above relation between the electro-osmotic volume flow and the streaming potential is that it establishes a criterion for accepting an apparent electro-osmotic flow as a genuine one. That is, $(\mathcal{J}_v/I)$ must be corroborated by the corresponding determination of $(\Delta E/\Delta p)$.

It is possible to analyse electro-osmotic flow further by considering three different models (see Dainty, Croghan & Fensom, 1963) which try to describe possible interactions between ions and water in the pore fluid.

Helmholtz–Smoluchowski model. This picture of electro-osmosis is probably the most familiar and yet the least helpful for discussing this phenomenon in artificial and biological membranes. It stems from a theory which treats the case of electro-osmosis in wide capillaries (Overbeek, 1952).

Consider once more our porous membrane. The interface between the pore walls and the solution will be the site of an electrical double layer of positive and negative charges. Let us assume that the pore walls have a negative charge then the adjacent fluid will have a net positive charge due to an excess of cations. When an electric field is applied across the pore in a direction parallel to its surface the pore fluid, by virtue of its electrical charge, moves by laminar flow through the pore. According to this model there is a radial variation in the fluid velocity and the velocity is zero not at the pore wall but at a short distance from the wall in the 'slipping plane'. It is possible, therefore, to calculate the velocity of the liquid at a large distance from the wall. This will be the electro-osmotic velocity, provided that the thickness of the pore fluid outside of the double layer is very much larger than the thickness of the double layer itself. This proviso must be emphasized since the fluid velocity in the double layer is smaller than the calculated electro-osmotic velocity. The thickness of the double layer in angströms is approximately equal to $3/\sqrt{}$(ionic strength), where the ionic strength of the solution is given by $0.5 \sum c_i z_i^2$; c_i is the molarity of an ion i. For example, the thickness of the double layer for a 1 mM NaCl solution will be about 100 Å. Thus, the Helmholtz–Smoluchowski model almost certainly cannot be applied to electro-osmosis across cell membranes but it may be relevant to electro-osmotic flow through wide channels, such as the intercellular spaces in epithelia, and even larger channels such as blood vessels in animals and xylem and phloem in plants.

It is possible to express L_p, β and κ' in terms of the parameters of the Helmholtz–Smoluchowski model but the application of these new coefficients to transport across artificial and biological membranes is fraught with difficulties (see Dainty et al., 1963). For example, consider the apparently straightforward determination of the membrane conductance κ'. On the assumption that all of the current passes through the aqueous pores then κ' is given by

$$\kappa' = \frac{A_p \theta}{\Delta x} \qquad 3.61$$

where θ (Ω^{-1} cm^{-1}) is the specific conductance of the pore. However, because of the accumulation of ions in the double layer there exists a so-called surface conductance pathway in the pore. This component can be as large as the conductance through the rest of the pore fluid and its existence means that θ is exceedingly difficult to estimate. The hydraulic conductivity can, of course, be obtained from Poiseuille's law, whereas the derivation of β comes from the calculation of the electro-osmotic velocity (Overbeek, 1952; Dainty *et al.*, 1963).

Schmid model. Contrary to the previous model which could only be applied to pores with radii larger than the double layer, the model of Schmid (1950) is applicable only when the pore radius is much smaller than the double layer. In fact, the pore radius must be only a few angströms. In this case the ions within the pore fluid possess only charges of opposite sign to those of the pore wall. Moreover, it is assumed that the distribution of these counter ions in the pore is uniform. We may visualize that an applied electrical field exerts a force on the counter ions and that the resultant force experienced by the pore fluid is analogous to a hydrostatic pressure gradient. It is possible, therefore, to estimate β for this model by making the questionable assumption that Poiseuille's law applies to such narrow pores. If there are $\bar{X}$ (equiv. cm^{-3}) electrical charges per unit volume of pore fluid due to counter ions, then the force per unit area exerted on the pore fluid by an electric field ΔE is $\bar{X}F\Delta E$ joules cm^{-3}. This is equivalent to a hydrostatic pressure of $10\bar{X}F\Delta E$ atm. If we can apply Poiseuille's law to this situation then the volume flow will be

$$\mathcal{J}_v = \frac{A_p r^2 10 \bar{X} F \Delta E}{8\eta_w \Delta x} \qquad 3.62$$

and since $\Delta E = I\Delta x/A_p\theta$ we can express β by the relation

$$\beta = \left[\frac{\mathcal{J}_v}{I}\right]_{\Delta p} = \frac{r^2 \bar{X} F}{8\eta_w \theta} 10 \qquad 3.63$$

which, in principle, could be used to estimate the pore radius from β when $\bar{X}$ and θ are known. For this model both L_p and κ' are given by the corresponding expressions for these coefficients in the Helmholtz–Smoluchowski model. Again one meets difficulties in trying to estimate θ, the specific conductance of the pore fluid, but for a different reason since Dainty *et al.* (1963) have estimated that the electro-osmotic drag of ions which could occur, would increase the apparent conductance of the membrane. Thus, the danger is that κ' and hence θ may be overestimated.

In comparison to that of Helmholtz–Smoluchowski the Schmid model seems more applicable to electro-osmosis in biological membranes and possibly in most artificial membranes since it is valid only for small pore radii much less than 30 Å or so.

Frictional model. Dainty et al. (1963) employed the frictional model of Spiegler (1958) to obtain the phenomenological equations for the flows $\mathcal{J}_v$ and I in relation to the forces Δp and ΔE. These authors discussed the frictional model for the case of narrow pores containing only counter ions i and water w. If f_{iw} is the frictional force between one mole of cation and the water in the pore when the relative velocity of the cation with respect to the water is 1 cm sec^{-1}, then the force on 1 mole of cations due to the water is $f_{iw}(u_i - u_w)$; the velocities of i and w are given by u_i and u_w. Similarly the force on one mole of cations due to the membrane m is $f_{im}u_i$. The flows of i and w also involve their velocities since they are given by (concentration × velocity). When the phenomenological equations were written to take account of the frictional coefficients Dainty et al. (1963) were able to obtain a set of Onsager coefficients involving the frictional coefficients f_{iw}, f_{wm} and f_{im}.

The practical coefficients L_p, β and κ', which have been cited above, can be rewritten to include f_{iw}, f_{wm} and f_{im} since Dainty et al. (1963) have obtained the coefficients L_{11}, L_{12} and L_{22} employed in the phenomenological equations 3.53 and 3.54 and we know that L_p, β and κ' are described by equations 3.55, 3.56 and 3.57 respectively.

For example, the electro-osmotic permeability, β, is given

$$\beta = \frac{1}{F} \frac{(c_i/c_w)f_{iw} + c_i \bar{V}_i f_{wm}}{c_i[(c_i/c_w)f_{iw} + f_{wm}]} \qquad 3.64$$

where $\bar{V}_i$ is the partial molar volume of the counter ion. Using this expression Dainty et al. (1963) obtained the number of moles of water transported electro-osmotically per mole of counter ion. This electro-osmotic transport ratio is

$$\frac{\mathcal{J}_w}{\mathcal{J}_i} = \frac{f_{iw}}{\dfrac{c_i}{c_w}f_{iw} + f_{wm}} \qquad 3.65$$

and it reveals what the maximal efficiency of electro-osmotic transport of water can be. The maximum will be achieved when the frictional force between the water and the pore wall is much less than that between the counter ions and water; that is, $f_{wm} \ll f_{iw}$. Under these conditions equation 3.65 shows that (J_w/J_i) will be equal to (c_w/c_i) and equation 3.64 shows that β becomes $1/Fc_i$. Thus, optimally the pore fluid will move as an entity during electro-osmosis. In general, of course, the ratio of water to ion transport during electro-osmosis will lie in the range from 0 to (c_w/c_i).

Summary

Several features of solute and water transport can be explained by postulating that the membrane is porous and methods are available for establishing the size of the equivalent pore radius All of the phenomena which underlie such determinations stem from interactions between solutes and water in a common transport pathway although it is known that a limited amount of interaction can occur in membranes without aqueous pores. The theoretical approaches outlined in the preceding discussion have been successful in varying degrees. Nevertheless, any success which they enjoy in yielding estimates of the equivalent pore radius, which are mutually compatible, is all the more surprising when one considers some of the unlikely assumptions on which they are based. For example, it has been assumed that Poiseuille's law applies to pressure-driven volume flow in extremely narrow pores. Another weakness is the assumption that solute and water transport occurs exclusively through the porous channels; this particular condition surely does not apply to biological membranes. In this connexion, nevertheless, the criterion for membrane porosity, which rests on the magnitudes of the reflexion coefficient, is still valid since Dainty & Ginzburg (1963) have shown that it holds even for a membrane permitting solute and water transport through a non-porous route in parallel with the porous route.

4

SOME EXPERIMENTS ON ARTIFICIAL MEMBRANES

Unstirred layers	104
Unstirred layer thickness	106
Practical implications	109
Water transport	114
Diffusion	114
Osmosis	118
Comparison of water permeabilities	124
Effect of temperature on water permeability	127
Interactions between solute and water molecules	130
Restricted motion of solutes	130
Reflexion coefficient	134
Frictional coefficients	138
Solvent drag	141
Composite membranes	144
Cellulose acetate membranes	145
Asymmetrical double-membrane system	148

THE study of water and solute transport in artificial membranes *may* help to improve our understanding of the corresponding problems in cell membranes and this optimistic aim is embodied in the following description of some experiments on artificial membranes. Before we can begin to discuss the permeability characteristics of artificial membranes, however, we ought to re-examine the assumption that the membrane actually rate-controls the movement of substances across itself.

The theoretical treatment of membrane transport, which has been outlined in the previous chapters, has assumed that the solutions on both sides of the membrane are so well stirred that there are no local gradients of concentration at the membrane's surfaces. This condition cannot be satisfied in practice since at the interface between an artificial or a biological membrane and the bathing medium there is a Nernst diffusion layer—commonly called an *'unstirred layer'*. Let us take a look at the sort of influence the unstirred layer exerts on the apparent permeability characteristics of the membrane.

Unstirred layers

The concept of unstirred layers was developed originally by Noyes & Whitney (1897) and later by Nernst (1904) in their experimental studies of heterogeneous reactions. According to the theoretical views of Nernst, there is a layer of static fluid at the interface of a solid body and the bathing liquid. It should be stressed that the solute concentration within this layer is not necessarily identical to that in the bulk solution and, indeed, it is a function of position. Actually the fluid within such a layer is not stationary and the unstirred layer is a region of slow laminar flow parallel to the solid–liquid interface in which convection cannot play an important role in solute transport. Diffusion is the primary mode of transport within the unstirred layer. The thickness of the layer, δ, is actually indeterminate but it can be defined for convenience as an operational quantity given by

$$\left(\frac{dc}{dx}\right)^{\text{int}} = (c^b - c^{\text{int}})/\delta \qquad 4.1$$

where $(dc/dx)^{\text{int}}$ is the concentration gradient at the interface, c^b is the bulk concentration of the solute and c^{int} is the concentration at the interface (see Fig. 4.1). Hydrodynamic studies have demonstrated that δ is related to the hydrodynamic boundary layer where the latter is defined by the velocity gradient at the solid–liquid interface. The hydrodynamic boundary layer is invariably larger than δ.

In the original treatment of this problem Nernst assumed that δ was a constant for a given state of fluid motion. Current theory (Levich, 1962) predicts, however, that δ is a function not only of the diffusion coefficient of the solute but also of the physical properties and the velocity of the solution. Under a given regime of motion, therefore, the effective thickness of the unstirred layer may not be identical for different substances.

Whether or not such diffusion layers limit the rate of transport of a certain substance across a membrane depends upon the membrane's permeability for the substance. Such an unstirred layer can be envisaged as an equivalent membrane in series with the actual membrane and the equivalent permeability coefficient, P_s^δ, of the unstirred layer for a solute s may be defined by

$$P_s^\delta = \frac{D_s}{\delta} \qquad 4.2$$

where D_s (cm^2 sec^{-1}) has its usual meaning. Clearly it is possible that the transfer of some solutes across a given membrane may be wholly or partially rate-controlled by diffusion in the unstirred regions adjacent to the membrane.

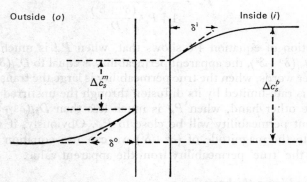

FIG. 4.1. A possible concentration profile for a permeating solute in the solutions adjacent to the membrane (Dainty & House, 1966a: Fig. 1).

A possible instantaneous concentration profile during the permeation of s across a membrane is illustrated in Fig. 4.1. The operational thicknesses, δ^o and δ^i, of the unstirred layers are defined by equation 4.1. Under these circumstances the flux of solute, J_s, is given apparently by

$$J_s = P_s \Delta c_s^b \qquad 4.3$$

where P_s is the apparent permeability of the membrane to the solute. The 'true' membrane permeability, P_s^t, is given by

$$J_s = P_s^t \Delta c_s^m \qquad 4.4$$

Under steady-state conditions, where Δc_s^b, Δc_s^m and J_s remain constant, P_s is related to P_s^t by

$$\frac{1}{P_s} = \frac{1}{P_s^t} + \frac{\delta^o}{D_s} + \frac{\delta^i}{D_s} \qquad 4.5$$

where the solute diffusion coefficients in the outer and inner solutions have been taken equal to D_s. Thus, the presence of unstirred layers may produce a significant discrepancy between the values of P_s and $P_s{}^t$ for a particular solute.

Even although the bulk solutions on either side of the membrane may be well-stirred, there are still zones of fluid adjacent to the membrane where convective transport of material is not possible. It is instructive to consider the effect of that restriction on diffusion across the membrane. Equation 4.5 can be re-arranged to yield

$$P_s = \frac{P_s{}^t}{1 + P_s{}^t \frac{(\delta^o + \delta^i)}{D_s}} \qquad 4.6$$

Inspection of equation 4.6 shows that, when $P_s{}^t$ is much larger than $D_s/(\delta^o + \delta^i)$, the apparent permeability is equal to $D_s/(\delta^o + \delta^i)$. In other words, when the true permeability is large the transport of solute is rate-limited by its diffusion through the unstirred layers. On the other hand, when $P_s{}^t$ is much less than $D_s/(\delta^o + \delta^i)$ the apparent permeability will be close to $P_s{}^t$. Obviously, if we can estimate the magnitude of $(\delta^o + \delta^i)$ we can use equation 4.6 to obtain the 'true' permeability from the apparent value.

Unstirred layer thickness

Many attempts have been made to obtain the thickness of the unstirred layers at membrane surfaces. It must be stressed, however, that the unstirred layer does not have an exact thickness and and so the precision of certain estimates of its size is deceptive.

Some workers (Stewart & Graydon, 1957; Peterson & Gregor, 1959; Mackay & Meares, 1959) have obtained estimates of δ from studies of ionic fluxes across ion-exchange membranes. It can be shown that the solute flux is a function of both δ and the membrane thickness (e.g. Peterson & Gregor, 1959) and so δ can be found by measuring the solute fluxes across membranes of different thickness. Moreover, δ has been obtained for different stirring conditions (Table 4.1).

Ginzburg & Katchalsky (1963) measured the diffusional permeability to water of a cellulose membrane (Sylvania wet gel) at different stirring speeds. They observed that $(1/P_d)$ declined to a limiting value as the stirring rate of the bathing solutions was increased from 0 to 2000 revolutions per minute. From the minimal value of $(1/P_d)$ and their estimate of $(1/P_d{}^t)$, where $P_d{}^t$ is the 'true' permeability to labelled water, they calculated by equation 4.5 that the minimal value of δ was about 12 μm on the reasonable assumption that δ had the same magnitude on both sides of the membrane. In this paper Ginzburg & Katchalsky (1963) suggested that δ could be estimated from measurements of the apparent solute permeability of two membranes, which were studied separately and then subsequently as a series-membrane arrangement. If the apparent permeabilities for the first and second membranes are P_1 and P_2 respectively, then we have

$$\frac{1}{P_1} = \frac{1}{P_1{}^t} + \frac{2\delta}{D_s} \qquad 4.7$$

$$\frac{1}{P_2} = \frac{1}{P_2{}^t} + \frac{2\delta}{D_s} \qquad 4.8$$

where $P_1{}^t$ and $P_2{}^t$ are the 'true' solute permeabilities of the first and second membrane and the thickness of the unstirred layer on either side of each membrane is δ. For two membranes in series (see page 75) we expect that the apparent permeability, P_{12}, of the composite series-membrane is

$$\frac{1}{P_{12}} = \frac{1}{P_1{}^t} + \frac{1}{P_2{}^t} + \frac{2\delta}{D_s} \qquad 4.9$$

Substituting equations 4.7 and 4.8 into equation 4.9, followed by some re-arrangement, gives

$$2\delta = D_s \left[\frac{1}{P_1} + \frac{1}{P_2} - \frac{1}{P_{12}} \right] \qquad 4.10$$

Hence, measurements of the apparent solute permeabilities P_1, P_2 and P_{12}, will yield an estimate of δ from equation 4.10. Everitt, Redwood & Haydon (1969) used this technique to estimate δ at the surface of Cellophane membranes. They found from the labelled water permeabilities conducted in the absence of stirring that $2\delta/D_w$ was $3 \cdot 12 \times 10^3$ cm^{-1} sec at 20°C; knowing D_w at 20°C one can calculate δ (see entry for Everitt *et al.*, 1969, Table 4.1).

Furthermore, there is the ingenious method of Peers (1956) for estimating δ. This method employs a cation-exchange membrane which separates two half-chambers containing a dilute salt solution (10 mM NaCl). Peers recorded the potential across the membrane during the passage of constant currents.

TABLE 4.1. Dependence of unstirred layer thickness on stirring conditions

'Membrane'	Thickness of unstirred layer (μm)										Reference
					Stirring rate (rev/min)						
	0	30	60	100	200	300	400	800	1200	2000	
Ion-exchange membranes Polystyrene-sulphonic acid					30	20	14	4	1		Peterson & Gregor (1959)
Homogeneous Phenol sulphonic + formaldehyde (Zeo-Karb 315; Permutit)		330*		10*							Stewart & Graydon (1956)
Cation-exchange membrane (C-10; Permutit)			65		50	47					Mackay & Meares (1959)
Sylvania wet gel	400									≥12	Peers (1956) Ginzburg & Katchalsky (1963)
Cellophane membrane	300										Everitt et al. (1969)
Contact lens	150						≤20				Green & Otori (1970)
Isolated cornea of rabbit	350						65				Green & Otori (1970)
Lipid membrane			55								Andreoli & Troutman (1971)
Lipid membrane				85 (stirring rate?)							Holz & Finkelstein (1970)

* Agitation of bathing solutions achieved by shaking; values refer to number of oscillations per minute.

At a certain voltage the current attained a limiting magnitude, i_{lim}, because the transport of sodium ions up to the cation-exchange membrane was ultimately limited by the diffusion of the accompanying chloride ions across the unstirred region near to the membrane. According to Peers the magnitude of i_{lim} is given by $FD_s c_s/\delta \tau_a$, where D_s is the diffusion coefficient of the salt, c_s is the salt concentration and τ_a is the anion transport number in the bulk solution. Hence, δ was estimated from the value of i_{lim} (see Table 4.1).

Andreoli & Troutman (1971) investigated the dependence of the apparent values of P_d and P_s for artificial lipid membranes on the viscosity of the bathing solutions. Since the diffusion coefficient of a solute is inversely related to the viscosity of the solvent, these authors argued that a plot of $(1/P_s)$ against the viscosity should be linear (cf. equation 4.5) and that δ can be obtained from the slope. Their experimental plots for labelled water, urea and glycerol penetration were in accord with this view and the unstirred layer thickness was quite similar in each case; in fact, they found that δ was about 55 μm on each side of the membrane.

Yet another method was devised by Green & Otori (1970) who measured the thickness of the unstirred layer adjacent to a contact lens by using a pachometer. This optical instrument is normally used to measure the thickness of corneas; Green and Otori found that there was a region adjacent to the lens (or the cornea) into which small latex spheres (0·2 μm diameter) moved by diffusional transport only. They measured the thickness of this region and demonstrated that its size was diminished by stirring the bathing solution (Table 4.1).

In addition to the estimates of the thickness of unstirred layers adjacent to artificial membranes, numerous estimates have been made for biological membranes. Those experiments will be discussed later within the physiological context of the particular investigations. At present, however, we can draw the conclusion that in all transport studies on artificial and biological membranes there will be unstirred layers in the range 1–500 μm at the membrane's surfaces. The magnitude of these layers will depend, of course, upon the degree of agitation of the bathing media.

Practical implications

There seem to be several areas in transport studies where

unstirred layers are important. For example, their presence may lead to erroneous estimates of the practical membrane coefficients ω_s, L_p and σ_s. Moreover, they may cause difficulties in solvent drag experiments, especially where the flux-ratio for labelled water is being determined, and yet another place where they cause trouble is in the interpretation of electrokinetic phenomena.

One way of evaluating the effectiveness of the unstirred layer in rate-limiting the transport of a substance across a membrane is to obtain an estimate of D/δ and compare this with the apparent permeability for the substance (see equation 4.5). For example, consider the passage of labelled water across an artificial membrane. Let us assume that the apparent diffusional permeability of the membrane to water is found to be 2×10^{-3} cm sec^{-1}. If the unstirred layer thickness at the surface of the membrane were 100 μm, then D_w/δ for labelled water would be about 2×10^{-3} cm sec^{-1}. The similarity between D_w/δ and the apparent permeability in this case raises the objection that the transport of labelled water across the membrane is rate-controlled by its movement in the unstirred layer. Another way of putting this is to say that, although the experiment was designed to measure the diffusional permeability of the membrane to water, the apparent permeability actually represents the equivalent permeability of the unstirred layer.

Do unstirred layers also affect the measurement of L_p? Several authors (e.g. Kuhn, 1951; Dainty, 1963a; Everitt & Haydon, 1969) have considered this problem and the common conclusion is that L_p will be underestimated. The argument supporting that view is briefly described below.

Consider a net inflow, J_v, of volume across a membrane separating two aqueous solutions of an impermeant solute (Fig. 4.2). The net volume flow at any point x in the outer unstirred region ($o < x < \delta°$) will produce a convective movement ($J_v c^x$) of solute up to the membrane, where c^x is the solute concentration at x, and this will be opposed by the diffusion of solute ($-D dc^x/dx$). In the steady-state these forces will balance each other to give

$$J_v c^x - D \frac{dc^x}{dx} = 0 \qquad 4.11$$

This equation can be integrated across the outer region ($o < x < \delta°$) to give

$$c_m{}^o = c_b{}^o \exp(J_v \delta°/D) \qquad 4.12$$

where $c_b{}^o$ and $c_m{}^o$ are the solute concentrations in the outer solution ($x = 0$) and at the outer surface of the membrane ($x = \delta°$) respectively.

A similar treatment for the inner unstirred region yields

$$c_m^i = c_b^i \exp(-\mathcal{J}_v \delta^i/D) \qquad 4.13$$

where c_b^i and c_m^i are the solute concentrations in the inner solution and at the inner surface of the membrane respectively. To simplify the treatment let us assume that $\delta^o = \delta^i = \delta$. Then the osmotic pressure difference $RT(c_m^i - c_m^o)$ exerted at the surface of the membrane is given by the difference between equations 4.13 and 4.12, namely

$$RT(c_m^i - c_m^o) = RT\left[c_b^i \exp\left(-\frac{\mathcal{J}_v \delta}{D}\right) - c_b^o \exp\left(\frac{\mathcal{J}_v \delta}{D}\right)\right] \qquad 4.14$$

In practice, we expect that $\delta = 10^{-2}$ cm, $D = 10^{-5}$ cm^2 sec^{-1} and $\mathcal{J}_v$ would be of the order of 10^{-5} cm^3 cm^{-2} sec^{-1} or less: thus $\mathcal{J}_v \delta/D$ is much

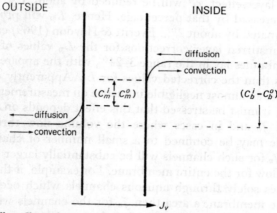

FIG. 4.2. The effect of unstirred layers on an osmotic experiment designed to measure L_p for the membrane. The volume inflow will 'sweep' solute up to the membrane at its outside surface and away from it at its inner surface by convection. To a certain extent these local concentration gradients will be opposed by solute diffusion in the unstirred layers adjacent to the membrane. The end result of this is that the effective osmotic gradient is less than the apparent gradient across the membrane.

less than 1 and $\exp(\mathcal{J}_v \delta/D) = 1 + \mathcal{J}_v \delta/D$. Using this approximation equation 4.14 can be reduced to the more useful form

$$RT(c_m^i - c_m^o) = RT(c_b^i - c_b^o) - RT\frac{\mathcal{J}_v \delta}{D}(c_b^i + c_b^o) \qquad 4.15$$

This relation shows that the actual osmotic gradient is less than the nominal osmotic gradient $RT(c_b^i - c_b^o)$ because of the unstirred layer term $RT\mathcal{J}_v\delta(c_b^i + c_b^o)/D$. The latter term arises in this case because of solute accumulation at the outer surface of the membrane and solute depletion at the inner surface.

The hydraulic conductivity will be underestimated by an amount which depends on J_v. This effect will also be present even when J_v is driven by a hydrostatic pressure gradient rather than an osmotic gradient provided, of course, that solutes are present in the bathing solutions. However, the influence of unstirred layers is likely to be small because of the size of $J_v\delta/D$. For example, consider equation 4.15; clearly c_b^i will be reduced at the membrane to a value $c_b^i(1-J_v\delta/D)$ whereas c_b^o will be enhanced at the membrane to $c_b^o(1+J_v\delta/D)$. If $\delta = 10^{-2}$ cm, $D = 10^{-5}$ cm^2 sec^{-1} and, say, $J_v = 10^{-5}$ cm^3 cm^{-2} sec^{-1} then $J_v\delta/D = 0\cdot01$. Under those conditions, which incidentally have been chosen to magnify the unstirred layer effect, c_b^i will be reduced by about 1% while c_b^o will be increased by that percentage. Hence, L_p will probably be underestimated by about 2%. Everitt & Haydon (1969) concluded that the unstirred layer corrections for the L_p values of artificial lipid membranes lay in the range 3–24% with the apparent values being less than the corrected values for L_p. Apparently unstirred layers exert an almost negligible influence on measurements of L_p. However, it must be stressed that this effect depends on the magnitude of J_v and in cases where the volume flow through the membrane may be confined to a small number of channels the value of J_v for such channels will be substantially larger than that nominal flow for the entire membrane. For example, if the volume flow passes solely through aqueous channels which occupy, say, 1% of the membrane's area then J_v for the channels will be 100 times that for the membrane; therefore, at the mouth of the channels the solute concentrations will be displaced by about 200% and not just 2% as they would for the whole surface. Hence, it is not safe to ignore this effect on L_p, especially for membranes which possess preferential routes for viscous water flow (see Chapter 9).

The presence of unstirred layers also interferes with the measurement of the reflexion coefficient. For instance, an investigator might hope to estimate σ_s by observing the volume flow generated by a difference in osmotic pressure, $RT\Delta c_s^m$, across the membrane. When there is no hydrostatic pressure gradient J_v is given by $-\sigma_s L_p RT\Delta c_s^m$. In practice, J_v must be estimated when the original Δc_s^m is established across the membrane, otherwise the permeation of s may alter significantly the magnitude of Δc_s^m. Because of the existence of unstirred layers the procedure for estimating the initial flow may become difficult and imprecise. Of course other methods for estimating σ_s are available; however, any estimate of the reflexion coefficient is liable to be erroneous if no account is taken of the possible discrepancy between the nominal value of $\Delta c_s{}^b$ for the bathing solutions and the actual value of $\Delta c_s{}^m$ itself.

Such discrepancies, with their concomitant errors in P_s, L_p and σ_s, are created by the presence of unstirred layers under certain circumstances. Fortunately, it is possible to correct for this source of error, although such corrections are made too infrequently.

It follows from what has been said about the influence of unstirred layers on determinations of P_s and L_p that measurements of flux ratios, especially of rapidly permeating solutes, will be subject to error and even more so in the presence of volume flow across the membrane. This effect could be of great importance in the interpretation of solvent drag experiments. For example, consider the determination of the unidirectional influx of labelled water or urea in the presence of volume inflow. The volume flow will create an accumulation of the labelled solute at the outer surface of the membrane and a depletion at the inner surface. Consequently one would expect an enhanced influx of solute as compared to that in the absence of volume flow. Now consider the unidirectional efflux. In the presence of volume inflow it will be reduced due to the local gradients which we already mentioned. The net result is that volume inflow *appears* to augment the unidirectional influx while hindering the efflux of solute. In ignorance of the unstirred layer effect one might conclude that such experimental evidence favoured solvent drag of the solutes through aqueous pores in the membrane.

Another area of membrane transport where unstirred layers may play an important part is in the observation of electrokinetic phenomena. For example, Barry & Hope (1969a) have shown theoretically that, when a current is passed across a membrane, there may be sufficiently large alterations in the ionic concentrations at the membrane surfaces to produce an observable osmotic volume flow in addition to any purely electro-osmotic volume flow. The local alterations in concentration are caused by the differences in the ionic transport numbers between the membrane and the bathing solutions. Such ionic gradients in the unstirred layers adjacent to the membrane will tend to be dissipated by diffusion from the unstirred layers into the bulk solution and into the membrane, and by convective movement of solute arising from the flow of water along the local osmotic gradients. Barry and Hope concluded that, in addition to the genuine electro-osmotic volume flow, j_o, there would be a transient osmotic component of the total volume flow. According to these authors the latter component should increase to a maximal value of $\sigma_s L_p(\tau_{m+}-\tau_{s+})I/F\omega_s$ where τ_{m+} and τ_{s+} are the transport numbers of cations in the membrane and adjacent solutions respectively on the assumption that the cations carry most of the current through the membrane. Thus the maximal volume flow, $\mathcal{J}_{vm}$, is given by

$$\mathcal{J}_{vm} = j_o + \frac{\sigma_s L_p \alpha I}{F\omega_s} \qquad 4.16$$

where $\alpha = \tau_{m+}-\tau_{s+}$. Since the chief aim of experiments on electro-osmosis is to obtain j_o, their analysis indicates that measurements of the volume flow should be performed very rapidly, that the bathing solutions should be well-stirred and, finally, that any transient behaviour of the volume flow should be noted and regarded possibly as anomalous.

Streaming potential experiments will also suffer from problems with unstirred layers. In this case pressure-driven volume flow will perturb the local ionic gradients at the membrane's surfaces and these alterations will generate a spurious transient potential across the membrane which complicates the interpretation of the experimental data (e.g. Schmid & Schwarz, 1952).

Water transport

Diffusion

Table 4.2 shows the spectrum of values which have been found for the diffusional permeability to water of certain artificial membranes. In several of these experimental studies no account has been taken of the unstirred layers at the membrane surfaces. In some cases, where the effect of stirring on the apparent value of P_d has been investigated, a significant degree of correction for unstirred layers was required. For example, Ginzburg & Katchalsky (1963) found that vigorous agitation of the solutions bathing a Silvania wet gel membrane produced a pronounced rise in the apparent diffusional water permeability (see Fig. 4.3). In order to estimate the limiting size, δ_l, of the unstirred layer thickness in

TABLE 4.2. Diffusional permeability to water of some artificial membranes

Membrane	$P_d \times 10^4$ (cm sec^{-1})	Water content %	Reference
Collodion I	30	–	Robbins & Mauro (1960)
II	14	–	Robbins & Mauro (1960)
III	2·7	–	Robbins & Mauro (1960)
Sylvania wet gel	5·6	84	Renkin (1954)
Sylvania wet gel	3·8	–	Durbin (1960)
Sylvania wet gel	19·6*	77	Ginzburg & Katchalsky (1963)
Cellophane	3·9	76	Renkin (1954)
Cellophane	4·0	–	Durbin (1960)
Visking dialysis tubing	11·2*	68	Ginzburg & Katchalsky (1963)
Visking dialysis tubing	4·5	66	Renkin (1954)
Visking dialysis tubing	3·5	–	Durbin (1960)
Cellulose acetate	6·4	60	Gary-Bobo et al. (1969)
Cellulose acetate	10·5	60	DiPolo et al. (1970)
Cellulose acetate	3·0	–	Hays (1968)
Phenol sulphonic acid cation-exchange membrane (PSA)	1·7	59	Lakshminarayanaiah (1967)
Polyethylene-styrene graft copolymer type AMF C-103	0·36	21	Lakshminarayanaiah (1967)
Cellophane	3·4	39	Thau et al. (1966)
Polyvinyl alcohol	1·0	31	Thau et al. (1966)
Cellulose acetate	0·18	10	Thau et al. (1966)
Tributyl phosphate (on paper)	0·105	6	Thau et al. (1966)
Polyethylacrylate (on paper)	0·025	2	Thau et al. (1966)
Triacetine	0·011	7	Thau et al. (1966)
Lipid	2·3	–	Hanai, Haydon & Taylor (1965)
Lipid	4·4	–	Huang & Thompson (1966)
Lipid	10·6*	–	Cass & Finkelstein (1967)
Lipid	21·3*	–	Everitt, Redwood & Haydon (1969)
Lipid	2·0	–	Holz & Finkelstein (1970)
Lipid (treated with nystatin)	12·0	–	Holz & Finkelstein (1970)
Lipid (treated with amphotericin B)	6·0	–	Holz & Finkelstein (1970)
Lipid	13·8*	–	Andreoli & Troutman (1971)
Lipid (treated with amphotericin B)	107·5*	–	Andreoli & Troutman (1971)

* Corrected for diffusion in unstirred layers.
In several other studies cited in the table unstirred layers were considered to be of no quantitative significance.

their experiments Ginzburg and Katchalsky rewrote equation 4.5 in the form

$$\frac{D_w}{2P_d} = \frac{D_w}{2P_d{}^t} + \delta_l \qquad 4.17$$

where P_d and $P_d{}^t$ are the apparent and 'true' diffusional permeabilities to water, and D_w has its usual meaning. When the data in Fig. 4.3 were replotted as $D_w/2P_d$ against stirring rate these authors noted that $D_w/2P_d$ decreased asymptotically to a minimum

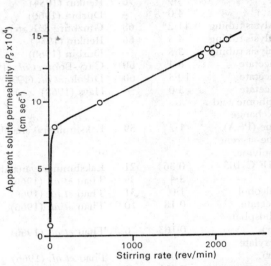

FIG. 4.3. The apparent permeability of a Sylvania wet gel membrane to labelled water as a function of the rate of stirring in the bathing solutions (Ginzburg & Katchalsky, 1963: Fig. 4).

of 100 μm and from this value they estimated that δ_l lay in the range 12 to 47·5 μm. On the basis of these experiments Ginzburg and Katchalsky allotted an arbitrary value of 25 μm to δ so that they could correct certain solute permeability coefficients for the existence of unstirred layers. Their data for the diffusional permeability to water of Visking dialysis tubing and Sylvania wet gel have also been corrected for that unstirred layer thickness.

Inspection of the data in Table 4.2 shows that there is a correlation between the water content of the artificial membranes and their diffusional permeability to water. The water content is usually expressed as the weight of water per gram of wet weight and it is a rough guide to the porosity of the membrane. Thus, we see that the more porous the membrane is the larger P_d becomes. This is underlined by the measurements of Robbins & Mauro (1960). These workers intentionally prepared collodion membranes of different permeabilities and they estimated by independent means that the average radius of the membrane pores ranged from 40 Å in group I to about 15 Å in group III. The measurements of Thau, Bloch & Kedem (1966) also illustrate the relation between diffusional permeability and the porosity of the artificial membranes. Essentially their data for the porous membranes (cellulose acetate, polyvinyl alcohol and cellophane) are compatible with the view that the self-diffusion of water in such membrane is determined by the total available area for water transport. In contrast, the water permeabilities of the so-called 'liquid membranes' (Triacetine, Tributylphosphate (TBP) and Polyethylacrylate (PEA)) are considerably smaller than those of the porous membranes.

In a dual sense the cellulose acetate membranes cited in Table 4.2 occupy a unique position in the study of the mechanisms of water permeation through artificial membranes. This type of membrane was designed by Loeb (1966) for desalination by reverse osmosis and it consists of a thin dense layer supported by a thick porous layer. Obviously this artificial membrane has great practical importance. In addition, cellulose acetate membranes behave not only as porous barriers for hydrophilic solutes but also like lipophilic ones for solutes with large partition coefficients (DiPolo, Sha'afi & Solomon, 1970). They represent, therefore, a crude analogue of cell membranes. For this reason, several studies have been made recently of their water permeability characteristics (Hays, 1968; Gary-Bobo, DiPolo & Solomon, 1969; DiPolo *et al.*, 1970), and these will be discussed later in the section on composite membranes (see page 145).

The water permeabilities of certain ion exchange membranes (Lakshminaranyanaiah, 1967) are also shown in Table 4.2. Again the discrepancy between the water permeabilities of these membranes can be qualitatively attributed to the differences in their water contents. Lakshminaranyanaiah also found that the presence of an additional supporting membrane increased the resistance to water diffusion across these ion exchange membranes.

Techniques have been developed for producing thin (100 Å)

membranes of phospholipids with certain additives. These membranes have some structural similarities to those of cell membranes (Mueller, Rudin, Tien & Wescott, 1962; Huang, Wheeldon & Thompson, 1964). Because they can be considered as useful models of natural plasma membranes, it is extremely pertinent to study their permeability characteristics for water. Numerous workers have estimated P_d for this type of membrane (see Table 4.2). The significant point about these data is that the permeability coefficients fall within the range of values reported for cell membranes (see Table 5.1). Two important points emerge from Table 4.2. First, the diffusional permeability to water is related to the water content and, secondly, in the cases in which unstirred layers have been corrected the 'true' permeability is larger than the apparent value.

Osmosis

At the beginning of a paper published in 1966 Longuet-Higgins and Austin wrote: *'few phenomena are so well understood thermodynamically, or so ill understood kinetically, as the osmotic flow of solvent through a semipermeable membrane'*. Their assessment seems very apt. Nevertheless, the kinetics of osmosis can be described clearly although not rigorously.

Experimental work on pressure-driven flow of solvent through artificial porous membranes indicates that the volume flow is only partially diffusional in character. Numerous terms have been applied to the predominant component of this kind of volume flow—bulk, laminar or quasi-laminar flow. Robbins & Mauro (1960) suggested that it should be called 'non-diffusional' flow and in view of the complexity of the kinetics of volume flow their expression seems the most appropriate. Complementary studies of the volume flow, driven by differences in osmotic pressure across porous membranes, also indicate that the flux is non-diffusional. A consideration of the mechanism of osmotic flow is extremely germane not only to the study of water transport in artificial and biological membranes as a whole but also to a basic understanding of the kinetic processes occurring within porous membranes. Mauro (1957, 1960, 1965), Ray (1960) and Dainty (1963a, 1965) have published essentially similar accounts of osmosis in porous membranes. The theoretical views of these workers can be summarized briefly in the following way.

Consider a uniform porous membrane separating pure water in the inner compartment from an ideal aqueous solution of an impermeant solute s in the outer compartment. Thus, the pores are filled with water alone. Further, let us assume that both compartments are at the same pressure, p (say, 1 atm). Volume flow can be halted either by applying an excess hydrostatic pressure, Δp, to the outer compartment or by constraining the inner solution so that a drop in the hydrostatic pressure, $-\Delta p$, of the inner compartment occurs. In both equilibrium conditions the chemical potentials for water in the outer and inner solutions are equal. Hence, we can equate $\mu_w{}^o$ and $\mu_w{}^i$ to give, for example

$$\bar{V}_w(p+\Delta p)+RT\ln(n_w{}^o) = \bar{V}_w p + RT\ln(n_w{}^i) \qquad 4.18$$

or

$$\Delta p = -\frac{RT}{\bar{V}_w}\ln(n_w{}^o) \qquad 4.19$$

since the mole fraction for pure water, $n_w{}^i$, is unity, i.e. $\ln(n_w{}^i) = 0$. For dilute solutions equation 4.19 can be rewritten as $\Delta p = RTc_s$, where c_s is the concentration of s in the outer solution. Figure 4.4

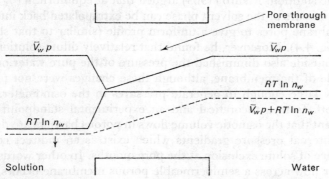

FIG. 4.4. The profiles of the chemical potential of water and its components through a narrow pore in a membrane. The upper solid lines show the pressure component ($\bar{V}_w p$) and the concentration component ($RT\ln n_w$) respectively. The lower interrupted line shows the total chemical potential profile ($\bar{V}_w p + RT\ln n_w$) (Dainty, 1965: Fig. 2).

shows the profiles of the thermodynamic variables within a typical pore. In the steady-state there must be a step in the profile of the mole fraction at the mouth of the pore since the pore contains only water molecules. If the assumption is made that the thermodynamic expressions, such as pressure, are valid for the microregion of the pore mouth, then the step in the mole fraction for water must be associated with an opposite step in the pressure profile (see Fig. 4.4). This follows from the fact that μ_w is continuous at the pore mouth. Therefore, the pressure just within the pore must fall from p by an amount Δp given by equation 4.19. This picture of osmosis predicts that there is an internal pressure gradient within the porous membrane and that it ought to be possible to observe states of tension within the adjacent water phase. Mauro (1965) confirmed that such an internal pressure gradient exists (see Fig. 4.5a); he used cellulose acetate membranes and the tension developed in the solvent phase was monitored with a displacement transducer. Figure 4.5b shows a typical transient record of the pressure of the water phase when an aqueous solution (0·02M) of the impermeant solute, polyethylene glycol, was placed in the outer compartment. The equilibrium state, which is attained, corresponds to a negative pressure difference of 2·2 atm; the discrepancy between the observed value and the theoretically predicted value RTc_s (0·4 atm) reflects the non-ideal behaviour of the solution. Mauro (1965) argued that at equilibrium ($J_v = 0$) the pressure of the solvent phase can be extrapolated back into the membrane pores to give a uniform profile (similar to that shown in Fig. 4.4). Moreover, he found that relatively dilute solutions on the outside also diminished the pressure of the pure water on the inside of the membrane, although these changes were not necessarily large enough to generate pressures in the osmometer.

Both from a theoretical and an experimental standpoint it is evident that the osmotic volume flows in porous barriers are driven by internal pressure gradients which exist as an indirect consequence of solute exclusion at the pore mouths. In other words, the water flow across a semipermeable porous membrane occurs by a single common mechanism whether it is driven by a difference in osmotic pressure or hydrostatic pressure.

On the basis of two quite independent models, Ray (1960) and Dainty (1965) have attempted to compute the magnitude of Δp at the pore mouth during osmosis.

Ray (1960) expressed the osmotic water flux in two ways; first, he obtained the water flux across the pore mouth by invoking a diffusional transfer of water down its concentration gradient and, secondly, he employed Poiseuille's law to describe the water flow

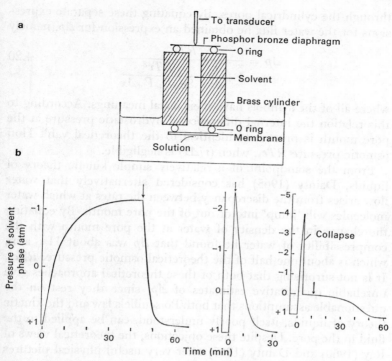

FIG. 4.5. **a** Hepp-type osmometer used to determine pressure in solvent phase. **b** Pressure responses in solvent phase due to presence of different solutions on the opposite side of the membrane. On the left is shown the response to a solution of polyethylene glycol (120 g/1000 ml: molecular weight, 6000–7500). The arrow indicates when the solution was replaced by distilled water and the relaxation began. The asymmetry in the 'rise' and 'fall' of pressure is an indication of the difficulty in establishing boundary conditions at the membrane with a viscous solution. The middle trace shows the pressure response to a 2M sucrose solution. The pressure was allowed to rise only to −2·4 atm and then to relax by replacing the solution with distilled water at the moment indicated by the arrow. On the right this experiment was repeated but allowed to attain higher negative pressure until cavitation set in. Although not indicated on this scale, pressure fell ('instantaneously') to +22 mm Hg with osmotic flow continuing at this constant pressure. Arrow indicates moment when solution was replaced with distilled water and relaxation of pressure began (Mauro, 1965: Figs. 3 & 4).

Copyright (1965, American Association for the Advancement of Science). Used by permission of AAAS.

through the cylindrical pores. By equating these separate expressions for the water flux he obtained an expression for Δp, namely

$$\Delta p = \frac{RTc_s}{1 + \dfrac{r}{\Delta x} + \dfrac{RTr^3}{8\eta_w D_w \bar{V}_w \Delta x}} \qquad 4.20$$

where all of the symbols have their usual meanings. According to this relation the internal difference of hydrostatic pressure at the pore mouth is practically identical to the theoretical van't Hoff osmotic pressure RTc_s, when $(r/\Delta x)$ is negligible.

From the standpoint of a relatively simple kinetic theory of liquids, Dainty (1965) has considered alternatively that water flow arises from the discrepancy between the rates at which water molecules will 'jump' into or out of the pore mouth. By equating the change in the density of water at the pore mouth with the compressibility of water he found that Δp was about 11 c_s atm, which is about one half of the theoretical osmotic pressure, RTc_s. It is not surprising that both of these theoretical approaches give unreliable quantitative estimates of Δp, since they rest on the questionable assumptions that both Poiseuille's law and the kinetic theory of liquids, itself poorly understood, can be applied to the fluid in the pore. Despite these objections, the theoretical views of Ray (1960) and Dainty (1965) offer very useful physical pictures of osmosis.

The foregoing discussion of osmosis in semipermeable porous membranes suggests that their water permeabilities, determined by applying either Δp or $RT\Delta c_s$ should be quantitatively different from their diffusional permeabilities to water. Comparison of the L_p values in Table 4.3 with those of P_d in Table 4.2 reveal that is the case except for artificial lipid membranes; of course, to compare these coefficients one must multiply L_p by $RT/\bar{V}_w$ (see Table 4.4 for a comparison).

No unstirred layer corrections of really significant size have been required in the determinations of L_p for these artificial membranes (e.g. Lakshminarayanaiah, 1967; Everitt & Haydon, 1969). The absence of serious errors in L_p contrasts markedly with the influence of unstirred layers on the estimates of the diffusional permeability for these membranes, but actually such a disparity is expected on theoretical grounds (see preceding section on *Unstirred layers*).

TABLE 4.3. *Hydraulic conductivities of some artificial membranes*

Membrane	$L_p \times 10^7$ (cm sec^{-1} atm^{-1})	Water content %	Reference
Collodion			
I	15800	–	Robbins & Mauro (1960)
II	560	–	Robbins & Mauro (1960)
III	70	–	Robbins & Mauro (1960)
Sylvania wet gel	1560	–	Durbin (1960)
Sylvania wet gel	1940	84	Renkin (1954)
Cellophane	400	–	Durbin (1960)
Cellophane	220	76	Renkin (1954)
Visking dialysis tubing	310	68	Ginzburg & Katchalsky (1963)
Visking dialysis tubing	106	–	Durbin (1960)
Visking dialysis tubing	95	66	Renkin (1954)
Cellulose acetate	120	–	Hays (1968)
Cellulose acetate	95	60	DiPolo et al. (1970)
Cellulose acetate	42	60	Gary-Bobo et al. (1969)
PSA ion-exchange membrane	140	59	Lakshminarayanaiah (1967)
AMF C-103 ion-exchange membrane	2·1	21	Lakshminarayanaiah (1967)
Cellophane	194	39	Thau et al. (1966)
Polyvinyl alcohol	9·4	31	Thau et al. (1966)
Cellulose acetate	0·32	10	Thau et al. (1966)
Tributylphosphate (on paper)	0·14	6	Thau et al. (1966)
Polyethylacrylate (on paper)	0·019	2	Thau et al. (1966)
Triacetine	0·017	7	Thau et al. (1966)
Lipid	12·3	–	Huang & Thompson (1966)
Lipid	6·4	–	Hanai et al. (1965)
Lipid	8·3	–	Cass & Finkelstein (1967)
Lipid	14·5*	–	Everitt & Haydon (1969)
Lipid	1·5	–	Holz & Finkelstein (1970)
Lipid (treated with nystatin)	29·2	–	Holz & Finkelstein (1970)
Lipid (treated with amphotericin B)	12·4	–	Holz & Finkelstein (1970)
Lipid	12·2	–	Andreoli & Troutman (1971)
Lipid (treated with amphotericin B)	290	–	Andreoli & Troutman (1971)

* Corrected for unstirred layer effect.
In several other studies cited in the table unstirred layers were considered to be of no quantitative significance.

The range of values for L_p which have been found for artificial membranes, again bears a qualitative relation to the water contents of such barriers just as it did for P_d. In the cases where the porosity of the barrier has been independently established, we also find a correlation between the degree of porosity and the respective values for the hydraulic conductivity (see Robbins & Mauro, 1960).

The hydraulic conductivities, like the diffusional permeabilities, of the artificial lipid membranes bear a strong quantitative resemblance to the values of these permeability coefficients for cell membranes (see Table 5.2).

Comparison of water permeabilities

It has been tacitly assumed that the hydraulic conductivity of artificial membranes is the same when it is determined by either filtration or osmotic flow of water. Although this point has not been examined exhaustively there is some experimental evidence (e.g. Durbin, 1960) upholding this reasonable assumption. Nevertheless, it is possible that L_p determined by osmotic gradients may be lower than that obtained from filtration simply because the osmotic gradients themselves will alter the degree of hydration of the membrane. For example, Ginzburg & Katchalsky (1963) found that L_p for Visking dialysis tubing decreased when the concentration of the bathing solutions was increased and they attributed this

TABLE 4.4. Values of $(L_p RT/\bar{V}_w P_d)$ for some artificial membranes

Membrane	$\left(\dfrac{L_p RT}{\bar{V}_w P_d}\right)$	Water content %	Reference
Collodion			
I	730	–	Robbins & Mauro (1960)
II	55	–	Robbins & Mauro (1960)
III	36	–	Robbins & Mauro (1960)
Sylvania wet gel	570	–	Durbin (1960)
Sylvania wet gel	490	84	Renkin (1954)
Sylvania wet gel	66*	77	Ginzburg & Katchalsky (1963)
Cellophane	140	–	Durbin (1960)
Cellophane	79	76	Renkin (1954)
Visking dialysis tubing	41	–	Durbin (1960)
Visking dialysis tubing	38*	68	Ginzburg & Katchalsky (1963)
Visking dialysis tubing	29	66	Renkin (1954)

Cellulose acetate	56	–	Hays (1968)
Cellulose acetate	12	60	DiPolo et al. (1970)
Cellulose acetate	8·7	60	Gary-Bobo et al. (1969)
Cellulose acetate			
CA–4	48	60	Gary-Bobo & Solomon (1971)
CA–3	3·2	15	Gary-Bobo & Solomon (1971)
CA–2	1·7	14	Gary-Bobo & Solomon (1971)
CA–1	1·0	13	Gary-Bobo & Solomon (1971)
Cellulose triacetate	1·7	13	Gary-Bobo & Solomon (1971)
PSA ion-exchange membrane	110	59	Lakshminarayanaiah (1967)
AMF C–103 ion-exchange membrane	8·0	21	Lakshminarayanaiah (1967)
Cellophane	80	39	Thau et al. (1966)
Polyvinyl alcohol	12·5	31	Thau et al. (1966)
Cellulose acetate	2·5	10	Thau et al. (1966)
Triacetine	2·1	7	Thau et al. (1966)
Tributyl phosphate (on paper)	1·8	6	Thau et al. (1966)
Polyethylacrylate (on paper)	1·1	2	Thau et al. (1966)
Lipid	≥4	–	Hanai et al. (1965)
Lipid	≥4	–	Huang & Thompson (1966)
Lipid	1·1*	–	Cass & Finkelstein (1967)
Lipid	0·94*	–	Everitt & Haydon (1969); Everitt et al. (1969)
Lipid	1·0	–	Holz & Finkelstein (1970)
Lipid (treated with nystatin)	3·3	–	Holz & Finkelstein (1970)
Lipid (treated with amphotericin B)	3·0	–	Holz & Finkelstein (1970)
Lipid	1·2*	–	Andreoli & Troutman (1971)
Lipid (treated with amphotericin B)	3·8*	–	Andreoli & Troutman (1971)

* Corrected for unstirred layer effect.
In several other studies cited in the table unstirred layers were considered to be of no quantitative significance.

effect to an increase in the viscosity of the solutions; however, it could also be explained by a decrease in the water content, and hence in the dimensions of the aqueous channels of the membrane. Cass & Finkelstein (1967) also observed in artificial lipid membranes that increasing the osmotic gradients (established with either glucose or sucrose solutions) led to a reduction in L_p, but the mechanism is obviously complex since similar effects were not found evidently with NaCl solutions. It seems, therefore, that in artificial membranes there is no serious discrepancy between the rates of filtration and osmotic flow for equivalent driving forces, although this may not be the case for biological membranes.

In contrast to the similarity between filtration and osmosis there is a marked difference between either of these mechanisms of volume flow and that of diffusion across some membranes. Probably the most informative study of viscous and diffusional flow of water through artificial membranes is that of Thau et al. (1966). These workers obtained estimates of $(L_p RT/\bar{V}_w P_d)$ for a number of artificial membranes (Table 4.4). The series of membranes included porous membranes, such as Cellophane, and 'liquid membranes', such as paper coated with polyethylacrylate (PEA). Although no unstirred layer corrections were applied to the permeability data, the results probably do not warrant such a correction because most of the membranes were relatively impermeable to water. It is notable that the 'liquid membranes' gave values of $(L_p RT/\bar{V}_w P_d)$ of up to 2·1. Since they do not contain pores as such, the degree of interaction of water molecules with each other, expressed by the ratio $(L_p RT/\bar{V}_w P_d)$, does suggest that there must be transport mechanisms employing group transfer of some kind which is different from viscous flow of water. Clearly the work of Thau et al. (1966) indicates that the value of $(L_p RT/\bar{V}_w P_d)$ must significantly exceed 2 before it can be accepted as proof of pores.

At the extreme end of the range of values cited in Table 4.4 are Collodion I (Robbins & Mauro, 1960) and Sylvania wet gel (e.g. Renkin, 1954) and this is consistent with their highly porous character. For example, Robbins and Mauro estimated that the average pore radius of Collodion I was about 40 Å whereas the value of $(L_p RT/\bar{V}_w P_d)$ yields a pore radius of about 100 Å from equation 3.36. In the intermediate part of the range of $(L_p RT/\bar{V}_w P_d)$ values we find the ion-exchange membranes (e.g. Lakshminarayanaiah, 1967) whereas at the other extreme there are the artificial lipid membranes. Because the latter are probably our best analogues of cell membranes their permeability characteristics deserve some mention.

The original comparisons of $(L_p RT/\bar{V}_w)$ and P_d for artificial lipid membranes revealed a significant disparity (e.g. Huang & Thompson, 1966). However, later work in which unstirred layer corrections were made to P_d values demonstrated that $(L_p RT/$

$\bar{V}_w P_d$) is indistinguishable from unity (e.g. Cass & Finkelstein, 1967). This evidence suggests that these lipid membranes are non-porous and consequently that the osmotic movement of water proceeds by diffusion rather than by viscous flow. Nevertheless, the water permeability characteristics of lipid membranes can be altered by the antibiotics nystatin and amphotericin B (Andreoli, Dennis & Weigl, 1969; Holz & Finkelstein, 1970) and by lysine vasopressin (Fettiplace, Haydon & Knowles, 1972). For example, Andreoli *et al.* (1969) found the amphotericin B increased P_d from 11×10^{-4} to 18×10^{-4} cm sec^{-1} and L_p from 17×10^{-7} to 400×10^{-7} cm sec^{-1} atm^{-1}. Holz & Finkelstein (1970) have confirmed the observations of Andreoli *et al.* (1969) but there is a quantitative discrepancy between their data. Andreoli *et al.* (1969) obtained a value of 30 for $(L_p RT/\bar{V}_w P_d)$ in the presence of amphotericin B, whereas Holz & Finkelstein (1970) found a value of 3. It seems that the disparity between these data could be due to an insufficient correction for unstirred layer effects on P_d in the earlier work of Andreoli and his colleagues. Holz & Finkelstein (1970) also showed that nystatin raised $(L_p RT/\bar{V}_w P_d)$ from unity to 3 by again increasing L_p predominantly. It is interesting to note also that Fettiplace *et al.* (1972) found that lysine vasopressin increased L_p for artificial lipid membranes by more than a factor of two; as yet there have been apparently no reports of the effect of vasopressin on P_d for those membranes.

Effect of temperature on water permeability

One way of tackling questions about the mechanisms of diffusion, osmosis and filtration in membranes is to examine the temperature coefficient of these processes. Such data are expressed in the form of an Arrhenius plot where a parameter, say P_d, is plotted against the reciprocal of the absolute temperature. When the slope of the graph is linear one can say that a single process is occurring and that it is characterized by an apparent activation energy, E_a (kcal mole^{-1}). This approach allows one, in principle, to obtain the activation energies of water diffusion and filtration across a membrane so that they can be compared with the corresponding values in free solution. Gary-Bobo & Solomon (1971) adopted this approach in their study of water transport across cellulose acetate membranes and their data will be discussed first to illustrate the general argument.

Gary-Bobo and Solomon used four types of cellulose acetate

membranes and a cellulose triacetate membrane in order to obtain a graded series of different membranes (see Tables 4.2, 4.3 and 4.4). The reason for the inclusion of the cellulose triacetate membrane was that it contained fewer polar groups than the others and accordingly it was less likely to participate in hydrogen bonding with water molecules in transit through the membrane. The apparent activation energies for water diffusion and filtration across these membranes are shown along with the corresponding values of $(L_p RT/\bar{V}_w P_d)$ in Table 4.5. Since the first cellulose acetate membrane (CA–1) is not porous it is not surprising that the value of E_a for diffusion is larger than that for self-diffusion (4·6 cal mole^{-1}, see page 21). In fact, Gary-Bobo and Solomon argued that the water molecules diffusing through CA–1 would participate

TABLE 4.5. Apparent activation energies for diffusion and filtration of water across cellulose acetate membranes

Membrane	$\left(\dfrac{L_p RT}{\bar{V}_w P_d}\right)$ at 20°C	Apparent activation energies (kcal mole^{-1})		
		Diffusion (P_d)	Filtration (L_p)	'Bulk flow' ($L_p - P_d \bar{V}_w/RT$)
Cellulose acetate				
CA–1	1·00	7·8	7·8	–
CA–2	1·71	7·7	5·9	4·3
CA–3	3·20	8·1	5·0	4·5
CA–4	48·4	5·7	4·7	4·7
Cellulose triacetate				
CTA	1·65	5·5	5·1	4·9

Modified from Gary-Bobo & Solomon, 1971.

in hydrogen bonds with alcoholic groups in CA–1 and that these bonds have a higher activation energy than hydrogen bonds between water molecules. E_a for filtration across CA–1 is identical to that for diffusion and this is what we expect for a non-porous membrane where filtration and osmosis proceed by diffusional movement of water molecules. The next two membranes CA–2 and CA–3 in the list also have high E_a values for diffusion. This, taken along with their low values of $(L_p RT/\bar{V}_w P_d)$, seems to indicate that water molecules still experience considerable frictional interaction with these membranes. Gary-Bobo and Solomon considered that the lower values of E_a for filtration suggested that a component of the flow was non-diffusional; they attempted to obtain this 'bulk-flow' component by subtracting $(P_d \bar{V}_w / RT)$ from the corresponding values of L_p throughout the temperature range. The apparent activation energies (Table 4.5) for this 'bulk-flow' component were practically identical to that for water viscosity (4·2 kcal mole^{-1}). This evidently confirms their assumption that filtration across CA–2 and CA–3 has both diffusional and viscous components. The most permeable cellulose acetate membrane CA–4 has activation energies for diffusion and filtration which are only a little greater than the corresponding values for water itself. As a good approximation, therefore, we may accept that water diffuses through aqueous channels in CA–4 just as it does during self-diffusion and that it is filtered across CA–4 by a Poiseuille-type mechanism. Finally, the cellulose triacetate membrane (CTA) can be compared with its closest relative in the cellulose acetate series, namely CA–2, since both have similar values of $(L_p RT / \bar{V}_w P_d)$. E_a for diffusion across CTA is significantly less than that for CA–2 and this perhaps reflects the lower power of CTA for forming hydrogen bonds with the diffusing water molecules. Again for CTA the value of E_a for filtration is lower than that for CA–2 and the estimated E_a for 'bulk-flow' is close to that for viscous water flow.

From the study of Gary-Bobo and Solomon it emerges that high E_a values for diffusion and filtration occur in non-porous membranes which have relatively low water permeabilities and where diffusion seems to be the sole mechanism of water transport. On this basis, therefore, one would expect that artificial lipid membranes would also have large activation energies for water flow. Both Price & Thompson (1969) and Redwood & Haydon (1969) have observed high values of E_a for osmotic water flow across lipid membranes. For example, Price & Thompson (1969) used two different sorts of lipid membrane which were prepared from different solutions of lecithin and cholesterol. The membranes manufactured from the relatively more concentrated solution of lecithin and cholesterol had a lower L_p (18×10^{-7} cm sec^{-1} atm^{-1}) at 25°C than that (23×10^{-7} cm sec^{-1} atm^{-1}) for membranes prepared from the dilute solutions, while E_a values for the former and latter were 13·1 and 12·7 kcal mole^{-1} respectively. Similar data ($L_p = 20 \times 10^{-7}$ cm sec^{-1} atm^{-1}; $E_a = 14.6$ kcal mole^{-1})

were obtained by Redwood & Haydon (1969). Both sets of authors computed theoretical values for E_a of about 12 kcal mole^{-1} on the premise that osmotic water flow was achieved by diffusion across the membrane. As pointed out by Price & Thompson (1969) it is conceivable, though less likely, that the large values for E_a arise because water molecules pass through a system of small pores in which the pore fluid is different from bulk water. In other words, it is simpler to assume that the high energy of activation for osmotic flow stems from water diffusion in lipid rather than from water movement through an aqueous route, particularly since the values of $(L_p RT/\bar{V}_w P_d)$ offer no convincing proof that such a route exists in lipid membranes under normal conditions.

Interactions between solute and water molecules

So far we have considered only the movement of water molecules across artificial membranes. However, when we include solute molecules in our discussion we immediately must acknowledge that solute and water molecules may interact in the membrane. There is quite a lot of evidence for such an interaction. For example, solvent drag occurs when water flow can exert a convective 'force' on solute motion in the membrane. Indeed, most of the evidence for solute–water interaction is taken as indices of the porosity of certain membranes. Some of this evidence is reviewed below.

Restricted motion of solutes

We have already considered in detail the theoretical approach to restricted diffusion of solutes across porous membranes (see Chapter 3). This was used to analyse restricted diffusion in porous cellophane membranes by Renkin (1954). In particular, Renkin obtained estimates of the area available for solute diffusion per unit path length of membrane; that is, he obtained $(A_{sd}/\Delta x)$ from the permeability coefficients for a number of test solutes, such as labelled water, urea and sucrose. Table 4.6 shows the relations which he found between $(A_{sd}/\Delta x)$ and the molecular size of the solutes in three different kinds of membrane. It is clear that $(A_{sd}/\Delta x)$ decreases as the molecular size increases and this decrease is most pronounced in Visking dialysis tubing which is the membrane with the lowest water content (see Tables 4.2 and 4.3). Certainly these data can be explained by postulating that the solute molecules move through a system of idealized cylindrical pores in the membrane. As we have discussed before, the fall in $(A_{sd}/\Delta x)$ is possibly a result of steric hindrance at the mouth of pores and also of frictional resistance at the pore wall. To account

TABLE 4.6. Values of $(A_{sd}/\Delta x)$ for diffusion of various solutes across porous cellulose membranes

Solute	Molecular radius (Å)	$(A_{sd}/\Delta x)$ (cm)		
		Visking dialysis tubing	Cellophane	Sylvania wet gel
Labelled water	1·97	19·0	16·6	23·6
Urea	2·70	17·2	18·3	22·6
Glucose	3·57	9·6	14·7	23·7
Antipyrine	3·96	11·9		24·6
Sucrose	4·40	6·6	11·4	21·1
Raffinose	5·64	5·1	9·9	20·1
Haemoglobin	30	0	0	2·4

Modified from Renkin, 1954.

for this Renkin used the Faxen treatment for hydrodynamic motion in porous channels and fitted the data with equation 3.8. Figure 4.6 shows that the latter equation offered a reasonable fit to the experimental values of $(A_{sd}/\Delta x)$ for an equivalent pore radius of 15 Å in Visking dialysis tubing. This value for the pore radius is also consistent with that required to explain restriction to solute movement during filtration in this membrane (Renkin, 1954); this point will be discussed later.

Numerous other studies of a similar kind have been performed on artificial membranes, particularly ion-exchange membranes (Lagos & Kitchener, 1960; Peterson & Gregor, 1959; Kawabe, Jacobson, Miller & Gregor, 1966) and thin mica membranes (Beck & Schultz, 1970). It is also interesting to note that a similar relation between $(A_{sd}/\Delta x)$ and molecular size, as that found by Renkin (1954), has been observed in thin lipid membranes treated with either amphotericin B or nystatin (Andreoli et al., 1969; Holz & Finkelstein, 1970). It will be remembered from the earlier discussion that both of these agents seem to create pores in the lipid membranes, at least on the basis of comparisons between $(L_p RT/\bar{V}_w)$ and P_d. Of course, it is not possible to test these

132 4. SOME EXPERIMENTS ON ARTIFICIAL MEMBRANES

experimental data properly by equation 3.8 since $(A_p/\Delta x)$ is not known accurately. Both Andreoli *et al.* (1969) and Holz & Finkelstein (1970) estimated that the pores occupied about 0·01% of the total membrane area, but the former workers suggested that the equivalent pore radius lay in the range 7–10·5 Å while the latter authors indicated that it was only about 4 Å.

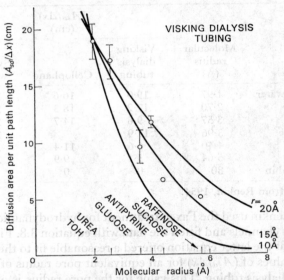

FIG. 4.6. Apparent diffusion area per unit path length for solute transport across Visking dialysis tubing. The experimental data can be fitted by equation 3.8 for an equivalent pore radius, r, of 15 Å: the other theoretical curves for $r = 20$ and 10 Å are also shown. The length of the bars show the standard errors of the means of the determined values of $A_{sd}/\Delta x$ (Renkin, 1954: Fig. 3a).

The characteristics of ultrafiltration of solute molecules through artificial membranes have also been examined in the light of the theoretical approach outlined in Chapter 3. In this case the frictional hindrance experienced by the solute is considered to be identical to that during diffusion but the steric hindrance term is different (cf. equations 3.6 and 3.10). Renkin (1954) measured the concentrations, c_2 and c_1, of various test solutes, in the filtrate and filtrand respectively. He expressed c_2 as a function of c_1 by

$$c_2 = \left(\frac{A_{sf}}{A_{wf}}\right) c_1 + \frac{dn}{dt} \cdot \frac{1}{\mathcal{J}_v} \qquad 4.21$$

where (dn/dt) is the number of moles of solute crossing unit area of membrane in unit time and the other symbols have their usual meaning. Solomon (1968) has attempted to reappraise Renkin's description of ultrafiltration in terms of irreversible thermodynamics and he has concluded that equation 4.21 is similar to the practical solute flux equation (2.46) namely $J_s = (1-\sigma_s)\bar{c}_s J_v + \omega_s \Delta \pi_s$, provided that $(1-\sigma_s)$ can be substituted for (A_{sf}/A_{wf}). This substitution can be made if the term $\omega_s \bar{V}_s/L_p$ in the expression 3.20 for the reflexion coefficient can be ignored. For some porous membranes, such as Visking dialysis tubing, Solomon (1968) argued that $\omega_s \bar{V}_s/L_p$ is negligible and consequently Renkin's experimental data can be fitted by a theoretical curve derived from equation 3.13 for (A_{sf}/A_{wf}) and equation 4.21 for the filtrate concentration. Figure 4.7 shows the results of Renkin's ultrafiltration study on Visking dialysis tubing where the so-called sieving coefficient, (c_2/c_1), is plotted against the filtration rate, J_v, for a number of solutes.

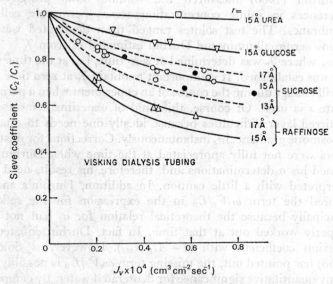

FIG. 4.7. The sieving coefficient of Visking dialysis tubing plotted as a function of the filtration rate. The theoretical curves were computed with the aid of equations 3.13 and 4.21. The good agreement between theory and experiment for the solutes tested indicates that the equivalent pore radius, r, is about 15 Å (Renkin, 1954: Fig. 5a).

Thus, the ultrafiltration study of Visking dialysis tubing indicates that the equivalent pore radius is about 15 Å. To a certain extent the excellent agreement between the latter value and that from the study of restricted diffusion in the same membrane may be regarded as somewhat fortuitous since these experiments were performed without any corrections for unstirred layers; moreover, the calculations employed molecular radii now regarded as slightly erroneous.

Reflexion coefficient

Several workers (Meschia & Setnikar, 1958; Durbin, 1960; Ginzburg & Katchalsky, 1963; Holz & Finkelstein, 1970) have measured the reflexion coefficients for certain solutes in various artificial membranes, but there has been apparently no really exhaustive study of the dependence of σ on the size of the test solute comparable to that carried out by Wright & Diamond (1969a,b) for a biological membrane.

Durbin (1960) measured the volume flows produced by differences in solute concentration across permeable cellulose membranes. The test solutes ranged from deuterated water to bovine serum albumin and Durbin estimated σ_s from $\mathcal{J}_v = -L_p \sigma_s \Delta\pi_s$ where $\mathcal{J}_v$ was determined at zero time, i.e. at the time when $\Delta\pi_s$ was established. It is important to obtain $\mathcal{J}_v$ at zero time since $\mathcal{J}_v$ will change during the course of an experiment when a permeant solute s is used. Of course, this kind of experiment meets with unstirred layer difficulties because ideally one needs to establish the osmotic gradient $\Delta\pi_s$ instantaneously. Corrections for unstirred layers were not fully appreciated at the time when Durbin performed his σ determinations and, therefore, his results need to be interpreted with a little caution. In addition, Durbin's analysis ignored the term $\omega_s \bar{V}_s/L_p$ in the expression for the reflexion principally because the theoretical relation for σ_s had not been properly worked out at that time. In fact, Durbin equated the reflexion coefficient with $(1-A_{sf}/A_{wf})$. However, as Solomon (1968) has pointed out, the missing term $\omega_s \bar{V}_s/L_p$ is possibly only of any quantitative significance for deuterated water. By comparing L_p and P_d values for Visking dialysis tubing, Cellophane and Sylvania wet gel membranes Durbin estimated that their equivalent pore radii were 23 Å, 41 Å and 82 Å respectively. For the particular case of Visking dialysis tubing Durbin compared the values of σ_s with those calculated from $(1-A_{sf}/A_{wf})$, where A_{sf}/A_{wf} was given by equation 3.13. Figure 4.8 shows the results of this comparison between experiment and theory and the agreement between these is very good when the pore radius is taken as 23 Å. Solomon (1968) has re-examined Durbin's data for Visking dialysis tubing and taken into account the term $\omega_s \bar{V}_s/L_p$ for each

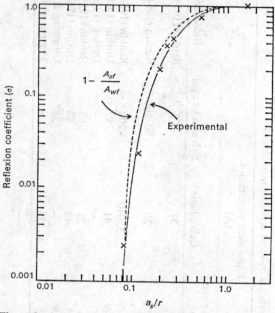

FIG. 4.8. The reflexion coefficient of Visking dialysis tubing as a function of the ratio of solute molecular radius, a_s, to the equivalent pore radius. The experimentally determined values of σ are indicated by the crosses and the solid line while the interrupted line shows the theoretical values obtained from $\sigma = 1 - A_{sf}/A_{wf}$ (see text) (Durbin, 1960: Fig. 3).

of the test solutes in the expression for σ_s. It turns out that after such corrections have been made the estimated pore radius becomes 22 Å as opposed to Durbin's estimate of 23 Å.

At this point it is pertinent to ask how consistent the estimates are of pore radii obtained by different methods. Table 4.7 contains such estimates for different cellulose membranes. Following the notation of Renkin (1954) the equivalent pore radius, r_p, is defined by $\sqrt{(8L_p \eta_w D_w/P_d)}$ where L_p is expressed in cm³ dyne⁻¹ sec⁻¹

TABLE 4.7. Estimates of equivalent pore radius in some artificial membranes

Method		Equivalent pore radius (Å)		Reference
	Visking dialysis tubing	Cellophane	Sylvania wet gel	
Restricted diffusion $r_p = \sqrt{(8D_w\eta_w L_p/P_d)}$	19	31	77	Renkin (1954)
analysis	15	30	80–100	Renkin (1954)
Molecular sieving during ultrafiltration $r_p = \sqrt{(8D_w\eta_w L_p/P_d)}$	15	35–40	~200	Renkin (1954)
	23	41	82	Durbin (1960)
	23	–	–	Durbin (1960)
Reflexion coefficient	19	31	83	Renkin (1954)
	23	43	89	Durbin (1960)
$r = 3\cdot 8 \sqrt{\left(\dfrac{L_p RT}{\bar{V}_w P_a} - 1\right)}$	22*	–	31*	Ginzburg & Katchalsky (1963)
	–	32	–	Thau et al. (1966)

* Corrected for unstirred layer effect.

(see Pappenheimer *et al.*, 1951). Both Renkin (1954) and Durbin (1960) obtained similar estimates of r_p for Visking dialysis tubing, Cellophane and Sylvania wet gel. When the pore radius is calculated from their data by equation (3.36) the values obtained differ little from their original estimates (see Table 4.7). Moreover, other estimates (e.g. Ginzburg & Katchalsky, 1963; Thau *et al.*, 1966) of P_d and L_p can be used in a similar calculation and yield essentially the same pore radii for the different membranes. The analysis of restricted motion of solutes during diffusion and ultrafiltration (Renkin, 1954) produces estimates of pore radius which are close, but not identical, to the previous values. Finally, the reflexion coefficient measurements of Durbin (1960) on these membranes seem to indicate pore radii which are compatible with all of the other estimates; in particular, Durbin's analysis of σ_s in terms of (A_{sf}/A_{wf}) for Visking dialysis tubing showed that his original estimate of r_p (23 Å) was correct. As far as can be judged, therefore, the estimates of pore radius from the reflexion coefficient, from $(L_p RT/\bar{V}_w P_d)$ values and from the hydrodynamic theory of restricted motion are quite consistent. The agreement is surprisingly good, in fact, especially when one considers that the underlying assumptions of these theoretical treatments probably fail at the molecular level.

There have been some criticisms of Durbin's study and also of Renkin's work on ultrafiltration. Lakshminarayanaiah (1967) has objected to the use of additional supporting membranes which were present during the filtration measurements but not during the diffusion measurements. His own work showed that the addition of a supporting barrier significantly reduced not only the apparent hydraulic conductivity but also the diffusional permeability of the membrane. It is difficult to assess the degree of significance of this objection without knowledge of the effects of mechanical deformation on membrane permeability and also of the magnitudes of the unstirred layers in those experiments. Another source of contention is that the molecular radii quoted by Renkin and Durbin are inaccurate. For water this is particularly cogent, since the radius was taken as 1·9–2·0 Å as opposed to the currently accepted value of 1·5 Å (see Chapter 1). Moreover, Soll (1967) has demonstrated that the effective molecular radius of a solute carried along in a bulk flow should be obtained by considering that the permeating molecule is cylindrical. That condition tends to give smaller molecular radii than those quoted by Durbin.

Despite these experimental and theoretical objections the studies of Renkin and Durbin are very informative and illustrate an interesting approach to the determination of membrane structure by means of transport measurements.

Some measurements of σ have been made on artificial lipid membranes treated with amphotericin B and nystatin. For example, in nystatin-treated membranes Holz & Finkelstein (1970) found σ for urea, ethylene glycol, glycerol, glucose, sucrose and NaCl; of these solutes only urea ($\sigma = 0.55$), ethylene glycol (0.67) and glycerol (0.78) have reflexion coefficients less than unity. Since the nystatin-treated membrane could discriminate between glycerol and glucose ($\sigma = 1$) Holz & Finkelstein (1970) concluded that the equivalent pore radius was about 4 Å; the basis of their argument was, of course, that an equivalent pore could exclude solutes, such as glucose with a radius of 4.2 Å, but not smaller solutes, such as glycerol with a radius of 3.1 Å. Their estimate of the equivalent pore radius from σ is close to that obtained from comparisons of $L_p RT/\bar{V}_w$ and P_d.

Frictional coefficients

Another way of expressing the interactions which solute and water molecules undergo is through the frictional coefficients describing the drag exerted on them. That is, f_{sw} describes the mutual drag between solute and water in the membrane and f_{sm} and f_{wm} describe the drag on solute and water exerted by the membrane. The membrane coefficients L_p, ω_s and σ_s can be written as functions of f_{sw}, f_{sm} and f_{wm} (e.g. Kedem & Katchalsky, 1961; Ginzburg & Katchalsky, 1963). Thus, it is possible to obtain conversely the frictional coefficients as functions of L_p, ω_s and σ_s. In fact, Ginzburg & Katchalsky (1963) determined L_p, ω_s and σ_s for Visking dialysis tubing and Sylvania wet gel and hence obtained the frictional coefficients for certain solutes and water in these membranes. Table 4.8 shows some of their data. The most striking feature is that f_{wm}, the coefficient representing friction between water and the membrane, is several orders of magnitude smaller than f_{sw} or f_{sm}. That is what we would expect for a highly porous membrane. Secondly, f_{sm} is significantly less than f_{sw} and hence again as we might expect for porous membranes the solute experiences more frictional resistance from water than from the membrane. Indeed, the solute suffers more drag from water in the

TABLE 4.8. Frictional coefficients* for solute diffusion in Visking dialysis tubing and Sylvania wet gel membranes and in water

Solute	Concentration × 10⁵ (mole cm⁻³)	Free water f_{sw}^0	Visking dialysis tubing f_{sw}	Visking dialysis tubing f_{sm}	Visking dialysis tubing f_{wm}	Sylvania wet gel f_{sw}	Sylvania wet gel f_{sm}	Sylvania wet gel f_{wm}
				(dyne sec mole⁻¹ cm⁻¹)				
Labelled water	–	101	328	–	–	114	–	–
Urea	50	172	660	65	8·3	282	4·6	1·7
Glucose	5	373	1890	230	8·5	780	30	1·7
Sucrose	5	500	3430	590	8·7	1130	67	1·7

* All of the frictional coefficients should be multiplied by 10¹³ to give them their actual numerical magnitudes. Modified from Ginzburg & Katchalsky, 1963.

membrane than it does in free solution, i.e. $f_{sw}^o < f_{sw}$ where f_{sw}^o is the frictional coefficient of solute in solution. We also should note that as the size of the test solute increases from labelled water to sucrose so also do the frictional coefficients, f_{sw} and f_{sm}, referring to the solute, thus demonstrating that the restriction to solute motion in the membrane is related to molecular radius. Finally, it should be emphasized that the frictional coefficients are simply an alternative way of describing the transport characteristics of the membrane and that they can be related to our earlier attempts. For example, Essig (1966) has shown that $(L_pRT/\bar{V}_wP_d)$ is equal to $(1+f_{w*w}/f_{wm})$ where w^* denotes labelled water. Table 4.8 shows that in both Visking dialysis tubing and Sylvania wet gel w^* suffers more drag from water than water does from the membrane. For Visking dialysis tubing (f_{w*w}/f_{wm}) is about 40 and this is consistent with the fact that $(L_pRT/\bar{V}_w)$ greatly exceeds P_d. Sylvania wet gel offers even less drag on water and labelled water and (f_{w*w}/f_{wm}) is about 67 which is compatible with the higher water content and larger equivalent pore radius of this membrane.

The frictional interpretation of membrane coefficients can also be applied to the movement of ions and water across membranes (e.g. Spiegler, 1958; Mackay & Meares, 1959) as we indicated in Chapter 3. Again it is possible to express the frictional coefficients for counter ions, coions and water in terms of the practical coefficients, such as L_p, τ_c' and β. The outcome of these studies in ion-exchange membranes is that the counter ions experience the largest frictional drag and, in particular, the drag exerted on them by the membrane is quite important (Mackay & Meares, 1959). On the other hand, the friction exerted on water molecules by the membrane is exceedingly small in comparison; for example, Spiegler (1958) estimated that the ratio of the frictional coefficient for the counter ion to that for water relative to the membrane is about 130 for a membrane composed of sodium polymethacrylate resin (Despic & Hills, 1956). Again the low values of f_{wm} for these ion-exchange membranes indicate their highly porous structure. Moreover, the low values of f_{wm} relative to the ion frictional coefficients means, for example, that the efficiency of electro-osmotic transport of water will be high if not at a maximum (see page 102).

For ion-exchange membranes estimates of β, expressed in mole Faraday^{-1} or water molecules per ion, lie in the range 10–100

(George & Courant, 1967; McHardy, Meares, Sutton & Thain, 1969; Tanny, 1973).

Solvent drag

Some evidence for solvent drag in cellulose membranes has already been discussed. In particular, Renkin (1954) described the ultrafiltration of certain test solutes by means of a sieving coefficient. However, these experiments need to be reappraised because of spurious solute fluxes that can arise from local solute concentration gradients in the neighbouring unstirred layers (see page 113). In an interesting paper Andreoli, Schafer & Troutman (1971) described experiments on solute fluxes across lipid membranes treated with amphotericin B to make them porous (Andreoli *et al.*, 1969). Andreoli *et al.* (1971) measured the net fluxes of urea, glycerol and meso-erythritol in the presence of net volume flow. The aim of their experiments was to establish whether the alterations in the solute fluxes, which were produced during osmotic volume flow, were due entirely to changes in the local solute concentrations at the membrane or to a genuine solvent drag effect in the membrane. Provided the size of the unstirred layers is known, it is possible to compute the solute fluxes under these separate circumstances. Andreoli *et al.* (1971) referred to the solute flux arising solely from the unstirred layer effect as J_{di} and that from solvent drag plus unstirred layer effect as J_i. Figure 4.9 shows the theoretical dependence of both J_i and J_{di} on the osmotic volume flow J_v as it is predicted by Andreoli *et al.* (1971) for urea, glycerol and meso-erythritol; the experimental values for the solute fluxes are also shown. It is clear that one can distinguish between unstirred layer effects and genuine solvent drag when J_v is sufficiently large. In the case of urea the apparent permeability is so large that it is probably dominated entirely by rate-control in the unstirred layers. However, when we consider the less permeant solutes we see that the solvent drag (plus unstirred layers) model is more successful at describing the solute flux than the simple unstirred layer model. For these solutes the flux clearly exceeds that expected from the unstirred layer although it might be contended the actual unstirred layer thickness (55 μm) used in their model is perhaps too small, and hence there is an underestimate of the diffusion flux J_{di}. Certainly the presence of solvent drag in lipid membranes treated with amphotericin B is compatible with the other evidence (Andreoli *et al.*, 1969) that these treated membranes are porous.

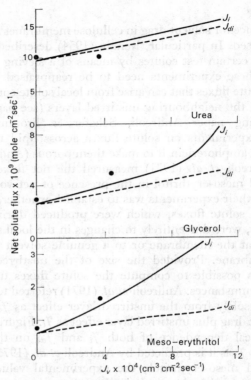

FIG. 4.9. The effect of osmotic water flow on net solute flux across an artificial lipid membrane. The points represent the observed fluxes in the absence and presence of water flow. The theoretical curves for the total solute flux, J_i, and the solute flux, J_{di}, arising solely from the unstirred layer effect show their dependence on the rate of osmotic flow (Andreoli *et al.*, 1971: Fig. 3).

Solvent drag has also been observed in a porous artificial membrane (Diaflo UM-3) by Franz, Galey & Van Bruggen (1968). These authors established an osmotic flow by maintaining a concentration gradient of sucrose across the membrane; actually sucrose is a permeant solute in this membrane and its reflexion coefficient is 0·02. They measured the unidirectional fluxes of inulin both before and after the establishment of osmotic flow.

In the absence of volume flow the unidirectional fluxes were equal but they became dissimilar in the presence of volume flow inasmuch as the inulin flux in the direction of volume flow was enhanced while that in the opposite direction was reduced. Although this is evidently satisfactory evidence for solvent drag on inulin movement, it could be due to an unstirred layer effect; the latter possibility was apparently not considered by Franz *et al.* (1968). These workers then performed an exceedingly interesting experiment; the osmotic flow due to the sucrose gradient was abolished by increasing the hydrostatic pressure of the hypertonic solution and again the inulin fluxes were determined. Surprisingly, the unidirectional fluxes were not equal under these conditions! In fact, the flux of inulin from the hypertonic solution to the dilute solution greatly exceeded that in the opposite direction and Franz *et al.* (1968) attributed the net flux of inulin to a frictional interaction between solutes referred to as 'solute drag'. They argued that the diffusion of sucrose down its concentration gradient exerted a drag on inulin and this interaction produced a net flux of inulin. This phenomenon occurs in free solution (Curran, Taylor & Solomon, 1967) and apparently in other artificial membranes (Galey, & Van Bruggen, 1970) and also in biological membranes (e.g. Ussing, 1966). However, it is possible that this apparent 'solute drag' may actually stem from solvent drag in a heteroporous membrane.

Our previous discussion of membranes composed of different elements in parallel demonstrated that volume circulation in the membrane could occur even when net volume flow is absent (see page 69). A similar conclusion has been reached, of course, by several workers (Sollner, 1945; Rapoport, 1966) as well as Kedem & Katchalsky 1963*b*). The anomalous flux ratios observed by Galey & Van Bruggen (1970) could be due to volume circulation and, in fact, Ussing & Johansen (1969) have referred to such a phenomenon as 'anomalous solvent drag' and they used a similar model to account for the anomalous transport of sucrose and urea across a biological 'membrane' bathed by hypertonic solutions.

Patlak & Rapoport (1971), in particular, have considered whether the data of Galey & Van Bruggen (1970) can be attributed solely to 'solute drag' or if heteroporosity of the membrane is responsible. Following their reasoning, let us assume that a membrane possessing both small and large pores is bathed on the outside by an aqueous solution of a permeant solute, say sucrose, and on the inside by pure water. Let us assume also that net volume flow is abolished by applying a pressure to the sucrose solution and that the sucrose reflexion coefficient for the small pores is larger than that for the large pores. Then there will be a circulation of volume flow in the membrane such that a volume efflux will occur through the small pores and an equal volume influx will occur through the large pores. If we now place an additional permeant solute, say inulin, at equal concentrations on both sides of the membrane then inulin, i, will be dragged outwards at a rate proportional to $(1-\sigma_i')$, where σ_i' is the inulin reflexion coefficient for the small pores; and it will be dragged inwards at a rate proportional to $(1-\sigma_i'')$, where σ_i'' is the reflexion coefficient for the large pores. Since $(1-\sigma_i'')$ is greater than $(1-\sigma_i')$, the net effect of volume circulation is that inulin will be dragged inwards in the same direction as the diffusion of sucrose. The same anomalous transport of inulin may also occur even when the osmotic volume efflux is allowed to occur since it arises basically because of the higher efficiency of solvent drag in the inwards direction and because some inflow of volume must always exist. Patlak & Rapoport (1971) treated this problem in some detail and their model of the heteroporous membrane also took account of possible interactions between solutes in the membrane. Their conclusion was that it was possible to explain the apparent 'solute drag' effect in some of the experiments of Galey & Van Bruggen (1970) simply on the sole basis of volume circulation in a heteroporous membrane, except in the case where dextran was used to generate osmotic volume flow. In the latter case, they concluded that it was necessary to accept also that dextran exerted some 'solute drag' on the tracer solute, inulin.

Composite membranes

Although all artificial membranes can be characterized by values of L_p, ω_s and σ_s it appears on closer inspection that some of them contain components with different transport properties from that of the entire membrane itself. This means that if one is to understand the behaviour of such a composite membrane one needs to study separately the properties of its components. The particular examples of composite systems which we shall discuss below are first the cellulose acetate membrane and secondly an asymmetrical double-membrane system possessing a central fluid compartment. These systems exhibit behaviour that can be identified with the permeability characteristics of some biological membranes.

Cellulose acetate membranes

Cellulose acetate membranes were developed by Loeb (1966) for desalination. They are composed of a dense thin 'skin' of about 0·25 μm supported by a highly porous layer of about 150 μm in thickness (Riley, Merten & Gardner, 1966). Desalination by reverse osmosis is achieved when a salt solution is filtered across the membrane since the skin has a low salt permeability; the sieving coefficient for salt is about 0·1 or less. The permeability coefficients of the porous layer can be determined after the skin has been removed from the membrane and thus the properties of the skin can be deduced from those of the entire membrane and the porous layer. This approach has been used to measure L_p, ω_s and σ_s of the skin and porous layers by Hays (1968), Gary-Bobo et al. (1969) and DiPolo et al. (1970).

Water permeabilities. Since the skin and porous layer of the cellulose acetate membrane are in intimate contact with each other the hydraulic conductivity L_p of the membrane is given by $L_p{}^s L_p{}^c/(L_p{}^s + L_p{}^c)$ where $L_p{}^s$ and $L_p{}^c$ are the hydraulic conductivities of the skin and porous layer respectively (see equation 2.115). Similarly the diffusional permeability to water P_d is given by $P_d{}^s P_d{}^c/(P_d{}^s + P_d{}^c)$. Hays (1968), Gary-Bobo et al. (1969) and DiPolo et al. (1970) measured L_p, $L_p{}^c$, P_d and $P_d{}^c$ and from these measurements obtained $L_p{}^s$ and $P_d{}^s$. Table 4.9 shows their data for the membrane and its component layers. Also shown in the table is a comparison of hydraulic conductivity and diffusional permeability for the membrane and its components. Presumably the disparities between the permeability values obtained by these authors reflect the differences in membrane preparations. Considering first the entire membrane we see $(L_p RT/\bar{V}_w P_d)$ ranges from about 9 to 56; however this ratio does not give an accurate guide to the porosity either of the skin or the porous layer. As we might expect from the known structures of the dense skin and the highly porous layer, the corresponding values of $(L_p RT/\bar{V}_w P_d)$ are vastly different from each other and also from that for the whole membrane. For the skin Hays (1968) estimated an equivalent pore radius of about 9 Å whereas the data of Gary-Bobo et al. (1969) and DiPolo et al. (1970) suggest that perhaps such a calculation is not justified for their membranes. In particular, the value of $(L_pRT/\bar{V}_wP_d)$ obtained by DiPolo et al. (1970) is just on the limit at which this criterion can be used for membrane porosity.

TABLE 4.9. Water permeabilities of intact cellulose acetate membranes and of the component layers of these membranes

Membrane	$P_d \times 10^4$ (cm sec^{-1})	$L_p \times 10^7$ (cm sec^{-1} atm^{-1})	$\left(\dfrac{L_pRT}{\bar{V}_wP_d}\right)$	Reference
Intact	3·0	120	56	Hays (1968)
	10·5	95	12	DiPolo et al. (1970)
	6·4	42	8·7	Gary-Bobo et al. (1969)
Porous layer	3·4	3200	1320	Hays (1968)
	12·0	450	53	DiPolo et al. (1970)
	7·0	255	50	Gary-Bobo et al. (1969)
Skin	29·6	127	5·9	Hays (1968)
	83·0	122	2·0	DiPolo et al. (1970)
	75·0	50	0·9	Gary-Bobo et al. (1969)

Consequently DiPolo et al. (1970) examined the reflexion coefficient of the skin layer to decide whether or not this component is porous.

Another interesting feature in Table 4.9 is that it is evident that although the skin layer rate-controls osmotic water flow, it is the porous layer that rate-controls the diffusional movement of water largely owing to its thickness. As Hays (1968) pointed out, however, the thickness of the porous layer is not the sole important factor since he found that the restricted diffusion coefficient for water in this highly porous layer is only about 25% of the self-diffusion coefficient, thus indicating that the tortuosity of the diffusion path is probably large and its area is small.

Solute permeabilities. Both Gary-Bobo et al. (1969) and DiPolo et al. (1970) have measured solute permeability coefficients for the membrane and its porous component. Vigorous stirring conditions were employed to reduce the effect of diffusion in the unstirred layers; similar conditions were also used during the measurements of P_d. Again the solute permeability for the skin component was computed from the permeabilities of the whole membrane and its porous layer.

The skin layer in the cellulose acetate membranes prepared by Gary-Bobo *et al.* (1969) was apparently a dense non-porous barrier and these authors investigated its permeability properties for certain alcohols and amides. They concluded that the permeability depended not only on the size of the solute molecule but also on its partition coefficient between solution and the skin. Moreover, they considered that the larger the number of hydrogen bonds a given solute can have the slower it will diffuse through the skin; this seems to be supported by their observations that the permeabilities of the amides were lower than those of the alcohols of a comparable size. The latter source of restriction to diffusion occurs evidently in the porous layer also, but to a lesser extent.

Reflexion coefficients. DiPolo *et al.* (1970) extended the study of solute permeability coefficients, originally performed by Gary-Bobo *et al.* (1969). In addition, the reflexion coefficients of the skin and porous layer were determined from $\sigma_s = (\sigma_s{}^s \omega_s/\omega_s{}^s) + (\sigma_s{}^c \omega_s/\omega_s{}^c)$, where the superscripts s and c again refer to the skin and porous layer; the preceding relation is the appropriate equation for a membrane with two components in series (cf. equation 2.137). The skin component of the cellulose acetate membranes prepared by DiPolo *et al.* (1970) was evidently of a slightly different character from that used in the study of Gary-Bobo *et al.* (1969); for example, its hydraulic conductivity was about double that of the earlier value (see Table 4.9). In fact, the ratio $(L_p RT/\bar{V}_w P_d)$ suggested it might be porous and consequently that interactions between water and solute within the skin might exert an influence on solute permeation in addition to the factors studied earlier by Gary-Bobo *et al.* (1969). Thus DiPolo *et al.* (1970) compared the estimated values of $\sigma_s{}^s$ for certain solutes with the corresponding values of $(1 - \omega_s{}^s \bar{V}_s/L_p{}^s)$ (Fig. 4.10). From our discussion (Chapter 3) about the significance of the reflexion coefficient it will be remembered that if it is less than $(1 - \omega_s \bar{V}_s/L_p)$ for a particular solute then the membrane must contain a common pathway for water and the solute. For all of the nine solutes studied in the skin it is clear that the reflexion coefficients lie below the line of identity between σ_s^s and $(1-\omega_s^s \bar{V}_s/L_p^s)$. Thus, there must be a porous route for solute permeation in addition to the diffusion pathway whose characteristics had been discussed previously by Gary-Bobo et al. (1969). One of the solutes (1-butanol; point 9 in

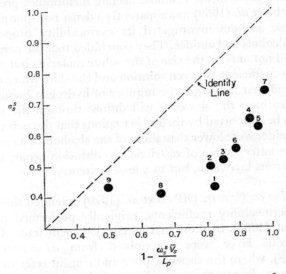

FIG. 4.10. Reflexion coefficient for the skin component of a cellulose acetate membrane as a function of $(1-\omega_s^s \bar{V}_s/L_p^s)$ (DiPolo et al., 1970: Fig. 3).

Fig. 4.10) lies quite close the the line of identity and this is compatible with the fact that this solute had an exceedingly high partition coefficient ($K_s = 0.95$) in the skin component and hence probably permeates mainly through the non-porous route.

Asymmetrical double-membrane system

The cellulose acetate membrane, which we have just discussed, is composed of two different permeability barriers in series. In that system one can assume quite safely that the resistances to water or solute movement are the sum of the resistances of the individual components of the membrane. However, if the components are separated by a central fluid compartment, that assumption may not hold and somewhat anomalous behaviour may result (see Chapter 2). For example, Curran & McIntosh (1962) measured net volume flow across a composite system possessing two different membranes placed in series and separated by a closed compartment. They observed that volume flow could proceed against its apparent driving force across the entire system under certain conditions and they suggested that this might be due to the different permeability characteristics of the membranes. Since many biological permeability barriers composed of two or more membranes in series are also capable of transporting water apparently against its own activity gradient (see Chapter 10) it is important to discuss this double-membrane system in a little more detail.

Following the suggestion of Curran & McIntosh (1962) that the asymmetry of the double-membrane system is responsible for net volume flow, Ogilvie, McIntosh & Curran (1963) set out to test that hypothesis experimentally. The experimental apparatus of Ogilvie *et al.* (1963) is shown in Fig. 4.11. It consisted of a Perspex chamber subdivided into three compartments by two artificial membranes, each supported on both sides by wire meshes. The first membrane between A and B was Visking dialysis tubing while the second membrane between B and C was Dupont wet gel. The hydraulic conductivities and the reflexion coefficients for the test

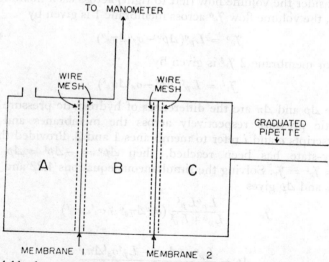

FIG. 4.11. Apparatus for measuring the pressure and volume flow in a three compartment system containing two dissimilar membranes in series (Ogilvie *et al.*, 1963: Fig. 1).

solute, sucrose, of each of those cellulose membranes were measured independently; they obtained $\sigma_s = 0.075$ and $L_p = 21 \times 10^{-5}$ cm sec^{-1} atm^{-1} for membrane 1 and $\sigma_s = 0.019$ and $L_p = 101 \times 10^{-5}$ cm sec^{-1} atm^{-1} for membrane 2. That is, the second membrane is relatively more permeable to water and sucrose than the first. The central compartment of the apparatus was connected to a manometer so that its pressure could be monitored, and the volume changes of compartment C were recorded with a graduated pipette. The solutions in B and C were stirred by rotating magnetic bars while that in A was stirred by an air stream. The experimental procedure consisted of filling A and C with distilled water and introducing sucrose solutions of different concentrations into B. Thereafter the hydrostatic pressure in B and the volume of C were recorded.

Even although the compartments A and C contained identical media, these authors observed that a net volume flow, J_v, occurred from A to C. Moreover, a hydrostatic passive difference was recorded between the central compartment and the outer ones A and C. Curran and his colleagues found that it was possible to account for the volume flow on the following basis.

Consider the volume flow (left to right) across each membrane. Then, the volume flow $J_v{}^o$ across membrane 1 is given by

$$J_v{}^o = L_p{}^o(\Delta p^o - \sigma_s{}^o \Delta \pi_s{}^o) \qquad 4.22$$

and for membrane 2 $J_v{}^i$ is given by

$$J_v{}^i = L_p{}^i(\Delta p^i - \sigma_s{}^i \Delta \pi_s{}^i) \qquad 4.23$$

where Δp and $\Delta \pi$ are the differences of hydrostatic pressure and osmotic pressure respectively across the membranes and the superscripts o and i refer to membranes 1 and 2. Provided that a steady-state has been reached, then $\Delta p^o = -\Delta p^i = \Delta p$ and $J_v{}^o = J_v{}^i = J_v$. Solving the simultaneous equations 4.22 and 4.23 for J_v and Δp gives

$$J_v = -\frac{L_p{}^o L_p{}^i}{L_p{}^o + L_p{}^i}(\sigma_s{}^o \Delta \pi_s{}^o + \sigma_s{}^i \Delta \pi_s{}^i) \qquad 4.24$$

and

$$\Delta p = \frac{L_p{}^o \sigma_s{}^o \Delta \pi_s{}^o - L_p{}^i \sigma_s{}^i \Delta \pi_s{}^i}{L_p{}^o + L_p{}^i} \qquad 4.25$$

Equations 4.24 and 4.25 enabled Ogilvie *et al.* (1963) to predict the magnitudes of J_v and Δp from all the measured values appearing in those equations. Table 4.10 summarizes the results of their study in which they noted that the presence of sucrose solutions in the central compartment gave rise to a net volume flow across the double-membrane system and that the observed and predicted values of both J_v and Δp agreed satisfactorily. The double-

TABLE 4.10. Experimental and theoretical values for net volume flow and hydrostatic pressure gradients developed in an artificial double-membrane system (Ogilvie et al., 1963)

Initial sucrose concentration in compartment B (mM)	Net volume flow (μl min^{-1})		Difference in hydrostatic pressure (mm Hg)	
	Observed	Calculated	Observed	Calculated
100	1·1	1·0	30	34
200	2·2	2·0	74	69
300	3·5	3·2	128	115
400	4·2	4·1	156	141
500	5·2	5·4	206	190

membrane model, therefore, accounts quantitatively for volume flow in this system. It is interesting to note that without knowledge of the properties of the individual membranes and the composition of the central compartment one would be forced to the conclusion that active water transport can occur even in this inanimate system! These authors stressed that the so-called double-membrane effect demanded the existence of a positive hydrostatic pressure in the central compartment and different values for the reflexion coefficients of the individual membranes. The latter condition is equivalent to saying that the overall 'membrane' must exhibit asymmetrical permeability characteristics.

The double-membrane model has been treated in detail by Patlak, Goldstein & Hoffman (1963) and Kedem & Katchalsky (1963c). Both sets of workers have derived equations which describe the flows of solute and water across a system composed of two membranes arranged in series. Both treatments are similar except that Patlak et al. (1963) have included active solute transport in their theoretical scheme. Consequently their conclusions seem more directly applicable to the discussion of active salt and water transport in some biological systems and consequently their model will be considered more fully in Chapter 10.

5

WATER PERMEABILITIES OF ANIMAL AND PLANT CELLS

Water transport in animal cells	152
Measurement of P_d	152
Does the cell membrane rate-control water exhange?	156
Measurement of L_p	162
Does the cell membrane rate-control osmotic flow?	169
Water transport in plant cells	177
Measurement of P_d	178
Measurement of L_p	179
Comparison of $L_p RT/\bar{V}_w$ and P_d values	184
Effect of temperature on water permeabilities	186

BEFORE one can begin to analyse the problems of the water relations of cells, one needs a quantitative assessment of how rapidly water moves into or out of the cells. Such a description may involve the water permeability of the cell provided, of course, that its plasma membrane is the rate-limiting barrier to transport. In principle, the water permeability can be measured in two ways. The diffusional permeability can be determined by studying the exchange of labelled water between the cell and its surroundings whereas the hydraulic conductivity can be obtained from the net water flux generated by a given gradient of osmotic or hydrostatic pressure.

Water transport in animal cells

Measurement of P_d

The diffusional permeability to water can be obtained by measuring the rate at which labelled water, w^*, enters or leaves the cell. For instance, if the cell is placed in a relatively large volume of solution containing a concentration $c_{w*}{}^o$, of labelled water, the rate of increase of the internal concentration, c_{w*}^i, is

$$\frac{dc_{w*}^i}{dt} = \frac{P_d A}{v}(c_{w*}^o - c_{w*}^i) \qquad 5.1$$

where A and v are the surface area and volume of the cell respectively. The solution of this equation is

$$\frac{c_{w*}^i}{c_{w*}^o} = 1 - \exp\left(-\frac{P_d A t}{v}\right) \qquad 5.2$$

provided that c_{w*}^o, P_d, A and v are independent of t and that c_{w*}^i is zero at $t = 0$. Equation 5.2 can be re-arranged to give

$$\ln\left(1 - \frac{c_{w*}^i}{c_{w*}^o}\right) = -\frac{P_d A t}{v} \qquad 5.3$$

Thus the diffusional permeability can be obtained from a plot of $(\ln -c_{w*}^i/c_{w*}^o)$ against time.

If the volume of the external medium is not large in comparison to the cellular compartment, then labelled water will equilibrate with the cellular water according to two-compartment kinetics (Paganelli & Solomon, 1957).

In the case of the influx of tritiated water across the red cell membrane Paganelli & Solomon (1957) used the following symbols and equations to describe the exchange process;

$$\frac{dP}{dt} = -kp + kq \qquad 5.4$$

$$P = pv_p \qquad 5.5$$

$$p_o v_p = pv_q + qv_q \qquad 5.6$$

where P is the quantity of labelled water in the external medium, k is a proportionality constant, p and q are the concentrations of labelled water in the medium and in the cell respectively and v_p and v_q are the volumes of the medium and the intracellular water and p_o is the value of p at $t = o$. Taking the boundary conditions; $p = p_o$ at $t = o$ and $p = p_\infty$ when $t = \infty$, equation 5.4. can be solved to yield

$$\frac{p}{p_\infty} - 1 = \left(\frac{p_o}{p_\infty} - 1\right) \exp\left(-\frac{kp_o t}{v_q p_\infty}\right) \qquad 5.7$$

Figure 5.1 shows the semilogarithmic plot of $(p/p_\infty - 1)$ against time for the uptake of tritiated water into human red cells (Paganelli & Solomon, 1957). The linearity of the relation between $\log(p/p_\infty - 1)$ and t substantiates the view that the exchange of labelled water in this system follows two-compartment kinetics.

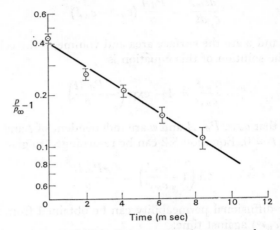

FIG. 5.1. The uptake of labelled water in human erythrocytes under isotonic conditions. The bars associated with the experimental points indicate ±S.D. and the curve has been drawn according to the appropriate theory of redistribution of a labelled substance in two compartments (see text) (Paganelli & Solomon, 1957: Fig. 5).

The half-time of the exchange of labelled water across the red cell membrane is about 4·2 msec. Paganelli and Solomon estimated from the rate constant, k/v_q (fraction of intracellular water exchanging in unit time), that the diffusional permeability of the red cell membrane is 53×10^{-3} cm sec^{-1}.

An alternative approach to measuring the rate of influx of labelled water is to follow the efflux of labelled water from a cell previously equilibrated with labelled water. Consider the case when the cell is washed in a very large volume of unlabelled solution. The rate of decrease in c_{w*}^i is given by

$$-\frac{dc_{w*}^i}{dt} = \frac{P_d A}{v} c_{w*}^i \qquad 5.8$$

and the solution of this is

$$\ln\left[\frac{c_{w*}^i}{(c_{w*}^i)_0}\right] = -\frac{P_d A t}{v} \qquad 5.9$$

where $(c_{w*}^i)_0$ is the initial concentration of labelled water in the cell at the beginning of the efflux experiment. Thus, a plot of ln $[c_{w*}^i/(c_{w*}^i)_0]$ against time should yield a straight line whose slope is related to the diffusional permeability of the cell to water.

These methods of measuring P_d rest on several assumptions.

First, it has been assumed that the labelled water, say tritiated water, is an ideal tracer for water. Wang, Robinson & Edelman (1953) found that the diffusion coefficient of $^1H^3H^{16}O$ in water is about 14% smaller than that of $^1H_2^{18}O$ which is probably the best tracer for water. Kohn (1965) has compiled a list of the values for the self-diffusion and isotopic diffusion coefficients of water and his table shows the small but significant differences between the different isotopes. In the case of water transport across a biological membrane, however, it is impossible to predict what sort of transport rates might result from the properties of different isotopic forms of water. Nevertheless, it is notable that King (1969), for example, found no significant difference between the permeability coefficients of frog skin for deuterated and tritiated water and, therefore, it seems unlikely that an 'isotope effect' is a serious source of error in determinations of P_d for biological membranes.

The second assumption is that the chemical and physical properties of the labelled water do not interfere with the normal behaviour of the cell membrane. For instance, tritiated water emits weak radiation which might damage the cell membrane; however, the radiation dose delivered to the cell is generally negligible (e.g. see Paganelli & Solomon, 1957).

The third assumption is that cells do not change in volume during the exchange experiment. Fortunately this assumption can be checked.

The fourth assumption is that the labelled water does not exchange with water elsewhere in the apparatus or, perhaps, with water vapour above the external medium. As in the preceding case this assumption can be checked; of course, this source of error could be troublesome only in the case of the influx experiment.

The final assumption is that both the cellular and the external medium are well mixed. It turns out that this condition is often the most difficult to satisfy and also that this source of error can be very serious. In other words, we have to deal with the errors produced by unstirred layers not only in the bathing medium but also in the interior of the cell.

All of these assumptions, with the exception of the unstirred layer difficulties, have usually been taken for granted and a number of estimates of P_d for animal cells have been obtained (Table 5.1).

TABLE 5.1. Apparent values of P_d for animal cells

Cell	$P_d \times 10^4$ (cm sec^{-1})	Reference
Dog erythrocyte	64	Villegas, Barton & Solomon (1958)
Human erythrocyte	53	Paganelli & Solomon (1957)
Human erythrocyte	53	Villegas *et al.* (1958)
Cow erythrocyte	51	Villegas *et al.* (1958)
Human erythrocyte	48	Barton & Brown (1964)
Dog erythrocyte	44	Rich, Sha'afi, Barton & Solomon (1967)
Human (foetal) erythrocyte	32	Barton & Brown (1964)
Squid axon	4·0	Nevis (1958)
Barnacle muscle	2·6	Bunch & Edwards (1969)
Squid axon	1·4	Villegas & Villegas (1960)
Frog (ovarian) egg	1·3	Prescott & Zeuthen (1953)
Crab muscle	1·2	Sorenson (1971)
Xenopus egg	0·90	Prescott & Zeuthen (1953)
Frog egg	0·75	Prescott & Zeuthen (1953)
Zebra fish (ovarian) egg	0·68	Prescott & Zeuthen (1953)
Zebra fish egg	0·36	Prescott & Zeuthen (1953)
Amoeba (*Chaos chaos*)	0·23	Prescott & Zeuthen (1953)
Amoeba proteus	0·21	Prescott & Zeuthen (1953)

These apparent values for the diffusional permeability are quite similar to the corresponding permeabilities of certain artificial membranes, particularly the artificial lipid membranes (see Table 4.2). Perhaps the actual range of P_d values for animal cells is larger than one might have expected; however, it must be stressed that most of these estimates have been obtained under conditions where the possible influence of rate-limiting diffusion of water either in the cytoplasm or in the external solution has not been checked. Thus, it must be emphasized that Table 5.1 contains apparent values of P_d for these cells and not necessarily the 'true' permeabilities of their cell membranes. Obviously this raises the doubt about whether the cell membrane actually rate-controls the exchange of labelled water across the cell's surface.

Does the cell membrane rate-control water exchange?

Apart from the cell membrane itself the main alternative sources of rate-control on the exchange of water molecules are diffusion in the cytoplasm and in the external unstirred layer.

The role of internal diffusion. Within the interior of a cell the exchange of labelled water is limited inevitably by diffusion and, therefore, it is important to know the speed of internal diffusion. For instance, Paganelli & Solomon (1957) estimated that in the erythrocyte the intracellular water would equilibrate relatively rapidly with the labelled water which had entered. Such equilibration is achieved by diffusion and it should reach 90% of its final value within 0·2 msec, which is substantially lower than the apparent half-time (4·2 msec) for exchange of water across the membrane of that cell. Their calculations are based on an internal diffusion path of 0·5 μm, however, and consequently the limitations set by internal mixing cannot be ignored completely in the exchange kinetics of cells larger than the erythrocyte. This point has been examined by several workers, notably Ling (1966) and Ling, Ochsenfeld & Karreman (1967) for the exchange of labelled water in the frog egg.

Ling and his co-workers considered that, in addition to the cell membrane, there are two other alternative sources of rate-control. These alternative mechanisms are the rate of diffusion in the cytoplasm itself and, moreover, the rate of adsorption and desorption of the labelled water on certain sites within the cell. Figure 5.2 shows the effects that these three sources of rate-control can exert on the influx of labelled water into a spherical cell. According to Ling et al. (1967), the influx resulting from membrane-limited diffusion would be given by equation 5.3 where (c_{w*}^i/c_{w*}^o) has been written as (M_t/M_∞), i.e. the uptake of labelled water, M_t, at any time divided by the total amount, M_∞, of labelled water in the system. Figure 5.2a shows that, when (M_t/M_∞) is plotted against $\sqrt{t}$ for the membrane-limited exchange, a sigmoidal curve is obtained. Ling (1966) coined the term 'influx profile' to describe the graphical relation between (M_t/M_∞) and $\sqrt{t}$. On the other hand, Fig. 5.2b shows the influx profile for the case where the cell membrane does not rate-limit the entry of labelled water; here the influx is rate-controlled by the rate of water diffusion in the cytoplasm. The relation between (M_t/M_∞) and $\sqrt{t}$ for this type of exchange is approximately linear over a wide range of $\sqrt{t}$ provided t is small. Finally in Fig. 5.2c there is

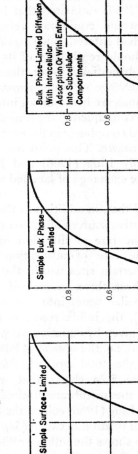

FIG. 5.2. The time course of the influx of a labelled substance into various systems with rate-limiting steps as shown on each graph. Each of these 'influx profiles' is obtained theoretically from the corresponding model system. The ordinate represents the uptake, Mt, of the labelled substance at time t as a fraction of the total amount of material, $M\infty$, in the system after an 'infinite' time. The abscissa represents the square root of time (Ling et al, 1967: Fig. 1).

a mechanism of rate-control which relies not only on diffusion in the cytoplasm but also on adsorption of water on to intracellular sites. The solution to the differential equations describing the latter cases were taken from Crank (1956).

Ling and his collaborators made the point that influx studies are generally not accurate enough for their 'influx profile' analysis since the influx into a single cell cannot be followed throughout the entire time course of its equilibration with the labelled medium. Consequently they preferred to observe the efflux of labelled water from single eggs (mature ovarian eggs from leopard frogs) or from clusters of several eggs because the efflux curves could be obtained quite accurately. The relation 5.9 governing the efflux can be rewritten as

$$1 - \frac{c_{w*}{}^i}{(c_{w*}{}^i)_o} = 1 - \exp\left(-\frac{P_d A t}{v}\right) \qquad 5.10$$

Since the right hand side of equation 5.10 is identical to that of equation 5.2 governing the influx of labelled water, Ling *et al.* (1967) argued that by subtracting the efflux values of $[c_{w*}{}^i/c_{w*}{}^i)_o]$ from unity they obtained the time course of the fractional uptake for the influx in the same cell. That manipulation of the experimental data by their so-called 'inversion method' allowed to them to examine 'influx profiles' indirectly obtained from the relatively more accurate efflux data. Fig. 5.3a shows the time course of the labelled water efflux from a cluster of five eggs; incidentally they found no difference between experiments employing small clusters of eggs or single eggs. On the basis of a membrane-limited exchange one would expect the efflux curve to be linear (equation 5.9). The relation, however, is curvilinear. By employing the 'inversion method' they obtained the corresponding 'influx profile' shown in Fig. 5.3b; in fact, the influx profile was fitted with a theoretical curve derived from 'a simple bulk-phase limited' model similar to that of Fig. 5.2b. In addition to this type of influx profile these workers also observed influx profiles which were described by bulk-phase limited diffusion plus adsorption (see Fig. 5.2c). In the latter cases they concluded that there was a slowly exchanging fraction (0·5–25%) of intracellular water which might be attributed to adsorption or influx of labelled water into some subcellular compartments. According to Ling and his co-workers, therefore, the exchange of labelled water is rate-controlled partially by its diffusion within the cytoplasm and to a certain extent its entry into some intracellular sites. Løvtrup (1963a) has also argued that the apparent values of P_d for single cells must be corrected for the effect of intracellular diffusion.

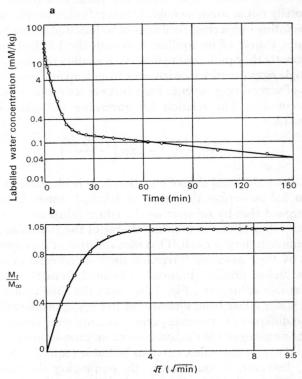

FIG. 5.3. Efflux of labelled water from frog eggs and the corresponding 'influx profile' for labelled water entry into the same eggs. In a the concentration of labelled water in a group of five eggs during an efflux experiment is plotted as a function of time. The corresponding 'influx profile' obtained theoretically is shown in b and it is based on a bulk phase-limited diffusion model for labelled water entry (Ling et al., 1967: Figs. 3 & 4).

It is important to recognize that the diffusion coefficient for water inside cells is not much less than that for self-diffusion. For example, values for the diffusion coefficient lie in the range $0 \cdot 48 - 2 \cdot 40 \times 10^{-5}$ cm^2 sec^{-1} (Løvtrup, 1963a; Ling et al., 1967;

Bunch & Kallsen, 1969) whereas the self-diffusion coefficient for water is about 2.5×10^{-5} cm^2 sec^{-1}.

Even although the diffusion coefficient for water inside cells is apparently not reduced by a great amount, the rate-limiting effect of diffusion in the cytoplasm might still be important, especially in large cells where the diffusion pathlength may be as big as the external unstirred layer. This particular point is illustrated in a study by Sorenson (1971) of the exchange of deuterated water in the large muscle fibres of a marine crab. Sorenson noted that the efflux of deuterated water from isolated muscle fibres was composed of a fast component and a slow component. It was concluded that the slow component possibly represented the exchange of water between intracellular organelles and the cytoplasm whereas the fast component probably represented the exchange of water across the cell membrane. Sorenson considered that the latter flux was rate-limited by permeability of the cell membrane rather than by diffusion of water in the cytoplasm. Taking an arbitrary, but reasonable, value for the diffusion coefficient (10^{-5} cm^2 sec^{-1}) Sorenson estimated that a muscle fibre with a radius of 600 μm would have a relatively short half-time of about 6.3 sec for water exchange compared to the actual half-time of 66 sec. Duplicate calculations based on the same theory and data, however, show that the half-time for cytoplasmic diffusion is about 22.7 sec. In view of this it would be unwise to ignore the possible rate-controlling influence of water diffusion in the cytoplasm. One way of resolving this difficulty would be to obtain an estimate of the diffusion coefficient for water inside these large cells.

External unstirred layers. Besides the inadequacy of internal mixing in labelled-water exchange experiments we also have to deal with the problem of external unstirred layers at the surface of the cell. The significance of unstirred layers in transport studies on biological membranes has been discussed by several workers (Osterhout, 1933; Jacobs, 1935; Teorell, 1936). To a large extent the value of those earlier papers was not appreciated fully until Dainty emphasized the importance of unstirred layers from both experimental and theoretical standpoints (e.g. Dainty, 1963a).

If we consider, for example, the attempts of Paganelli & Solomon (1957) and Villegas, Barton & Solomon (1958) to measure P_d for erythrocytes it turns out that the adequacy of mixing in the external medium is crucial. Paganelli & Solomon (1957) demonstrated that mixing of dissimilar media in their apparatus was complete in about 0·9 msec, which is significantly less than the half-time (4·2 msec) for the apparent exchange of labelled water across the cell surface. However, as Dainty (1963a) has argued, it is the effect of the unstirred layer on the exchange which may be important because that region at the surface of the cell persists even in strongly agitated media. If the unstirred layer thickness in the vicinity of the red cell were 100 μm, then the equivalent permeability (D_w/δ) of this layer to water would be about $2·5 \times 10^{-3}$ cm sec^{-1}. The similarity between (D_w/δ) and the apparent value of P_d, which is $5·3 \times 10^{-3}$ cm sec^{-1} (Paganelli & Solomon, 1957), makes it imperative that the magnitude of δ should be estimated under the conditions of these exchange experiments. Sha'afi, Rich, Sidel, Bossert & Solomon (1967) estimated that the magnitude of the unstirred layer at the surface of the human red cell during osmotic experiments had an average value of 5·5 μm. Their conclusion was that the unstirred layer probably exerts an almost negligible effect on the determination of P_d; in this case, therefore, the estimated value of P_d is a reliable estimate of the actual permeability of the membrane for water.

At present the conclusion must be that estimates of P_d which are larger or equal to (D_w/δ), say 20×10^{-4} cm sec^{-1}, must be regarded with caution unless it has been shown that unstirred layers are exceedingly small. This contention is particularly cogent for determinations of P_d performed in the absence of stirring, for example in the Cartesian Diver Balance (Prescott & Zeuthen, 1953).

Measurement of L_p

The hydraulic conductivity of a cell membrane can be determined by suddenly placing the cell in, say, a hypertonic solution and then recording the time course of shrinkage. Let us assume that the osmotic pressure gradient is due to presence of some impermeant solute i at a concentration c_i in the external solution. At the beginning of cellular shrinkage the volume flow across the membrane is $-L_p RT c_i$; moreover, the volume flow can be expressed as $(dv/dt)/A$ where v and A are the volume and area of the cell. Thus, a knowledge of the initial value of (dv/dt) and A

allows one to determine L_p. Several workers (e.g. Vargas, 1968a; Wallin, 1969) have obtained the initial values of (dv/dt) from continuous records of cellular volume during the time course of such osmotic experiments.

Apart from the kinetic analysis of the initial phase of an osmotic transient it is also possible to use the entire time course of the osmotic volume change to determine L_p. The latter approach, however, can be achieved if, and only if, we can assume first that cellular volume is inversely related to the osmotic pressure of the cell and secondly that there is a relationship between v and A that will hold during the osmotic experiment. The first of these assumptions is supported by considerable evidence and this feature of the water relations of cells will be discussed in Chapter 6. The second assumption can be satisfied for a number of cells but, of course, the particular relation between v and A depends upon the cellular geometry. For instance, in the case of the erythrocyte A does not change significantly during changes in cellular volume (Jacobs, 1932) while in the sea urchin egg v and A are related by its spherical geometry. Thus, when we can satisfy both of these conditions we can rewrite the volume flux equation as a differential equation involving cellular volume. This equation can be integrated to give the cellular volume as a function of time during the osmotic experiment. When the solution has been obtained it can be used to describe the experimental data on osmotic shrinkage or swelling provided that an appropriate value of L_p can be found. This procedure has been used, for example, by Lucké (1940) to analyse the osmotic swelling curve for sea urchin eggs (see Fig. 5.4).

In this typical experiment the sea urchin egg is removed from sea water at zero time and placed in a dilute sea water. Subsequently the cell swells owing to osmotic entry of water and the points signify the volumes at various times during the osmotic swelling curve. An estimate of L_p can be obtained from the initial slope of the swelling curve and this can be used in the theoretical relation describing the volume as a function of time. The agreement between theory and experiment is very good and it indicates that a reliable estimate of the hydraulic conductivity of the cell can be obtained by this method. Thus, this kind of kinetic analysis of osmotic swelling or shrinkage curves allows one to estimate L_p. It should be pointed out, however, that this theoretical approach fails to account for the osmotic behaviour of cells placed in solutions containing permeant solutes. That kind of discrepancy will be discussed in Chapter 6. It should also be stressed that this approach fails to describe the final stages of osmotic swelling in some cells; for example, Lucké (1940) argued that the final portion

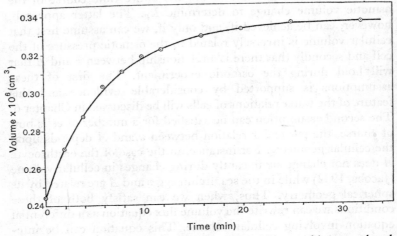

FIG. 5.4. A typical curve for osmotic swelling of sea urchin eggs placed in diluted sea water. In this case the egg was placed in 70% sea water and its volume was subsequently measured at certain intervals. The theoretical curve was calculated on the basis that during osmotic swelling the cellular volume is governed by the modified Boyle–van't Hoff law (see equation 6.2) and that the cell retains its spherical geometry (Lucké, 1940: Fig. 4).

of the swelling curve (not shown in Fig. 5.4) was probably influenced by the swelling of proteins and other macromolecules in the cytoplasm and presumably also by the swelling of subcellular organelles.

Table 5.2 shows some representative values of L_p for certain cells. Again, just as for the diffusional permeability to water, the hydraulic conductivities fall within a surprisingly wide range of values for animal cells. It might be considered that part of this variation stems from an additional source of rate-control of water flow other than that from the cell membrane itself. In fact, we could draw a close parallel between the sources of rate-control on diffusional permeability and those on hydraulic conductivity; however, the main source of such rate-control in the former case

TABLE 5.2. Apparent values of L_p for animal cells

Cell	$L_p \times 10^7$ (cm sec^{-1} atm^{-1})	Reference
Cat erythrocyte	250	Rich, Sha'afi, Barton & Solomon (1967)
Frog muscle	166	Hodgkin & Horowicz (1959)
Dog erythrocyte	147	Rich et al. (1967)
Horse erythrocyte	127	Blum & Foster (1970)
Cow erythrocyte	120	Rich et al. (1967)
Ehrlich ascites tumour	107	Hempling (1960)
Frog muscle	94	Zadunaisky, Parisi & Montoreano (1963)
Human erythrocyte	91	Sha'afi, Rich, Sidel, Bossert & Solomon (1967)
Human (foetal erythrocyte	86	Sjölin (1954)
Crab muscle	72	Sorenson (1971)
Frog (ovarian) egg	65	Prescott & Zeuthen (1953)
Crayfish muscle	49	Reuben, Girardier & Grundfest (1964)
Chick heart fibroblast	46	Dick (1959b)
Crayfish axon	24	Wallin (1969)
Eel erythrocyte	24	Blum & Foster (1970)
Human leucocyte	23	Shapiro & Parpart (1937)
Zebra fish (ovarian) egg	21	Prescott & Zeuthen (1953)
Squid axon	8·1	Villegas & Villegas (1960)
Chicken erythrocyte	6·1	Farmer & Macey (1970)
Rabbit leucocyte	4·8	Shapiro & Parpart (1937)
Squid axon	4·7*	Vargas (1968a)
Sea urchin egg	1·7	Lucké, Hartline & McCutcheon (1931)
Aplysia neuron	1·4	Austin, Sato & Longuet-Higgins (1966)
Xenopus egg	1·2	Prescott & Zeuthen (1953)
Frog egg	0·96	Prescott & Zeuthen (1953)
Amoeba proteus	0·43	Mast & Fowler (1935)
Zebra fish egg	0·33	Prescott & Zeuthen (1953)
Chaos chaos	0·27	Prescott & Zeuthen (1953)

* This value was obtained when water flowed along an osmotic gradient: however, when an hydrostatic pressure gradient was employed instead, the hydraulic conductivity was approximately 100 times larger.

is the unstirred layer and we know that this effect exerts a minor influence on the determination of L_p (see page 112). Although the unstirred layer effect cannot be ignored completely, it is likely to be important only where the velocity of osmotic flow is high. This means that errors due to unstirred layers may arise in the most permeable cells, but since they will produce underestimates of the 'true' L_p values it is clear that this kind of error cannot explain the wide range of hydraulic conductivities. Of course, the osmotic flow of water into and out of cells may be rate-limited by other kinetic effects which will be discussed later.

All of the L_p values in Table 5.2 were obtained by measuring the water flow arising from osmotic gradients and most of the values are similar to those found in artificial membranes, particularly the lipid membranes. Vargas (1968a), however, measured the hydraulic conductivity in experiments where water flow was driven either by an osmotic or a hydrostatic pressure gradient. Contrary to what one might expect, the values of L_p he obtained in these different experiments were not identical. Vargas used both perfused and intact axons of the squid, *Dosidicus gigas*. In one set of experiments a hydrostatic pressure was applied to the cell's interior by a capillary manometer and the rate of water transport across the axolemma was monitored by the displacement of the meniscus in the capillary. Alternatively L_p was measured by making the internal or external medium hyperosmotic. Vargas found that the hydraulic conductivity determined in the hydrostatic pressure experiments was significantly larger than that observed in the osmotic pressure experiments (see Table 5.3). At first glance one might speculate that the solutes used to establish the osmotic gradients have small reflexion coefficients; however, some simple arithmetic shows that the reflexion coefficients for the experimental solutes, such as sucrose, would need to be about 0·01, which is substantially lower than the values obtained by Villegas & Barnola (1961) for the squid axon.

The remarkable difference between the hydraulic conductivities determined in the separate osmotic and filtration experiments of Vargas is difficult to explain. Vargas considered three possibilities.

First, the axolemma may be a porous structure and these pores may increase in size when pressure is applied to the axoplasm. Two experimental facts, however, argue against this. The hydraulic conductivity was evidently independent of the pressure

TABLE 5.3. Hydraulic conductivities of the axon membrane of the squid *Dosidicus gigas*

Driving force for volume flux	Experimental technique	$L_p \times 10^7$ (cm sec^{-1} atm^{-1})
Hydrostatic pressure	Perfused	1030
	Intact	330
Osmotic pressure	Perfused	4·7
	Intact	5·0
	Perfused*	2·3

* Hypertonic solution inside.
Modified from Vargas, 1968a.

gradient and, secondly, neither membrane conductance nor permeability to glycerol was enhanced markedly during the filtration experiments.

The second possibility is that the osmotic gradients employed in the osmotic experiments alter, say, the degree of hydration of the axolemma and this in turn produces a change in L_p. In this case one might expect the hydraulic conductivity to be reduced as the mean osmolarity of the bathing media is increased. Such a mechanism has been proposed to account for apparent changes in the hydraulic conductivity of erythrocytes (Rich, Sha'afi, Romualdez & Solomon, 1968) and of the rabbit gall bladder (Diamond, 1966a) although in both of these instances there are rival explanations which are discussed in Chapters 6 and 9 respectively. In principle, Vargas' data on the squid axon could be explained on that basis. Indeed, the decrease in L_p which he observed in the osmotic experiments, where the interior of the axons was perfused with hypertonic solutions, seems entirely compatible with that mechanism. Nevertheless, the disparity between the results of the filtration and osmotic measurements seems too large to be explained by the relatively small increase in the mean osmolarity of the bathing media employed in the latter experiments. Probably the most telling evidence against this second mechanism is that Vargas found that it was possible to counterbalance a given osmotic gradient with a hydrostatic pressure gradient so that net volume flow was abolished. Under these circumstances the required hydrostatic pressure gradient was predicted on the basis of the L_p derived from the filtration experiments and not the osmotic ones. This indicates apparently that the

L_p determined during osmosis is a serious underestimate of the 'true' hydraulic conductivity.

Vargas favoured the third hypothesis, namely that the axolemma was a heteroporous membrane possessing a few large pores and numerous small pores. If the small pores exclude the osmotic solute, say, sucrose, but the large pores do not, then a given osmotic gradient will generate a smaller volume flow than expected simply because a certain fraction of the membrane has an exceedingly low reflexion coefficient for sucrose. On the other hand, the filtration of water across such a membrane is unaffected by such considerations. Furthermore, Vargas argued that if the axolemma contains n small pores of radius r and n' large pores of radius r' then the ratio (L_p^h/L_p^o) is given by

$$\left[1 + \frac{n'}{n}\left(\frac{r'}{r}\right)^4\right]$$

where L_p^h and L_p^o are the observed hydraulic conductivities in the filtration and osmosis experiments respectively. It should be noted that the ratio will always exceed unity and may be quite large when $r' \gg r$ even although n' is less than n. Although this is an attractive model there is no evidence in favour of it. Nevertheless, the postulated heteroporosity of the axolemma could be appraised by other experiments designed in the light of theories of membranes in series (see Chapter 2); for example, perhaps the axolemma, like the artificial membranes studied by Galey & Van Bruggen (1970), exhibits anomalous solvent drag (see Patlak & Rapoport, 1971).

The disparity between L_p^h and L_p^o is remarkable and puzzling. Moreover, it is not confined to single cells because a similar sort of discrepancy has been found for capillary walls (see Chapter 8) and for certain epithelia (see Chapter 9). The origin of this effect, however, is probably different in these individual instances. Certainly the important point about these measurements and those on the squid axon, in particular, is that they cast doubt on the osmotic method of determining L_p. Such doubt can only be dispelled by further experimental comparisons of filtration and osmosis in cells and by a careful reappraisal of the nature of hydraulic volume flow under these circumstances.

In addition to the questions raised about the characteristics of volume flow during filtration and osmosis across cell membranes, we should also attend to the claim that the membrane itself does not actually rate-control osmotic volume flow into and out of the cell. In other words, do the apparent values of L_p cited in Table 5.2 really reflect the hydraulic conductivities of cell membranes or do they reflect, even partially, other sources of rate-control on water flow?

Does the cell membrane rate-control osmotic flow?

It has been assumed so far that the cell membrane is the main barrier to the osmotic entry or exit of water, and on this basis the hydraulic conductivity of the membrane has been determined. However, this approach to swelling and shrinkage of cells has been criticized from two independent standpoints. On the one hand, Troshin, Ling and Cope have maintained that changes in cell volume can be attributed to alterations in the hydration of intracellular proteins; for summaries of these views see Troshin (1966), Ling (1962) and Cope (1967b). These authors are aligned in their disbelief of the classical membrane theories of solute and water transport. On the other hand, Dick (1959a, 1964, 1966) has suggested that what rate-controls the osmotic swelling or shrinking of cells is not only the water permeability of the membrane but also the rate of water diffusion inside the cell.

Let us examine the claims of the 'sorptionists' first. Cope (1967b), for example, has suggested that the changes in cellular volume occurring during osmotic experiments are governed by the Bradley isotherm (Bradley, 1936) for water adsorption on cell proteins; experimental studies have shown that the adsorption of water on to proteins, such as casein, ovalbumin and silk, is described by the Bradley isotherm (Hoover & Mellon, 1950) and this has been confirmed for the hydration of wool and collagen (Ling, 1965). In fact, Cope postulated that the rate-limiting phenomenon during osmosis in cells is the adsorption or desorption of water in 'multiple polarized layers' of water molecules forming the hydration crust of proteins. His treatment yields a relation for the cellular volume as a function of the total concentration of all solutes in the external medium; this equation is approximately equivalent to that derived from membrane theory (see equation 6.2). Cope's analysis is based on the concept that the solutes in the bathing solution decrease its vapour pressure and that consequently there is a decrease in the adsorption of water on to intracellular proteins since the cell membrane is alleged to offer no restriction to the movement of solutes and water. In the first instance Cope applied his theoretical description to experimental data where it was certain that changes in cellular volume occurred in the absence of solute adsorption on to intracellular proteins; this condition was satisfied, for example, for experiments where sucrose was used as an external solute. The agreement between his theory and the data on swelling of muscle, erythrocytes and eggs is good. Secondly, Cope considered the osmotic experiments which employed solutes, such as sodium ions, which are capable of being complexed by cellular macromolecules (e.g. Cope, 1965, 1967a). After correcting his original relation between cellular volume and the external concentration for the effect of solute adsorption, Cope found that the modified equation was compatible with some experimental data on erythrocytes and muscle cells exposed to different concentrations of KCl or NaCl.

The theoretical work of Cope is interesting because it offers an alternative description of osmotic phenomena in cells to that of the 'classical' membrane theory. Furthermore, it takes account of solute adsorption within cells, which is often ignored by the protagonists of the membrane view. Nevertheless, this type of theoretical approach requires a great deal more knowledge about the behaviour of water on cellular macromolecules than is currently available, and even in the relatively simple form proposed by Cope (1967b) it has not been thoroughly tested from an experimental standpoint.

Dick (1959a,c, 1966) has concluded from his reviews of the magnitudes of the hydraulic conductivities of animal cells that the cell membrane may not be the sole source of rate-control of osmotic flow in all cells. Recalling the hydraulic conductivities of animal cells shown in Table 5.2, we can see that even within this small sample there is an enormous range of values from about $0 \cdot 2$ to 250×10^{-7} cm sec^{-1} atm^{-1}. If we are to accept that these water permeabilities reflect accurately some properties of the individual cell membranes, then these membranes must have more intrinsic variation than electron micrographs often suggest. This is quite possible, of course. Dick has suggested, however, that most of these hydraulic conductivities have been underestimated because of the influence of the slow rate of diffusion of water within the cytoplasm. To support his case Dick presented the correlation

(Fig. 5.5) which exists between the apparent hydraulic conductivity of certain cells and their surface-to-volume ratios. Obviously large cells, such as muscle, will have a smaller surface-to-volume ratio than small cells, such as erythrocytes. This might mean that the rate of osmotic water flow into or out of the larger cells is rate-limited by internal diffusion of water rather than by the water permeabilities of their membranes. This appears to be supported

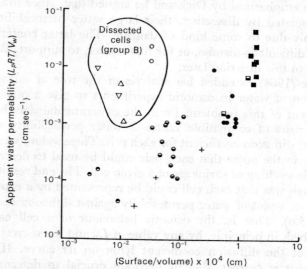

FIG. 5.5. Relationship between the apparent water permeability and the surface to volume ratio for a variety of cells. The units of the apparent L_p (cm sec^{-1} atm^{-1}) have been converted to cm sec^{-1} by the factor $RT/\bar{V}_w$. A correlation between water permeability and surface to volume ratio exists except for several dissected cells (group B, Table 4, Dick, 1966): the significance of this correlation is discussed in the text. The symbols represent; ◓ amphibian and fish eggs: + protozoa: ◖ marine invertebrate eggs: ● mammalian leukocytes and ascites tumour cells: ◳ chick heart fibroblasts: ■ mammalian erythrocytes: △ cephalopod axons: ▽ amphibian and fish eggs: ○ frog muscle fibres (Dick, 1966; redrawn from Fig. 1 of Dick, 1959c).

by the correlation in Fig. 5.5 but, as Dick also observed, 'the correlation may be fortuitous'!

The relation between the apparent permeability and the surface-to-volume ratio of the cells makes sense only if the diffusion coefficient of water in the cell's interior is extremely low; in fact, it must be substantially lower than the self-diffusion coefficient of water for Dick's argument to hold. If this is so, then the swelling and shrinking of cells in response to osmotic gradients might be rate-controlled, particularly in the large cells, by intracellular diffusion of water rather than by its rate of transport across the cell membrane. In Fig. 5.5 a certain group of cells was found to have larger water permeabilities than one might expect from the general pattern emphasized by Dick and he argued that, since those cells were isolated by dissection, their large water permeabilities are probably due to some kind of damage. The latter contention is rather difficult to dismiss, or for that matter, to support. We shall return to this question later.

Dick (1964) extended his analysis of the role of intracellular diffusion of water in osmotic experiments to give a quantitative treatment of this problem. From the experimental data he computed pairs of compatible values of water permeability and the internal diffusion coefficient for each cell. These values were compatible in the sense that each pair could be used to describe the osmotic swelling or shrinkage of a given cell. The end result of this approach was that each cell could be represented by a continuous curve on a plot of water permeability against diffusion coefficient (Fig. 5.6). That is, the osmotic behaviour of a cell could be described, in principle, by any value of L_p and the corresponding value of the diffusion coefficient lying on its curve. If Dick's argument is correct, then it becomes crucial to determine the diffusion coefficient for water inside cells, especially the large ones, so that we can assign the appropriate value of L_p to the cell. We can only judge how valid Dick's arguments are by looking at the magnitudes of the diffusion coefficient which are required to explain the apparent correlation between L_p and the surface-to-volume ratio in Fig. 5.5. It must be remembered that the latter correlation is the corner-stone of Dick's argument about the possible important role of intracellular water diffusion in osmotic experiments.

Dick argued that, if all of the cells examined have the same water permeability, then the curves in Fig. 5.6 should intersect at a common point indicating that permeability value; if the cells have similar rather than identical water permeabilities then the curves ought to pass through a common domain rather than a common locus. Fig. 5.6 shows that the curves for the majority (19 out of 23) pass through a square region, indicating a range of diffusion coefficients from 0.08 to 2×10^{-8} cm^2 sec^{-1} and of hydraulic conductivities from 2 to 50×10^{-7} cm sec^{-1} atm^{-1}. There are several points to note about these values. First, the conventional treatment of osmotic phenomena, which ignores the influence of intracellular water diffusion, yields a wider range of water permeabilities for these cells than Dick's analysis does; indeed the ratio of the extreme values of L_p drops from about 1000:1 to 23:1 when account is taken of diffusion of water in the cell's interior. Secondly, Dick's range of water permeabilities for those cells clusters about the

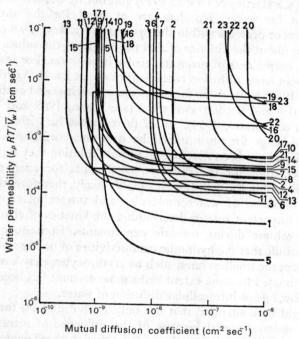

FIG. 5.6. Values of the water permeability ($L_p RT/\bar{V}_w$) and of the mutual diffusion coefficient (D) in the cytoplasm which are mutually compatible for describing osmotic flow in various cells. Each line represents the appropriate range of pairs of values for $L_p RT/\bar{V}_w$ and D in a particular cell designated by the adjacent number (see Dick (1964) for details). The heavy square outlines the smallest range of common values that can describe the osmotic behaviour of all cells except the dissected cells, numbers 20–23. The latter include axons, eggs and muscle fibres (Dick, 1964: Fig. 2).

observed hydraulic conductivities of artificial lipid membranes (see Table 4.3). Finally, even although the internal diffusion coefficients of water in the expected range are about 1000 to 300,000 times smaller than the self-diffusion coefficient of water, these values are still quite acceptable. This is so because the coefficient which determines water diffusion inside the cell during osmotic experiments is not the self-diffusion coefficient but *the mutual diffusion coefficient*. The latter diffusion coefficient, representing the rate at which water mixes with the intracellular macromolecules, is determined by the rate of movement of the slower components (see Hartley & Crank, 1949) and not by the self-diffusion coefficient for water. No data are at hand for the diffusion coefficients of proteins within the cytoplasm; however, we can get some idea about the kinetics of that process from the values of the diffusion coefficients of macromolecules in solution. For instance, nucleic acid has a diffusion coefficient of about 2×10^{-8} cm^2 sec^{-1} (Jordan, 1960) while those for tobacco mosaic virus and ovalbumin are 5×10^{-8} and 9×10^{-8} cm^2 sec^{-1} (see Kohn, 1965 and Wang, Anfinsen & Polestra, 1954). All of these values lie within Dick's predicted range for the intracellular diffusion coefficients. Since these values have been determined in bulk solution they represent the upper limit for the diffusion coefficients for such macromolecules within the cell. It seems highly likely, therefore, that the mutual diffusion of water molecules and intracellular proteins plays an important role in determining the kinetics of changes in cellular volume during osmotic experiments. Consequently we may conclude that the hydraulic conductivities of nearly all animal cells except the smallest ones, such as erythrocytes, may have been underestimated to some extent because no account has been taken of the slow rate of intracellular diffusion of water.

It might be contended that not only L_p but also P_d for single cells is underestimated because of the influence of intracellular diffusion. This is true, but the water diffusion coefficients which one must use for the sake of correction are different in the two instances. For the corrections to L_p we need to know the mutual diffusion coefficient whereas for corrections to P_d we need the larger self-diffusion coefficient of water in the cytoplasm. Determinations of the rate of diffusion of labelled water in the cytoplasm of certain cells have given values in the range $0.5-2.4 \times 10^{-5}$ cm^2 sec^{-1} (Løvtrup, 1963a; Ling et al., 1967; Bunch & Kallsen, 1969).

That these values for the self-diffusion coefficient of water are substantially larger than those for the mutual diffusion coefficient is expected, since the former relate to the exchange of labelled water molecules with other water molecules in the cytoplasm whereas the latter relate to the mutual exchange of water and macromolecules in the cytoplasm. Several workers have erroneously assumed that these coefficients should be identical in magnitude (Edelman, 1961; Bunch & Kallsen, 1969; Kushmerick & Podolsky, 1969; Sorenson, 1971).

As we have seen earlier, a group of values for L_p in the dissected cells are apparently in discord with Dick's analysis of the kinetics of osmotic phenomena. He has offered several reasons for the discrepancies between the values for 'free-living' and 'dissected cells' (Dick, 1966, 1970). First, there may be a genuine difference in structure of the 'dissected cells' (ovarian eggs and squid axons) compared to the 'free-living' ones. In this connexion, Dick noted that the apparent ratios $(L_p RT/\bar{V}_w P_d)$ for the dissected cells were significantly larger than those for other cells; however, the actual validity of this argument is difficult to assess since the values of P_d were uncorrected for unstirred layers. Secondly, the dissection of ovarian eggs and squid axons may have raised the water permeability or the coefficient for internal diffusion. If this reason is valid one might expect that some leakage of solutes would have occurred during the osmotic experiments. Indeed, Hill (1950b) reported that such leakage did occur, although the recent work by Freeman, Reuben, Brandt & Grundfest (1966) has demonstrated that both lobster and squid axons behave as reliable osmometers when changes are made in the external NaCl concentration. Of course, it is possible that the water permeability of these cells is more sensitive to the disturbance produced by the dissection than is the sodium permeability. Finally, in the case of at least one kind of dissected cell—the amphibian ovarian oocyte—account must be taken of the actual surface area rather than the apparent geometrical surface area used in Dick's theory. In these cells there are numerous microvilli on the surface which increase the area by a factor of 2—10; thus, the apparent values of L_p, after correction for this increase in surface area, are about 50–10% of the apparent ones (Dick, Dick & Bradbury, 1970). Fig. 5.7a shows the apparent values for L_p of the oocytes as a function of their size and 5·7b gives the values of L_p, corrected for the actual surface area.

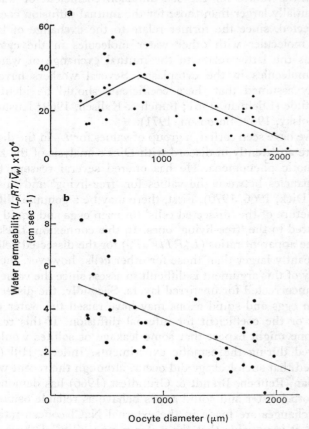

FIG. 5.7. (a) Apparent water permeability ($L_p RT/\bar{V}_w$) calculated for the membranes of oocytes of different sizes. Each point represents one oocyte. The permeability coefficients were calculated assuming that the oocyte was a smooth sphere and without taking account of microvilli at its surface. The coefficients rise at first with increasing cell size, reach a peak in oocytes about 800 μm in diameter, and then decline. The oocytes shown in this figure were all isolated from the ovary by dissection.

(b) Water permeabilities ($L_p RT/\bar{V}_w$) corrected for the effects of microvillar development and concomitant changes in the surface area of oocytes. There is a significant downward trend with increasing cell size (Dick *et al.*, 1970: Figs. 6 & 11).

Dick *et al.*, (1970) concluded that the apparent value of L_p for the oocyte is related to the variation in the area of the membrane. Despite this evidence for the influence of membrane rate-control it was concluded that intracellular diffusion of water is also probably important since the magnitude of the apparent L_p, corrected for the microvillar surface, decreases as the oocyte increases in size (Fig. 5.7b). Dick *et al.* (1970) calculated that the actual value of L_p lies in the range $1 \cdot 5 – 22 \times 10^{-7}$ cm sec^{-1} atm^{-1} and the internal diffusion coefficient for water is in the range $6 – 100 \times 10^{-8}$ cm^2 sec^{-1}.

According to the views of Dick, the rate-control of osmotic water movement in the cytoplasm cannot be ignored, except in very small cells, otherwise an underestimate of the actual L_p may result. Taken in its wider context it would seem that Dick's argument probably does not apply to osmotic water transport across tissues, such as capillary walls or epithelial layers, for in these instances the transcellular route for water transport is relatively short and consequently the significance of mutual diffusion of water and intracellular macromolecules must be slight. Moreover, this view has not been applied to plant cells as yet in order to see if their values of L_p have been underestimated too. We shall discuss this point briefly in the next section.

Water transport in plant cells

As an experimental object for transport studies the vacuolated plant cell is complex; certainly it seems more complicated than its distant relation—the animal cell. The central aqueous vacuole of the 'typical' plant cell is enclosed by a membrane—the tonoplast. This plasma membrane is one of a pair which bounds the cytoplasm; the outer membrane of this pair is termed the plasmalemma and it makes intimate contact with the plant cell wall which behaves as a heterogeneous ion-exchange membrane under considerable tension. It is important to realize at the outset that the entire plant cell is under considerable hydrostatic pressure, often called the *turgor pressure* or *wall pressure*. Under equilibrium conditions it is this large internal hydrostatic pressure of several atmospheres which counterbalances the adverse osmotic pressure gradient tending to drive water into the vacuole. A surprising feature of this system is that such a high internal hydrostatic pressure does not force the plasmalemma membrane through the apparently open network of the plant cell wall. In considering the water relations of plant cells, as opposed to animal cells, one comes up against the additional complications that the physical properties of the cytoplasm and the cell wall may alter significantly during swelling or shrinkage and in turn these alterations may affect the rate of water flow between the vacuole and the cell's environment.

Measurement of P_d

Exceedingly little work has been done on the kinetics of the exchange of labelled water between single plant cells and their external media. A study by Dainty & Hope (1959) on the giant internodal cells of *Chara* demonstrated that the exchange of labelled water was rate-controlled by diffusion of water in the cell wall and external and internal unstirred layers. Earlier work on *Tolypellopsis* (now called *Nitellopsis*) by Wartiowaara (1944) is possibly also open to the interpretation that unstirred layers rate-control the exchange of labelled water. It is true that Wartiowaara did attempt to correct for unstirred layers but his method of doing so is not entirely free from criticism; after such corrections he found that P_d for *Nitellopsis* was 4.4×10^{-4} cm sec^{-1}. A similar question mark should also probably accompany the estimate of 25×10^{-4} cm sec^{-1} for *Nitella mucronata* (Collander, 1954). Gutknecht (1967) measured P_d in the internally perfused cells of a marine alga, *Valonia*, and he found a mean value of 1.22×10^{-4} cm sec^{-1}. Upon completion of an individual measurement of P_d Gutknecht perfused the internal vacuole with sea water which induced the protoplast to form aplanospores. This process left the cell wall intact and subsequently the P_d for the cell wall and its associated unstirred layers was measured. By treating the cell as a series of permeability barriers it was possible from these two measurements to obtain the actual value of P_d for the series barrier, composed of tonoplast, cytoplasm and plasmalemma, commonly referred to as the protoplast. Gutknecht found that the diffusional permeability of the protoplast was 2.36×10^{-4} cm sec^{-1}. This value is quite close to the P_d values for some animal cells, such as the squid axon, and those for artificial lipid membranes. Apparently the value of P_d for *Valonia* is the only reliable estimate we have for single plant cells. The lack of data on the exchange of labelled water is probably an accurate reflection of the emphasis that plant physiologists have placed on the osmotic properties of plant cells rather than the more general aspects of their water relations.

Measurement of L_p

As mentioned earlier, the plant cell is under considerable hydrostatic pressure (turgor pressure) and this additional parameter plays an important role in the osmotic relations of plant cells, whether we are considering the plant cell in a state of equilibrium or when it is swelling or shrinking during an osmotic experiment.

When a plant cell is bathed by a solution of some impermeant solute, the volume flux is given by

$$J_v = L_p(\Delta p - \Delta \pi) \qquad 5.11$$

where Δp and $\Delta \pi$ represent the differences in hydrostatic and osmotic pressure between the vacuole and the external medium. This relation offers a method of determining L_p. It must be stressed that the hydraulic conductivity measured in this way is that of the barrier—tonoplast, cytoplasm, plasmalemma and cell wall in series. The same problem is met, of course, in the measurements of P_d for plant cells. It might be thought that the cell wall can be safely ignored. However, this is not entirely the case since Barry & Hope (1969b), for example, have noted that the hydraulic conductivity of the cell wall of *Chara australis* (now called *Chara corallina*) is 400×10^{-7} cm sec^{-1} atm^{-1} which is only about four times the L_p for these cells. Clearly the plant cell wall does offer some resistance to osmotic water flow. Moreover, if Dick's argument about the role of the mutual diffusion coefficient (e.g. Dick, 1966) in rate-controlling osmotic water movement in animal cells can be applied to plant cells, then we cannot afford to ignore the cytoplasmic barrier either. Thus, the values of L_p, which are cited for plant cells cannot be accepted unquestionably as the effective conductance of either the tonoplast or the plasmalemma.

Once again, just as in animal cells, we can estimate L_p from the initial slope of an osmotic swelling or shrinkage curve and also from the entire time course of such responses. Philip (1958), for example, has published a revealing account of the swelling and shrinkage of turgid plant cells placed in different solutions of impermeant solutes. His treatment takes into account the elastic properties of the plant cell wall. Indeed, this is an important point because the elasticity of the wall exerts its influence on the volume changes happening during osmotic experiments. According to Philip's treatment, the changes in cellular volume, v, during swelling or shrinkage are governed by a rearrangement of equation 5.11, namely

$$\frac{dv}{dt} = L_p A (\Delta p - \Delta \pi) \qquad 5.12$$

where again A is the area of the cell; it is usually satisfactory to assume that A is constant. Following essentially similar arguments to those applied to osmotic studies on animal cells, Philips obtained a relation describing the volume as a function of time during the experimental period. Only in this case the expression includes the elasticity of the cell wall, a feature absent, of course, from animal cells. Again the theoretical relation is not quoted here, principally for the same reasons that it was omitted in our previous discussions of animal cells. That is, the equations are cumbersome and one cannot easily intuit from them how cellular volume varies with time. From Philip's treatment it can be deduced that the half-time, $t_{1/2}$, for osmotic shrinkage or swelling is given by

$$t_{1/2} = \frac{0 \cdot 693 \, v^o}{AL_p(\epsilon_w + \pi^o)} \qquad 5.13$$

where π^o and v^o are the osmotic pressure and volume of the plant cell when the turgor pressure is zero and ϵ_w is the elastic modulus of the cell wall. Provided the plant cell remains turgid throughout the osmotic swelling or shrinkage which is a prerequisite of Philip's analysis then equation 5.13 shows that the kinetics of the volume changes are dictated not only by the hydraulic conductivity of the plant cell but also by the elastic properties of the plant cell wall. For example, if either L_p or ϵ_w is small, then the time course of the osmotic volume change will be slow. This means that osmotic experiments on plant cells must be accompanied by independent measurements of the elastic modulus of cell wall. Philip also considered the interesting case where swelling and shrinkage were induced by concentrations of permeant solutes instead of the impermeant ones used in determinations of L_p. This work will be discussed in Chapter 6.

The preceding treatment assumes that during the osmotic experiments the plant cell maintains its usual turgor pressure. At this point it should be noted, however, that a number of measurements have been made on plasmolysed cells (Stadelmann, 1963). This method has been employed for a wide range of cells and the values for L_p lie in the range 0.5–20×10^{-7} cm sec^{-1} atm^{-1}. Unfortunately this approach is open to several objections and the main one is, of course, that the permeability of the membranes in the plasmolysed cells may be substantially different from that in the normal turgid plant cell. In contrast, some measurements of L_p have been made on cells exhibiting their full turgor pressure. Kelly, Kohn & Dainty (1963), for example, have applied the theoretical analysis of Philip to the volume changes in *Nitella translucens* in order to determine L_p. They succeeded in measuring the very small changes in volume of this large cell in osmotic experiments which were designed to minimize the loss of turgidity; the degree of shrinkage was limited by placing the cells in solutions which were insufficiently concentrated to produce plasmolysis. In their study the elastic modulus of the cell wall, which is an important determinant of osmotic water flow, was also measured. Actually there is an additional complication over the elasticity of the cell wall in *Nitella* because it is a cylindrical cell and the longitudinal elastic modulus may not be equal to the radial elastic modulus. Finally, Kelly *et al.* (1963) made allowances for the finite rate of diffusion of the external solutes at the cell's surface during the changes in the external osmotic pressure. Their examination of the osmotic shrinkage and swelling of *Nitella* gave values for L_p in the range 100–300×10^{-7} cm sec^{-1} atm^{-1} which are in good agreement with other estimates obtained by a different method described below.

The most reliable estimates of L_p for single plant cells have been obtained by observing transcellular osmosis in the giant internodal cells of the *Characeae*. A relatively simple account of transcellular osmosis was published by Osterhout (1949*a,b*) and later a more detailed theory was given by Kamiya & Tazawa (1956). Unfortunately neither of these analyses considered the effect of changes in the turgor pressure of the cell during the experiment, but the later treatment of transcellular osmosis (Dainty & Hope, 1959; Dainty & Ginzburg, 1964*a*) does take account of turgor pressure changes. Moreover, it dictates the necessary experimental conditions for the determination of L_p by transcellular osmosis. In this method the cell is placed in a watertight dividing wall between two compartments (see Fig. 5.8). One compartment is closed and contains a capillary so that its volume can be monitored. The entire apparatus is mounted in a thermostatic water bath after the compartments have been filled with water and the experimental solution respectively. Since the solution in the right-hand chamber can be changed easily, different osmotic gradients can be established across the cell and their concomitant water flows can be recorded quite conveniently. L_p was estimated from the relation

$$\mathcal{J}_v = L_p \frac{A_n A_x}{A_n + A_x} \Delta\pi \qquad 5.14$$

where $\mathcal{J}_v$ is the initial rate of volume flow across the cell, $\Delta\pi$ is the difference in osmotic pressure between the compartments and A_n and A_x are the areas of the cell in the left- and right-hand compartments. Dainty & Hope (1959) observed that the water flux across the cell decreased with time due to the build up of a concentration gradient of solute in the vacuole; such a gradient would tend to reduce the effective driving force on the water movement across the cell. The relatively more complicated aspects of this method of measuring L_p by transcellular osmosis have been discussed fully by Dainty & Ginzburg (1964a). Transcellular osmosis gave values of about 100×10^{-7} cm sec^{-1} atm^{-1} for the hydraulic conductivity of both *Chara corallina* (Dainty & Hope, 1959; Dainty & Ginzburg, 1964a) and *Nitella translucens* (Dainty & Ginzburg, 1964a). Similarly Kamiya & Tazawa (1956) found L_p lying in the range 120–300×10^{-7} cm sec^{-1} atm^{-1} for *Nitella flexilis*.

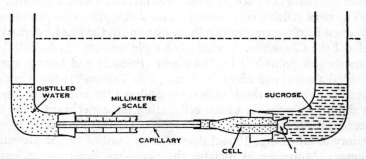

FIG. 5.8. Apparatus for recording the rate of transcellular osmosis across a giant plant cell. The cell was sealed with Vaseline into a split Perspex taper (t) containing a central 1·6 mm diameter hole (Dainty & Hope, 1959: Fig. 2).

Table 5.4 gives values for the hydraulic conductivities of some plant cells. The methods which have been described for measuring these permeabilities invariably employ osmotic gradients as the driving force for water flow. In view of the astounding difference between the estimates of L_p obtained from separate filtration and osmotic experiments on the squid axon (Table 5.3) it is pertinent to ask if this disparity also occurs in plant cells. According to Vargas (1968a) it does. He cited unpublished measurements of Villegas on *Valonia ventricosa* in which L_p obtained from osmotic experiments was $1 \cdot 2 \times 10^{-7}$ cm sec^{-1} atm^{-1} whereas that from filtration experiments was 230×10^{-7} cm sec^{-1} atm^{-1}. This is a remarkable state of affairs and one which certainly needs to be confirmed again experimentally. It is interesting to note incidentally that Villegas' estimate for L_p by osmotic methods is quite close to that obtained similarly by Gutknecht (1967) on internally perfused *Valonia* cells (Table 5.4).

TABLE 5.4. Apparent values of L_p for plant cells. Values for the thickness of the cytoplasmic layer in these cells are also given

Cell	$L_p \times 10^7$ (cm sec^{-1} atm^{-1})	Thickness of cytoplasmic layer (μm)	Reference
Nitella flexilis	120–300	–	Kamiya & Tazawa (1956)
Nitella axillaris	–	7	Diamond & Solomon (1959)
Nitella translucens	107	20	Dainty & Ginzburg (1964a,b)
Chara corallina	93	–	Dainty & Hope (1959)
Chara corallina	101	–	Dainty & Ginzburg (1964a)
Chara globularis	–	10–20	Gaffey & Mullins (1958)
Nitellopsis obtusa	19	14	Palva (1939): Holm-Jensen, Krogh & Wartiowaara (1944)
Valonia ventricosa	1·8	12	Gutknecht (1967)

Again, as we observed for animal cells there is a surprisingly wide range of L_p values even for the relatively meagre number of plant cells that have been examined. According to the views of Dick (1966) this range of L_p values might be markedly reduced if corrections were made for the influence of internal diffusion of water molecules. Actually Dick's argument cannot be applied in its original form to plant cells because of the presence of the large central vacuole. However, one way of testing it would be to plot the apparent values of L_p for these cells against the reciprocal of the thickness of their cytoplasmic layers on the grounds that mutual diffusion of water and macromolecules can occur presumably only in the cytoplasm. Values for the thickness of the cytoplasmic layers are also given in Table 5.4. Despite the wide range of permeabilities the thickness of the cytoplasm is quite uniform from cell to cell and there is no correlation between L_p and the reciprocal of cytoplasmic thickness. It seems, therefore, that Dick's argument about the role of intracellular diffusion of water probably does not apply to osmotic phenomena in this group of plant cells.

Comparison of $L_p RT/\bar{V}_w$ and P_d values

Before proceeding with the discussion of the comparisons that have been made between the diffusional permeability and the hydraulic conductivity in some cells, it is wise to keep two cardinal points in mind. First, measurements of P_d are subject to errors arising from external and internal unstirred layers and also estimates of L_p are subject to errors mainly due to the possible influence of mutual diffusion of water and large molecules inside cells. Secondly, the work of Vargas (1968a) raises an even more serious doubt over the values of L_p determined in osmotic experiments. As yet we cannot assess the full significance of his finding simply because of lack of experimental data. It is possible, for example, that all of the values of L_p quoted in Table 5.2 are really underestimates of the actual hydraulic conductivities. Because of these complications it could be contended quite correctly that we have no right to compare P_d and $L_p RT/\bar{V}_w$. Nevertheless, I shall do so, but the reader should remember that appropriate corrections to P_d will tend to lower the size of $(L_p RT/\bar{V}_w P_d)$ while new determinations of L_p by filtration measurements may tend to raise the ratio quite substantially.

There have been several comparative studies of P_d and L_p in cells. Table 5.5 contains the values of $(L_p RT/\bar{V}_w P_d)$ which have been derived from those investigations. At this point it should be recalled that work on artificial membranes (Thau et al., 1966) has revealed that some non-porous artificial membranes have water permeability ratios which lie between 1 and 2. Thus, we may conclude that a number of cells in Table 5.5 probably have non-porous membranes. It is perhaps significant too that cells, such as the human erythrocyte and the alga *Valonia*, which have been studied carefully to avoid errors from unstirred layers, have low values of $(L_pRT/\bar{V}_wP_d)$. Parallel studies on artificial lipid membranes, where it is relatively easy to check on errors, have demonstrated that the different water permeabilities are identical. The exception to that particular rule occurs, however, when these lipid membranes are treated with certain antibiotics that apparently make them porous; that is, $(L_pRT/\bar{V}_wP_d)$ attains values around 3.

Three cells in Table 5.5. have very large values of $(L_pRT/\bar{V}_wP_d)$. The first of these is the large muscle fibre of the marine crab, *Chionoecetes bairdi*. In estimating P_d for this cell Sorenson (1971) made corrections for the external and internal unstirred regions

TABLE 5.5. Apparent values of $(L_pRT/\bar{V}_wP_d)$ for some cells

Cell	$\left(\dfrac{L_pRT}{\bar{V}_wP_d}\right)$	Reference
Crab muscle	81	Sorenson (1971)
Frog (ovarian) egg	70	Prescott & Zeuthen (1953)
Zebra fish (ovarian) egg	43	Prescott & Zeuthen (1953)
Lobster nerve	20	Nevis (1958)
Squid axon	7·8	Villegas & Villegas (1960)
Dog erythrocyte	6·3	Villegas, Barton & Solomon (1958)
Dog erythrocyte	4·6	Rich, Sha'afi, Barton & Solomon (1958)
Cow erythrocyte	3·1	Villegas *et al.* (1958)
Human erythrocyte	2·4	Villegas *et al.* (1958)
Human erythrocyte	2·4	Sha'afi, Rich, Sidel, Bossert & Solomon (1967)
Xenopus egg	1·8	Prescott & Zeuthen (1953)
Frog egg	1·7	Prescott & Zeuthen (1953)
Amoeba (*Chaos chaos*)	1·6	Prescott & Zeuthen (1953)
Zebra fish egg	1·3	Prescott & Zeuthen (1953)
Marine alga (*Valonia ventricosa*)	1·0	Gutknecht (1967)

but possibly (see page 161) he has underestimated the importance of water diffusion in the cytoplasm of these large cells (radius ~ 600 μm). For that reason I feel that the estimated ratio of water permeabilities for those muscle cells is probably much larger than the actual ratio. The other two cells with high permeability ratios are from a study by Prescott & Zeuthen (1953) who measured P_d by monitoring the exchange of deuterated water in cells placed in a Cartesian diver balance. Because of the inadequacy of stirring in that method the P_d values are likely to be seriously underestimated. Under the circumstances no reliance can be placed on the values of $(L_p RT/\bar{V}_w P_d)$ for any of these cells.

A number of other cells, including the lobster axon, squid axon and dog erythrocyte, also have values of $(L_p RT/\bar{V}_w P_d)$ which greatly exceed 2 and, hence, their membranes may be porous. Here, as in other instances one could contend that sources of rate-control other than cell membranes dictate the kinetics of exchange and flow of water molecules into and out of these cells. These difficulties just underline the fact that we cannot rely solely on the value of $(L_p RT/\bar{V}_w P_d)$ as an index of membrane porosity especially in the case of cell membranes. Nevertheless, this criterion remains a useful analytical tool to be used conjointly with other methods, of course.

Effect of temperature on water permeabilities

Several workers have studied the temperature dependence of water movement across cell membranes in order to gain some insight into the nature of the membrane and, in particular, the physical state of water in the membrane. The temperature dependence of the water permeabilities can be expressed by means of the well-known Q_{10}. This quantitative approach allows one to compare the temperature dependence of, say, P_d with that for the self-diffusion coefficient, D_w, for water. In fact, it is more common to refer to the activation energy, E_a, of a given process like diffusion than it is to quote its Q_{10}.

Consider, for example, the self diffusion of water. This process can be described by a single diffusion coefficient—D_w—whose temperature dependence obeys the classical Arrhenius equation, namely

$$D_w = D_o \exp(-E_a/RT) \qquad 5.15$$

which can be re-written as

$$\ln(D_w) = \ln(D_o) - \frac{E_a}{RT} \qquad 5.16$$

where D_o is a constant termed the frequency factor. E_a can be obtained from the slope of the linear relation between $\ln(D_w)$ and T^{-1} and it can easily be shown that it is related to the Q_{10}; at 20°C Q_{10} is approximately equal to $\exp(0.056\,E_a)$ where E_a is expressed in kcal mole^{-1}. A similar treatment of the viscosity of water can also be given and the final expression is similar to equation 5.16.

Another way of describing the temperature dependence of D_w and η_w for water is through the transition state theory of rate processes (see Glasstone, Laidler & Eyring, 1941). Consider again, for the sake of an example, the self-diffusion of water. According to this theory a water molecule diffuses by going through a transition state during the molecular jumping process. The formation of the transition state is associated not only with the activation energy but also with an entropy change. The end result of this theoretical approach, as applied to the self-diffusion and viscosity of water, is that the temperature dependence of both of these processes is governed by expressions similar to equation 5.16. Thus, the temperature dependence of certain transport processes, such as diffusion, involves both the activation energy and the entropy of activation, although it is customary for most physiologists to concentrate almost exclusively on the activation energy.

Table 5.6 shows some representative values of E_a for osmotic flow and diffusion of water in animal and plant cells. The general pattern in this table is that the larger activation energies are found in the relatively impermeable cells and *vice versa*. Danielli & Davson (1935) have offered explanations for this. They envisaged that the activation energy was like a barrier to the molecular motion and to move over this barrier each molecule had to acquire an energy greater than E_a. According to this scheme, the lower the value of the water permeability then the higher the corresponding value for the activation energy will be. In addition to these general accounts of the dependence of water permeability upon temperature we really need a more detailed explanation for the values of E_a. This gap is beginning to be filled. Price & Thompson (1969), for example, have attempted to calculate the size of the activation energy for water transport across artificial lipid membranes. The values they obtained lay somewhere in the range 4–16 kcal mole^{-1} depending on the particular model chosen; for example, they estimated that the activation energy for water diffusion across a lipid membrane ought to be 10.5–12.4 kcal mole^{-1}.

Most of the estimates of E_a for osmotic flow and diffusion of water in animal and plant cells exceed the corresponding activation energies for viscous flow and self-diffusion of water. Obviously these disparities illustrate the different nature of water transport

TABLE 5.6. Estimates of activation energies for osmotic and diffusional flows of water across cell membranes

Cell	Water permeability × 10^4		E_a (kcal mole^{-1})	Reference
	$\dfrac{L_p RT}{\overline{V}_w}$ (cm sec^{-1})	P_d		
Sea urchin egg	3·2	—	13–17	McCutcheon & Lucké (1932)
Chicken erythrocyte	8·3	—	11·4	Farmer & Macey (1970)
Ehrlich ascites tumour cell	37	—	9·6	Hempling (1960)
Nitella translucens	140	—	8·5	Dainty & Ginzburg (1964a)
Sheep (foetal) erythrocyte	9·3*	—	7·6	Widdas (1951)
Barnacle muscle	2·6	—	7·5	Bunch & Edwards (1969)
Human erythrocyte	—	33	6·0	Vieira, Sha'afi, & Solomon (1970)
Squid nerve	—	4·0	3–5	Nevis (1958)
Dog erythrocyte	—	56	4·9	Vieira *et al.* (1970)
Cow erythrocyte	160	—	4·0	Farmer & Macey (1970)
Dog erythrocyte	200	—	3·7	Vieira *et al.* (1970)
Human erythrocyte	120	—	3·3	Vieira *et al.* (1970)

* Approximate estimate (W. F. Widdas, *personal communication*).

in cells as opposed to that in bulk water. As we have noted elsewhere, E_a for self-diffusion of water is about 4·5 kcal mole^{-1}. This energy, for example, is equivalent to that required to rupture a hydrogen bond between water molecules. In some cells, however, the energy of activation is considerably larger than that for self-diffusion presumably because the membranes are not highly porous. Parallel studies on artificial membranes show quite convincingly that the structure of the membrane is intimately related to the activation energy for water transport. This point is exemplified clearly in the study (Gary-Bobo & Solomon, 1971) of cellulose acetate membranes (see page 128).

It seems that the structure of cell membranes must be a central factor in any description of the activation energy for water transport, although it is also possible that the rate of water diffusion in the cytoplasm could be very temperature dependent (e.g. Dick, 1966). According to Stein (1967) the activation energy for water permeation across lipid membranes should be considerably larger than 4·5 kcal mole^{-1} since he considers that the rate-limiting step in transfer is the disruption of a certain number of hydrogen-bonds by which the water molecule is held in the water lattice. Stein considers that this number is about 3 or 4 and, therefore, one might expect that E_a should lie between 13·5 and 18 kcal mole^{-1}. In this connexion it is interesting to recall that Price & Thompson (1969) have found that E_a lies in the range 12·7–13·1 kcal mole^{-1} for permeation through a homogeneous lipid barrier. One might expect, therefore, that cell membranes should exhibit similar large values for E_a provided, of course, that the rate of water transport is controlled by the membrane and not by, say, diffusion in unstirred layers where E_a would attain values of about 4·5 kcal mole^{-1}. Although Price & Thompson (1969) concluded that the high values of E_a for lipid membranes occur because water must first dissolve in the lipid and then diffuse across it, they were unable to exclude the alternative possibility that the large values for E_a were due to water transport through narrow aqueous pores in the membrane. The latter alternative seems to be more appealing to most workers probably because other experimental evidence apparently suggests that cell membranes are porous, but it should be stressed that water diffusion across a lipid membrane without pores will also be associated with a large activation energy. Most investigators, however, seem to have favoured the notion that cell membranes contain 'quasi-crystalline water' probably sited within narrow pores. Hempling (1960), for example, has suggested that the high energy of activation and entropy of activation for water transfer in ascites tumour cells (see Table 5.6) may indicate that the structural order of water in narrow membrane pores is higher than that of bulk water. It is a difficult job even to speculate about the nature of water in such narrow pores let alone to calculate what E_a might be for water transport through them. Price & Thompson (1969), for example, have used the plausible argument that the upper limit for E_a is 13–16 kcal mole^{-1} on the basis that the activation energies for self-diffusion of water molecules in ice have values in that range (Dengel & Riehl, 1963; Itagaki, 1964).

If water permeation across cell membranes proceeds through narrow pores containing highly ordered water then the diameter of the pores may have a crucial effect on the size of the activation energy. Vieira, Sha'afi & Solomon (1970) have examined that possibility by measuring E_a for both P_d and L_p in dog and human erythrocytes (see Table 5.6). These cells were chosen deliberately because the equivalent pore radius for the dog erythrocyte is about 6 Å according to Villegas et al. (1958) and that for the human erythrocyte is about 4·4 Å (Paganelli & Solomon, 1957; Sidel & Solomon, 1957). For the moment we shall accept these particular values for the equivalent pore radius (but see Chapter 6) so that we can retrace the views of Vieira et al. (1970) about the different activation energies in these cells.

The E_a values for osmotic flow in both cells are close to that for the viscosity of water. Does this mean that the osmotic flows of water across their membranes are genuinely viscous in character? If that were so, then maybe Poiseuille's law describes the volume flow and we can write the hydraulic conductivity as $n\pi r^4/8\eta_w \Delta x$ (cf. equation 3.30). Since perhaps η_w is the only parameter in this expression for L_p that will be influenced by temperature, we might expect that $L_p\eta_w$ will be independent of temperature. In other words, if we find that $L_p\eta_w$ for a given membrane is invariant over a certain temperature range then we can conclude that osmotic flow is achieved by bulk flow through pores rather than by diffusion. In fact, Vieira et al. (1970) observed that $L_p\eta_w$ was practically independent of temperature for both dog and human erythrocytes. Therefore, osmotic flow through both sets of different equivalent pores in these cells is indistinguishable from viscous flow of water in bulk. This seems a somewhat surprising conclusion, especially when one recalls how much emphasis is placed by some workers on the 'quasi-crystalline water' in such narrow pores. Of course, some studies of L_p in artificial membranes also have demonstrated that bulk water flow occurs through quite narrow pores. For instance, Madras, McIntosh & Mason (1949) have noted that $L_p\eta_w$ is independent of temperature in their experiments on Cellophane membranes with pore radii of about 15 Å.

Let us now turn to the activation energies for water diffusion across the erythrocytes to see what they can tell us about water transport through such narrow pores.

In the dog erythrocyte E_a for water diffusion is 4·9 kcal mole^{-1} while that for human erythrocytes is larger, i.e. 6·0 kcal mole^{-1}. Since the former activation energy is close to that for self-diffusion it seems that water molecules diffuse through the relatively large pores in that membrane just as they do in water itself. Again this suggests, at least in the case of the dog erythrocyte, that such small pores do not contain 'quasi-crystalline water'. On the other hand, the activation energy for water diffusion across the human red cell membrane is higher than that for self-diffusion and these authors ascribed this disparity to possible interactions between water molecules and the membrane. However, it is hard to understand why such interactions do not also influence E_a for osmotic water flow across this membrane. Indeed, the activation energy for water diffusion in the human erythrocyte is a little difficult to reconcile with the low value of E_a for osmotic flow. Probably part of this difficulty is caused by the tacit assumption that water transport occurs solely through systems of small pores about which we know very little. Even the question of their existence is not entirely free from controversy.

At present we seem to have two ways of accounting for the large activation energies encountered in studies of water transport. Either we can accept that membranes contain some form of 'highly organised' water possibly forming an aqueous route through them or that water molecules cross them by first dissolving in the lipid phase and subsequently diffusing across that phase. It may well be that both mechanisms operate side-by-side in the same membrane.

We know from experiments on the effects of temperature that it exerts a strong influence on the kinetics of water exchange and flow in some cells. Moreover, it has wider consequences on, for example, the metabolism and the water content of cells. In the next chapter we shall discuss the power of cells to withstand different temperatures, particularly sub-zero ones.

6
WATER RELATIONS OF CELLS

Osmotic relations of cells	193
Erythrocytes	196
Egg cells	199
Skeletal muscle	202
Smooth muscle	207
Axons	209
Intracellular aspects of water relations	212
Organelles	212
Pinocytosis	218
Contractile vacuole	219
Cytoplasmic streaming	221
Interrelations of solute and water transport	223
Osmosis in the presence of permeant solute	223
Osmotic behaviour—index of solute permeability	230
Osmometric method for determining solute permeability	232
Does the cell membrane rectify volume flow or is its hydraulic conductivity dependent upon osmolarity?	236
Electro-osmosis	242
Action potentials and water flow	248
Reflexion coefficients	252
Water relations of cells at low temperatures	257
'Minimum cell volume' hypothesis	259

IN this chapter we shall concentrate on some selected aspects of the water relations of cells. Attention will be focused first on the view that cells and their organelles can be treated as osmometers. Then we will go on to consider how this osmotic behaviour breaks down when permeant solutes are present, and this leads us on naturally to a discussion of coupling between solute and water transport in cells. The latter coupling may show itself, for example, as a volume flow accompanying action potentials in axons or as anomalous osmosis in erythrocytes. The final facet of water relations, which is on our agenda, is the power of cells to withstand cold temperatures.

Osmotic relations of cells

In a sense we have already dealt with some aspects of the osmotic relations in the preceding chapter where it was indicated that the osmotic swelling or shrinkage curves for cells can be described mathematically. The basis of this theoretical approach is to assume that the cell membrane can be characterized by a certain hydraulic conductivity and then for the particular cellular geometry one can predict how cellular volume should change with time during an osmotic experiment. For instance, this approach has been used successfully for the erythrocyte (Sidel & Solomon, 1957), the sea urchin egg (Lucké & McCutcheon, 1932) and the axons of the squid and cuttlefish (Hill, 1950b). This sort of description of osmotic swelling and shrinkage of cells is based on the relation (Lucké & McCutcheon, 1932) between cellular volume and osmotic pressure, namely

$$\pi(v-b) = \pi^o(v^o-b) \qquad 6.1$$

where v^o is the volume of the cell placed in an isotonic Ringer solution with an osmotic pressure π^o and v is the corresponding volume of the cell in equilibrium with a solution of osmotic pressure π. The parameter b is the 'non-solvent' volume or the 'dead-space' of the cell. This relation which is a modification of the Boyle–van't Hoff law (i.e. πv = constant) has been quite useful for describing osmotic phenomena in cells and, of course, it is one of the basic assumptions of the kinetic description of osmotic swelling and shrinkage curves. Equation 6.1 can be re-written to give the volume v at any osmotic pressure π in the form.

$$v = \frac{\pi^o}{\pi}(v^o-b)+b \qquad 6.2$$

Thus, if the volume is measured after the cell has come into equilibrium with a number of solutions of different osmotic pressure then v may be plotted against π^{-1}. According to the modified Boyle–van't Hoff relation (6.2) there should be a linear relation between v and π^{-1} and the intercept when π^{-1} equals zero should yield b; b can also be obtained from the slope $\pi^o(v^o-b)$ since both π^o and v^o are known. The value of b indicates that part of the cellular volume which does not participate in the osmotic swelling or shrinkage. For that reason it has been customary to compare b with the non-aqueous volume of the cell which is given by the difference between v^o and the volume of water, v_w^o, in the cell under isotonic conditions. One might expect that b should be identical to $(v^o - v_w^o)$; however, that is not invariably the case and we shall discuss the disparity later.

It is as well to remember that the foregoing view of the equilibrium conditions of cells in solutions of different osmotic pressure rests on certain assumptions.

It is considered, for example, that under equilibrium conditions the internal osmotic pressure of the cell is identical to the external osmotic pressure. Robinson (1953, 1954) challenged that assumption and he proposed that water was being continually excreted from the cells by an active mechanism. Robinson's arguments rested on the observations that the depression of freezing point for certain tissue homogenates exceeded the serum values and, further, that cellular swelling occurred when the metabolism of certain mammalian tissues was depressed by either metabolic inhibitors, anoxia or cooling. Conway, Geoghegan & McCormack (1955) showed, however, that the apparent hypertonicity of the cytoplasm was due to the rapid breakdown of substances, such as glycogen; this work agrees with earlier findings by Conway & McCormack (1953) that tissue homogenates, which were rapidly cooled, were, in fact, isotonic and not hypertonic. In the latter experiments the tissues were quickly frozen with liquid oxygen to prevent autolysis. Moreover, it was also shown that the cellular swelling following impaired metabolism was caused by the entry of an isotonic fluid containing principally sodium chloride (Conway & Geoghegan, 1955; Leaf, 1956). Further work (Buckley, Conway & Ryan, 1958; Applebloom, Brodsky, Tuttle & Diamond, 1958; Maffly & Leaf, 1959) confirmed the conclusion that the intracellular fluid of certain cells was indeed isotonic to their extracellular fluids. Subsequently Robinson (1960, 1965) recanted his hypothesis that mammalian cells employ active water transport. It is interesting to note, however, that Sigler & Janáček (1969) have reopened the question of the intracellular osmolarity and they suggest that there may be a genuine difference between the intracellular and external osmolarities in the case of the frog oocyte. For example, they claim that the internal osmolarity of some oocytes exceeds the external osmolarity and that possibly this reflects certain properties of the cytoplasm which are akin to those of elastic gels. Since their explanation implies the existence of hydrostatic pressures it will be necessary to substantiate their argument by direct measurement of such pressures within the cytoplasmic phase. Apart from this claim, which admittedly needs experimental support, there is nothing else to deflect us from accepting that at equilibrium there is no difference in osmotic pressure across the membranes of animal cells. Of course, when we consider plant cells we see that such a gradient does exist but that it is counterbalanced by the hydrostatic pressure difference.

Another tacit assumption is that the cell membrane is truly semi-permeable and consequently no account is usually taken of the possible net flux of solute across the membrane during osmotic flow. In general, cells are much more permeable to water than to solutes (see Table 6.1). In the erythrocyte and the alga *Nitella*, the ratios of the permeabilities are quite large, even for small solutes, such as urea. Nevertheless, the fact that the ratios are large does not mean that we can ignore the effect of solute transport. Instead, what we need to know are the solute reflexion coefficients. Probably, it can be accepted quite safely that the reflexion coefficients of

TABLE 6.1. Comparison of water and solute permeabilities in the human erythrocyte and in the alga, *Nitella mucronata*

Permeant solute	Hydraulic conductivity $\left[\dfrac{\text{Solute permeability}}{\left(\dfrac{L_p RT}{\overline{V}_w P_s}\right)}\right]$	
	Erythrocyte*	*Nitella*†
Methanol	–	3·4
Urea	30[b]	8500
Acetamide	300–400[a]	168
Propionamide	620[a]	143
Methyl urea	1180[a]	3480
Glycerol	2450[b]	34,8000
Thiourea	10,200–28,000[a]	3090

* The value taken for the hydraulic conductivity was that of Sidel & Solomon (1957). The solute permeabilities were those of (a) Ørskov (1947) and (b) Jacobs *et al.* (1935)

† The water and solute permeabilities were measured by Collander (1949) and Collander (1954) respectively.

most osmotic solutes are close or equal to unity. For instance, Zadunaisky *et al.* (1963) found the σ for both mannitol and sucrose in skeletal muscle is 1·0.

Although we commonly refer to the cell membrane as being relatively freely permeable to water in comparison to solutes, it is interesting to note what is meant quantitatively by the term 'freely permeable'. Robinson (1965), for example, has noted that: 'The pressure of a column of water 33 ft high, acting on a square centimetre of a membrane with a permeability of 10^{-6} ml cm^{-2} sec^{-1} atm^{-1} should force 1 ml of water through the membrane in about 12 days. It is only because cells are so minute that they can reach equilibrium with their surroundings rapidly through membranes with this low order of permeability. Perhaps instead of speaking of cell membranes as "freely permeable to water" we should rather marvel that so thin a layer can be so nearly waterproof.'

It is now certain that the relative impermeability to solutes does not invariably hold in osmotic shrinkage and swelling experiments on some cells, and therefore the theoretical equations must be modified to take account of solute flow.

Finally we come to the most important assumption of all, namely that the cell membrane, and nothing else, rate-controls the osmotic water flow into or out of the cell. This question was discussed in the preceding chapter when we considered the measurement of L_p for cells and it was concluded that mutual diffusion of water and large molecules in the cytoplasm may also play a role in dictating the kinetics of osmotic swelling or shrinkage of cells. Nevertheless, if we confine our analysis to the steady-state or equilibrium conditions of osmotic experiments then that objection will not apply.

Erythrocytes

Figure 6.1 shows some results of osmotic experiments on human erythrocytes (Cook, 1967). In this graph there is linear relation between cellular volume and the reciprocal of the osmolarity of the bathing solution in accord with the modified Boyle–van't Hoff equation 6.2. Only about 55% of the cellular volume apparently responds to changes in osmotic pressure whereas the water content is about 75% of the cellular volume. In Cook's experiments, for example, it was found that v_w^o exceeded $(v^o - b)$ by about 17%

of the isotonic volume v^o. Ponder (1948) expressed the discrepancy between v_w^o and $(v^o - b)$ by means of a ratio 'R' given by

$$'R' = \frac{v^o - b}{v_w^o} \qquad 6.3$$

In erythrocytes and in other cells 'R' is invariably less than unity. The fact that 'R' does not equal unity has been used as an argument for non-solvent water within cells (see Chapter 1) simply on

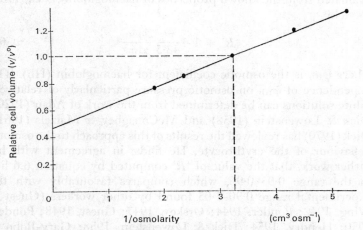

FIG. 6.1. The relative volume (v/v^o) of human erythrocytes as a function of the reciprocal of the osmotic pressure of the bathing medium. The interrupted line shows the relative volume under isotonic conditions (Cook, 1967: Fig. 1).

the grounds that a fraction of the cellular water content is not available for the osmotic response. There are, however, several other reasons for the low values of 'R'.

Dick (1959a, 1966) has argued that equations 6.1 and 6.2 are incorrect since they assume that the osmotic coefficient of the intracellular solute is independent of the concentration of the solution. According to Dick and others (e.g. Nobel, 1969) equation 6.1 should be re-written as

$$\pi(v - b) = \varphi R T n_s \qquad 6.4$$

where φ is the osmotic coefficient of the intracellular solute and n_s is the number of gram molecules of solute in the cell. Thus, it can be shown that Ponder's 'R' is given by

$$'R' = 1 - \frac{\pi^o}{\varphi^o} \cdot \frac{\Delta\varphi}{\Delta\pi} \qquad 6.5$$

where φ^o is the value of the osmotic coefficient at $\pi = \pi^o$ (see Dick, 1966). Since cells contain many proteins and their osmotic coefficients increase significantly with concentration, equation 6.5 predicts that 'R' < 1. For the erythrocyte, in particular, 'R' can be evaluated from the known properties of haemoglobin (Dick, 1966) to yield

$$'R' = 1 - \frac{\pi^o}{43\cdot 3} \frac{\Delta\varphi_{Hb}}{\Delta\pi} \qquad 6.6$$

where φ_{Hb} is the osmotic coefficient for haemoglobin (Hb). The dependence of φ_{Hb} on osmotic pressure particularly of relatively dilute solutions can be determined from the work of Adair (1929), Dick & Lowenstein (1958) and McConaghey & Maizels (1961). Dick (1970) has reviewed the results of this approach to the osmotic behaviour of the erythrocyte. He finds, in agreement with his earlier work, that the value of 'R' computed by equation 6.6 lies in the range 0·95–0·96, which compares favourably with the experimental range 0·90–1·05 found by other workers (Guest & Wing, 1942; Ponder, 1944; Ørskov, 1947; Guest, 1948; Ponder, 1950; Hendry, 1954; Dick & Lowenstein, 1958; Gary-Bobo & Solomon, 1968).

The conclusion from Dick's analysis is that erythrocytes placed in relatively dilute solutions behave as 'perfect osmometers' when account is taken of the expected variations in φ_{Hb}. Unfortunately this argument cannot be extended successfully to the case where erythrocytes are placed in hypertonic solutions, since the behaviour of φ_{Hb} under these circumstances is difficult to assess. For the cellular shrinkage experiments Dick obtained theoretical estimates of 'R' lying in the range 0·87–0·92 whereas the experimental values of 'R' fall in the range 0·67–1·02 (Ørskov, 1947; Ponder & Baretto, 1957; Olmstead, 1960; White & Rolf, 1962; Le Fevre, 1964; Savitz, Sidel & Solomon, 1964; Cook, 1967, Gary-Bobo & Solomon, 1968). The agreement between theory and experiment is quite good in view of the dubieties over the osmotic coefficient of haemoglobin in concentrated solutions.

Thus, when account is taken of the variations in the osmotic coefficient of haemoglobin the behaviour of the erythrocyte in both hypotonic and hypertonic solutions resembles that of a perfect osmometer governed by the modified Boyle–van't Hoff relation (Lucké & McCutcheon, 1932). Unfortunately Dick's analysis involving the osmotic coefficient of the intracellular proteins cannot be applied readily to other cells because extensive data on the osmotic coefficients of their intracellular proteins are not at hand.

Egg cells

A good deal of work has been done on the water relations of the large spherical cells of invertebrates, amphibia and fish.

Of the invertebrate eggs that have been examined, probably the work on the sea urchin, *Arbacia punctulata*, is the best known. Lillie (1916) studied osmotic phenomena in fertilized and unfertilized sea urchin eggs and the analysis was extended by Lucké (e.g. Lucké, Hartline & McCutcheon, 1931; McCutcheon & Lucké, 1932). There have also been similar osmotic studies of eggs of the annelid, *Chaetopterus pergamentaceus* (Lucké, Hartline & Ricca, 1939) and of the molluscs, *Cumingia tellenoides* (Lucké, Hartline & Ricca, 1939) and *Ostrea virginica* (Lucké & Ricca, 1941). In particular, McCutcheon, Lucké & Hartline (1931) and Lucké, Larrabee & Hartline (1935) tested the compatibility of the osmotic behaviour of the unfertilized eggs of *Arbacia* with the modified Boyle–van't Hoff relation (6.1). They found that the theoretical relation matched the experimental data provided that the osmotic 'dead-space' b was about 12% of the isotonic cell volume. Other experiments by Lucké & Ricca (1941) confirmed that the egg cells of *Ostrea*, *Cumingia* and *Chaetopterus* also obeyed the theory although the values of b were much larger than that for *Arbacia*. The mean value of b for *Ostrea*, for example, was about 45% of the isotonic volume.

Whereas the eggs of those invertebrates cited above do seem to behave as 'perfect osmometers' containing an osmotically inactive volume, the eggs of certain frogs (De Luque & Hunter, 1959; Hunter & De Luque, 1959; Løvtrup, 1960; Berntsson, Haglund & Løvtrup, 1964, 1965; Merriam, 1966) exhibit more complex osmotic behaviour. Why that is so remains unclear. No doubt a large contribution to the complexity stems from the use of solutions containing permeant salts. In some recent work Sigler & Janáček (1969) have tried to exclude that difficulty by adding an impermeant solute—lactose—to their Ringer solutions. They studied the effect of hypertonicity on the size of the oocytes of both *Rana temporaria* and *Rana esculenta* and found that their water contents did not obey the relation 6.1. The experimental curves were not linear and, in fact, they resembled the corresponding curves that were obtained for different gels manufactured from agar and gelatine. Sigler and Janáček also pointed out that the non-linear osmotic behaviour could not be due to a constant osmotically inactive volume in the oocytes. They noted, moreover, that the intracellular osmolarity of the oocytes of *R. esculenta* exceeded the external osmolarity by about 100 m-osm under the equilibrium conditions of their osmotic experiments. This is a remarkable finding. Even more remarkable is their observation that conversely the internal osmolarity of oocytes of *R. temporaria* is less than that of the bathing solutions in similar osmotic conditions. Other workers (Holtfreter, 1943; Løvtrup, 1960; Berntsson et al., 1965) have suggested that amphibian embryos, like plant cells, are under hydrostatic pressure and that the surface coats of the oocytes behave like plant cell walls and constrain high (positive) internal pressures. However, the data of Sigler & Janáček (1969) exclude such an analogy because 'negative' hydrostatic pressures must be postulated for the interior of oocytes of *R. temporaria*. On the contrary, they suggested that the osmotic behaviour of the oocytes reflected the properties of elastic gels presumably associated with the cytoplasm. In principle, it should be possible to substantiate their model by determining the hydrostatic pressure of the cytoplasmic phase under different osmotic conditions. Methods have been developed by plant physiologists to ascertain the turgor pressure of plant cells (for example, Green & Stanton, 1967) and probably such methods could be applied to this problem in the oocyte. Clearly we need this additional evidence before we can accept such wide latitudes of internal hydrostatic pressure in these egg cells. Nevertheless, the views of Sigler & Janáček (1969) also make one wonder how safe it is to ignore the small hydrostatic pressure gradients which exist normally across the surfaces of other cells. In comparison to experimental gradients of osmotic pressure such hydrostatic pressure differences look really negligible. For instance, Cole (1932) found that the internal pressure of *Arbacia* eggs exceeded that outside by 2·6 mm H_2O and Rand & Burton (1964) observed a corresponding value of 2·3 mm H_2O for the erythrocyte. Expressed in the more conventional units of osmotic pressure these pressures are about 2×10^{-4} atm, roughly 10,000 times smaller than the pressure gradient that has been postulated in the frog oocyte (Sigler & Janáček, 1969). In view of the finding (Vargas, 1968a) that the hydraulic conductivity determined in filtration experiments on the squid axon is at least 100 times larger than that obtained from osmotic experiments (see page 166), perhaps we should at least keep an open mind on the potential importance of such apparently inconsequential pressures and not simply dismiss them by comparing them with osmotic gradients. After all we have no right to expect that equal hydrostatic and osmotic pressure gradients will invariably exert equal influences on water transport in cells.

Of the fish eggs that have been studied perhaps the trout egg is the classic example. Before these eggs are shed they are in osmotic equilibrium with the maternal blood. Each egg is enveloped by a membrane called the chorion and when it is shed into fresh water it swells over the first hour or so due to water uptake. The increase in volume is due to the formation of a perivitelline fluid (Bogucki, 1930) between the chorion and a cytoplasmic membrane referred to as the vitelline membrane (Gray, 1932). Thus, the egg cell becomes suspended in the perivitelline fluid and during this period the entire egg becomes extremely impermeable to water. Eggs in this state are said to be 'water-hardened'. Gray (1932) has shown that this change in water permeability does not occur in the chorion since it remains permeable to water even in the 'water-hardened' state of the egg, and he concluded that the properties of the vitelline membrane are responsible for the low permeability. This view of the decline in the water permeability was confirmed by Krogh & Ussing (1937) using deuterated water to study the exchange of water between the egg and its surroundings. In fact, Krogh and Ussing concluded that the diffusional permeability of the egg was too low to be measured accurately and this ties in with the osmotic experiments of Gray (1932) who found the 'water-hardened' egg did not alter in volume even when it was placed in salt solutions with an osmotic pressure eight times that of Ringer solution. Thus, the trout egg becomes effectively isolated, in an osmotic sense, shortly after being shed into fresh water and it remains so for more than 12 days when it again becomes permeable to water at an advanced stage of embryonic development. Clearly the inception and maintenance of the 'water-hardened' state is a distinct advantage to trout eggs, as it is to the eggs of other freshwater fish too, because it protects them against osmotic swelling and potential rupture.

Can we say anything quantitatively about the water permeability of the 'water-hardened' egg? Certainly Gray's osmotic experiments cannot help us in this respect but more recent work on labelled-water exchange in the trout egg does yield a quantitative estimate of its diffusional permeability. For instance, Kalman (1959) has monitored the exchange of tritiated water between the egg yolk and the external medium and Potts & Rudy (1969) have taken this further to estimate P_d. These workers found that the P_d for eggs shortly after shedding was 6×10^{-6} cm sec^{-1}; during this initial period it was found that P_d occasionally rose to about 2×10^{-5} cm sec^{-1}. Those values represent the limits of the water permeability of the eggs in their relatively permeable state. Even so, trout eggs must be considered to be relatively impermeable to water when compared to erythrocytes or other cells (see Table 5.1). When the eggs had become 'water-hardened' it was found that P_d dropped to values less than or equal to 4×10^{-7} cm sec^{-1}. Assuming that one can equate the diffusional permeability and the hydraulic conductivity one can estimate that L_p for the 'water-hardened' egg is about 3×10^{-10} cm sec^{-1} atm^{-1} (cf. values in Table 5.2). In practical terms, therefore, we would expect that a huge osmotic gradient, say 100 atm, would generate a volume flow of only 3×10^{-8} cm^3 cm^{-2} sec^{-1} or $0 \cdot 1$ μl. cm^{-2} hr^{-1}. By biological standards the trout egg is truly watertight!

Skeletal muscle

The osmotic responses of muscle cells have been examined by numerous workers (e.g. Loeb, 1897; Overton, 1902; Hill, 1930; Boyle & Conway, 1941; Shaw, 1958; Dydyńska & Wilkie, 1963; Reuben, Lopez, Brandt & Grundfest, 1963; Reuben, Girardier & Grundfest, 1964; Blinks, 1965; Gainer, 1968; Lang & Gainer, 1969). In particular, interest has been focused on the question of whether the volume of muscle fibres is governed by the modified Boyle–van't Hoff relation. Even in the early studies (e.g. Overton, 1902) it was suggested that a certain fraction of cellular water was 'bound' and hence incapable of taking part in osmotic responses. Initially, it appeared that muscle cells behaved as good osmometers but a closer look at their osmotic responses revealed deviations from ideal behaviour both in hypertonic and hypotonic conditions.

For instance, Dydyńska & Wilkie (1963) observed that the sartorius muscle of the frog evidently did not shrink in very hypertonic solutions to the extent predicted by theory (equation 6.2). In addition to estimating the cellular volume by measuring the width of the cells these authors also determined the fibre water by subtracting the extracellular water (sucrose space) from the total water content in the muscle. These measurements showed that, in contrast to the cellular volume, the fibre water did alter in volume as if it were confined in a perfect osmometer. Consequently Dydyńska and Wilkie suggested that part of the sucrose space might be within the muscle fibres themselves.

On the other hand, Reuben et al. (1963) found that isolated single fibres of the frog semitendinosus muscle apparently did not swell to the extent predicted by theory when they were placed in dilute Ringer solutions. The same result was noted by Grieve (1963) who estimated the fibre water content of frog sartorius muscle. Reuben et al. (1963) suggested that the non-linear relation between cellular volume and osmotic pressure was due to the fact that volume changes were also accompanied by movements of solute; but Grieve (1963) suggested that muscles placed in hypotonic solutions developed a hydrostatic pressure that limited swelling.

The question of such discrepancies from ideal osmotic behaviour was re-examined by Blinks (1965) who estimated the volume of isolated fibres of the anterior tibial muscle of the frog. In order to estimate volume, Blinks measured both the length of an individual fibre and its cross-sectional areas at regular intervals along the fibre's length. It turns out that this comprehensive estimate of cellular volume is much more reliable than the corresponding value derived from fibre diameter alone. The results of Blink's osmotic experiments are shown in Fig. 6.2, where v/v^o has been plotted against π^{-1}. Separate regression lines have been drawn for both hypotonic and hypertonic conditions and there is no significant difference between their slopes. Thus, the osmotic behaviour of this muscle is virtually ideal, and according to Blinks the inaccuracies (stemming from measurements of fibre diameters) in other studies were probably large enough to account for the non-linear osmotic behaviour reported, for example, by Dydyńska & Wilkie (1963) and by Reuben et al. (1963).

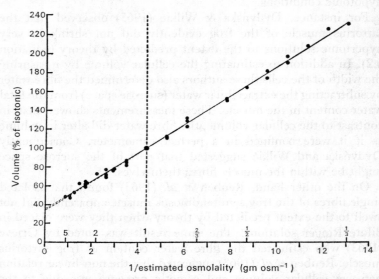

FIG. 6.2. The influence of osmotic pressure on the volume of isolated single muscle fibres. *Rana temporaria*, anterior tibial muscle. Results of nine experiments. All volumes expressed as percentages of volume of same fibre in isotonic solution. The abscissa represents the reciprocal of the external osmotic pressure: figures above the axis indicate relative tonicity and the vertical interrupted line identifies isotonic solution. Separate regression lines for points in hypotonic and hypertonic solutions, both constrained to go through the point of 100% volume in isotonic solution (Blinks, 1965: Fig. 4).

The intercept on the ordinate axis in Fig. 6.2 gives b as about 33% of the (isotonic) cellular volume and this means that the osmotic volume of the cell is apparently about 67% of its full volume. On the other hand, Blinks argued that the aqueous fraction of cellular volume was about 80%. This discrepancy can, of course, be expressed as an appropriate value of 'R' which can be interpreted in various ways. For example, the difference between these fractions is 13% and it could be attributed to fibre water that it is inaccessible to solute and thus not free to participate in the osmotic response of the fibre. Alternatively, the apparent disparity may be due to the fact that the osmotic coefficients of certain solutes in the cytoplasm increase with increasing osmotic pressure just as haemoglobin does in the erythrocyte (see Dick, 1966). However, there is a more plausible explanation for the disparity and it relates to the way in which the actual measurements were made.

It will be recalled that Dydyńska & Wilkie (1963) found that the muscle fibre water behaved as if it were confined in an ideal osmometer. However, when these data are compared with those of Blinks for cellular volume there is some disagreement, as noted by Blinks. In particular, the slopes of the linear relations are different and so also are the intercepts (Fig. 6.3). According to Dydyńska and Wilkie the osmotic dead space is about 20% of the cellular volume,

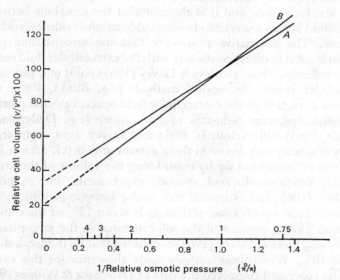

FIG. 6.3. The relationship between the volume of muscle cells and the osmotic pressure of the medium. The ordinate represents the volume expressed as a percentage of its value under isotonic conditions, i.e. $(v/v^o) \times 100$, while the abscissa represents the reciprocal of the relative osmotic pressure, i.e. (π^0/π): above the horizontal axis are shown the corresponding values of (π/π^0). Line A, from Blinks (1965), is the line of best fit to the observed changes measured by an optical technique. Line B, from Dydyńska & Wilkie (1963), describes the behaviour of a perfect osmometer possessing a non-osmotic component equal to 20% of its cell volume at π^0. The discrepancy between the intercepts on the ordinate axis is discussed in the text (Birks & Davey, 1969: Fig. 2).

while Blinks found b equal to 33%. Both sets of workers were aware of the fact that the sucrose space measurements from which the extracellular fluid space is obtained might be misleading in the sense that part of the sucrose space might lie within the muscle cell. Birks & Davey (1969) have considered the possibility that the sarcoplasmic reticulum behaves like an extracellular fluid compartment. Electron-microscopic studies on muscle have shown where the extracellular space proper makes contact with the sarcoplasmic reticulum. In particular, there are invaginations of the cell membrane which form a system of transverse tubules (e.g. Andersson-Cedergren, 1959; Frazini-Armstrong & Porter, 1964) open to the extracellular space. Although the transverse tubules are distinct from the sarcoplasmic reticulum they do make contact with the reticulum, and it is thought that the junctions between these fluid spaces are relatively permeable to small solutes, such as sucrose. The alternative possibility that the sarcoplasmic reticulum itself is in direct continuity with the extracellular fluid seems quite unlikely. Thus, as Birks & Davey (1969) point out, estimates of cellular volume by optical methods (e.g. Blinks, 1965) will include a fraction of the extracellular fluid space, i.e. sarcoplasmic reticulum, whereas estimates of fibre water (e.g. Dydyńska & Wilkie, 1963) will exclude it. Birks and Davey went on to argue that the discrepancy between these measurements (cf. data in Fig. 6.3) can be accounted for by considering the volume of the sarcoplasmic reticulum in such osmotic experiments. For example, Peachey (1965) has estimated that under isotonic conditions the volume of the sarcoplasmic reticulum is about 13% of the cellular volume. Taking account of the solid material in the sarcoplasmic reticulum, Birks & Davey (1969) concluded that its fluid volume is about 10%. When these authors made allowance for this extracellular space and compared the data of Dydyńska & Wilkie (1963) and Blinks (1965) they found that these results were mutually compatible provided that the sarcoplasmic reticulum swells in hypertonic conditions. The postulated swelling could certainly be due to the entry of the osmotic solute (sucrose) and concomitant water movement into this compartment from the outside, and indeed, these workers showed with electron microscopy that it did swell in hypertonic solutions in a manner predicted by the physiological results.

It might be contended that the transverse tubules are an alternative site for the extracellular fluid compartment that swells under hypertonic conditions. Apart from the controversial question about whether or not transverse tubular swelling actually occurs in these osmotic experiments (see Birks & Davey, 1972) the main argument against its possible significance is that the transverse tubules occupy a very small fraction of the cellular volume. For instance, Huxley (1964) estimated that the transverse tubular system in frog sartorius muscle fibres makes up less than 0·5% of the fibre's volume, whereas Birks & Davey (1972) put this fraction at only 0·2%.

Apart from these osmotic studies described above, it is interesting to note that other experiments suggested that a fraction of the muscle fibre's volume is extracellular. For instance, there is a discrepancy between the measured intercellular ion concentrations and those predicted by Donnan equilibrium. A number of workers (e.g. Simon, Shaw, Bennett & Muller, 1957) have stated that the disparity arises because there is an extracellular fluid compartment within the apparent confines of the cell, and Harris (1963) calculated on the basis of Donnan equilibrium that such a compartment must occupy about 15% of the total water content of the muscle fibre. This value is in good agreement with that derived from osmotic studies and strengthens the view that such an extracellular compartment exists in muscle.

We see, therefore, that skeletal muscle apparently behaves like a perfect osmometer after account has been taken of the complexity of its fluid compartmentation. Indeed, it appears that as muscle fibres are shrinking in hypertonic media their extracellular fluid compartments are swelling due to the entry of solutes and water.

Smooth muscle

A number of osmotic studies have been performed on smooth muscle cells (Meigs, 1912; Meigs & Ryan, 1912; Bozler, 1959, 1961a,b 1962, 1965; Brading & Setekleiv, 1968). Meigs concluded from a comparative osmotic study of frog skeletal and smooth (stomach) muscle that the former behaved as an osmometer with a semipermeable membrane whereas the distribution of water within the latter seemed to be controlled by properties of the cytoplasm rather than the cell membrane. Using frog stomach muscle Bozler came to a similar conclusion. On the other hand the relatively recent work of Brading and Setekleiv demonstrated that the osmotic responses of the smooth muscle of guinea-pig taenia coli, placed in hypertonic solutions, are governed by the modified Boyle–van't Hoff relation. These workers estimated cellular volume by subtracting the extracellular space (sorbitol space) from the wet weight of the tissue and they found that 'R' was unity although their estimate rested upon the assumption that the entire dry weight of the tissue originates solely in the cells.

Although most of the data, which have been obtained in osmotic experiments on smooth muscle cells of amphibians and mammals, can be explained on the basis that the smooth muscle cell is an osmometer (see Brading, 1970), certain experimental results have been considered to deviate from membrane theory (Bozler, 1961a,b). In relatively more recent work Bozler (1965) determined the changes in volume of the fibre water of frog stomach, sartorius and cardiac muscles placed in different NaCl Ringer solutions. The volume of the fibre water was estimated indirectly either by subtracting the size of the extracellular space from the wet weight or from the alteration in the concentration of radioactive dextran in the bathing solution. Bozler considered that the data on striated muscle were consistent with the behaviour of an osmometer provided allowance was made for the loss of electrolyte in hypotonic conditions. In contrast he found that smooth and cardiac muscles in hypotonic Ringers did not swell to a degree predicted by the 'osmotic laws' (i.e. equations 6.1 and 6.2) although in hypertonic Ringers they did shrink in accordance with such theoretical relations. He suggested that the swelling of smooth and cardiac muscle cells was limited as in a gel and, further, that the interiors of those cells were under hydrostatic pressure. This view is somewhat similar to those expressed by Ling (1962) and Ernst (1963) for skeletal muscle and by Sigler & Janáček (1969) for frog oocytes. The small degree of swelling observed in hypotonic solutions, however, might be caused by considerable solute leakage from the fibres; Bozler estimated the leakage of potassium ions from these types of muscle and he found that it was relatively small in sartorius but significantly larger in stomach and cardiac muscle. He concluded, however, that the leakage of potassium from the fibres was insufficiently large to account for their poor enlargement in hypotonic conditions. Of course, in addition to the loss of potassium there might also be loss of other solutes, such as amino acids, and this possibility has not been completely elimiated by Bozler's experiments. In fact, other independent lines of evidence are required before one can accept that the swelling of smooth muscle is 'limited by a gel structure'. In view of the errors that can arise in estimations of cellular volume it seems safer to doubt that the swelling of these muscles in dilute media is actually limited by a postulated gel structure. Certainly it is imperative to re-examine osmotic phenomenon in smooth and cardiac muscle with the particular aims of obtaining accurate time courses for swelling and shrinkage and also an adequate picture of solute transport in those muscle cells during osmotic experiments.

Axons

Apparently very little work indeed has been done on the osmotic behaviour of nerve axons. Hill (1950b) studied the kinetics of osmotic swelling and shrinkage of large axons of the squid and cuttlefish and observed that they could be described by suitable relations derived from the modified Boyle–van't Hoff relation provided that b lay in the range 20–30% of isotonic cellular volume. Freeman, Reuben, Brandt & Grundfest (1966) published a report of some osmotic experiments performed on squid and lobster giant axons and also Shapiro (1966) has described the osmotic properties of the frog sciatic nerve. The latter investigation employed the entire nerve trunk and its weight was measured with a torsion balance. Since it is difficult to interpret Shapiro's data quantitatively in terms of the properties of individual nerve fibres it will not be discussed here except to note that the osmotic responses of the nerve trunk apparently obeyed the modified Boyle–van't Hoff relation.

Freeman *et al.* (1966) estimated the changes in fibre volume from enlarged photomicrographs; they measured the total diameter of the fibres and also the diameter of the axon itself, although it was somewhat difficult to establish optically the boundary between the axon and its connective tissue sheath. The overall diameter of the lobster axons did not shrink significantly in hypertonic salt solutions whereas the axon itself did shrink. In contrast the diameter of the entire fibre increased in diluted media and during this swelling the connective tissue sheath decreased in thickness. The data from these swelling and shrinkage experiments are shown in Fig. 6.4; in this figure part **a** shows the quantitative aspects of the relative changes in volume of the axon proper and of the

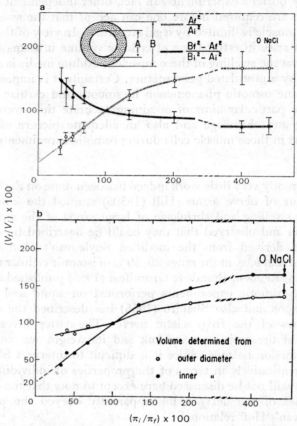

FIG. 6.4. The relationship between the volume of the lobster axon and the external osmotic pressure. The ordinate represents the relative volume, i.e. v_f/v_i, and the abscissa shows the reciprocal of the relative osmotic pressure, i.e. π_i/π_f. Both ratios, are converted to percentages. In this case v_f and v_i are identical to v and v^0, referred to in the text (see equation 6.2); the same identity exists between π_f and π_i and π and π^0 respectively. In **a** the osmotic relations of the axon (○—○) and of the connective tissue compartment (●—●) are both shown: the bars indicate the S.E. of the mean values. The inset shows the method of calculating the two volumes. In **b** the osmotic relations of the axon (●—●) are again shown but in this case for comparison with those of the whole fibre (axon plus connective sheath) (Freeman et al., 1966: Figs. 4 & 5).

connective sheath, whereas part **b** shows the relations between cellular volume and osmotic pressure for the axon itself and for the entire fibre. In Fig. 6.4a Freeman and his co-workers plotted the relative volume change of the axon itself, i.e. (A_f^2/A_i^2), and of the connective tissue sheath, i.e. $(B_f^2 - A_f^2)/(B_i^2 - A_i^2)$, against the relative osmotic pressure (π_i/π_f) where the subscripts i and f denote the initial (isotonic) conditions and the final (equilibrium) condition in the new osmotic pressure and B and A signify the diameters of the entire fibre and the axon respectively. These data demonstrate that in hypertonic media the sheath expands while the axon itself shrinks whereas in hypotonic media the sheath shrinks as the axon expands. In Fig. 6.4b the osmotic pressure-volume relations for the entire fibre and for the axon alone are shown and, of course, the difference between these curves is attributed to the swelling and shrinkage of the sheath. For the lobster axon proper it appears that the osmotic 'dead space' is about 20% (while for the squid axon it is 40% of the cellular volume). Figure 6.4b also shows that the lobster axons did not swell to the predicted extent in hypotonic solutions and it is not clear why this is so. Perhaps there is considerable solute leakage under those conditions. Although exposure of lobster axons to hypertonic media did not interfere with the generation and conduction of action potentials, an irreversible loss of excitability occurred in axons exposed to dilute (25%) Ringer solutions. Since these axons are unable to withstand such hypotonic media without impaired function we could conclude that departure from ideal osmotic behaviour under such conditions is possibly to be expected and does not signify a genuine deviation from the 'osmotic laws'.

The osmotic behaviour of squid axons was similar to that of lobster axons. Again, for example, it was found that the axon and the connective sheath behaved differently in response to an osmotic challenge. In both the squid and lobster axons Freeman *et al.* (1966) noted that the sheath did not constitute a significant barrier to water movement since the alterations in the axonal volume occurred rapidly and were apparently independent of the swelling or shrinkage of the sheath. These data confirm the view of Villegas & Villegas (1960) that the sheath does not represent a serious barrier to the transport of substances from the external medium up to the axolemma.

Intracellular aspects of water relations

In dealing with the water relations of cells we cannot simply assume that the cell itself is our basic unit, for it contains a nucleus and organelles all of which will exchange water with the cytoplasm. In some instances it may turn out that the kinetics of water transport in such subcellular compartments influence the transport rate between the cell and its surroundings. For example, Sorenson (1971) observed that the efflux of labelled water from isolated single muscle fibres could be described by a model involving two compartments in series and one of these compartments is presumably a subcellular one.

In addition to the transport properties of organelles there are other facets of cellular function which have a bearing on water relations. In particular, pinocytosis, cytoplasmic streaming and the formation of contractile vacuoles all could play a part in rate-controlling not only water transport within cells but also transport between the cell and its environment.

For the sake of continuity we shall discuss the osmotic relations of organelles first and then move on to potentially important transport phenomena like pinocytosis.

Organelles

Nucleus. The earliest report of osmotic studies on the nucleus is apparently that of Hamburger (1906) who observed the changes in the volume of both epithelial cells and their nuclei. According to that account the nuclei behaved as osmometers and, furthermore, so also did isolated nuclei. Later work (Beck & Shapiro, 1936, Kamada, 1936; Churney, 1952) on the eggs of certain invertebrates confirmed Hamburger's conclusions about the osmotic behaviour of nuclei. These investigations suggested that the nuclear membrane is a semipermeable barrier (although perhaps not in the same way as the cell membrane is) and in hindsight this seems a surprising conclusion particularly since electron micrographs of the nuclear membrane reveal evidently the presence of large pores of about 1000 Å in diameter which occupy about 20–30% of the surface area. Watson (1959) coined the term *pore complex* to describe these apparent perforations of the membrane and we shall return to the question of their structure after we have discussed the behaviour of the nuclear membrane.

It is now accepted that the nuclear membrane is permeable to water and salts but not to macromolecules. Some earlier work Anderson, 1953; Holtfreter, 1954; Merriam, 1959) on isolated nuclei seemed to be compatible with the view that their membranes were permeable to large molecules. However, MacGregor (1962) has discounted those claims on the grounds that these studies were not designed to limit nuclear swelling and that such swelling, by itself, increases the permeability to large molecules, presumably due to a disruption of the membrane's structure. MacGregor's views are in line with the evidence that injection of large molecules, such as albumin, into the cytoplasm of oocytes could produce nuclear shrinkage (Harding & Feldherr, 1959); in fact, the osmotic pressure–volume relation for the nuclei in that kind of experiment is linear. On the other hand, when Harding and Feldherr injected hypertonic salt solutions, nuclear swelling occurred probably due to the relatively rapid entry of salt and water into the nucleus from the cytoplasm. Thus, the nuclear membrane is not truly semi-permeable but may be regarded as such if there is an osmotic gradient arising solely from large molecules. This conclusion is supported further by the results of osmotic experiments on isolated nuclei of amphibian oocytes (Battin, 1959; Hunter & Hunter, 1961). Moreover, Feldherr (1964) confirmed the relative impermeability of the nuclear membrane to macromolecules by noting the failure of colloidal gold particles to enter the nucleus from the cytoplasm. However, there are exceptions to the rule that the nuclear membrane is impermeable to macromolecules. For example, Davies (1961) demonstrated the rapid transport of haemoglobin from the nucleus to the cytoplasm of frog erythrocytes and Feldherr (1969) has shown that colloidal gold enters the *Amoeba* nucleus.

How do the observations that the nuclear membrane is permeable to ions and water but not to, say, some proteins fit in with the presence of large pores in the membrane? Electron microscopical studies of the 'pore complex' indicate that it simply cannot be regarded as a perforation. Indeed, its structure seems quite complicated. Apparently its circumference is octagonal and not circular (Gall, 1964) and it has a central plug of material (Dawson, Hossack & Wyburn, 1955). Between the central plug and the periphery there is possibly a diaphragm (Afzelius, 1955) or an array of struts (Abelson & Smith, 1970). It is not difficult to imagine, therefore, that the 'pore complex' may greatly hinder the passage of macromolecules. However, it is impossible to conclude what hindrance this structure might offer to water and ions. It is interesting to note that Agutter (1972) has shown that the structure of the 'pore complex' is destroyed by high ionic strengths (e.g. 0·25 M NaCl) and, hence, the evidence that the membrane is permeable to salts may be doubted on the grounds that the experimental conditions disrupted the 'pore complex'.

In summary, it seems that the nucleus behaves like an osmometer provided that the osmotic pressure changes are achieved by altering the concentrations of macromolecules in the cytoplasm. However, when faced with a hypertonic solution of an electrolyte or non-electrolyte the nucleus swells and the nature of this response is poorly understood. Indeed, the study of the transport properties of the nucleus, in general, is in its infancy.

Mitochondria. Several workers (Dianzani, 1953; Raaflaub, 1953; Tedeschi & Harris, 1955; Bentzel & Solomon, 1967; Packer, Wrigglesworth, Fortes & Pressman, 1968; Harris & van Dam, 1968; Stoner & Sirak, 1969) have studied the osmotic properties of mitochondria placed in different solutions of electrolytes, such as NaCl, or non-electrolytes, such as sucrose. In general, the osmotic behaviour of the mitochondrion is governed by the Boyle–van't Hoff relation with the values of b lying in the range 0·40–0·85. For instance Tedeschi & Harris (1955) found $b = 0·75$ whereas the volume of solids is only about 30–40% of the mitochondrial volume. In fact, all the published work after 1955 indicates that b lies in the range 0·75–0·85. In accounting for this large osmotically inactive volume Bentzel & Solomon (1967) proposed that there are two fluid compartments in the mitochondrion. The first occupies about 70% of the total water in the mitochondrion and it is not bound apparently by a semipermeable membrane since its volume does not respond to changes in external osmotic pressure and, moreover, sucrose enters it relatively easily. The second compartment occupying 30% or so of the mitochondrial fluid is inaccessible to sucrose and does behave as an osmometer. Its 'R' value is only 0·49. Consequently Bentzel and Solomon concluded that about 50% of this small osmotic compartment is made up of non-solvent water but, just as for the corresponding data in cells, the value of 'R' may be attributed alternatively either to solute movements or to large osmotic coefficients for the macromolecules in such a compartment. There is no clear correlation between the size of the functional fluid compartments, which Bentzel and Solomon have proposed, and the anatomical compartments within the mitochondrion.

The osmotic behaviour of the mitochondrion is largely what we would expect from an organelle possessing a typical unit-membrane. Other factors apart from osmotic pressure influence the mitochondrial volume and water content. These changes in volume have been called non-osmotic and they fall into two classes (Lehninger, 1964).

First, we have the small amplitude changes referred to as Phase I. In this instance the volume of the mitochondrion is altered by about 20–40% due to changes in the rate of respiration. The latter is closely coupled to the production of ATP and we find, for example, that the addition of inorganic phosphate in the absence of ADP generates mitochondrial swelling which can be reversed when respiration is stimulated by the addition of ADP. This kind of change in volume is both reversible and rapid. It is not clear how these changes in volume are brought about. Perhaps, for example, the alterations in respiration rate produce temperature gradients which move water by thermo-osmosis. Apart from that highly speculative hypothesis it has also been suggested that ATP helps to maintain the normal volume possibly by inducing contraction of the mitochondrial membrane (Raaflaub, 1953).

Secondly, there are large amplitude changes or Phase II. In this type of non-osmotic volume alteration the mitochondrion may swell to about double its normal volume and this type of swelling probably results from the accumulation of solute and accompanying amounts of water (e.g. Chappell & Crofts, 1965). The swelling occurs over a longer period (10–15 min) than Phase I and it can be induced by a number of substances including calcium ions, glutathione and hormones such as vasopressin and insulin. The mechanism of this kind of non-osmotic swelling, as well as that of the Phase I volume change, is ill defined, but certainly it depends on metabolism because it is abolished, for example, by 2,4-dinitrophenol.

It is interesting to note that similar, but not identical, sorts of non-osmotic volume changes occur in chloroplasts and these are equally difficult to understand.

Chloroplasts. Like the mitochondrion, the chloroplast is an organelle bounded by a unit-membrane. The similarity does not stop there for the chloroplast, as well as the mitochondrion, undergoes changes in its volume which are both osmotic and non-osmotic in origin. For example, Itoh, Izawa & Shibata (1963) and Packer (1963) have shown that chloroplasts experience volume changes in response to light. On the other hand, Nishida (1963) and Tolberg & Macey (1965) have demonstrated that volume changes can also be induced by osmotic gradients.

Considering first the osmotic volume changes, one sees that we can crudely visualize the chloroplast as a semipermeable sac which swells or shrinks as a result of osmotic water flow. However, its osmotic behaviour is not entirely ideal since Mercer, Hodge, Hope & McLean (1955) found that the osmotic swelling of chloroplasts from *Nitella* and spinach was not reversible. Provided the chloroplasts are incubated in the cold, however, osmotic swelling is actually reversible (Tolberg & Macey, 1965). Moreover, Tolberg and Macey also found that the size of the osmotically active compartment in the chloroplast is larger for chloroplasts incubated in NaCl solutions than for those in sucrose solutions, probably because of salt entry into some fraction of the chloroplast volume. In this connexion it is interesting to note that Gross & Packer (1967) have observed that the osmotic behaviour of the isolated grana from chloroplasts is also governed by the modified Boyle-van't Hoff relation and that it is independent of the nature of the external osmotic solute. Thus, in the chloroplast some part of its volume, other than that occupied by the grana, is accessible to NaCl but not sucrose. Obviously the chloroplast cannot be described as a perfect osmometer and account must be taken of solute movements into and out of it during osmotic experiments.

Let us now turn to the volume changes which are apparently non-osmotic in origin. It is known that light generates two kinds of volume changes in chloroplasts. On the one hand, light may produce relatively slow swelling which is irreversible; on the other, it may cause a rapid shrinkage which is fully reversible in the dark. In fact, we can draw an analogy between these volume changes and the Phase II and I changes in mitochondria. The light-induced swelling of chloroplasts may double their volume which is comparable in size with Phase II changes in mitochondria. Moreover, it is not dependent on ATP formation since Packer, Siegenthaler

& Nobel (1965) have shown that its time course is unaffected by the conditions for ATP synthesis or hydrolysis or, for that matter, by the presence of uncoupling agents. The mechanism of the high-amplitude swelling remains obscure, although it has been established that it requires co-factors, such as phenazine methosulphate, which promote electron flow (Packer *et al.*, 1965). In contrast, a great deal more seems to be known about the light-induced shrinkage of chloroplasts. The mechanism of shrinkage is somehow coupled to energy-dependent reactions in the chloroplast. In particular, uncouplers of photophosphrylation, such as ammonium chloride, abolish the shrinkage response to light (see Packer, 1966). Moreover, the maximum shrinkage is obtained when both light and ATP are available to furnish energy. This sort of evidence along with many other lines of corroborative support has suggested that the shrinkage mechanism requires energy. Again, however, the mechanism of this non-osmotic volume change still eludes us. Several workers have argued by analogy with mitochondria that chloroplasts have active mechanisms for ion uptake and, furthermore, that these are stimulated by light. Some evidence for this proposal has been gathered (e.g. Nobel & Packer, 1965) although it is certainly a difficult task to establish conclusively without knowing the membrane potential that active transport of ions does occur in small organelles like chloroplasts. There does seem to be a strong correlation between the conditions that favour maximal accumulation of ions, such as calcium, sodium and phosphate, and those for maximal shrinkage. However, the presence of such a correlation does not immediately indicate a possible mechanism for the light-induced shrinkage. In comparison the Phase I swelling in mitochondria is also correlated with stimulated ionic accumulation but the correlation here immediately leads to the attractive view that an osmotic gradient is established by solute uptake which inevitably draws water into the organelle (Rasmussen, Chance & Ogata, 1965).

Packer (1966) has drawn an interesting parallel between the mitochondrion and the chloroplast. He makes the point that in both organelles the high-amplitude volume changes do not require energy and are manifested as irreversible swelling. In contrast, changes in volume that require energy differ in these organelles. In the mitochondrion swelling occurs whereas in the chloroplast there is shrinkage when energy is supplied. Furthermore, he draws the intriguing, but perhaps unwarranted, conclusion that such changes in volume concur with the physiological demands of the chloroplast and the mitochondrion, especially since photosynthesis uses water molecules while respiration forms them.

Pinocytosis

Small vesicles with diameters of about 600 Å have been observed in a wide variety of cells including amoeba, leucocytes, capillary endothelium and certain epithelial cells. Apparently each vesicle is formed by the engulfment of a small droplet of the external fluid as the cell membrane invaginates. Lewis (1931) coined the term *pinocytosis* to describe this process.

In the early studies of pinocytosis in the freshwater amoeba, *Amoeba proteus*, it was concluded that pinocytosis was a concomitant of cell shrinkage (Mast & Doyle, 1934), and consequently that it probably fulfilled an important role in regulating cellular volume. Of course, the main objection to that view is that pinocytosis can be induced, for example, in the amoeba after it has undergone enlargement (Chapman-Andresen & Dick, 1961). Several workers, notably Schumaker (1958), Brandt & Pappas (1960) and Chapman-Andresen (1962), have studied the environmental conditions which give rise to pinocytosis in amoeba. It has been noted that the presence of uncharged molecules in the bathing solution do not induce pinocytosis whereas certain electrolytes may do so. Of the charged molecules which generate pinocytosis, certain cationic polyelectrolytes and smaller electrolytes, such as methylene blue, become adsorbed on to the cell surface. In contrast, it has been found that some amino acids and inorganic salts, which do not adsorb significantly on to the surface, also evoke pinocytosis. It seems, therefore, that certain solutes can stimulate the cell membrane to invaginate and form vesicles, but the mechanism remains obscure. Since most of these experimental solutions were hypertonic, the induction of pinocytosis might seem, at first glance, to result from the reduction in cell volume. To settle this point Chapman-Andresen & Dick (1961) studied the relation between pinocytosis and alterations in cell volume of the amoeba *Chaos chaos*. As we have mentioned earlier, these authors made the important observation that in certain NaCl solutions pinocytosis occurred after cellular volume had increased. Even although pinocytosis probably does not play a major role in regulating the volume of these cells, there is some experimental evidence (Brandt, 1958; Schumaker, 1958; Brandt & Pappas, 1960; Chapman-Andresen & Holter, 1964; Gosselin, 1967) that it is an important mechanism of solute transport, especially in the absorption of large molecules. We shall return to the important question of whether pinocytosis plays a significant part in the transport of water and solutes across capillary walls and across epithelia in Chapters 9 and 10.

Contractile vacuole

In certain protozoa, algae and sponges we find an intriguing mechanism of fluid transport. These organisms contain fluid-filled structures, the so-called contractile vacuoles, and the expulsion of water *via* the contractile vacuole constitutes an important osmo-regulatory mechanism. In the freshwater protozoa, for instance, the internal osmotic pressure exceeds that of the environment and this has been substantiated by measurements of electrical conductivity (Gelfan, 1928), vapour pressure (Picken, 1936) and freezing-point depression (Schmidt-Nielsen & Schrauger, 1963). Kitching (1934, 1936), for example, confirmed the earlier data by estimating the external osmotic pressure which would just cause the cells to shrink. In 1938 Kitching published a review of his experiments which clearly demonstrated that the contractile vacuole excreted just enough water to balance the osmotic influx. Moreover, his experiments demonstrated that the rate of fluid output from the vacuole was reduced considerably when the external pressure was raised, and that the organism swelled when metabolic poisons, such as cyanide, reduced the vacuolar excretion of fluid. All of these data point conclusively to the fact that the contractile vacuole is an excretory 'organ'. Kitching (1952) has suggested that the extrusion of the vacuolar contents to the exterior could be due to a relatively small difference in hydrostatic pressure, while a secretory process is probably responsible for the formation and growth of the vacuole. Later Kitching (1954) modified his earlier views on the emptying of the vacuole to the exterior to include the possible contractile properties of the vacuolar wall. He cited evidence of Mast (1938) that the wall surrounding the vacuole in *Amoeba proteus* is relatively thick (0·5 μm) and of Schmidt (1939) that the birefringence of this structure disappears during expulsion of fluid. From these observations Kitching concluded that the vacuolar wall contained, in addition to a cell membrane, a structural layer of protein capable of contraction (Schmidt, 1939). The contractile vacuole in *Paramaecium* has been examined in the electron microscope by Schneider (1960) and this revealed the presence of a system of fibrils associated with the vacuole. From the similarity between their appearance and that of the cilia on the surface of *Paramaecium* it could be concluded that they have a contractile function.

In this earlier work there was only indirect evidence for the view that fluid excretion by the contractile vacuole helped to maintain a relatively high osmotic pressure inside the cell. In fact, until the osmolarity of the vacuolar fluid was determined it was not possible to assess fully its osmoregulatory rule. Relatively recent experiments by Schmidt-Nielsen & Schrauger (1963) and Riddick (1968) have shown that the osmotic pressure of vacuolar fluid is less than half that of the cytoplasm. Now we know, therefore, that the vacuole is continually bailing out a dilute fluid and thus helping to maintain the relatively high osmotic pressure of the cytoplasm. Although the expulsion of a hypotonic vacuolar fluid confirms the osmoregulatory function of the contractile vacuole it does raise some questions about the elaboration of this fluid compartment. In particular, what is the driving force that moves water into the vacuole against an osmotic pressure gradient? For instance, is the vacuolar fluid elaborated by active water secretion or by a passive process such as filtration? Let us consider whether filtration could be responsible. From the rate of growth of the vacuole in the giant amoeba *Chaos chaos* (Riddick, 1968) one can calculate that the volume flux is about 10^{-6} cm^3 cm^{-2} sec^{-1} (see page 425). Let us assume that the vacuolar membrane has a hydraulic conductivity equal to that of the most permeable cell membrane—the erythrocyte. Then $L_p = 2 \cdot 5 \times 10^{-5}$ cm sec^{-1} atm^{-1} and the required difference in hydrostatic pressure is 4×10^{-2} atm or about 40 cm H$_2$O. Such a large pressure gradient surely does not exist inside the amoeba between the cytoplasm and the vacuole. Of course, that argument does not completely exclude filtration since it might be argued that the permeability of the vacuolar membrane is considerably larger than that of the erythrocyte. If filtration is the driving force for fluid transport into the vacuole, then several puzzling questions remain. What is the osmolarity of this filtrate? If it is isotonic to the cytoplasm, as one might expect, there must be an opposing reabsorption of a hypertonic fluid so that the vacuolar fluid remains hypotonic to the cytoplasm. This 'answer' seems to raise even more puzzling questions about the vacuolar membrane and its ability to transport water and solutes. It seems more likely to me that filtration plays no part in the elaboration of the vacuolar fluid and that it is powered possibly by two opposing active systems for moving ions and water. Our ignorance of the way that the vacuolar membrane transports ions and water, however, precludes us from drawing a convincing picture of the formation of the vacuolar fluid. Nevertheless, we can make a tentative guess about the mechanism on the basis of what is known about the ionic composition of the cytoplasm and the vacuolar fluid. Since this intriguing problem is so closely tied up with the general question of coupling between active salt and water transport I would prefer to discuss it under that heading (see page 422) in Chapter 10.

Cytoplasmic streaming

In our previous discussions about self-diffusion of water inside cells we tacitly assumed that there was no convective motion in the cytoplasm. This was also taken for granted in dealing with osmotic water transport where the mutual diffusion of water and macromolecules might exert a rate-limiting influence (Dick, 1966). The cytoplasm, however, is not stationary. Indeed, cytoplasmic streaming has been observed in a wide variety of cells including protozoa, nerve axons and most plant cells. The range of velocities is quite large. For example, in the slime mould *Physarum polycephelum* cytoplasmic streaming may achieve rates as high as 1350 μm sec^{-1} whereas in the alga, *Nitella*, the velocity of flow is about 50 μm sec^{-1} (see MacRobbie, 1971). Most of the evidence about axoplasmic flow in nerves indicates that the velocity is quite low, i.e. about 0·01 μm sec^{-1} (e.g. Taylor & Weiss, 1965). These values have been obtained mainly by monitoring the transport of labelled proteins along the axon. It is interesting that some authors (e.g. Karlsson & Sjöstrand, 1968; Fernandez, Burton & Samson, 1971) have found both slow ($\sim$ 0·01 μm sec^{-1}) and fast (0·1–1·0 μm sec^{-1}) transport rates for proteins along axons. The maximal rate of axoplasmic flow seems to be about 6 μm sec^{-1} (e.g. Lasek, 1968).

In principle, the movement of the cytoplasm in cells, even if it were proceeding at low velocities of 0·01 μm sec^{-1}, might still exert profound effects on the kinetics of solute and water transport. Although cytoplasmic streaming is too slow to reduce the internal unstirred layer it could affect the rate of water exchange in the cytoplasm, particularly if the exchange has to occur over relatively large distances. It might be contended too that the bulk flow of cytoplasm would effectively reduce the rate-controlling effect of mutual diffusion of water molecules and macromolecules during osmotic experiments. For cytoplasmic streaming to be effective in this connexion the diffusion or osmotic flow would need to be driven over a relatively large distance. We can get some insight into the relative efficiencies of diffusion and convection as modes of transport by the following argument. Consider a water molecule diffusing over a distance x cm. The time, t sec, required for diffusion is given by $x^2 = 2D_w t$. On the other hand a different time t' would be required for convective transport on the assumption that we can ignore diffusional flow; t' is given by x/v where v is speed of bulk flow. Over a certain pathlength l the speed of convection will equal that for diffusion and l will be given by $l = 2D_w/v$. What significance does this have for water transport in cells? Consider, first, self-diffusion of water molecules in the cytoplasm and let us assume that the speed of cytoplasmic streaming is 1 μm sec^{-1}. Then, if D_w is, say, 10^{-5} cm^2 sec^{-1} l will equal 0·1 cm or 1000 μm. This means that cytoplasmic streaming could only improve the rate of water exchange in cells if the pathlength is larger than 1000 μm. Thus, we may conclude that cytoplasmic streaming probably will have no effect on water exchange experiments on cells. On the other hand, if we accept Dick's (1966) argument about osmotic experiments, then the appropriate mutual diffusion coefficient for water is perhaps 10^{-8} cm^2 sec^{-1} or less. In this instance the pathlength for equality between diffusion and convection is about 1 μm. Thus, it is conceivable that the mixing of the cytoplasm by its bulk transport could reduce the rate-limiting effect of mutual diffusion, especially in large cells. Unfortunately we cannot really refine these rather crude arguments about the significance of cytoplasmic streaming at present and we must await some experimental evidence on this question. Irrespective of whether or not cytoplasmic flow is important in the water relations of cells it certainly performs a useful physiological role in transporting proteins and other substances in some cells, notably nerves.

Interrelations of solute and water transport
Osmosis in the presence of permeant solute

In the previous chapter we discussed the results of osmotic experiments on single cells and it was assumed that the osmotic gradients were established by impermeant solutes. It is intuitively obvious that the results of such experiments will be altered if, instead of impermeant solutes, permeant ones are used to create osmotic gradients. Several workers have tried to analyse that situation. Jacobs (1952), for example, tried to derive the appropriate theory for osmosis in the presence of permeating solute and he described the rate of change of cellular volume by

$$\frac{dv}{dt} = P_w A \left[\frac{a_i + a_s}{v} - c_s^o - c_i^o \right] \qquad 6.7$$

where v is the volume of water in the cell at any time t, A is the surface area of the cell, P_w is the water permeability given by $L_p RT$, a_i and a_s are the amounts of impermeant solutes, i, and permeant solute, s, in the cell at time t, and c_i^o and c_s^o are the concentrations (assumed constant) of i and s in the external solution. Johnson & Wilson (1967) have modified Jacob's equation to include the reflexion coefficient σ_s for the permeant solute. Their equation is

$$\frac{dv}{dt} = L_p RTA \left[\frac{a_i + a_s \sigma_s}{v} - \sigma_s c_s^o - c_i^o \right] \qquad 6.8$$

In order to solve this equation one needs to know how a_s varies with time. Of course, a_s can be expressed as vc_s^i, where c_s^i is the internal concentration of s. In his original treatment of this problem, Jacobs described the transport of s by

$$\frac{d(vc_s^i)}{dt} = P_s A [c_s^o - c_s^i] \qquad 6.9$$

where P_s is the permeability of the cell membrane for s. Again Johnson and Wilson modified this equation for solute flux to account for the reflexion coefficient of s to yield

$$\frac{d(vc_s^i)}{dt} = P_s A [c_s^o - c_s^i] + (1 - \sigma_s) \bar{c}_s \frac{dv}{dt} \qquad 6.10$$

(cf. equation 2.46) where $\bar{c}_s$ is the mean concentration of s across the membrane. With the aid of equations 6.8 and 6.10 these authors obtained an expression for v as a function of time which could be used to describe the kinetics of volume change during an osmotic experiment. For example, consider a cell that is suddenly transferred from a solution of an impermeant solute to a solution with a higher concentration of a permeant solute. In such an experiment the volume of the cell will decrease initially due to the efflux of water and subsequently its volume will increase towards a new value owing to solute and water entry. Johnson and Wilson showed that the kinetics of these changes in cellular volume could be described adequately by their theoretical treatment provided suitable values of L_p, P_s and σ_s have been chosen. In other words, such a description of the cellular volume change offers an indirect method of obtaining the practical coefficients (L_p, P_s and σ_s) for the cell membrane. The values of these coefficients can be checked against independent estimates to see if a self-consistent picture of solute and water transport across the cell membrane has been obtained. For instance, consider the results of applying this analysis to the osmotic experiments of Stewart & Jacobs (1932) on the sea urchin egg. Fig. 6.5 shows the experimental data of Stewart & Jacobs (1932) for the volume changes of fertilized (F) and unfertilized (U) sea urchin eggs placed in a sea-water medium containing 0·5 M ethylene glycol. The results are expressed in terms of the relative cell volume which is the volume at any time divided by the initial volume. Both sets of data show that there is a transient shrinkage followed by a relatively slow swelling presumably due to the entry of ethylene glycol and water. Of course, the time courses of these osmotic responses are different because of the different permeability characteristics of the fertilized and unfertilized cells. For example, Lucké (1940) reported that the fertilized eggs are about twice as permeable to water as the unfertilized ones and this explains why the initial shrinkage due to osmotic water loss is more rapid in the former than in the latter (Fig. 6.5). Moreover, Stewart & Jacobs (1932) claimed that fertilized cells are almost three times as permeable to ethylene glycol as the unfertilized ones and this explains why the secondary swelling is faster in the fertilized cells. Of course, these arguments are only qualitative and also they ignore the reflexion coefficient of the cells for ethylene glycol. In contrast to this rather crude view of these osmotic transients the analysis of Johnson and Wilson gave a quantitative description for certain chosen values of L_p,

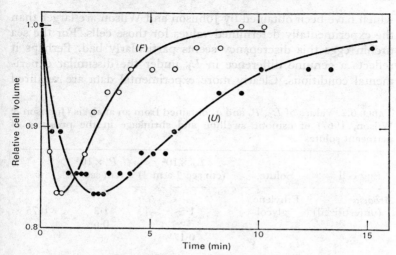

FIG. 6.5. Osmotic shrinkage of sea urchin eggs placed in a medium containing a permeant solute. The experimental data are those of Stewart & Jacobs (1932) for eggs, both unfertilized (U) and fertilized (F), immersed in sea water containing 0·5 M ethylene glycol at zero time. The theoretical curves describing the relative cell volume (v/v^0) as a function of time were derived on the basis that ethylene glycol permeates the egg cell membrane (Johnson & Wilson, 1967: Fig. 2).

P_s and σ_s. Their theoretical curves are also shown in Fig. 6.5 and the values of L_p, P_s and σ_s employed to give good agreement between theory and experiment are shown in Table 6.2. As can be seen from the table, these authors also used this approach to describe similar osmotic transients in other egg cells. In the case of the sea urchin the values for L_p and P_s for the fertilized eggs are larger than those for the unfertilized ones whereas the estimated reflexion coefficients are the same. The corresponding estimates of L_p, P_s and σ_s for other eggs are also shown in the table. The theoretical fit of the experimental data was good in all cases except for *Ostrea* eggs placed in diethylene glycol. In addition to these estimates of the membrane coefficients some independent values for L_p and P_s have been included in the table. In particular, the values for P_s in the sea urchin eggs are about twice those obtained by Stewart & Jacobs (1932) and this seems relatively good agreement when it is considered that the analysis of Stewart and Jacobs did not include the reflexion coefficient. All of the values for L_p

which have been obtained by Johnson and Wilson are larger than the experimentally determined values for those cells. For the sea urchin eggs this discrepancy seems particularly bad. Perhaps it reflects a genuine difference in L_p under the dissimilar experimental conditions. Clearly more experimental data are required

TABLE 6.2. Values of L_p, P_s and σ_s obtained from an analysis (Johnson & Wilson, 1967) of osmotic swelling and shrinkage in the presence of permeant solutes

Egg cell	Solute	$L_p \times 10^6$ (cm sec^{-1} atm^{-1})	$P_s \times 10^5$ (cm sec^{-1})	σ_s
Arbacia (unfertilized)	Ethylene glycol	1·5	1·3	0·75
		–	0·6[a]	–
		0·17[b]	–	–
Arbacia (fertilized)	Ethylene glycol	3·7	3·3	0·75
		–	1·6[a]	–
		0·34[b]	–	–
Ostrea	Diethylene glycol	6·8	1·0	0·8
		1·0[b]	–	–
	Glycerol	1·8	5·0	1·0
Chaetopterus	Ethylene glycol	1·0	2·5	0·8
		0·75[b]	–	–

References: (a) Stewart & Jacobs (1932): (b) Lucké (1940).

and it would be interesting to have independent estimates of the reflexion coefficients so that a more convincing picture of these transients could be drawn.

An experimental study of the interrelations between solute flow and osmotic water flow has been performed by Sigler & Janáček (1969, 1971) on frog oocytes. In the first of these papers they reported that the swelling and shrinkage of oocytes placed in different concentrations of lactose—a poorly permeating molecule—were dictated not only by the osmotic gradient but also by the hydrostatic pressure gradient across the cell membrane. They postulated that there was a difference in hydrostatic pressure between the cell's interior and the external medium and claimed that it was a consequence of gel properties of the oocyte cytoplasm. However, their evidence for the existence of a difference in hydrostatic pressure is indirect since they inferred its magnitude from the difference in the measured osmotic pressures of the cytoplasm and the external medium. If their contention is correct, then a modification of the preceding equations for the volume flow is required. In fact, Sigler & Janáček (1971) did modify the theory to include the effect of hydrostatic pressure on the osmotic transients of frog oocytes placed in different concentrations of mannitol. These experimental solutions were either hypotonic or hypertonic to the usual incubation medium. The upper half of Fig. 6.6. shows the time courses of the changes in cellular volume. When the typical oocyte was transferred to a hypotonic mannitol solution (○—○) its volume initially increased due to osmotic water entry and then decreased towards its original value due to a relatively slow passive efflux of KCl and water. The time course of solute efflux is shown in the lower half of Fig. 6.6. When the oocytes were transferred to different hypertonic solutions of mannitol, biphasic responses were again observed (●—●, □—□, Fig. 6.6) consisting of an initial shrinkage followed by a slow swelling of the cells. They considered that the slow swelling occurred because mannitol and water entered the oocytes from the concentrated external solutions; the lower half of Fig. 6.6. shows the time course of mannitol permeation in these osmotic experiments. Berntsson, Haglund & Løvtrup (1965) obtained similar experimental data to that shown in Fig. 6.6; these workers, however, attributed the phenomenon to the tension of the vitelline membrane bounding the oocyte. On the other hand, Sigler and Janáček favoured an explanation similar to that offered by Johnson & Wilson (1967) for such osmotic transients. In attempting to describe the time courses of the swelling and shrinkage of frog oocytes, Sigler and Janáček adopted essentially the treatment of Jacobs (1932) (see equations 6.7 and 6.9) and they assumed that the reflexion coefficient for mannitol was unity. Moreover, as we indicated earlier they also included a term expressing the difference in hydrostatic pressure that might exist in these experiments (see Sigler & Janáček, 1969). The value of this hydrostatic pressure difference across the membrane was assumed to be constant and taken as 1 atm. The agreement between their theoretical model and the experimental results in both hypotonic and hypertonic solutions is good. From their analysis they obtained estimates of both the hydraulic conductivity and the mannitol permeability of the oocyte membrane. In the hypertonic media the

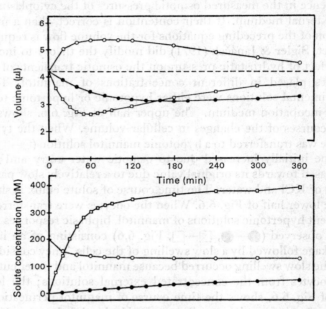

FIG. 6.6. The osmotic relations of frog oocytes placed in solutions containing permeant solutes. The upper part shows the time course of the change in cellular volume during swelling and shrinkage experiments. The lower part shows the theoretical estimates of the intracellular concentration of the permeant solute in these experiments. The separate experimental conditions are designated by the symbols; ○—○, hypotonic solution containing permeant KCl: ●—●, slightly hypertonic solution containing permeant mannitol: □—□, a strongly hypertonic solution containing mannitol (Sigler & Janáček, 1971: Fig. 4).

oocytes had a water permeability of about 8×10^{-7} cm sec^{-1} atm^{-1} while in the hypotonic medium it was about 32×10^{-7} cm sec^{-1} atm^{-1}. The permeability coefficient for mannitol also was dependent apparently on the osmolarity of the bathing solution; it increased from 10^{-6} to 10^{-5} cm sec^{-1} when the hypertonicity was increased. This interesting work on the oocyte poses several questions. For example, what gives rise to the changes in water and solute permeabilities in the different media? What is the reflexion coefficient for mannitol? Probably the assumption that it is close to unity is quite sound, for if we take $\sigma_s = 1 - \omega_s \bar{V}_s/L_p$ (cf. equation 3.16) and insert Sigler and Janáček's values for ω_s (i.e. P_s/RT) and L_p it turns out that σ_s virtually identical to 1. However, the reflexion coefficient for mannitol may include an additional term arising from interaction of mannitol and water in the membrane (see equation 3.20).

The cellular volume changes which are shown in Fig. 6.6 for the oocyte look superficially like readjustments designed to maintain cellular volume in the face of adverse osmotic gradients. It would be wrong, however, to conclude that the cell is regulating its volume and it seems more likely that this apparent regulation stems from passive entry or exit of solute. A similar explanation for apparent volume regulation has been given for crab muscle (Shaw, 1958; Lang & Gainer, 1969), fish erythrocytes (Fugelli, 1967) and duck erythrocytes (Kregenow, 1971a,b).

Osmotic transients have also been recorded in plant cells (e.g. Werth, 1961) placed in solutions of permeant solutes. Again one can describe the time course of the volume changes by taking into account the movement of the permeant solute into or out of the cell. For example, Philip (1958) has derived a theory of osmotic transients which is based on lines identical to the arguments described above.

Unfortunately Philip has assumed that the reflexion coefficients for each permeating solute is unity and, hence, his analysis is incorrect in principle. Dainty (1963a) has fully discussed Philip's theory and concluded that it is satisfactory provided the solutes penetrate rather slowly. In other words, it is possible to ignore the reflexion coefficient, as Sigler & Janáček (1971) did for the frog oocyte, if P_s is small enough to make $(1 - P_s \bar{V}_s/RTL_p)$ approach unity. When the reflexion coefficient is less than unity, as it would be for rapidly permeating solutes, the theory of Philip must be modified (see Dainty, 1963a).

Osmotic experiments performed in the presence of permeant solutes are characterized by transient changes in cellular volume. From the time course of the volume change we conclude that not only the water permeability but also the permeability and the reflexion coefficient for the solute are important in dictating the features of the osmotic response. Although these osmotic transients at first seem troublesome they have been used by some workers (e.g. Collander, 1950) to determine the permeability characteristics of cell membranes for certain solutes. This avenue of investigation has taken two forms which will be discussed more fully in the next section. First, some workers have used the osmotic transient as a way of deciding qualitatively whether or not a given solute is permeant. Secondly, the osmotic transient may be used to assess quantitatively the solute permeability coefficients as we have already discussed for the egg cells of the sea urchin (Johnson & Wilson, 1967) and frogs (Sigler & Janáček, 1971).

Osmotic behaviour—index of solute permeability

As an example of this approach to solute permeability let us consider the work of Freeman *et al.* (1966) who compared the osmotic behaviour of the lobster giant axon with that of the squid. One of the particular aims of their experiments was to establish whether or not the squid axon was relatively permeable to chloride ions. This interest stemmed from the discrepancy between independent conclusions from electrophysiological studies that chloride was relatively permeant in the axolemma (Hodgkin & Katz, 1949; Hodgkin, 1951) and the converse of this (Grundfest, Kao & Altamirano, 1954; Baker, Hodgkin & Meves, 1964).

Figure 6.7 shows the responses of lobster axons to exposure to hypertonic media. Whereas the lobster axon shrank in hypertonic NaCl solution, it was found that the hypertonic solution containing KCl produced an initial shrinkage followed by a swelling. Upon returning to the original medium the axon which had been in hypertonic NaCl swelled to its original volume, whereas the axon which had been in hypertonic KCl swelled initially and then its volume slowly decreased. Thus, lobster axons apparently behave as reliable osmometers towards NaCl solutions but they behave anomalously in the presence of KCl. The data for lobster axons placed in KCl solutions are similar to earlier observations on muscle fibres of the frog (Boyle & Conway, 1941; Reuben *et al.*, 1963) and crayfish (Reuben *et al.*, 1964). In Fig. 6.7, therefore, the initial shrinkage is due to an osmotic efflux of water and the subsequent swelling probably results from the entry of KCl and water into the axon. This view is compatible with the finding of Freeman *et al.* (1966) that the substitution of KCl for NaCl in the external medium produced cellular swelling even although the external osmotic pressure was unaltered. A similar substitution in the presence of an impermeant anion (propionate), however, produced no change in volume. These observations support the notion that lobster axons are permeable to KCl but not to NaCl. In contrast to these findings the squid axon did not swell when the external

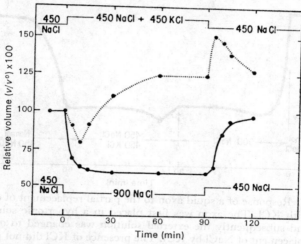

FIG. 6.7. Responses of a pair of lobster axons to solutions made hypertonic by addition of NaCl or KCl. The solid line refers to the experiment where NaCl was added and the dotted line to that with KCl addition. Cellular volume is expressed as a percentage of that under isotonic conditions, i.e. $(v/v^o) \times 100$ (Freeman et al., 1966: Fig. 7).

concentration of KCl was increased by a suitable replacement of a fraction of the external NaCl concentration (Fig. 6.8). Under those conditions some swelling should have occurred if the membrane were permeable to KCl. This indicates that the squid axon, in contrast to the lobster axon, is not permeable to KCl. Since the electrophysiological evidence (Curtis & Cole, 1942) offers convincing proof that the squid axon is permeable to potassium ions we are led to the conclusion that the axon's permeability to chloride ions must be quite low. Although this argument is only a qualitative one it is, nevertheless, a powerful one. Moreover, this kind of osmotic study illustrates a useful way of examining the solute permeability of cells. It also has had a significant impact on the study of solute permeabilities of plant cells (e.g. Collander, 1950) and within the last decade it has been used in conjunction with electrophysiological analysis to study the permeability characteristics of frog muscle (Reuben et al., 1963), crayfish muscle (Reuben

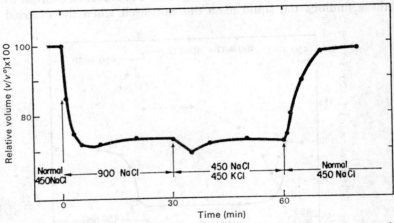

FIG. 6.8. Response of a squid axon to the partial replacement of external NaCl with KCl. The axon was first placed in a hypertonic solution of NaCl and subsequently the external solution was changed to one with 50% replacement of NaCl by KCl. The presence of KCl did not produce a pronounced swelling, thus indicating that KCl (as well as NaCl) is not a permeant salt. Cellular volume is expressed as a percentage of that under isotonic conditions (Freeman et al., 1966: Fig. 15).

et al., 1964) and lobster muscle (Gainer & Grundfest, 1968). Furthermore, it now turns out that this osmotic method for studying solute permeability can be developed further with the aid of the phenomenological description of transport processes to yield reliable estimates of solute permeability coefficients.

Osmometric method for determining solute permeability

Analyses of the relations between solute and water flows, which have been described in the preceding pages, have ignored several practical difficulties. For instance, it was assumed by Johnson & Wilson (1967) among others that the water permeability of the cell membrane was not dependent on the osmotic pressure of the medium. In fact, it seems that some cells swell faster under a given osmotic gradient than they shrink under an equivalent, but opposite, gradient; the origin of this phenomenon will be discussed later (see page 236). Therefore, the hydraulic conductivity of cells cannot be regarded as invariant during osmotic transients induced by permeant solutes. Sha'afi, Rich, Mikulecky & Solomon (1970) took account of this in their analysis of the volume changes occurring in erythrocytes placed in hypertonic solutions of permeant solutes.

Sha'afi *et al.* (1970) observed that the volume of erythrocytes exposed to a sudden increase in the concentration of a permeant solute, such as urea, decreased, and after passing through a minimum returned to its original value. In an earlier study Jacobs (1933) had reasoned that the initial time course of the volume change and the magnitude of the minimal volume reflected the permeability of the permeant solute; and this interpretation is basically similar to that of Sha'afi *et al.* (1970) who expressed the flows of solute and solvent in terms of the phenomenological coefficients. From their theoretical treatment these authors concluded that the solute permeability, P_s, for the permeant solute s is given by

$$P_s = \frac{v'_{\min} \left(\dfrac{d^2 v'}{dt^2}\right)_{\min}}{A^2 L_p \Delta \pi_i} \qquad 6.11$$

where the subscript min denotes the value of a given parameter at the time when the cellular volume, v, is at a minimum, i.e. when $\mathcal{J}_v = 0$; v' is the osmotically active volume, *i.e.* $(v-b)$; $\Delta \pi_i$ is the difference in osmotic pressure due to impermeant solutes across the red cell membrane and it was determined independently in practice from calibration curves of the relative cell volume against the external osmolarity. Figure 6.9 shows the result of an experiment where cellular volume was measured after exposure to a hypertonic solution of urea; from the theoretical fit (derived from an arbitrary sum of two exponentials) for the time course of the cellular volume change it was possible to obtain v' and the second derivative of v' at the minimum. Finally, corrections were applied for the apparent dependence of L_p on the osmolarity of the bathing medium (Rich, Sha'afi, Romualdez & Solomon, 1968) so that the appropriate value for the hydraulic conductivity was used in equation 6.11. This procedure for determining solute permeability from the minimum of the volume curve is referred to as the '*minimum method*' and, in this case, it was used to measure urea and formamide permeabilities. The values of P_s which these authors obtained for those solutes agreed very well with the corresponding values determined by tracer methods; for instance, the '*minimum method*' gave the urea permeability as $3 \cdot 5 \times 10^{-4}$ cm sec^{-1} in good agreement with $4 \cdot 05 \times 10^{-4}$ cm sec^{-1} obtained from flux

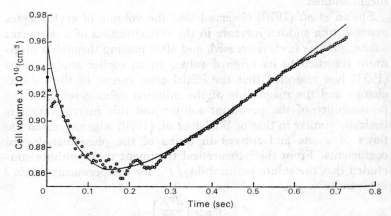

FIG. 6.9. Time course of the change in volume of human erythrocytes placed in a hypertonic solution containing urea (Sha'afi *et al.*, 1970: Fig. 3).

measurements. The '*minimum method*' of Sha'afi *et al.* (1970) is a rather striking example of the merit of the osmometric approach to estimating solute permeability; it is undeniably sound since it makes use of that part of the experimental curve where the net volume flow is zero, and this is in accord with the definition of the solute permeability coefficient (see page 51).

These workers took the analysis of the urea permeability of erythrocytes a stage further by estimating P_{urea} when $\mathcal{J}_v$ was non-zero. The dependence of P_{urea} on $\mathcal{J}_v$ is shown in Fig. 6.10 and unexpectedly it was found that urea permeability was larger when volume efflux occurred than when there was a volume flow into the cell. It might be contended that such a dependence of urea permeability on $\mathcal{J}_v$ results from disturbance of the solute concentration profile by the volume flow. Intuitively one might expect, for example, that volume inflow might enhance the concentration profile across the membrane and, hence, give rise to an over-

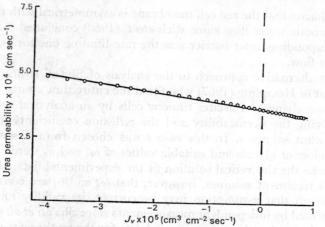

FIG. 6.10. Dependence of urea permeability on the volume flow across the membrane of human erythrocytes. The regression line for the experimental data is also shown (Sha'afi et al., 1970: Fig. 5).

estimate of P_{urea}. By the same reasoning one would expect an underestimate of P_{urea} to be obtained during volume efflux. Sha'afi et al. (1970) computed what effect such a proposal would exert on the apparent urea permeability and they found that its dependence on J_v would be quite weak. Moreover, according to their analysis and to intuition the slope of the relation ought to be positive not negative as observed in Fig. 6.10.

Sha'afi et al. (1970) concluded that the apparent dependence of P_{urea} on J_v does not stem from inadequacies in the phenomenological equations used in their treatment. On the other hand, they suggested that dependence of P_{urea} on volume flow indicates that the erythrocyte membrane does not behave as a homogeneous permeability barrier. Indeed, they found that their results were compatible with the existence of two separate permeability barriers—a relatively impermeable outer barrier in series with a permeable inner one—in the cell membrane. Kedem & Katchalsky (1963c) have described the behaviour of such a series-array of permeability barriers and Sha'afi et al. (1970) showed that their analysis accounted for the dependence of P_{urea} on J_v provided that the urea permeabilities of the outer and inner barriers were 5×10^{-4} and 15×10^{-4} cm sec^{-1} respectively. This is an extremely interesting conclusion and it seems to be consonant with an earlier conclusion that the red cell membrane is asymmetrical with regard to osmotic water flow since Rich *et al.* (1968) concluded that the corresponding outer barrier was the rate-limiting one for osmotic water flow.

An alternative approach to the analysis of Sha'afi *et al.* (1970) is that of Hempling (1967) who fitted the entire time course of the volume changes of ascites tumour cells by an analytical solution involving the permeability and the reflexion coefficients for the permeant solutes s. In this case s was chosen from a series of homologous glycols and suitable values of ω_s and σ_s were chosen to make the theoretical solution fit the experimental data. Hempling's treatment assumes, however, that ω_s and σ_s are constants. Although that assumption may be correct, its validity must be confirmed by independent measurements since Sha'afi *et al.* (1970) have shown that their experimental data for the erythrocyte can be fitted by a similar procedure and, in particular, that the required value of σ_s is grossly different from independent estimates of σ_s.

In conclusion it can be seen that the analysis of osmotic changes in cells has been used qualitatively to assess the permeability of plant and animal cell membranes and also quantitatively to estimate the transport coefficients (ω_s, σ_s and L_p) of these membranes. It is a useful way of approaching some aspects of cell permeability and no doubt it will be used in conjunction with other methods to explore further the properties of cell membranes.

Does the cell membrane rectify volume flow or is its hydraulic conductivity dependent upon osmolarity?

In both animal and plant cells an apparent dependence of L_p upon the direction of volume flow has been observed. Lucké & McCutcheon (1927), for example, observed that for a certain difference in osmotic pressure the swelling of cells placed in hypotonic solutions was much faster than the rate of shrinkage in hypertonic solutions. That is, it appeared that the osmotic permeability of the cell membrane was reduced when the osmolarity of the bathing medium was increased and *vice versa*. There are, however, other explanations for such a phenomenon and, in fact, the notable alternative model is that the cell membrane possesses an ability to rectify the flow of volume across it. In order to simplify the discussion of this topic I have chosen to focus attention on two types of cell which exhibit this phenomenon.

Characean Cells. Kamiya & Tazawa (1956), using the method of transcellular osmosis, found that the osmotic efflux of water (exosmosis) from *Nitella flexilis* was less than water influx (endosmosis) for a given osmotic pressure gradient. Dainty & Hope (1959) also used transcellular osmosis to estimate the hydraulic conductivity of *Chara corallina* and they also found that the endosmotic water permeability was apparently larger than the exosmotic permeability. These authors concluded, however, that the discrepancy between the permeability values was spurious and that it stemmed from the fact that the effective driving force on the water was different in the two cases. They argued that the apparent decrease in L_p with increasing external osmolarity (exosmosis) arose from the 'sweeping away' of the solute at the external surface by the water efflux; such an effect is dependent upon the presence of external unstirred layers and would produce an effective reduction in the osmotic pressure gradient at the cell's surface. Dainty (1963*b*) has given a quantitative analysis of the role of unstirred layers in such transcellular osmosis measurements. In a later publication (Dainty & Ginzburg, 1964*a*) however, it was noted by Dainty that his original estimate of the effect of unstirred layers in exosmosis was probably too large since he had used the thickness, δ, of the unstirred layer for steady-state conditions rather than that existing during the experiment. The latter value of δ is very difficult to estimate since it is probably time-dependent. Dainty & & Ginzburg (1964*a*) repeated the endosmosis and exosmosis experiments on *Chara corallina* and *Nitella translucens* and they noted that there was a genuine difference between the endosmotic and exosmotic water permeabilities in these cells. They concluded that 'the effective decrease in L_p with increase in osmotic pressure could be then interpreted either as a direct effect of the solution on the membrane and/or in terms of the rectification phenomenon'. Thus, Dainty and Ginzburg have left the conclusion about the actual mechanism of this phenomenon as an open question; however, they did allude to other evidence (Dainty & Ginzburg, 1964*b*) that the permeability of these cells to urea was reduced when the external osmotic pressure was increased.

Erythrocyte. Rich *et al.* (1968) have studied the influence of the osmolarity of the medium on the water permeability of erythrocytes. Figure 6.11 shows the relation between the final equilibrium osmolarity of the medium on L_p of human and dog erythrocytes. Both Rich *et al.* (1968) and Blum & Forster (1970) have shown that the change in L_p induced by a given change in osmolarity is very rapid; apparently the time required for L_p to change to its new value is within the range of 10–50 msec. Rich and co-workers obtained estimates of L_p from their quantitative descriptions of the time courses of cellular volume change and each theoretical

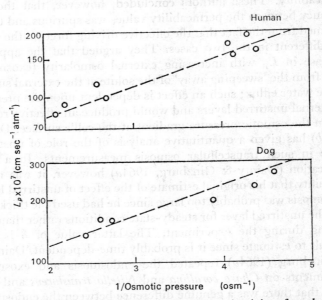

FIG. 6.11. The dependence of the hydraulic conductivity of erythrocytes on the reciprocal of the osmotic pressure of the bathing medium. The results of single experiments on human and dog erythrocytes are given and L_p is expressed logarithmically on the ordinate. The interrupted lines are the regression lines for the experimental data (Rich *et al.*, 1968: Fig. 3).

curve was obtained for an appropriate value of L_p related to the osmolarity of the bathing solution. On the other hand, in a series of experiments designed to assess whether or not there was rectification of volume flow, these workers measured L_p during swelling and shrinkage of human and dog erythrocytes. They found for human erythrocytes that the L_p values for swelling were 2·20, 2·00, 1·67 and 1·62 × 10⁻⁵ cm sec⁻¹ atm⁻¹ at the corresponding external osmolarities of 202, 207, 234 and 233 m-osm respectively.

In contrast the L_p values for shrinkage were 1.10 and 0.93×10^{-5} cm sec^{-1} atm^{-1} at 330 and 396 m-osm. By a process of normalization, which rested on the assumption that there was intrinsically an effect of the bathing osmolarity on L_p, they concluded that there was no rectification of volume flow. The corresponding data for swelling of dog erythrocytes were 2.92, 2.25 and 2.24×10^{-5} cm sec^{-1} atm^{-1} at osmolarities of 230, 271 and 288 m-osm whereas for shrinkage L_p had the values of 2.39, 2.16, 2.12 and 2.05×10^{-5} cm sec^{-1} atm^{-1} at osmolarities of 290, 322, 335 and 376 m-osm respectively. Again the conclusion was drawn that these values did not support the concept that volume flow is rectified in the erythrocyte membrane. Perhaps their conclusion, which is based on so few experimental results and a rather dubious method of comparison of hydraulic conductivities, is not warranted.

Although Rich *et al.* (1968) claimed that there was no rectification of water flow in the cell membrane of the erythrocyte, they inferred from the apparent rapidity of the change in L_p that the inner and outer facing surfaces of the red cell membranes are not symmetrical'. Indeed, they argued that the outer surface of the membrane was probably the rate-limiting barrier for osmotic water flow. If this picture of the erythrocyte membrane is correct, it prompts the question: would such an asymmetrical double-membrane not be expected to rectify volume flow? In principle, the answer is yes (see page 73). In fact, Sha'afi *et al.* (1970) proposed later on the basis of their measurements of urea permeability that the erythrocyte membrane behaved as an asymmetrical double-membrane and they concluded that the apparent rectification of volume flow observed by Rich *et al.* (1968) in the erythrocyte possibly stems from its asymmetrical water permeability characteristics.

Other evidence in favour of rectification of volume flow comes from the work of Farmer & Macey (1970) who have described a technique which determines the dependence of L_p upon the external osmolarity. Briefly, their method involves a sudden small perturbation of the bathing osmolarity which generates a small change in cellular volume. They have shown that the time course of the volume change ought to be exponential and, in particular, for the erythrocyte the time constant ought to be $[v^o(1-b)/\pi^o L_p A]/r^2$, where v^o is the cell volume under isotonic conditions and r^o and π are the isotonic and the experimental (final) osmotic pressures respectively. The time constant was obtained on the assumption that both L_p and the area A were independent of π. For the erythrocyte A can be assumed to be constant. On the other hand we cannot safely make any assumption about whether or not L_p is independent of the external osmotic pressure and, in fact, Farmer and Macey used the time constant and its dependence on π as an index of the dependence of L_p on π. That is, they argued that if L_p is independent of π a plot of the time constant against π^{-2} should be linear. Indeed, these authors did find such a linear relationship and, moreover, they noted that the slope for the shrinkage experiments was larger than that for the swelling measurements. Their measurements on erythrocytes indicated that L_p for osmotic water efflux was $1 \cdot 16 \times 10^{-5}$ cm sec^{-1} atm^{-1} and that for water influx was $1 \cdot 80 \times 10^{-5}$ cm sec^{-1} atm^{-1}. They compared their estimates of L_p with the values obtained by other workers for the water permeabilities of erythrocytes (see Table 6·3). According to Farmer and Macey there is evidence for rectification of water flow in cow, dog and human erythrocytes, although they concluded that the ratio of the permeabilities for chicken erythrocytes was not significantly different from unity.

The method of Farmer and Macey is a valuable tool for investigating the question of rectification of water flow across membranes of intact cells; however, their particular conclusions are based, in the case of human erythrocytes, on a small number of measurements and there is considerable scatter (see Fig. 5, Farmer & Macey, 1970).

At present it seems that the most satisfactory conclusion is that of Dainty & Ginzburg (1964a), that we cannot really differentiate properly between volume rectification and the effects of the external osmolarity on L_p. It is obviously important to resolve this question because the answer may tell us whether it is the hydration of the cell membrane that is important in controlling L_p or if the cell membrane is an asymmetrical structure.

It would be interesting to know if the diffusional permeability to water is influenced by the osmolarity of the bathing medium. Villegas et al. (1958) have reported that P_d decreased as the volume of the erythrocyte increases and clearly this is diametrically opposed to the changes in L_p that occur in similar conditions. However, Veatch (personal communication in Sha'afi et al., 1970) found that the addition of 0·3 M urea to the external medium

TABLE 6.3. Comparison of water permeabilities of erythrocytes for both swelling (L_p^{in}) and shrinkage (L_p^{out})

Species	Method	$(1-b)$	$(L_p^{in}) \times 10^5$ (cm sec^{-1} atm^{-1})	$(L_p^{out}) \times 10^5$	$\dfrac{L_p^{in}}{L_p^{out}}$	Reference
Cow	Perturbation	0·54	1·84	1·18	1·52	Farmer & Macey (1970)
	Stop-flow	0·52	—	1·14		Rich et al. (1967)
	Flow-tube	(0·35)	(1·14)			Villegas et al. (1958)
		(0·52*)	(1·71*)			
Dog	Stop-flow	—	2·32†	1·47†	1·59	Rich et al. (1968)
	Stop-flow	0·57	—	1·47		Rich et al. (1967)
	Flow-tube	(0·70)	(2·94)		1·61	Villegas et al. (1958)
		(0·57*)	(2·37*)			
Human	Perturbation	0·60	1·67	1·18	1·39	Farmer & Macey (1970)
	Stop-flow	—	1·63†	0·90†	1·82†	Rich et al. (1968)
	Stop-flow	—		0·90(22°)	1·32	Sha'afi et al. (1967)
	Flow-tube	(0·46)	(0·94)			Sidel & Solomon (1957)
		(0·57*)	(1·19*)			
Chicken	Stop-flow	—	1·67	0·98	1·71	Blum & Foster (1966)
	Perturbation	0·46	0·061	0·057	1·07	Farmer & Macey (1970)

* Recalculated by Farmer & Macey (1970) using a more acceptable value for $(1-b)$.
† Reinterpreted by Farmer & Macey (1970) allowing for rectification of volume flow.
Values located between the columns (L_p^{in}) and (L_p^{out}) were determined from both swelling and shrinkage experiments.
Temperature is 23–26°C except where noted (after Farmer & Macey, 1970).

produced no significant change in P_d from that recorded in isotonic solution, although it is stated by Sha'afi *et al.* (1970) that these experiments were not designed to study the influence of osmolarity on P_d. We must await such an experimental study with interest, for it would throw some light on the apparent contradictory effects of osmolarity on L_p and P_d as revealed by the work of Rich *et al.* (1968) and Villegas *et al.* (1958).

Electro-osmosis

In view of the importance of electrokinetic measurements it is surprising to find that relatively little experimental work has been done on such phenomena. The existence of electro-osmosis across cell membranes would constitute strong evidence for aqueous channels within the membrane. Furthermore, the quantitative nature of electro-osmotic flow and of the corresponding streaming potentials would give us some idea of the size and charge distribution of membrane pores. Until recently most workers have been content to speculate about the sort of physiological roles that electrokinetic phenomena might play in transport rather than to attack the problem rigorously from both experimental and theoretical standpoints. Now, thanks to the efforts of several groups we have a foundation for the further examination of electro-osmosis in cells.

The poor progress in the study of electro-osmosis in single cells in comparison to that made on artificial membranes can be attributed to the practical difficulties of performing the appropriate measurements on living systems. To achieve some success in this one needs to work with giant cells, such as the squid axon, or the alga, *Nitella*. The experimental difficulties are compounded by the fact that there are several sources of error and the most important of these arises from the presence of unstirred layers at the surface of the membrane (e.g. Barry & Hope, 1969*a,b*; see Chapter 3).

The phenomenological equations describing electrokinetic phenomena in membranes (Staverman, 1952) assume that the bathing solutions contain no unstirred layers at the membrane's surfaces. However, as Barry & Hope (1969*a*) have shown, the flow of current will generate local concentration changes in the unstirred layers and consequently water, over and above the genuine electro-osmosis, will flow across the membrane in response to these local osmotic gradients. This 'transport number effect', as Barry and Hope call it, is a troublesome source of error which may explain other phenomena, such as transient alterations in the membrane potential during the passage of constant currents. In addition, Barry and Hope make the important point that the 'transport number effect' may perform an important physiological function as a link between electrical currents and other phenomena, such as osmotic water flow. Indeed, such an effect between action potentials and water flows in plant cells has been established (Barry, 1970a,b) and will be discussed in detail later.

The most thorough experimental examination of electro-osmosis in any cell has been that of Barry & Hope (1969b) on *Chara corallina*. Fig. 6.12 shows their experimental apparatus for measuring volume flows across the cell during current passage. Because of the additional influence of the plant cell wall in contributing to the 'transport number effect', Barry and Hope measured volume flows across both isolated cell walls and intact cells in order to correct for the influence of the cell wall. In both sets of measurements they found that the initial rates of volume flow, attributable to pure electro-osmotic flow, were smaller than

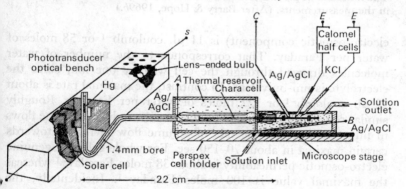

FIG. 6.12. A schematic representation of the apparatus used for measuring current-induced volume flows in whole *Chara* cells and cell walls. E refers to the input connexions to the electrometer used to measure membrane potential: A, B and C, connexions to a current generator: s, the stabilized light source supply and a, the main amplifier in the circuit used to measure the current output of the phototranducer. The overall sensitivity of this system is extremely good since it could detect volume changes of the order of 10^{-6} cm^3 (Barry & Hope, 1969b: Fig. 2).

the final rates because of the additional osmotically induced water flows arising from the 'transport number effect'. Table 6.4 shows the estimated electro-osmotic fluxes and also the maximal rates of volume flow recorded in both isolated cell walls and intact cells. For cell walls the average instantaneous rate of flow (or pure

TABLE 6.4. Average values of electro-osmotic and maximal rate coefficients for *Chara* cells and *Chara* cell walls in different solutions

External medium	Electro-osmotic flow μl coulomb^{-1}	Maximal rate of flow μl coulomb^{-1}
Chara cells		
Artificial pond water	7·6 (2)	18·5 (2)
1·0 mM NaCl }{ 0·1 mM KCl	7·9 (6)	16·4 (6)
0·1 mM KCl	6·0 (7)	16·5 (15)
1·0 mM KCl	7·3 (4)	20·5 (2)
Chara cell walls		
0·1 mM KCl	11 (3)	23 (3)
1·0 mM KCl	10 (4)	24 (4)
10·0 mM KCl	7 (2)	14 (2)

The numbers in parentheses show the number of cells or cell walls used in the measurements. (After Barry & Hope, 1969b.)

electro-osmotic component) is 11 μl. coulomb^{-1} or 58 moles of water per Faraday. That corresponds to the number of water molecules dragged through the cell wall by one ion, since the electrolyte is uni-univalent. In contrast, the maximal rate is about 23 μl. coulomb^{-1} or 122 moles of water per Faraday. Roughly similar results were obtained on the current-induced volume flows in intact cells; again the rate of volume flow increased towards certain maxima in about 70–150 sec. For whole cells the genuine electro-osmotic permeability is about 38 moles Faraday^{-1} whereas the maximal value is 100 moles Faraday^{-1}. Incidentally, the magnitudes of the coefficients, which have been quoted by Barry and Hope, are independent of the direction of the current.

Several other workers (Blinks & Airth, 1957; Fensom & Dainty, 1963; Fensom, Ursino & Nelson, 1967; Fensom & Wanless, 1967; MacRobbie & Fensom, 1969) have studied electro-osmosis in *Nitella*. The values for the electro-osmotic permeability which have been obtained by these workers with the exception of Blinks and Airth who concluded that electro-osmosis does not occur in *Nitella* 'under natural or applied potentials of moderate value' agree well with the maximal values obtained by Barry & Hope (1969b) for intact cells. For instance, Fensom & Dainty (1963) noted that the volume flow did not immediately attain a steady value when the current was applied and that the final value of the electro-osmotic permeability was 100 moles Faraday^{-1}; similar features have also been reported by Mackay & Meares (1959) for electro-osmosis in cation-exchange membranes. Fensom and Dainty took the maximal value of the apparent electro-osmotic flow as the appropriate rate. Now, however, the theoretical and experimental work of Barry & Hope (1969a,b) demonstrates that the initial rate, not the maximum, reflects electro-osmosis and that the 'true' electro-osmotic permeability for *Nitella* is about 40 moles Faraday^{-1}.

Actually the electro-osmotic efficiency seems quite high at its value of 40 water molecules per cation. It might be contended that this electro-osmotic flow is merely the transfer of the primary hydration shell of the ions carrying the current. If this were so, then such a volume flow would require no frictional interaction to occur between ions and water within the membrane, or in other words, we would not need to postulate the presence of aqueous pores in the membrane. The values of the primary hydration numbers for NaCl and KCl, however, are only about seven (or less) water molecules for each pair of ions (see Robinson & Stokes, 1959). Thus, the relatively large magnitude of the electro-osmotic permeability of *Nitella* demands the presence of aqueous pores.

The next question which these electrokinetic measurements pose is an important one raised by Fensom & Dainty (1963) and later by Briggs (1967) among others. They have asked whether or not the electrokinetic properties of the cell wall, rather than the cell membranes, can account for the overall electro-osmotic flow across the intact cell. Fensom and Dainty have attempted to answer this by considering the alternative current pathways across *Nitella* placed in the transcellular-flow apparatus (see Fig. 5.8) and they concluded that the lowest resistance pathway is apparently across the cell wall and cell membranes rather than solely through the cell wall itself in a longitudinal direction within the sealing plug between the two chambers. Indeed, they argued that only about 1–10% of the applied current ought to flow longitudinally through the cell wall in these experiments and consequently the electro-osmotic permeability of the cell would need to be 400–4000 moles Faraday^{-1} to account for the observed volume flows if the cell membranes were electrokinetically inert. When the latter values are compared to the electro-osmotic permeabilities of about 10 moles Faraday^{-1} for isolated cell walls (Barry & Hope, 1969*b*) we are forced to conclude that the cell membranes of *Nitella* are the source of the electro-osmotic flow. Barry & Hope (1969*b*) have concluded too that the cell membranes are responsible for about 80% of the measured electro-osmotic permeability of the intact cell. Of course, in *Nitella* and *Chara* we have two cell membranes in series—plasmalemma and tonoplast—and we are unable to assess their individual contributions to electro-osmosis as yet.

In any membrane where electro-osmosis has been observed, there should also be the converse electrokinetic phenomenon—streaming potentials. The latter phenomenon arises from the frictional interaction between a pressure-driven flow of water and ions within aqueous channels in the membrane. The sign of the streaming potential depends on both the direction of water flow and the electrical charge on the walls of the aqueous channels. Invariably that side of the membrane towards which water is flowing acquires a charge opposite to that of the membrane. Such streaming potentials are linearly related to the hydrostatic or osmotic pressure difference driving the water flow and they are related quantitatively to the electro-osmotic permeability by equation (3.60), namely

$$-\left(\frac{\Delta E}{\Delta p}\right)_{\Delta \pi, I} = \left(\frac{J_v}{I}\right)_{\Delta \pi, \Delta p}$$

See page 98 for the basis for this equality.

The above relation constitutes a criterion for the existence of genuine electro-osmosis in a given membrane. Unfortunately the relation between electro-osmosis and streaming potential has been examined in very few tissues indeed. In this connexion it is really surprising that streaming potentials have not been recorded in *Nitella* since this cell has proved to be an important preparation for the study of electro-osmosis. Of course, before such experiments on streaming potentials are attempted it is wise to estimate from the electro-osmotic measurements what the likely size of such potentials would be. Dainty (1963*a*) has considered this point and he concluded that the maximal streaming potential is given by $\Delta E = \Delta p/Fc_i$, where c_i is the concentration of the counter ion. Let us suppose that $\Delta p = 10$ atm (or 1 joule cm^{-3}) and $c_i = 10^{-3}$ equiv. cm^{-3}, then the streaming potential, ΔE, will be about 10 mV; although such a large difference in pressure could be employed in plant cells but not in animal cells, an equivalent osmotic pressure gradient could be employed in both with relatively more ease and safety. The calculation shows that in *Nitella* one should be able, in principle, to record streaming potentials and it is curious that such measurements have apparently not been reported.

In animal cells the study of electro-osmosis has not been as thorough as that in *Chara* by Barry & Hope (1969*a*,*b*). Despite this, there have been estimates of both electro-osmotic permeability and streaming potentials in the giant axons. Stallworthy & Fensom (1966) detected an apparent electro-osmotic permeability of 28 moles Faraday^{-1} for axons of the squid *Illex illecebrosus*, and later Stallworthy (1970) obtained an estimate of 16 moles Faraday^{-1} for the giant axons of *Loligo forbesi*. It is possible that the actual magnitudes of these estimates are erroneous because of two sources of difficulty in these experiments. First, the axoplasm itself may be the source of the electro-osmosis rather than the axolemma. Stallworthy (1970) did demonstrate that the axoplasm was a source of electro-osmosis but that its contribution to the overall electro-osmotic permeability was indeed small compared to that of the axolemma. There is obviously a parallel here between the electrokinetic role of the plant cell wall and the axoplasm in these studies. The second source of difficulty is similar to that discussed by Barry & Hope (1969*a*,*b*) for plant cells and relates to the transient nature of the electro-osmotic flow. For instance, Stallworthy did report that the electro-osmotic flow induced by a steady current increased towards a maximum (Fig. 3, Stallworthy, 1970) and he used these maximal flows to estimate the electro-osmotic permeability. Clearly, these observations support the view that there is a 'transport number effect' in the squid axon as well as in *Chara* and the electro-osmotic permeability coefficients should be corrected for this error. In common with most of the similar measurements on *Nitella* the electro-osmotic permeabilities are overestimates of the true values for the cell membranes.

Measurements of streaming potentials have been reported by

Vargas (1968b) for the giant axon of the squid *Disodicus gigas*. He found that the size of the streaming potential was 4×10^{-2} mV per cm of water (applied Δp) and only $2 \cdot 5 \times 10^{-4}$ mV per cm of water when an osmotic pressure difference was applied. This is a striking difference. Vargas accounted for that large discrepancy in terms of the difference in the water flow produced by these driving forces. He had found in earlier work (Vargas, 1965, 1968a) that L_p determined by Δp was approximately 100 times larger than that determined by $\Delta \pi$ (see page 166). Accordingly he argued that the size of the water flows in the streaming potentials experiments will depend on whether Δp or $\Delta \pi$ is applied and for the sake of comparison each streaming potential should be divided by the corresponding value of L_p. That procedure yielded values of 4×10^5 mV/(cm sec^{-1}) and $5 \cdot 4 \times 10^5$ mV/(cm sec^{-1}) for $\Delta \pi$ and Δp experiments respectively. This finding seems to reinforce Vargas' arguments about the different nature of the water flows in the hydrostatic and osmotic pressure experiments. Vargas also noted that the streaming potentials produced by Δp, for example, had a fast phase followed by a slow phase and he described the latter component as a transient diffusion potential. Such a time course for the streaming potential is expected on the basis of the 'transport number effect' since there will be (as a result of water flow) a transient change in the concentrations of electrolytes at the surfaces of the axolemma which will generate a slow transient diffusion potential on top of the streaming potential. Vargas corrected his data for the slow component in order to obtain a more reliable estimate of the true streaming potential occurring during the initial fast phase of the response to pressure. After such corrections the streaming potential became approximately 10^{-2} mV per cm of H_2O. (Both electro-osmosis and streaming potentials have been recorded apparently in epithelial tissues and these experiments will be discussed in Chapter 9.)

The existence of electrokinetic phenomena in certain excitable plant and animal cells means that some electro-osmosis may occur during the propagation of action potentials in these cells. Several workers have studied this question and it will be discussed below.

Action potentials and water flow

The excitability of nerve has been interpreted by several authors (e.g. Teorell, 1958, 1959a,b, 1962, 1966; Kobatake &

Fujita, 1964a,b) in terms of electrokinetic models. Teorell (1961), in particular, has suggested that electro-osmosis occurs during the action potential of *Nitella*.

According to Teorell's model the concentration profile across the membrane is initially linear and the electro-osmotic flow, which occurs during the passage of the stimulating current, decreases the concentration to yield a concave profile. Such a drop in concentration within the membrane will increase membrane resistance which will in turn increase the shift in membrane potential produced by the current pulse. That series of events will increase the electro-osmotic flow across the membrane until an adverse hydrostatic pressure builds up to exactly counter-balance the electro-osmotic flow. When that point is reached the diffusion of solute will ensure that the concentration profile in the membrane relaxes back to the original linear profile. Consequently the electro-osmotic flow and the membrane potential should both decrease. Teorell (1962), for example, has shown that the predictions of his model are consistent with the behaviour of certain artificial membranes. Such systems, however, contain large pores (7000 Å) and, thus, they cannot be considered as satisfactory analogues of cell membranes. Besides that objection, there are other ones to the application of Teorell's model to excitable cell membranes (e.g. see Barry, 1970b).

The model of Kobatake & Fujita (1964a,b) is rather similar to that of Teorell with the exception that the former theory envisages that the increased pressure which results from the electro-osmotic flow is the overall restoring force rather than the diffusion of solute. Again this model requires very large pores for its operation.

It is indeed interesting that such electrokinetic models can be used to explain the excitable characteristics of cell membranes; however, it seems that their main strength lies in the implication that excitable behaviour may be intimately connected with a transient water flow across the membrane. As plausible explanations for excitable phenomena they fall short on several counts. For example, these models do not predict the transient increase in anion conductance which occurs during the action potential of *Chara corallina* (Hope & Findlay, 1964).

Apart from the theoretical approaches outlined above, which constitute rather unsatisfactory attempts to account solely for action potentials in terms of electrokinetic mechanisms, there have been several experimental attempts to measure the volume flows accompanying action potentials in some cells.

Hill (1950a) investigated the effect of repetitive stimulation on the opacity of a nerve trunk from the spider crab, *Maia squinado*. He found that the response to stimulation consisted of basically a large decrease in opacity preceded often by an initial increase in opacity. In a parallel set of experiments, where the external osmotic pressure was altered, he found that the opacity of the nerve trunk decreased when the fibres swelled and *vice versa*. His conclusion that stimulation of the nerve trunk produced an increase of fibre size was confirmed in other experiments with single giant axons from *Sepia officinalis*. In the latter study he measured the increase in the volume of the axon resulting from repetitive stimulation. In order to explain the increase in axonal volume Hill postulated that in addition to the exchange of sodium and potassium occurring during the action potential there is also a net influx of NaCl; both the ionic exchange and the net influx were suggested by Hill to be responsible for the concomitant water entry. However, Keynes & Lewis (1951) were unable to measure the relatively large chloride influx postulated by Hill.

The electrokinetic models described above predict that volume flows and pressure changes should accompany action potentials. Obviously it is important to try to substantiate experimentally such changes during excitation and the giant algal cells, such as *Chara corallina*, are extremely useful preparations for that purpose. Their outstanding features are their large size, cylindrical geometry and slow action potentials of about 2–3 sec duration. Fensom (1966) reported preliminary measurements of the volume flow occurring during the propagation of an action potential in *Nitella*; the volume flow was measured by the technique used for transcellular osmosis experiments. Fensom found that during each action potential a volume change of about 0·2 nl (or 2×10^{-7} cm^3) occurred at one of the cells. Barry (1970a) has measured similar volume flows in *Chara corallina* and he has found that when an action potential occurs at some point in the cell there is an associated volume efflux of about 0·88 nl cm^{-2} sec^{-1}. Barry (1970a,b) also showed that Fensom's data could be expressed in the same way and were equivalent to an efflux of about 0·9 nl cm^{-2} sec^{-1}. Some representative records (Barry, 1970a) of the volume changes and action potentials in *Chara* are shown in Fig. 6.13. In this experiment it must be remembered that it is the volume of the solution bathing one end (A) of the cell that is being recorded. The cell is stimulated at the opposite end (B) and a volume efflux occurs at B which draws water and solute through the cell from end A.

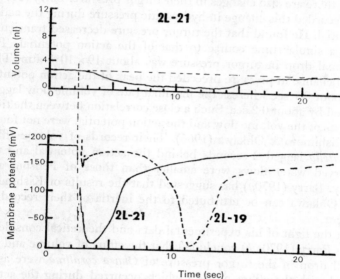

FIG. 6.13. Typical changes of volume and membrane potential during an action potential in a cell of *Chara corallina*. One end (B) of the cell was stimulated directly with a current pulse and the membrane potential was also recorded at B between the vacuole and the external solution. The volume of solution bathing the opposite end (A) of the cell was monitored over part of its length by a technique previously described (Barry & Hope, 1969b: see also Fig. 6.12). This recording (2L–21) shows a typical biphasic volume flow resulting from the transmission of an action potential from B to A. The transmitted action potential has been simulated by superimposing another recording (2L–19) in which end A had been stimulated initially. Thus, at first there was a volume efflux at end B accompanying the action potential there. Then, after a few seconds, a volume efflux occurs at A when the action potential reached that region (Barry, 1970: Fig. 8).

Thus, the volume V decreased as volume inflow occurs at end A to match the efflux at B. Later the action potential reaches end A and the corresponding volume efflux takes place across end A to restore the volume V to its original value. Thus, the propagation of an action potential in *Chara* and *Nitella* is accompanied by a circulating volume flow with volume efflux occurring in the active region and volume influx in the inactive region of the cell. No net change in cellular volume results from excitation. In addition to the volume flows, which are induced by action potentials in these plant cells, there are also changes in their turgor pressure. Barry (1970a) has recorded this change in hydrostatic pressure during the action potential. He found that the turgor pressure decreased transiently with a similar time course to that of the action potential. The maximal drop in turgor pressure was about 19×10^{-3} atm. This small change in pressure preceded the peak of the action potential by about 0·1 sec whereas the maximal rate of volume flow lagged behind by about 0·2 sec. Such a close correlation between the time courses of the volume flow and the action potential were not found by Kishimoto & Ohkawa (1966). Their records of volume efflux in *Nitella* lagged 2 seconds behind the action potential and the observed water flows were smaller than those of Fensom and Barry. Barry (1970a) has suggested that the results of Kishimoto and Ohkawa can be attributed to the inertia of their recording system.

In the light of his experimental data and theoretical considerations Barry (1970a,b) concluded that the efflux of volume and the small drop in the turgor pressure of *Chara corallina* were associated with the efflux of KCl which occurred during the action potential. During excitation there is an increase in KCl permeability and consequently the frictional coupling between the movement of both potassium and chloride ions and water generates a volume efflux. To a small extent also there is a local osmotic gradient set up by the efflux of KCl which pulls water osmotically out of the cell in the active region. On the basis of that evidence and other arguments Barry considered that his observations did not imply obligatorily an electrokinetic mechanism for the action potential but rather that the volume efflux and turgor pressure changes were concomitants of an increase in KCl permeability.

Reflexion coefficients

Another way of examining the degree of interaction between solutes and water in cell membranes is to compare the magnitude of the reflexion coefficient with that of solute and water permeabilities. Consequently there have been several attempts to measure the value of σ_s for various solutes in some animal and plant cells. That σ_s can be used to establish the degree of interaction between solute and water is evident from relation 3.20 where it can be seen that a knowledge of σ_s, ω_s and L_p will determine the magnitude of the frictional interaction term $(K_s{}^c f_{sw}{}^c/f_{sw}{}^c + f_{sm}{}^c)$. Unfortunately several workers who have determined σ_s for such a purpose have not strictly adhered to that approach and their conclusions must be treated with some caution.

There are several ways in which σ_s can be measured. Actually the first method for cell membranes was that of Goldstein & Solomon (1960). They determined the rate of swelling or shrinking of erythrocytes placed suddenly into different concentrations of a given non-electrolyte; the initial rates of volume change were obtained by extrapolation to zero time. Examination of the data revealed that at a particular concentration, c_s, of the permeant non-electrolyte the cells would neither shrink nor swell and, therefore, they obtained σ_s from $c_i = \sigma_s c_s$, where c_i is the internal concentration of solute at zero time. This method has been used also by Dainty & Ginzburg (1964d) for measurements of σ_s for the 'membranes' of *Nitella translucens* and *Chara corallina*. They also employed another method which compared the initial rates of transcellular osmosis generated first by a certain concentration of an impermeant molecule (sucrose) and then by an identical concentration of the test solute. For reasons that will be discussed later, they considered that the first method—the so-called 'nul method'—was more accurate than the second one.

Estimates of σ_s for a variety of solutes have been obtained for human erythrocytes (Goldstein & Solomon, 1960), dog erythrocytes (Rich *et al.*, 1967), squid axons (Villegas & Barnola, 1961), frog muscle fibres (Zadunaisky *et al.*, 1963) and plant cells (Dainty & Ginzburg, 1964d). Some comparative values are given in Table 6.5. In particular, Goldstein & Solomon (1960) assumed that their test solutes must pass only through aqueous pores since they had low solubility in lipid. The validity of this assumption has been questioned by many workers, notably Diamond & Wright (1969) and Macey & Farmer (1970). Numerous investigators have interpreted data on σ_s in terms of an equivalent pore theory. Their approach rests on the assumption that σ_s can be expressed by

$$\sigma_s = 1 - \frac{A_s}{A_w} \qquad 6.12$$

where A_s is the pore area available for transport of solute and A_w is the area for water transport (see page 83). In order to assess the pore radius, r, from the reflexion coefficient it has been

TABLE 6.5. Reflexion coefficients of some cell membranes

Permeant solute	Solute molecular radius (Å)	Erythrocyte[a]	Squid axon[b]	Muscle[c]	Chara corallina[d]
Methanol	1·83[b]	–	0·35	–	0·30
Formamide	1·96[b]	–	0·44	0·65	1·00
Ethanol	2·13[b]	–	0·63	–	0·27
Urea	2·17[b]: 2·03[c]	0·62	0·70	0·82	1·00
Thiourea	2·18[a]	0·85	–	–	–
Ethylene glycol	2·24[a]	0·63	0·72	–	1·00
Acetamide	2·27[a]	0·58	–	–	–
Propionamide	2·13[a]	0·80	–	–	–
Methyl urea	2·37[a]	0·80	–	–	–
Malonamide	2·57[a]	0·83	–	–	–
Propylene glycol	2·61[a]	0·85	–	–	–
Glycerol	2·77[b]: 2·74[a]	0·88	0·96	0·86	–
Mannitol	–	–	–	1·00	–
Sucrose	–	–	–	1·01	–
Estimated pore radius (Å)		4·2	4·3	4·0	–

References: (a) Goldstein & Solomon (1960): (b) Villegas & Barnola (1961): (c) Zadunaisky *et al.* (1963): (d) Dainty & Ginzburg (1964d).

common practice to replace the term (A_s/A_w) in the above relation by (A_{sf}/A_{wf}) expressed by equation 3.13 to give

$$\sigma_s = 1 - \frac{\left[2\left(1-\frac{a_s}{r}\right)^2 - \left(1-\frac{a_s}{r}\right)^4\right]}{\left[2\left(1-\frac{a_w}{r}\right)^2 - \left(1-\frac{a_w}{r}\right)^4\right]}$$

$$\times \frac{\left[1 - 2\cdot104\left(\frac{a_s}{r}\right) + 2\cdot09\left(\frac{a_s}{r}\right)^3 - 0\cdot95\left(\frac{a_s}{r}\right)^5\right]}{\left[1 - 2\cdot104\left(\frac{a_w}{r}\right) + 2\cdot09\left(\frac{a_w}{r}\right)^3 - 0\cdot95\left(\frac{a_w}{r}\right)^5\right]} \quad 6.13$$

Provided that the molecular radii, a_s, of the test solutes are known, then theoretical values of σ_s can be evaluated for various values of r.

With this procedure one can generate a theoretical curve to fit the experimental data for some appropriate value of r. Figure 6.14 shows the results of this approach applied to reflexion coefficient measurements on the erythrocyte. Thus the degree of interaction between solutes and water occurring in the erythrocyte membrane

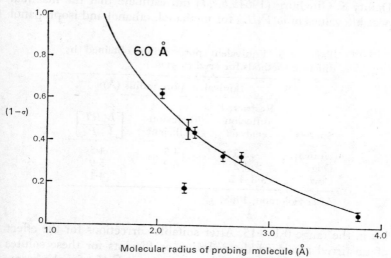

FIG. 6.14. The dependence of $(1-\sigma)$ for dog erythrocytes on the molecular radius of the test molecule. The theoretical curves were obtained from equations 3.13 and 3.14: in this case it was argued that σ can be identified with $(1-A_{sf}/A_{wf})$ because the term $\omega_s \bar{V}_s/L_p$ is negligible. The bars on the experimental points indicate the S.E. of the mean. All of the test solutes with the exception of ethylene glycol have reflexion coefficients which yield an equivalent pore radius of 6 Å (Rich et al., 1967: Fig. 4).

can be expressed in terms of an equivalent pore radius. Table 6.6 shows some of the values for the equivalent pore radius of certain erythrocytes that have been obtained from measurements of σ_s. The corresponding values for r, obtained either from a comparison of $L_p RT/\bar{V}_w$ and P_d or from analysis of restricted diffusion, are also shown. These data are in remarkably good agreement especially when one considers the questionable assumptions underlying the applications of these methods to cell membranes. There are, of course, several objections to the use of equation 6.13 for the reflexion coefficient (see for example

Diamond & Wright, 1969). For example, in equation 6.13 for σ_s the term $\omega_s \bar{V}_s/L_p$ has been omitted (cf. equations 3.20 and 3.21). However, several workers (Hoshiko & Lindley, 1964; Rich et al., 1967, Solomon, 1968) have argued that $\omega_s \bar{V}_s/L_p$ may be ignored legitimately since it is generally negligible. On the other hand, Dainty & Ginzburg (1964a,b,c,d) did estimate that the maximal possible values of $\omega_s \bar{V}_s/L_p$ for methanol, ethanol and isopropanol

TABLE 6.6. Equivalent pore radius obtained by different methods for erythrocytes

Species	Equivalent pore radius (Å)		
	Restricted diffusion analysis	Reflexion coefficient	$\left[\dfrac{L_p RT}{\bar{V}_w P_d}\right]$
Human	> 3·5	4·3	4·5
Dog	> 4·2	6·2	5·9
Cow	3·8–4·2	–	4·1

After Solomon, 1968.

lay in the range 0·1–0·15. After suitable corrections for the effect of unstirred layers on the reflexion coefficients for these solutes Dainty and Ginzburg found that $\sigma_s < 1 - \omega_s \bar{V}_s/L_p$ for the membranes of *Nitella translucens* and *Chara australis*. Their study is probably the most thorough of any performed on animal or plant cells since they evaluated the effects of unstirred-layer corrections and also applied the most satisfactory expression for σ_s, namely equation 3.20, to their data. From their work it may be concluded that certain solutes, such as methanol, permeate the cell membranes via pores. It must be noted, however, that the relation for the reflexion coefficient was derived for a single lipid–pore model of the cell membrane, whereas the characean cells contain two membranes—tonoplast and plasmalemma—in series.

Aside from the experimental difficulties which attend the measurement of σ_s and also, of course, ω_s and L_p, there are serious theoretical doubts about the validity of quantitative estimates of equivalent pore radii based on the magnitudes of σ_s. Apart from the obvious objection that the hydrodynamical treatment of solute and water transport in small aqueous pores is inadequate, there is another objection which has its roots in the fundamental assumptions upon which equation 3.20 and others, such as 3.18 are based. In their derivation it has been assumed that the continuous treatment of membrane transport can be applied to a discontinuous membrane system. This is a sound approach provided that we are dealing with a homogeneous membrane. However, if the membrane is porous we cannot satisfactorily describe the hydrostatic pressure as a continuous function across the membrane since there will be a discontinuity at the mouth of the pore, as Mauro (1965) has shown both theoretically and experimentally. Thus, the continuous treatment of irreversible thermodynamics is not really equipped to deal rigorously with a discontinuous system like a porous membrane.

Water relations of cells at low temperatures

The tolerance of animal and plant cells to low temperatures, especially those below the freezing point of water, is an intriguing problem. It raises several questions about the water relations of such tolerant cells at those temperatures. In particular, how do cells avoid, or rather prevent, ice formation and if, for example, extracellular ice does occur is it necessarily lethal to cells? Actually the problem of cold tolerance boils down to the main question of what factor causes injury to cells during freezing and thawing. It may seem that the answer is self-evident, namely that the cells are damaged mechanically by ice crystals. This proposition is in accord with the fact that there is considerable distortion of cells and tissues during the formation of extracellular ice. Nevertheless, it is currently accepted that the cellular damage occurring during freezing and thawing does not arise from disruption of cells by ice (see e.g. Meryman, 1970). Indeed, we must seek alternative explanations for cellular death at low temperatures.

Apart from the obvious suggestion that extracellular ice causes cell damage there is the equally obvious one that ice crystals inside the cells are lethal. It is questionable, however, that ice actually does form invariably inside cells except during very rapid rates of cooling in laboratory experiments or in cryosurgery. Certainly, when intracellular freezing does happen, it is lethal both in animal cells (Mazur, 1966) and plant cells (Levitt, 1966). Mazur (1963), in particular, has developed a theory that is based on the view that intracellular ice crystals are responsible for cell death at low temperatures. His model describes how various parameters influence the probability of intracellular freezing. In this connexion the water permeability of the cells is certainly of crucial importance because as the external temperature drops extracellular ice will form and this will raise the external osmotic pressure. Thus, water will leave the cell under the osmotic driving force at a rate determined by its hydraulic conductivity. Basically, Mazur's view is that if sufficient water remains in a cell during cooling then ice will form spontaneously in the super-cooled intracellular fluid. Clearly, the hydraulic conductivity of the cell membrane will dictate the rate at which water can be driven from the cell and if this rate is less than that demanded by the cooling rate of the cell then a substantial fraction of intracellular water will become supercooled. Mazur's theory also takes account of the temperature dependence of L_p. For example, Jacobs, Glassman & Parpart (1935) have measured the hydraulic conductivity of human erythrocytes between 0° and 30°C and Mazur used these and other data in his theoretical derivation of the fraction of supercooled water in a number of cell types including yeast, erythrocytes and sea urchin eggs. The theory meets a serious difficulty because it requires the temperature dependence of L_p at subzero temperatures and there are apparently no data of that kind in the literature. Thus, Mazur was forced to obtain the temperature coefficient by an extrapolation process. Of course, it is quite possible that L_p decreases with temperature much more steeply below 0°C than it does above it and this would exert an important influence on Mazur's theoretical curves for the kinetic loss of water during freezing and perhaps consequently on some of his conclusions.

Mazur's analysis shows that the most permeable cells (erythrocytes) can avoid freezing by a relatively rapid process of dehydration whereas the less permeable cells remain highly hydrated and supercooled; consequently they experience intracellular freezing. Thus, according to Mazur if the cooling rate is not too rapid, say < 10°C per min, water is able to leave cells, particularly the permeable ones, fast enough to raise the internal osmotic pressure and consequently to lower the freezing point of the intracellular fluid. Mazur's theory is an ingenious explanation of the intriguing ability of cells to withstand low temperatures without damage. Of course, his model assumes that cell death results solely from intracellular ice formation, and the cooling rates required to produce intracellular ice formation are necessarily very rapid (>100°C per min). Cellular injury, however, can also arise when the cooling rates are comparatively slow, say 1°C per min, and when extracellular ice but not intracellular ice is formed.

The latter kind of freezing injury lies outside Mazur's scheme and an alternative explanation for its origin has been derived by Meryman (1968) and Williams (1970). They have put forward their so-called 'minimum cell volume' hypothesis which will now be described.

'Minimum cell volume' hypothesis

This model suggests that the mechanism of freezing injury is due to cellular dehydration and the inevitable increase in the intracellular osmotic pressure that occurs at low temperatures. It must be remembered, of course, that the growth of extracellular ice produces an increase in the external osmotic pressure and consequently there is an osmotic water efflux from the cell. According to the 'minimum cell volume' model there is a limit to the extent of cellular dehydration that can occur during freezing. In fact, there is a minimum volume to which the cell may shrink without injury. This limit seems surprisingly constant from one type of cell to another. Before going on to discuss the 'minimum cell volume' model it is interesting to briefly trace the development of this idea.

Levitt (1966), for example, suggested that freezing injury in plant cells occurs when there has been enough cellular dehydration to reduce the distance between neighbouring intracellular proteins to such an extent that new disulphide bonds are formed between them. According to this hypothesis these bonds are stronger than others in the protein molecules and when the cell thaws and gains water these molecules are probably broken.

On the other hand, Lovelock (1953*a*) concluded from experiments on erythrocytes that freezing injury resulted from the steep rise in the external salt concentration that occurs as extracellular ice is formed. He observed that erythrocytes suffered haemolysis during freezing, particularly when the extracellular salt concentration exceeded 0·8 M NaCl. In accord with this view of freezing injury was Lovelock's observation that adding glycerol to the external medium afforded cryoprotection to the erythrocytes by a simple anti-freeze mechanism. That is, glycerol enters the erythrocytes and lowers the freezing point of both extracellular and intracellular fluids. Thus, in the presence of glycerol the amount of extracellular ice at any given temperature is less and so the external salt concentration is also reduced. Moreover, Lovelock (1953b) demonstrated that freezing haemolysis was invariably associated with an external salt concentration of 0·8 M irrespective of the temperature. Lovelock did not reach a definite conclusion about how high concentrations of salt damaged the erythrocytes except to suggest that they probably denatured the cell membrane. Although this proposal seems to agree with the observation that phospholipids and cholesterol are released from erythrocytes incubated in high salt concentrations (Lovelock, 1955), several other pieces of evidence are at variance with Lovelock's 'lyotrophic' concept. Mazur (1966), for example, has contended that during such freezing and thawing experiments haemolysis occurs within seconds whereas when erythrocytes are suspended in high salt concentrations phospholipids do not appear in the external medium until several minutes have elapsed. Furthermore, Meryman (1968, 1971) has pointed out that erythrocytes suffer injury at approximately the same freezing temperature when they are placed in either salt or non-electrolyte solutions. Of course, the latter evidence by itself is not definitive because it can be argued that the cells suspended in non-electrolyte solutions experience membrane denaturation due to high concentrations of intracellular electrolytes at low temperatures. Meryman (1968), however, has shown that erythrocytes can be suspended in 3 M ammonium chloride—a permeant salt—and no cellular injury evidently occurs. (Incidentally, because of its power to equilibrate readily between the exterior and interior of erythrocytes, ammonium chloride acts as a cryoprotective agent just like glycerol). This experiment demonstrates, therefore, that high intracellular concentrations of electrolytes are not necessarily lethal to the erythrocyte. Thus, we can conclude that neither the high concentration of electrolytes outside nor inside the erythrocyte is, by itself, responsible for cellular injury.

In 1968 Meryman concluded that the haemolysis of erythrocytes which results from slow freezing is 'the result of damage to the membrane caused by the development of an osmotic pressure gradient across the cell membrane greater than that which can be compensated by cell volume change'.

In short, the 'minimum cell volume' hypothesis proposes that there is a progressive increase in the resistance to cellular shrinkage occurring as more extracellular ice is formed. Finally, the cell attains its minimum tolerable volume and damage to the cell membrane possibly results from the residual osmotic gradient across it. Inseparable from that possibility is the alternative one that it is the dehydration or reduction in the size of the protoplasm that causes cellular injury.

Certain experimental facts favour the 'minimum cell volume' model. For example, Meryman's (1968, 1971) work on the effects of osmotic stress on erythrocytes demonstrates that high external osmotic pressures are inevitably accompanied by cellular shrinkage and that at a certain unique limit the membrane becomes permeable to solute and there is a concomitant cellular swelling. This was found in erythrocytes bathed in either concentrated salt or non-electrolyte solutions and the minimum volume at which membrane integrity was impaired was identical in both cases although the external osmotic pressures were significantly different. According to Meryman the same disturbance in membrane permeability is produced at low temperatures ($< -2.7°C$) and that as the erythrocytes are thawed out haemolysis takes place.

Other experiments which also confirm that there is a relation between the loss of a fixed fraction of cellular water and the injury induced by freezing are those of Williams & Meryman (1965) on human erythrocytes and of Williams (1970) on the clam, *Venus mercenaria*, and the mussel, *Mytilus edulis*. According to their studies erythrocytes undergo freezing injury at $-2.7°C$, equivalent to an external osmotic pressure of 1500 m-osm, whereas the clam and mussel show injury at -6 and $-10°C$, equivalent to 3200 and 5400 m-osm. Williams and Meryman estimated the relative amounts of ice and water in these preparations by a calorimetric method, and noted that at the point of freezing injury approximately 64% of the cellular water in each case had been removed to form extracellular ice. Because these measurements demonstrate that the common feature is the volume of cellular water that remains at the point of injury, they support the minimum volume hypothesis. Moreover, Williams (1970) observed that after *Mytilus* had been adapted to 150% sea water the cells were able to withstand temperatures down to $-15°C$ before injury was incurred. Nevertheless, freezing injury was still associated with a 64% loss of cellular water.

The 'minimum cell volume' hypothesis has been quite successful at emphasizing the importance of cell volume in the water relations of cells at subzero temperatures. Of course, it is not clear exactly whether the dehydration of the cytoplasm or the cell membrane is of crucial importance or, indeed, if it is the compression of the cell which is actually injurious. The relative importance of these factors cannot be distinguished at present. Another of the model's features is the curious one that the cellular water that must remain in the cell to prevent freezing injury need not be osmotically active. Apparently the significant fact is the volume and not the nature of the cellular water at low temperatures. Most of the work supporting this model has been done on animal cells but Williams (*personal communication*) has recently marshalled evidence in favour of a similar mechanism operating in plant cells. However, plant cells seem to have evolved several additional mechanisms to reduce the water loss that occurs at low temperatures. For example, they may accumulate small solutes which act as cryoprotective agents just as glycerol, for example, does in the erythrocyte; moreover, during periods of osmotic stress some plant cells increase their permeabilities to small solutes and consequently solute influx and accompanying water flow minimize the cell's approach to its minimum tolerable volume. It is of interest too that some woody species manufacture glycoproteins which substantially reduce the chemical potential for water and hence its osmotic efflux from cells at low temperatures. A similar group of substances have been found in the blood sera of antarctic fish (e.g. De Vries & Wohlschlag, 1969; De Vries, Komatsu & Feeney, 1970) and apparently afford these fish cryoprotection by basically the same mechanism.

7

FLUID DYNAMICS IN THE EMBRYO

Amphibian embryos	263
Chick embryo	271
Mammalian embryos	276
Blastocoel	277
Other extra-embryonic cavities	282
Conclusion	285

THE elaboration of the transient extracellular fluid compartments in embryos poses numerous intriguing problems connected with transport phenomena. The formation of these cavities occurs during the early stages of development, and interference with their growth can have profound effects on the embryo. In addition to the fluids contained within these extracellular compartments there are others—plasma, lymph and interstitial fluids—which are formed within the body of the embryo itself. So little is known about the creation of the latter cavities that it is impossible to delineate particular experimental problems to be tackled. Therefore, in this chapter attention has been focused on the areas of fluid dynamics in the embryo where some progress is being made.

Amphibian embryos

During their development amphibian embryos absorb a relatively large amount of water. In fact, this uptake is responsible for the increase in volume of the embryo before the onset of feeding. Numerous workers have attempted to correlate the water influx with the osmotic pressure of the embryo. Such analyses of water absorption, however, are complicated by the fact that two fluid compartments are formed transiently in the extracellular space during the initial period of development (Fig. 7.1). The first of these compartments is the blastocoel. This cavity increases in size and its growth is followed by the formation of another cavity produced by an infolding of the embryonic surface during gastrulation. The latter invagination produces the second cavity,

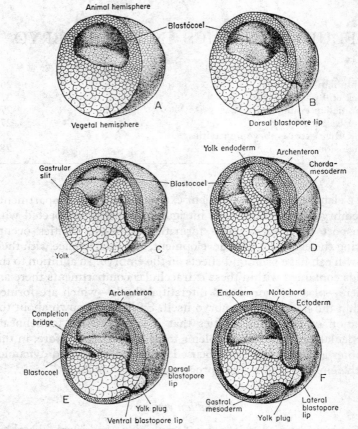

FIG. 7.1. Diagrammatic representation of six successive stages in the development of *Rana pipiens* from blastula to late gastrula. The formation of the blastocoel and archenteron cavities are shown (Rugh, 1951: figure redrawn from Fig. 47 of Huettner, 1941).

From Rugh: *The Frog, Its Reproduction and Development.* Copyright (1951, McGraw-Hill Book Co.). Used by permission of McGraw-Hill Book Company.

namely the archenteron, whose growth is accompanied by a decrease in the size of the blastocoel. Later the contents of the archenteron are released into the perivitelline space and subsequently the growth of the embryo is largely accounted for by a progressive increase in the intracellular fluid compartment (Tuft, 1965). Thus it is essential to measure the osmotic pressures of the blastocoel and archenteron cavities, as Tuft (1962) has done, rather than the osmotic pressure of the entire embryo (cf. Backmann & Runstrom, 1909, 1912; Krogh, Schmidt-Nielsen & Zeuthen, 1939). Moreover, it is now evident that many of the early estimates of water uptake were slightly erroneous since the increase in the embryo's volume was estimated from its diameter and the volume was assumed to be spherical. A more satisfactory method of measuring the volume is based on determinations of both the density and the embryo's reduced weight (i.e. of the embryo in water). For example, Tuft (1962) placed *Xenopus* embryos in linear density gradients of colloidal thorium oxide in order to estimate their density during their development and he measured the reduced weight by means of the Cartesian diver balance.

According to Tuft (1962) the increase in embryonic volume has four phases. From the early stages of cleavage until gastrulation there is an initial phase of rapid increase in volume which is followed by a second phase during gastrulation when there is a decrease in the rate of water influx. During the third phase there is much more rapid increase in volume which is terminated by the collapse of the archenteron. Finally, uptake of water is resumed but now it accumulates relatively rapidly in the cells. Figure 7.2 shows that in contrast to the increase in the volume of the whole embryo the cellular volume increases very slowly at the beginning. After the collapse of the archenteron, however, the increase in the size of the entire embryo can be attributed almost entirely to the increase in cellular volume.

In contrast to the observations on *Xenopus* there is in the sturgeon embryo an initial increase in the volume of the embryonic cells; however, within 10 hours of fertilization the growth of the cavities accounts entirely for the change in volume of the sturgeon embryo (see e.g. Zotin, 1965). The archenteron collapses at a much later state (50 hours) than in *Xenopus* and then the cellular volume begins to increase. Thus, during the initial stages of the development of both *Xenopus* and sturgeon embryos there are similar changes in the rate of water uptake from the environment and a common transient pattern of water distribution.

It seems, therefore, that a satisfactory account of the water relations of those embryos must explain the successive formation of the blastocoel and archenteron, the increasing rate of water flux into these cavities, the low rate of water uptake into the cells during initial development and the relatively faster swelling of the cells

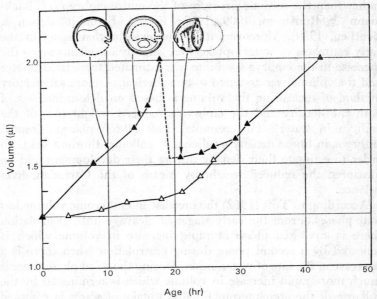

FIG. 7.2. Increase in the volume of the *Xenopus* embryo and its cells during development. The volume of the intact embryo (▲) increases during the formation of the blastocoel and subsequently of the archenteron. When the archenteron collapses at about 18 hours the volume decreases abruptly and then increases again as the embryo elongates. These stages of development are shown schematically above the graph. The volume of the cells (△) increases slowly throughout this period, but begins to increase more rapidly when the embryo elongates. The difference between the total volume and that of the cells can be attributed to the volume of the cavities—blastocoel and archenteron (Tuft, 1965: Fig. 3 slightly modified).

after the collapse of the archenteron. Several theories have been put forward to account for the formation and decay of the blastocoel and archenteron.

Løvtrup (1960), for example, studied the exchange of deuterated water in axolotl embryos in the Cartesian diver balance. Although he found that the rate of water exchange in the embryos was not constant throughout their development, there was no correlation between the rate of exchange and the net water influx into the embryo. It is not surprising that the rate of labelled water exchange is not an accurate index of the water uptake in these experiments because the former values are likely to be erroneous due to large unstirred layers in the Cartesian diver balance; moreover, water diffusion and net osmotic water flow may be rate-controlled by different barriers in the embryo. In discussing his data on water uptake in the embryo, Løvtrup suggested that its surface coat acted as a mechanical barrier, like the plant cell wall, and that this constraint limited osmotic swelling. He argued that the rapid influx of water occurring at neurulation was caused by the loss of this surface coat. It is difficult to accept this theory since it rests on the assumption that the embryo must be under considerable hydrostatic pressure. This is not so. Another serious criticism of Løvtrup's model is that the water uptake enters the cavities and not the cells as his model suggests. Subsequently Løvtrup (1965a) modified his original hypothesis by postulating that each embryonic cell has a surface coat with similar mechanical properties to that surrounding the entire embryo. In his model there is obviously an analogy, perhaps unintentional, between the embryo and a piece of plant tissue. According to Løvtrup the cells become smaller as development proceeds and the intracellular pressure rises to force water into the cavities. Finally, Løvtrup (1965b) proposed that the fluid movement from the blastocoel to the archenteron was driven by a hydrostatic pressure gradient. There is no evidence for such a difference in pressure as yet and, in fact, it would probably need to be several atmospheres to enlarge the archenteron at the observed rate.

Zotin (1965) has suggested that the increase in volume of the blastocoel arises from the secretion of glycogen granules into it from the neighbouring cells. A similar mechanism for the formation of the blastocoel in echinoderm embryos was originally published by Monné & Hardé (1951). According to Zotin's hypothesis the exhaustion of glycogen secretion into the blastocoel leads to a pronounced drop in the osmotic water uptake into the embryo during gastrulation and fluid is subsequently transferred from the blastocoel to the archenteron by a hydrostatic pressure gradient. Moreover, Zotin envisaged that the archenteron underwent a further growth stage which was produced by glycogen secretion and concomitant water transfer. Although glycogen and other polysaccharides are present in the cells and blastocoels of sturgeon and amphibian embryos, Zotin's theory seems questionable on several points. First, it does not account quantitatively for the water flux into the blastocoel; furthermore there is no evidence for the proposed pressure gradient between blastocoel and archteron.

Tuft (1962, 1965) has published another model for water uptake in the developing embryo. His model assumes that the distribution of water in the embryo is the result of water movement which is not only driven by osmotic gradients but also by some mechanism dependent upon the metabolic energy of the cells. He called the latter type of water flux 'energy-coupled flow' and suggested that each cell is actively pumping out the fluid which enters it down the osmotic gradient. Moreover, he postulated that the rate of active water transport is not uniform over the whole cell, but that the surface facing the vegetative pole has a larger active water flux. Actually in the uncleaved egg these two opposing active flows are assumed to be equal since the egg volume is constant (Fig. 7.3). According to Tuft's hypothesis the polarities of the 'energy-coupled' water flows are transferred to the daughter cells in the ectoderm and endoderm. Thus, if we consider first the ectodermal cells at the animal pole of the blastula, then we can see that there will be a net passive influx of water from the outside and a net active efflux of water from them into the blastocoel. On the other hand, across the endodermal cells at the vegetative pole there will be a net passive influx of water from the blastocoel and a net active efflux to the outside. According to this scheme, therefore, fluid ought to accumulate in the blastocoel because the area of the ectoderm is larger than that of the endoderm (Fig. 7.3). Moreover, the increase in the ectodermal area during the early stages of development will cause an increase in the rate of water accumulation. Throughout gastrulation, however, the endodermal area increases, and subsequently it lines the archenteron cavity as a consequence of the invagination of the vegetal surface. Thus, water is transported across the endoderm into the archenteron. A novel feature of Tuft's model is that it explains the observed transfer of fluid from the blastocoel to the archenteron: during the invagination of the vegetal surface the blastocoel becomes delineated from the archenteron by a layer of endodermal cells which transport water in an 'outwards' direction, i.e. from blastocoel to archenteron. When the archenteron reaches its maximal size it is deformed by the elongation of a notochord of the embryo and, as a result of this, the blastopore opens and the archenteron discharges its fluid to the outside.

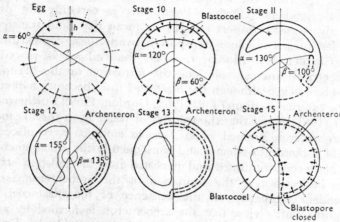

FIG. 7.3. Illustration of the mechanism of water regulation in the embryo of *Xenopus laevis* according to Tuft (1962). The dotted arrows indicate the direction of osmotic water flow whereas the solid arrows show the direction of water flow achieved by an active transport mechanism. When the passive and active water flows are in the same direction this indicates a net flow of water across the cells. It is assumed that the active water flow does not occur uniformly in all cells but rather that it is confined predominantly to the endodermal surface (interrupted line). Thus, according to this model the formation of the blastocoel results from both passive and active water flow (see stage 10): later the endodermal surface invaginates to line the archenteron and consequently the active water efflux is now directed towards the lumen of the archenteron (see stage 15). In the latter case the blastocoel acts as a source of water which is actively pumped into the archenteron (Tuft, 1962: Fig. 10).

Since Tuft's model requires no more than a uniform rate of water transport across the ectoderm and endoderm during development, and since these fluxes of water are similar to those accompanying active salt transport across amphibian epithelia, such as frog skin and toad bladder, this description of water distribution in amphibian embryos is worthy of further experimental attention. For example, it would be interesting to know the solute fluxes between the outside and the blastocoel and between the blastocoel and the archenteron. Certain experimental evidence is compatible with this account of water uptake in the amphibian embryo. Measurements of the freezing-point depressions of the blastocoel and archenteron fluids show that fluid is transferred from the former to the latter against an apparent osmotic gradient (Gordon, 1969); this is certainly not proof of active water transport but it may mean that fluid transfer is coupled to active solute transport between these compartments (cf. Chapter 10). Moreover, when the embryo is placed in concentrated sucrose solutions the formation of the blastocoel still occurs even although the blastocoel fluid becomes hypotonic to the external medium (Tuft, 1965; Gordon, 1969). Furthermore, Tuft (1961) found that the transfer of water from the blastocoel to archenteron stopped when *Xenopus* embryos were placed in 2 mM β-mercapto-ethanol and consequently the enlargement of the blastocoel continued and persisted into the tail-bud stages. This indicates that the formation of the archenteron is not dependent ultimately on the existence of the blastocoel and supports Tuft's view that the archenteron is formed by a net efflux of fluid across the invaginated endoderm. Such observations suggest strongly that the blastocoel and archenteron cavities are formed by fluid secretion, perhaps dependent on active solute transport. On the other hand, Tuft considers that the 'energy-coupled' water flow itself is genuinely active—a view which is not proved by the evidence, especially since little seems to be known about the ionic relations of the embryonic cells. Recently Slack & Warner (1973) have suggested that the blastocoel enlarges due, first, to osmotic flow into the cells from outside and, secondly, passive water movement coupled to active sodium from the cells into the intercellular fluid. Perhaps a similar mechanism accounts for the archenteron's growth.

This brief summary seems to underline just how little is known about the mechanism of fluid movement in the blastocoel and archenteron of amphibian embryos. Of course, other fluid cavities are formed as the embryo develops. For example, the coelomic cavity first appears during late gastrulation and like the archenteron it seems to be formed apparently as the result of an increase in the surface area of a cell layer; unfortunately the rate of fluid accumulation in the coelomic cavity has not been measured. Beyond gastrulation it becomes progressively more difficult to pinpoint precisely which differentiated layer of cells is elaborating a particular fluid compartment.

Chick embryo

The avian ovarian egg is fertilized in the first part of the reproductive tract and then albumen, shell membrane and the shell are successively added during its passage through the oviduct. The egg finally appears as a sealed system which is capable only of gaseous exchange with its environment. In contrast to the amphibian and mammalian embryo the avian embryo does not depend upon the external medium to supply fluid. The chick embryo, for instance, has a large store of water awaiting the requirements of its development; the yolk and the albumen contain about 9 and 31 cm^3 of water respectively. Only a small fraction (25%) of this store of fluid is lost by evaporation through the shell.

During the development of the chick embryo four extra-embryonic cavities are formed. In contrast to the amphibian egg, where there is a complete cleavage of the egg during the formation of the blastula, the cleavage of the bird's egg is different. The latter, of course, contains a very large yolk mass and its protoplasm is confined to a small region, called the blastodisc, at the egg's animal pole where cleavage and blastula formation occur. The yolk mass does not participate in this and many cell divisions lead to the formation of a blastula which lies at the animal pole. The blastula is basically a sphere which is flattened down into a sheet called the blastoderm overlying part of the yolk. This cell mass is actually separated from the underlying yolk by a fluid cavity called the sub-blastodermic fluid or liquefied yolk. This cavity is probably equivalent to the blastocoel observed in amphibian eggs and into this space moves the free edge of the blastoderm. The invagination carries a layer of cells—prospective endoderm—into the sub-blastodermic cavity. Subsequently both the upper layer of the blastoderm, namely the ectoderm, and the endoderm grow over the yolk mass (Fig. 7.4a). In particular, the extra-embryonic endoderm, referred to as the splanchnopleure forms a *yolk sac* which almost completely envelopes the yolk. The extra-embryonic ectoderm, referred to as the somatopleure, begins to grow over the embryo (Fig. 7.4b). At the same time the allantois grows out as

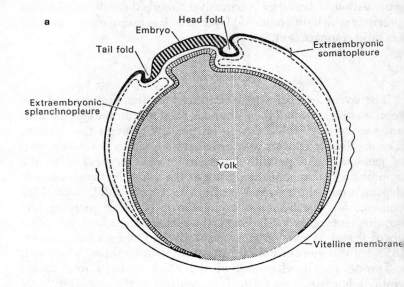
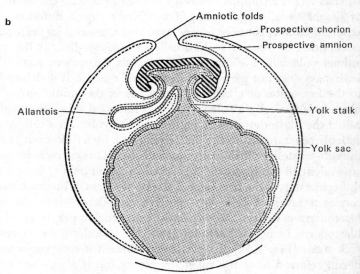

Fig. 7.4

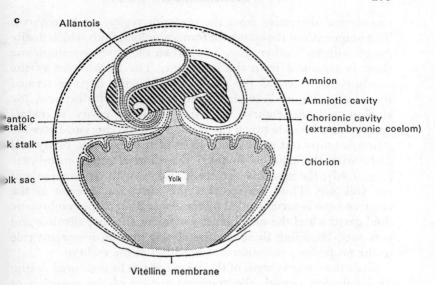
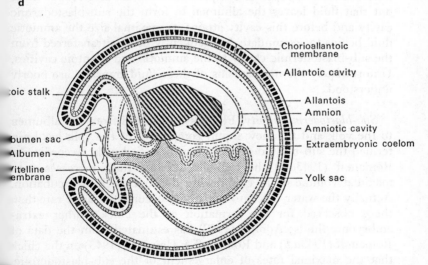

FIG. 7.4. The growth of the extra-embryonic membranes and cavities in the early stages of development of the chick. The body folds delineating the embryo from the extra-embryonic regions are shown diagrammatically in **a** and the later sequential stages of development are illustrated in **b**, **c** and **d** (Torrey, 1971: Figs. 10.11–10.14).

endodermal derivative from the posterior region of the embryo. The outgrowth of the ectoderm forms amniotic folds which finally merge with each other (Fig. 7.4c) when the outer somatopleuric sheet is separated from the inner one. The former part of the somatopleure is called the *chorion* whereas the latter is termed the *amnion*. Figure 7.4c shows the amniotic cavity between the amnion and the embryo and the chorionic cavity or extraembryonic coelom between the amnion and the chorion. Later the allantois grows out until it fills most of the chorionic cavity and fuses with the chorion to form the chorioallantoic membrane (Fig. 7.4d); the allantois also tends to fuse with the amnion and the yolk sac. Thus, we see that the early development of the chick embryo is accompanied by the elaboration of four embryonic fluid cavities and the membranes—amnion, chorion, allantois and yolk sac—bounding these cavities clearly play an important role in the protection, excretion and nutrition of the embryo.

Since the water content of the avian egg can be measured during its incubation period, the transient nature of the partition of fluid between its separate compartments can be assessed. It turns out that fluid leaves the albumen to form the sub-blastodermic cavity and before this cavity attains its maximal size the amniotic fluid begins to accumulate. Subsequently fluid is transferred from the sub-blastodermic cavity to the amniotic and allantoic cavities. Unfortunately the mechanisms of these fluid transfers are poorly understood.

Sub-blastodermic cavity. Fluid is transferred from the albumen to the yolk and this new liquid component of the yolk has been termed the sub-blastodermic fluid or liquefied yolk. According to Romanoff (1943a,b) the sub-blastodermic fluid increases to a maximal volume of 15 cm^3 during the first week of incubation. Actually the water flux into this cavity occurs at a faster rate than those observed for the formation of the chick's other extraembryonic fluids; Adolph (1967) has estimated from the data of Romanoff (1943a,b) and Romanoff & Hayward (1943) on the chick that the maximal rates of enlargement of the sub-blastodermic, amniotic and allantoic cavities are equivalent to fluid influxes of 5×10^{-6}, $2 \cdot 8 \times 10^{-6}$ and $1 \cdot 8 \times 10^{-6}$ cm^3 cm^{-2} sec^{-1} (or $1 \cdot 8$, $1 \cdot 0$ and $0 \cdot 66$ μl cm^{-2} hr^{-1}). These rates of net water flow are similar to those accompanying active salt transport across a number of epithelia (see Chapter 10, Table 10.1). The mechanism of fluid transport into the sub-blastodermic cavity is not known, although Yamada (1933) suggested that the driving force for water flow was an osmotic pressure difference between the yolk and the albumen. New (1956) has shown, however, that in the absence of osmotic gradients an isolated preparation of the blastoderm can still transport fluid in the right direction and at the appropriate rate to account for the normal accumulation of the sub-blastodermic fluid.

The composition of the sub-blastodermic fluid has been analysed by Howard (1957) from the second day to the eleventh day of incubation. During this period the osmolarity remained constant and identical to that of the amniotic fluid. Howard found that the potassium concentration always exceeded its value in the other extracellular fluids and ranged from an initial value of 13 mM to a final one of 43 mM on the eleventh day. As the potassium concentration rose the sodium concentration fell, so the total concentration of both ions in the sub-blastodermic fluid remained constant. Unfortunately it is impossible to decide whether this increase in potassium level is caused by the secretion of the blastoderm or by an efflux of this ion from the yolk. The analysis of the ionic relations of the sub-blastodermic fluid is hindered by a serious lack of experimental data and also by another feature—during the period between the seventh and fifteenth day of incubation the sub-blastodermic volume decreases in size at a rate slightly less than its growth rate (Romanoff, 1943a,b). Clearly it is imperative to obtain more information about the characteristics of solute and water transport across the blastoderm before an explanation can be offered for the growth and subsequent disappearance of the sub-blastodermic fluid.

Amniotic cavity. The volume of the amniotic cavity is regulated apparently by the amnion, although the regulatory mechanisms are not understood. Amniotic fluid is elaborated between the sixth and thirteenth days of incubation in the chick; its rate of formation, expressed as cm^3 per day, is 1·2 whereas the sub-blastodermic fluid is formed at five times that rate (Romanoff & Hayward, 1943). The maximal volume (about 4 cm^3) of the amniotic cavity is maintained until the seventeenth day and then it decreases in size. Romanoff (1952) has suggested that the amniotic fluid drains into the alimentary tract where it is absorbed. The composition of this fluid volume is distinctly different from that of the adjacent compartments (Harsh & Green, 1963). It has a higher chloride concentration and lower potassium concentration than the latter, but again we need a thorough study of ion and water transport into this cavity before a comprehensive picture of its development can be drawn. It is certainly free of protein until the fourteenth day, when it makes a connexion with the albumen sac and there is a subsequent rise in its protein concentration.

Extra-embryonic coelom. Apparently the formation of this cavity has not been studied thoroughly and, therefore, there is not even quantitative information about either its rate of growth or its composition.

Allantoic cavity. The accumulation of the allantoic fluid lags behind the growth of the amniotic fluid. The former increases from the fifth day until it attains its maximal size (7 cm^3) on the twelfth day of incubation. The maximal rate of its formation is 2·3 cm^3 per day, which is double that of the amniotic fluid and about half that for the sub-blastodermic fluid. The fluid in the allantoic cavity may be derived from the mesonephroi since the ureters empty into this compartment. After the twelfth day water is reabsorbed from the allantoic fluid and certain substances, notably uric acid, are concentrated in it. The allantoic fluid, therefore, serves initially as a reservoir for excretory products but later it is actually a source of water which is absorbed into the blood and lymphatic systems of the allantoic membrane to be redistributed to other compartments in the embryo.

Mammalian embryos

The mammalian embryo, like the avian embryo, has four extra-embryonic cavities which increase in size and subsequently decay. In the mammal, however, cleavage of the egg and subsequent cell divisions lead to a spherical arrangement of cells called the blastocyst. Its outer shell is termed the trophoblast and a small inner cluster is referred to as the inner cell mass. The central fluid-filled cavity is the blastocoel and although the blastocyst appears at first glance to be similar to an amphibian blastula it is more similar to the avian blastoderm. Eventually the blastocyst makes contact with the endometrium of the uterus and subsequently the development of the extra-embryonic membranes occurs. This is similar to that in the chick with the concomitant formation of an extra-embryonic coelom and amniotic and allantoic cavities. Of the four cavities that are formed most is known about the blastocoel and, therefore, we shall devote relatively more time to it.

Blastocoel

In the rabbit embryo, for example, blastulation begins at about 84 hours *post coitum*. At this time the blastocyst consists of a central blastocoel surrounded by a single layer of trophoblast cells except where they join with the inner cell mass. The latter cells develop into the embryoblast, or embryo proper, and they occupy a minor fraction of the entire surface area of the blastocyst.

Daniel (1964) has estimated the rate of accumulation of the blastocoel fluid in the rabbit blastocyst from measurements of its diameter at certain times *post coitum* (Table 7.1). The volume of the blastocyst increases from about 0·002 μl between the third and fourth day to about 2500 μl on the tenth day. Since the cells occupy less than 1% of the blastocyst volume this increase can be directly attributed to the accumulation of the blastocoel fluid. It is worth emphasizing that the net fluid flux into the blastocyst increases progressively from the fourth to the tenth day, thus indicating that the trophoblast layer becomes relatively more efficient at absorbing fluid as time goes on. It would be exceedingly interesting to have the corresponding data on solute transport into the blastocoel during that period.

The electrolyte composition of the blastocoel fluid has been analysed (Lewis & Lutwak-Mann, 1954) both before and after implantation in the uterine wall. In the former case the concentrations of sodium and chloride ions are significantly less than those on maternal serum; however, both potassium and bicarbonate concentrations exceed the serum values. In fact the ionic concentrations in the blastocoel are rather similar to those in uterine fluid. Smith (1970) has analysed the blastocoel fluid in the rabbit blastocyst and estimated the net transport rates of sodium, potassium, chloride, bicarbonate and water into this compartment. His data reveal that there is a net flux of sodium

TABLE 7.1. Time course of the increase in volume of the rabbit blastocyst and the associated water influx

Time (post coitum) (day)	Number of conceptuses studied	Average diameter (cm)	Average volume (μl)	Average surface (cm^2)	Accumulation rate (μl hr^{-1})	Net fluid influx (μl cm^{-2} hr^{-1})
1	10	0·016	0·0021	0·00080	—	—
2	6	0·016	0·0021	0·00080	—	—
3	14	0·016	0·0021	0·00080	—	—
4	12	0·027	0·0103	0·00229	0·00225	1·0
5	19	0·102	0·556	0·0327	0·068	2·1
6	18	0·280	11·5	0·246	0·8	3·2
7	10	0·501	65·8	0·789	5·6	7·1
8	8	0·870	344·5	2·38	18	7·6
9	7	1·190	880	4·45	42	9·4
10	6	1·690	2530	8·96	96	10·7

Modified from Daniel, 1964.

and chloride but not potassium and bicarbonate ions. The values for the internal concentrations of those ions are in agreement with those of Lewis & Lutwak-Mann (1954). Smith found that low temperature, acetazolamide and ouabain substantially decrease the net influx of sodium, chloride and water. Both of the latter agents are selective, but not exclusive, inhibitors of the active chloride and sodium transport respectively. Thus, Smith has concluded that the primary event in the formation of the blastocoel is the active influx of sodium and/or chloride ions and that there is a concomitant water influx. To decide about the nature of ion transport into the blastocoel we need to know the electrical gradient and also the unidirectional fluxes of the ions. The potential of the blastocoel relative to the outside medium has been measured by Cross & Brinster (1969) and Cross (1971). In particular, Cross (1971) found that the blastocoel was about -11 mV and that when it was perfused with a solution identical to the external medium the potential was of a similar sign but smaller (about -6 mV). Under his conditions of perfusion the current necessary to short-circuit the potential was equivalent to 0.15 μ equiv cm^{-2} hr^{-1}. It can be inferred, of course, that this short-circuit current is equal to the active ion influx (cf. Ussing & Zerahn, 1951); in this connexion it would be interesting to know the unidirectional fluxes of sodium and chloride ions because it seems quite probable that there is at least an active chloride influx (and possibly a passive sodium influx). The net fluxes of sodium and chloride ions estimated by Smith are about six times larger than the short-circuit current recorded by Cross.

One point of interest in Smith's (1970) data is that the ratio of the net influx of salt to that of water is approximately *isotonic* to the culture medium under all of the experimental conditions except at 0°C, when it becomes somewhat hypotonic to the culture medium. The osmolarity of the absorbed fluid is a parameter which is worthy of further study since it is intimately involved in the current concepts of how passive water flow is coupled to active salt transport across epithelia (see Chapter 10). Some workers have examined the osmolarity of the blastocoel fluid. For example, by freezing blastocysts rapidly *in situ* Brambell (1954) found that he could subsequently obtain blastocoel fluid for analysis by removing the trophoblast layer. It turns out that the blastocoel contains only about 3% protein of a similar nature to that in plasma of the adult rabbit (see also Lutwak-Mann, 1959). In fact, the protein concentration in the blastocoel is never larger than half that in plasma and, therefore, it contributes a negligible component to the total osmotic pressure of the blastocoel fluid. The measurements that have been made of the freezing-point depressions of the blastocyst fluid, uterine fluid and blood plasma in the rabbit (Lutwak-Mann, 1960; Tuft & Böving, 1970) show that the net flux of water into the blastocyst occurs against its chemical potential gradient between blood plasma and blastocoel (Fig. 7.5). Between the fourth and sixth days *post coitum* the difference in the chemical potential, $\Delta\mu_w$, for water between the uterine fluid and blastocoel increases from zero to a value which would account, at least *partially*, for

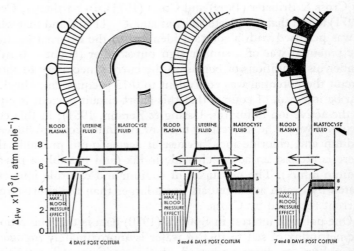

FIG. 7.5. Diagrammatic representation of the chemical potential gradient for water between maternal plasma and the uterine and blastocyst fluid in the rabbit. The direction of the expected passive water flow is indicated by a single arrow and that of the observed water flow by a double arrow. The difference in the chemical potential, $\Delta\mu_w$, between these fluids and plasma was determined by measuring the freezing-point depressions of plasma and uterine and blastocyst fluids. The drawings above the graph illustrate the sources of the fluid samples and the nature of the anatomical changes during three successive stages of blastocyst growth and implantation. The small circles represent blood vessels: the parallel lines indicate the columnar uterine epithelium: stipple tone identifies lemmas, and black shows the tissue elements of blastocyst (Tuft & Böving, 1970, Fig. 4).

some of the water influx. In particular, on the fourth day the water uptake from the uterine lumen into the blastocoel occurs in the absence of an osmotic gradient whereas during the fifth and sixth days the net influx of water continues to increase in size (Daniel, 1964; Tuft & Böving, 1970) although the chemical potential gradient for water remains small and invariant. Finally, on the eighth day the chemical potential gradient for water is opposed to water entry; nevertheless, the influx of water persists and continues to increase in size. The foregoing interpretation of these data, derived from freezing-point measurements, rests on the assumption that the solutes in the plasma and uterine fluid do not penetrate easily into the blastocyst. Tuft & Böving (1970), however, have argued that, if the concentration of impermeant solutes in the plasma were less than that in the blastocoel, then the plasma concentration of permeant solutes would be larger than that in blastocoel. This ought to be so, since the freezing-point depression for plasma is larger than that for blastocoel fluid. One might expect to see, therefore, an increase in the freezing-point depression of the blastocoel fluid due to the entry of the permeant solutes. No such increase is found. Indeed, between the sixth and eighth days *post coitum*, when the size of the blastocyst increases thirty-fold, the freezing point depression of the blastocoel fluid remains absolutely constant. According to the work of Tuft and Böving the large expansion in the volume of the rabbit blastocyst is caused by the active absorption of a solution which is *hypotonic* to plasma. They conclude that the flow of water into the blastocoel is active and that it cannot be a passive flow coupled to an active influx solute since this would produce a blastocoel fluid which was either isotonic or hypertonic to plasma (cf. Diamond & Bossert, 1967). Since very little is known about the ionic relations of the trophoblast cells and the mechanisms of ion uptake it is unwise to conclude that the influx of water into the blastocoel must be active. The fact that the absorbed fluid in the blastocyst is hypotonic to plasma is neither proof of active fluid transport nor incompatible with some current views on the coupling of active salt transport with passive water movement (see Chapter 10).

The regulation of the blastocoel cavity may be influenced by the endometrium after the implantation of the blastocyst. Some specialization of endometrial structure, and perhaps function, occurs at the sites of implantation. Thus, once the blastocyst has implanted, the regulation of the volume and composition of the blastocoel cavity is under the control not only of the foetal cells but also, perhaps, of the maternal cells since both are generally interposed between plasma and blastocoel.

The rabbit blastocyst is undoubtedly an attractive preparation for the study of ion and water transport and, moreover, since it absorbs a hypotonic fluid in contrast to most other tissues it presents a challenge to the currently accepted model of water transport (see '*Standing-gradient osmotic flow*', page 452).

Other extra-embryonic cavities

Amniotic cavity. As we have seen in the chick embryo, fluid accumulates within the enclosure delineated by the amnion. Adolph (1967) has surveyed the range of amniotic fluid volumes which have been found in some mammalian species and he has also estimated the rates of fluid influx which are compatible with their growth curves (Table 7.2). It is evident that, despite the enormous range of amniotic volumes, the maximal rates of formation expressed as the rate of fluid accumulation per unit area of amnion lie within very narrow limits. Since the composition of the amniotic fluid is approximately similar to plasma, although it is somewhat hypotonic in the rabbit (Davies & Routh, 1957), pig (McCance & Dickerson, 1957) and rat (Barker, 1961), the relatively uniform rate of formation cannot be due to osmotic entry of water. Nor is there any evidence for filtration of fluid into the amniotic cavity. Hence the narrow range of water influx values in Table 7.2 may reflect the relative uniformity of the maximal secretory activity of the amnions of these mammals. It is interesting too that the fluid transport rates are the same size as those observed in epithelial tissues (see Table 10.1).

Although several workers (Davies & Routh, 1957; McCance & Dickerson, 1957; Barker, 1961) have found that the amniotic fluid is slightly hypotonic to foetal and maternal plasma they also noted quite wide variations of the ionic concentrations. For instance, in the rat Barker (1961) found the ranges were 103–210 mM, 3–27 mM, 72–138 mM, 6–32 mM for sodium, potassium chloride and bicarbonate ions respectively; he noted also that the protein concentration varied from 0–2·6%, which is significantly lower than that in plasma. Unfortunately no thorough investigation of ion transport across the amnion seems to have been carried out such a study would help to unravel the mechanisms underlying the formation of the amniotic cavity.

TABLE 7.2. The volume of the amniotic cavity and the estimated rate of fluid influx into it in different species

Species	Maximal volume of amniotic cavity (cm^3)	Net fluid influx ($\mu l\ cm^{-2}\ hr^{-1}$)	Reference
Mouse	0·12	0·83	McCafferty (1955)
Golden hamster	0·3	0·83	Purdy & Hillemann (1950)
Rat	0·4	1·25	Barker (1961)
Rabbit	2·5	0·83	Daniel (1964)
Guinea pig	5	0·83	Ibsen (1928)
Cat	9	1·67	Wislocki (1935)
Pig	120	1·25	Wislocki (1935)
Human	120	1·25	Harrison & Malpas (1953)
Sheep	300	7·9	Needham (1931)
Cow	1700	2·5	Bergmann (1921)

Modified from Adolph (1967).

After the size of the amniotic cavity in each species has attained its maximum, some variation in its volume occurs, and later in pregnancy its volume decreases. The intermediate fluctuations in the amniotic fluid volume have been attributed to the net effect of entry of pulmonary fluid (Reynolds, 1953, 1964), entry of urine (Potter, 1961) and loss of fluid into the alimentary system of the foetus (Liley, 1963). The quantitative aspects of these mechanisms have not been explicitly described.

Extra-embryonic coelom. This cavity, which lies between the amnion and chorion can attain a volume of about 3 cm^3 in the rabbit (Dickerson & McCance, 1957). Again it has been found that, in common with the blastocoel and amniotic fluid, this cavity increases in size and then diminishes. There is exceedingly little information, however, about its formation not only in the rabbit but also in any other mammal.

Allantoic cavity. Some quantitative studies of the allantoic volume have been made. The maximal volumes, which have been found, are 2700 cm^3 in cattle (Bergmann, 1921), 700 cm^3 in sheep

(Needham, 1931), 180 cm^3 in pigs (Wislocki, 1935), 6 cm^3 in cats (Wislocki, 1935) and 0·5 cm^3 in rabbits (Davies & Routh, 1957). The rates of formation of the allantoic cavity are generally larger than those for the amniotic cavity and lie within the approximate range 3 to 10·4 μl cm^{-2} hr^{-1} (cf. Table 7.2 and also Table 10.1). There is considerable difference in the age at which this cavity begins to develop in different species. For instance, in the sheep it reaches its maximal volume towards the end of gestation (Needham, 1931), whereas in the pig the volume is maximal in the early stage of gestation and remains so until birth (Wislocki, 1935). In general, however, the allantoic cavity like the blastocoel, extra-embryonic coelom and amniotic cavity increases to a maximum and then declines.

During early development the composition of the allantoic fluid in the pig is somewhat similar to that of the amniotic cavity; but a disparity between them arises as development proceeds (McCance & Dickerson, 1957) when the allantoic fluid becomes significantly hypotonic. It is known that foetal urine enters the allantoic cavity in a number of species (e.g. pig, rabbit and sheep). However, the hypotonicity of the allantoic cavity in pigs and rabbits cannot be due to the entry of foetal urine since the latter fluid is quite similar to plasma (McCance & Stannier, 1960).

Some work has been done on isolated allantoic membranes. Crawford and McCance (1960) studied the transfer of fluid and salt across the chorio-allantoic membrane of the pig. Their experiments were performed without the use of isotopic tracers but measurements of the potential and short-circuit current were made. They found that there was no net flux of sodium chloride and water across the allantoic epithelium (separated from the chorionic epithelium) when it was bathed on both sides by isotonic salines. The entire chorio-allantoic membrane, consisting of allantoic and chorionic epithelia separated by a stromal layer, did transport sodium from the foetal surface to the maternal side; the net sodium flux occurred in the absence of a concentration gradient and against an electrical gradient (maternal side positive). The short-circuit current was approximately equal to the net flux of sodium. Unfortunately no measurements of water flow were made in these particular experiments; in fact, Crawford and McCance assumed that there was no net transfer of water across the isolated chorio-allantoic membrane when both sides were bathed by isotonic salines. Indeed they suggest that the active transport of sodium ions out of the allantoic compartment makes it hypotonic during the later stages of development.

Conclusion

Our ignorance of the processes that trigger, enhance and, finally, block the formation of the extra-embryonic cavities is enormous. Moreover, there is practically no information about the transport characteristics of the membranes which bound these cavities. For instance, there are no quantitative estimates of the water and solute permeabilities of these membranes; nor do we know how these permeabilities change with the age of the embryo. Not enough work has been done on isolated embryonic membranes, such as the blastoderm, and this is very unfortunate since it would facilitate quantitative work on water relations. Evidently the fluid cavities which have been examined are generated primarily by the movement of water in the apparent absence of any driving force, and these cavities contain either isotonic or hypotonic fluid. It is conceivable that temperature gradients within the embryo might shift fluid transiently from one site to another, but this remains an entirely open question. Although water moves into some embryonic cavities against its chemical potential gradient this does not constitute proof for active water transport (see definition on page 65. The extent and nature of active solute transport in the embryo must be assessed before active transport of water can be accepted as the main mechanism of cavity formation. Thus, we require knowledge not only of the passive and active transport characteristics of the embryonic membranes but also of the temporal dependence of these characteristics before a satisfactory picture of this feature of development becomes available.

In spite of an enormous lack of rigorous experimental data, several conclusions can be drawn about the regulation of the embryonic fluids. First, the accumulation of a new fluid volume may result directly either from the formation of a new secretory membrane or from the change in position of an existing membrane. Secondly, the movement of fluid into certain embryonic cavities is not driven by its own chemical potential gradient and it is dependent upon metabolic energy; nevertheless, there is no unequivocal evidence for active water transport. Thirdly, the maximal rates of water transport into certain embryonic cavities of vertebrates are quite similar. Finally, chemical analyses of the fluids in certain embryonic cavities reveal that the fluid which is secreted into them is either isotonic or hypotonic to plasma.

Why does a particular epithelium in the embryo develop the power to secrete or absorb water and electrolytes in a preferential direction? What are the mechanisms of fluid secretion? Is the polarity of fluid transfer transmitted to all of the tissues derived from an embryonic membrane? Unfortunately we cannot really begin to answer these fascinating questions at present.

8

TRANSPORT ACROSS THE CAPILLARY WALL

The structure of capillaries	288
Continuous capillaries	290
Fenestrated capillaries	292
Discontinuous capillaries	293
Water transport	293
Filtration and osmosis	293
Exchange of labelled water	300
Water permeabilities and aqueous channels	301
Solute transport	302
Solute permeabilities	302
Vesicular transport	309
The route of solute and water transport	311
Continuous capillaries	311
Fenestrated and discontinuous capillaries	316

IN this chapter we shall make the huge leap from the preceding discussion of embryos and single cells to some problems of fluid transport in the intact animal. Our eventual aim in subsequent chapters will be to discuss water movement in epithelial tissues, and in order to fill the gap between single cells and the secretory and absorbing organs of animals some aspects of the circulatory system supplying these tissues ought to be described. It turns out that this task is more than a routine one for it presents us with both interesting and formidable problems concerning the transport of solutes and water across biological barriers. In this connexion the actual barrier to be studied in higher animals is the capillary wall. Although the capillaries are continuous with the arterioles and venules of the circulatory system, they are distinguished from these vessels by their complete lack of associated smooth muscle cells. To speak of capillaries as if they represented a fairly uniform group of structural and functional units is a mistake. This becomes especially clear when one examines particularly their structure. For this reason I have chosen to discuss first their structural details rather than their transport properties.

The structure of capillaries

For classifying capillaries Majno (1965) has modified a scheme proposed by Bennett, Luft & Hampton (1959) on the basis of the structure of the components of the capillary wall. The capillary wall is a composite barrier and it contains three layers in series—endothelium, basement membrane and finally the adventitial layer. The latter contains fibres and cells called pericytes which are encapsulated by the basement membrane in some capillary walls (Bruns & Palade, 1968a). According to Majno one can distinguish broadly three groups of capillaries from the standpoint of their endothelial structure. Table 8.1 outlines the classification and the

TABLE 8.1. Classification of capillaries according to Majno (1965)

Type	Name	Site
I	Continuous capillaries (continuous endothelial sheet: Fig. 8.1a,b)	Striated muscle, myocardium, lung, central nervous system, smooth muscle of digestive and reproductive system, and subcutaneous and adipose tissue, placenta
II	Fenestrated capillaries (intracellular openings: Fig. 8.1c,d)	Endocrine glands, sites of production or absorption of fluids (renal glomerulus, ciliary body of eye, choroid plexus, exocrine pancreas, salivary glands, intestinal villus), counter-current systems (rete mirabile of renal medulla, fish eye and swim bladder)
III	Discontinuous capillaries or Sinusoids (intercellular gaps, Fig. 8.1e,f)	Liver, spleen, bone marrow

sites where continuous, fenestrated and discontinuous capillaries are found. Within any given group one finds considerable variation of the structure of capillaries existing in different organs Figure 8.1 shows schematically the sort of variation which is found

THE STRUCTURE OF CAPILLARIES

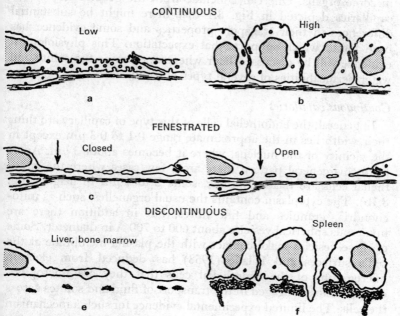

FIG. 8.1. Classification of capillary vessels according to the continuity of the main filtration barrier (the endothelial sheet). Three main types are distinguished (continuous, fenestrated, discontinuous) and, for each, two main varieties are given. Little detail is included because there are large variations from organ to organ: in fact almost each organ can be said to have its own type of capillary vessels. The scheme, according to Majno (1965), is based on information derived mainly from mammals. **a, b**; The endothelium has no recognizable openings. The low variety **a** is found in striated muscle, myocardium, central nervous system, smooth muscle of digestive and reproductive systems, and sub-cutaneous and adipose tissue. The high variety **b** is typical of the post capillary venules of lymph nodes and thymus: a similar endothelium is also found in the large arteries when contracted. **c, d**; The endothelium has intracellular fenestrae (*arrows*), either closed **c** as in endocrine glands, choroid plexus, ciliary body and intestinal villus, or open **d** as probably in the renal glomerulus. **e, f**; The endothelium has intercellular gaps. These vessels are also referred to as 'sinusoids'. They are typical of liver, bone marrow, and spleen: in each of these sites they differ in structural detail (Majno, 1965: Fig. 1).

in some organs. One can conclude from the general features of capillaries depicted in Fig. 8.1 that there might be substantial differences in their transport properties and some evidence has been found which supports that expectation. This physiological evidence will be discussed later when we examine the solute and water permeabilities of several types of capillaries.

Continuous capillaries

In general, the endothelial cells of this type of capillary are thin; their width lies in the approximate range 0·1 to 0·3 µm except in the vicinity of the nucleus, where it becomes about 3 µm. Apart from this general pattern there are a few cases where the endothelial cells are relatively thick, some 2 to 4 µm in height (Fig. 8.1b). The cytoplasm contains the usual organelles, such as mitochondria, granules and fine fibres, but in addition there are numerous spherical vesicles about 600 to 700 Å in diameter. Some of these vesicles make contact with the plasma membrane at the cell's surfaces and Palade (1953) has deduced from electron micrographs of the endothelial cells that they are micropinocytotic vesicles engaged in the transport of fluid and solutes across the cells. The limited experimental evidence for such a mechanism will be discussed later.

The nature of the intercellular junctions between endothelial cells has been the subject of several electron microscopical studies (e.g. Muir & Peters, 1962; Luft, 1965; Bruns & Palade, 1968a,b). In particular, Muir & Peters (1962) concluded that the intercellular spaces were closed invariably by 'tight junctions' or *zonulae occludentes* in the neighbourhood of the luminal surface of the capillary wall. At this point it is interesting to digress and consider the types of junctions that have been found between epithelial cells. Corresponding studies of intercellular junctions in epithelia have revealed several different types. For instance, Farquhar & Palade (1963) found three types of junction in the intestinal epithelium of mammals. In the immediate vicinity of the lumen the individual epithelial cells are joined by 'tight junctions'; below this they usually found the 'intermediate junctions' or *zonulae adhaerens*; finally, there are the desmosomes or *macula adhaerens*. Figure 8.2 shows these types of junctions diagrammatically. Since Farquhar and Palade did not observe the desmosomes in all of their sections they concluded that, in contrast to the other types, this junction does not form a continuous belt around the epithelial cells. On the other hand the 'tight junction' between epithelial cells represents apparently a region of fusion of the cell membranes where the intercellular space between the adjacent cell membranes disappears completely over a distance of about 0·4 μm. Accordingly it has been assumed that such 'tight junctions' are efficient seals against the passage of solutes and water across epithelia. The significance of the so-called 'tight junctions' in epithelia will be discussed in detail in Chapter 9. In the intermediate junctions, however, the intercellular space does not

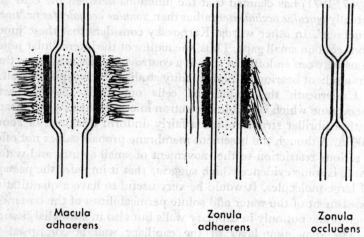

Macula adhaerens   Zonula adhaerens   Zonula occludens

FIG. 8.2. Schematic diagram of some intercellular junctions between epithelial cells. The desmosome or *macula adhaerens* occur between a wide variety of epithelial and endothelial cells. It is restricted to an oval area of the opposing cell surfaces. The *zonula adhaerens* differs from the *macula adhaerens* mainly because it is more extensive, has a less highly ordered cytoplasm border and has no widening of the intercellular space. This kind of junction may exist as a complete girdle around the apical region of the cell. The *zonula occludens* is a five-layered junction where the intercellular space is occluded by the fusion of the opposed plasma membranes. These junctions encompass the entire girth of the cells.

In some epithelia there is a characteristic arrangement of these junctions; the luminal *zonula occludens*, followed by an intermediate *zonula adhaerens* and finally a *macula adhaerens*. Variations from this arrangement have been noted: for example, in the ependyma the luminal junction is usually a *zonula adhaerens* rather than a *zonula occludens* (Brightman & Palay, 1963: part of Fig. 22).

disappear and certain workers have speculated that the cells are held mechanically together at these regions. According to Farquhar and Palade the endothelium of the capillaries exhibits principally 'tight junctions', with the 'intermediate junctions' being either absent or weakly developed. The question of the nature of the intercellular junction between the endothelial cells is quite important because the intercellular spaces may represent an important transport route across the capillary wall. In this respect it is interesting to note that Bruns & Palade (1968a,b), for example, have argued that the intercellular spaces in the wall of continuous endothelium are blocked by 'tight junctions' whereas Karnovsky (1967, 1971) has claimed that the junctions between the cells are actually *maculae occludentes* rather than *zonulae occludentes* or 'tight junctions'. In other words, Karnovsky considers that these junctions contain small gaps. Thus, the nature of the intercellular junctions between endothelial cells is a controversial issue and one that has a salient bearing on permeability studies of the capillary wall.

Underneath the endothelial cells one finds the basement membrane which is visible in electron micrographs as a membrane with a fibrillar structure and a fairly uniform thickness of about 500 Å. Although the basement membrane probably does not offer a serious restriction to the movement of small solutes and water, there is some evidence which suggests that it impedes the passage of large molecules. It would be very useful to have a quantitative assessment of the water and solute permeabilities of the basement membrane not only in capillary walls but also in epithelial tissues.

The remaining layer of the capillary wall is comprised of pericytes which are found in a wide variety of shapes. Invariably these cells do not form a complete layer around the wall but they have some processes which wrap themselves around the vessel. These cells have a comparable thickness to the endothelial cells and they also contain vesicles (Donahue & Pappas, 1961).

Fenestrated capillaries

In common with the continuous capillaries the fenestrated capillaries are comprised of a continuous basement membrane and endothelial cells which are interconnected by 'tight junctions'. In contrast, however, the endothelial cells are thinner than those in the continuous capillaries and they contain relatively few vesicles. Their other characteristic feature is fenestrae, which are circular perforations of the endothelial cells. Diameters cited for the fenestrae lie in the range 200 to 1200 Å (see Table 2, Manjo, 1965). In some cases the fenestrae are shuttered by thin diaphragms (less than 500 Å in thickness). There are several conflicting accounts of the nature of this diaphragm; for instance, it might be due to the apposition of two plasma membranes or alternatively it might be an artefact. Certainly on the basis of these structural features one could speculate that fenestrated capillaries ought to be more permeable to water and some solutes than the continuous capillaries. The experimental evidence bears that out.

Discontinuous capillaries

The endothelial cells of the sinusoids or discontinuous capillaries are separated by large gaps of 1000 to 10,000 Å in width. Moreover, the basement membrane is also discontinuous or in some cases absent. The presence of a loose association between adjacent endothelial cells is exemplified clearly in the sinusoids of the bone marrow, where both white and red blood cells have been observed in interstitial positions between endothelial cells. Of the three classes of capillaries we can see that the relatively open structure of the discontinuous capillaries ought to render them the most permeable type.

Water transport

Filtration and osmosis

The Starling hypothesis (Starling, 1896) is a suitable starting point for this discussion of fluid transport across the capillary wall. In the language of irreversible thermodynamics the hypothesis can be expressed as

$$J_v = L_p(\Delta p - \Delta \pi) \qquad 8.1$$

where J_v is the net volume flux ($cm^3\, cm^{-2}\, sec^{-1}$) across the capillary wall in response to the net effective 'driving force' composed of the hydrostatic pressure difference and the osmotic pressure difference. When both of these 'forces' are expressed in atmospheres the units of the hydraulic conductivity of the capillary wall become $cm\, sec^{-1}\, atm^{-1}$. In reality, Δp arises from the difference between the capillary blood pressure and the hydrostatic pressure of the interstitial fluid and $\Delta \pi$ arises from the difference between the osmotic pressure of the plasma proteins and that for the proteins in the interstitial fluid. Strictly speaking, equation 8.1 should include the reflexion coefficients for the particular protein molecules which may penetrate some capillary walls.

With the Starling hypothesis as a background, several investigators have measured fluid transfer in either perfused capillary beds or in single capillaries. Landis (1927, 1928) used single capillaries of the amphibian mesentery whereas other workers (Hyman, 1944; Pappenheimer & Soto-Rivera, 1948; Pappenheimer, Renkin & Borrero, 1951; Renkin & Zaun, 1955) have employed capillary beds in the perfused limbs of several mammals. In particular, the observations of Landis on single capillaries could be explained by equation 8.1, since he found that the net water transport across the capillary wall was a linear function of the capillary fluid pressure (Fig. 8.3). He noted that, provided

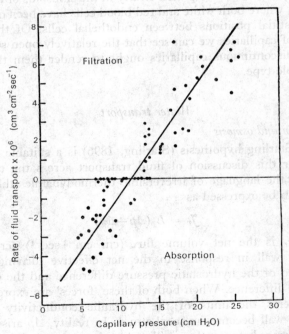

Fig. 8.3. Fluid transport across the capillary wall as a function of the capillary pressure (Landis, 1927: Fig. 10).

the capillary pressure exceeded 12 cm of water, fluid filtration occurred whereas below that pressure there was an absorption of fluid into the capillary lumen. Actually, within a certain range (9–12 cm H_2O) of capillary pressure there was no net flux because the difference in hydrostatic pressure was counterbalanced by the osmotic pressure of the plasma proteins. The so-called colloid osmotic pressure of plasma proteins in the frog is about 10 cm H_2O according to White (1924) and this fits well with the data of Landis. From the plot shown in Fig. 8.3 Landis obtained the filtration coefficient or hydraulic conductivity of the capillary wall; his average value for L_p was $5·8 \times 10^{-4}$ cm sec^{-1} atm^{-1} (Landis, 1927). Wind (1937) reported rather similar values in the range $3·3$–$8·7 \times 10^{-4}$ cm sec^{-1} atm^{-1} for capillaries in the toad mesentery but he noted that the value of L_p increased towards the upper limit during the course of the experiments. Estimates of L_p for a variety of capillary membranes are given in Table 8.2, although it should be stressed that most of these values are indirect estimates for capillary beds. These estimates have been obtained on the basis that all capillaries in a bed have identical L_p values and, indeed, that each capillary itself is uniformly permeable to water along its entire length. Notwithstanding those assumptions and the difficulties associated with estimates of the area of the capillary bed, the values for L_p in Table 8.2 are, in general, significantly larger than the corresponding values for cell membranes (Table 5.2) and for epithelia (Table 9.2). One obvious inference from those comparisons is that the route of pressure-driven flow across some capillary walls, particularly the fenestrated capillaries, is probably an extracellular one rather than a transcellular one. That assertion may apply to all capillary walls but supporting evidence is certainly necessary, particularly for the least permeable capillaries cited in Table 8.2. Indeed, such corroborative data arises from studies of solute transport which will be discussed later.

The range of L_p values for capillaries is large but the range of water permeabilities for cell membranes is even larger. Dick (1959a, 1966) has argued that the latter range is due to the rate-controlling influence of mutual diffusion of water and intracellular macromolecules rather than to a genuine variation in the water permeability of cell membranes. The crux of Dick's argument is that it is the relatively long diffusion distances in the

TABLE 8.2. Hydraulic conductivities of some endothelial walls

Tissue		$L_p \times 10^7$ (cm sec^{-1} atm^{-1})	Reference
Blood–brain barrier	(rabbit)	3	Fenstermacher & Johnson (1966)
Corneal endothelium	(cat)	9·3	Rhee et al. (1971)
Corneal endothelium	(rabbit)	14*	Green & Green (1969)
Corneal endothelium	(rabbit)	160*	Mishima & Hedbys (1967)
Skeletal muscle	(human)	69	Landis & Gibbon (1933)
Skeletal muscle	(cat)	220	Pappenheimer & Soto-Rivera (1948)
Skeletal muscle	(dog)	250	Pappenheimer & Soto-Rivera (1948)
Skeletal muscle	(rat)	600	Renkin & Zaun (1955)
Cardiac muscle	(rabbit)	880	Vargas & Johnson (1964)
Blood–brain barrier	(cat)	1060†	Coulter (1958)
Mesentery	(toad)	3300–8700	Wind (1937)
Mesentery	(frog)	5800	Landis (1927)
Mesentery	(rabbit)		
arterial side		2100–8300	Zweifach & Intaglietta
venous side		17,000–26,000	(1968)
Vasa recta	(rat)	306	Morgan & Berliner (1968)
Glomerulus	(frog)	22,800	Pappenheimer et al. (1951)
Glomerulus	(dog)	58,000	Unpublished work of Verney and Rushton cited by Pappenheimer et al. (1951). (See also Verney, 1950)

* Large unexplained disparity.
† This value was obtained by elevating the pressure of the cerebrospinal fluid and monitoring its volume: however, Fenstermacher & Johnson (1966) have criticized that method on the grounds that apart from fluid filtration across the brain capillaries there was also a significant, and possibly predominant, filtration across the arachnoid villi. Coulter's value, therefore, is probably a gross overestimate.

large cells which render them *apparently* less permeable to water and consequently that there ought to be a relation between the apparent value of L_p and the surface-to-volume ratio for cells. In the endothelial cells of the capillary wall, however, we meet relatively short diffusion distances in the range of say 0·2 to 0·5 μm, so that Dick's argument is inapplicable here. It seems more likely that the observed variation in the water permeabilities of capillaries reflect differences in the number and geometrical characteristics of pathways within the walls. That we are dealing here with Poiseuille-type bulk through aqueous channels is supported plausibly by the finding (Pappenheimer, 1953) that L_p for the capillary wall in the perfused hind limb changes with temperature in inverse proportion to the viscosity of water (cf. equations 3.30 and 3.31). Similar studies by Brown & Landis (1947) on single capillaries of the frog's mesentery revealed a similar trend, but due to a large variation in their data no quantitative test of the dependence of L_p on η_w is warranted.

Other evidence upholding Starling's hypothesis was obtained by Pappenheimer & Soto-Rivera (1948) who determined the osmotic pressure of the plasma proteins in perfusion experiments on hindlimb capillaries of cats and dogs. They found a strict equivalence between the effective osmotic pressure gradient across the capillary wall and the differences in hydrostatic pressure required to counter-balance it. These experiments, along with the classic study of Landis on single capillaries, substantiate the use of equation 8.1 to describe net fluid transport across the capillary wall, with the proviso that the reflexion coefficient of the plasma proteins can be safely ignored.

The conditions under which equation 8.1 is known to hold do not include, however, the cases where significant alterations in the total osmotic pressure may occur. Until recently no quantitative data were at hand about osmotic withdrawal of fluid from the capillary. Landis & Sage (1971), however, have reported such osmotic experiments on single capillaries of frog mesentery. They irrigated the capillaries with Ringer solution to serve as a control and at some predetermined time they permitted a new fluid, either hypotonic or hypertonic, to irrigate the preparation. The net water fluxes were determined from analysis of consecutive motion-picture frames of single capillaries undergoing osmotic transients. Fig. 8.4 shows their results expressed as the net rates of fluid transport recorded at given osmotic pressures of the perfusate. Also included in Fig. 8.4 are the fluid filtration and absorption rates obtained in previous experiments on single mesenteric capillaries of the frog; the small (hatched) bar gives the range 5–27 cm H_2O (Landis, 1927); the large (vertically shaded) bar gives the corresponding filtration rates in capillaries damaged by chemicals (Landis, 1927). The striking feature in Fig. 8.4 is that relatively small differences in hydrostatic pressures are about as effective as quite big osmotic gradients. Another way of expressing that is to calculate the hydraulic conductivity from the osmotic data in Fig. 8.4 and this gives L_p in the range $4 \cdot 0 – 7 \cdot 3 \times 10^{-6}$ cm sec^{-1} atm^{-1}. When those values for L_p are compared with the corresponding values determined in the filtration experiments

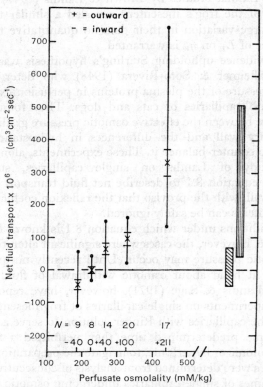

FIG. 8.4. Net fluid transport across walls of single capillaries as a function of the osmotic pressure of the bathing solution. Frog mesenteries were irrigated with Ringer's solutions having osmolalities ranging from 186 to 437 (mM/kg). For each osmotic pressure are shown total range (●), mean (×) and S.E. (−). N indicates number of experiments. The numbers above the horizontal axis denote the osmotic pressure difference (mM/kg) between the experimental solution and the preceding control Ringer's fluid. The shaded bars on the right signify the range of filtration rates observed in previous experiments where hydrostatic pressure rather than osmotic pressure was the driving force (Landis & Sage, 1971: Fig. 5).

they are found to be about 1/100 of the latter. It is apparently remarkable that some single animal cells (see page 167), capillaries and epithelia (see page 336) should all exhibit this anomaly even to the extent of a quantitative similarity between the ratios of the measurements. Probably the numerical identity of the ratios is completely fortuitous and it is unlikely that the discrepancy between the two coefficients stems from a similar source. In particular, since the mesenteric capillaries are so permeable to water it is fair to assume that their permeability for solutes, such as NaCl, is also high. Thus, in the description of the osmotic experiments of Landis & Sage (1971) we need to include the reflexion coefficient for NaCl. In other words, the effective osmotic gradients were probably less than the nominal ones employed. In order to get some idea of the underestimate in L_p which follows from ignoring σ, one can make a plausible guess about its magnitude; for instance, Taylor & Gaar (1970) estimate that σ for NaCl in pulmonary capillaries is about 0·02. Introducing a similar value into the data of Landis & Sage (1971) means that L_p becomes 50 times larger than its quoted value and only one half of the value obtained in the filtration experiments. Thus, the relatively low values of L_p found in the osmotic experiments arise because there is probably a rapid solute movement between the capillary lumen and the perfusate.

Implicit in the general approach to fluid transport in capillaries are the assumptions that the hydrostatic pressure and the protein concentration in the extravasular space are negligible and that each capillary has uniform permeability characteristics along its length. Those assumptions are surrounded by doubt, since, for example, there is good evidence (Guyton, Granger & Taylor, 1971) for the existence of interstitial pressure gradients in a variety of tissues. Moreover, Zweifach & Intaglietta (1968) have challenenged the long-standing view that a given capillary has a unique value for the filtration coefficient. They analysed the results of their osmotic transients, induced by intravenous injections of albumin, in order to obtain L_p for single mesenteric capillaries in the rabbit. Their data demonstrated that the venous side of single capillaries was more permeable to water than the arterial side; the former lay in the range $1·7$–$2·6 \times 10^{-3}$ cm sec^{-1} atm^{-1} whereas the latter had values within $2·1$–$8·3 \times 10^{-4}$ cm sec^{-1} atm^{-1}. In view of these results and other factors Zweifach & Intaglietta (1968) rightly stressed the hazards of attempts to estimate L_p for capillary beds in different organs either by direct means or by extrapolations from the behaviour of single capillaries. Thus, it is wise to regard the values for L_p cited in Table 8.2 as a rough guide only.

Exchange of labelled water

It is exceedingly difficult to obtain the 'true' rate of exchange of labelled water molecules in the capillary wall because the actual transmural concentration gradient is probably less than that estimated from the arterial concentration. In the capillary, just as in the single cell or epithelial tissue, the study of labelled water movement is bedevilled by the presence of rate-limiting influences at the boundaries of the preparation. On the basis of the values of L_p found for capillaries (Table 8.2) one might expect the corresponding values of P_d to lie in the range 4×10^{-4} to 8 cm sec^{-1}. It is important to note that even a small unstirred layer of, say, 2μm would rate-limit water exchange across any capillary with a permeability larger than 10^{-1} cm sec^{-1}. Irrespective of whether or not this argument is quantitatively correct it does indicate that in the capillary, which is a relatively permeable barrier, external sources of rate-control on water exchange are likely to be very troublesome. In particular, the efflux of labelled water from the capillary lumen is limited probably by the rate of blood flow itself. For instance, Johnson, Cavert & Lifson (1953) have claimed that the rate of blood flow over a wide range dominates the kinetics of labelled water exchange in both perfused cardiac and skeletal muscle. A similar state of affairs also occurs in the ventral sac of the cow, where the clearance of labelled water from the lumen is probably a 'useful index' of mucosal blood flow in that organ (Dobson, Sellers & Thorlacius, 1971). This is a particularly good example of turning what seems to be a disadvantage to one's advantage because it provides a way of monitoring mucosal blood flow, a parameter not easily measured by other means.

In a study of water and solute permeability properties of the rat kidney, Morgan & Berliner (1968) attempted to measure the diffusional and osmotic permeabilities of the vasa recta. They found that $P_d = 2 \times 10^{-3}$ cm sec^{-1} whereas $L_p = 3 \cdot 1 \times 10^{-5}$ cm sec^{-1} atm^{-1}. Unfortunately they did not determine to what extent their estimate of P_d was dependent upon the perfusion rate of 30 nl min^{-1}, and since the lumen of the vasa recta was about

18 μm their perfusion rate corresponds to a fluid velocity of about 10 cm min^{-1} in the capillary lumen. Because their estimate of P_d was obtained under rather poor stirring conditions it is probably an underestimate of the 'true' value for the diffusional permeability. In fact, it seems that there are no reliable values for P_d for any type of capillary and this is not surprising in view of the technical difficulties associated with such experiments.

Water permeabilities and aqueous channels

Apart from the temperature dependence of fluid filtration there is no direct evidence from the water flux measurements themselves that water flow across the capillary walls is quasi-laminar or Poiseuille type. Admittedly if one had accurate estimates for both L_p and P_d, the diffusional permeability to water, for a given capillary one might be able to assert whether or not it contained aqueous channels. This approach was tried by Pappenheimer et al. (1951) who obtained P_d indirectly. Using an 'osmotic transient' method, to be discussed later, they estimated the permeability of capillaries in cat skeletal muscle to various solutes. From the curvilinear relation between permeability and the square root of the molecular weight of the test solutes they extrapolated to a value for P_d on the basis that the behaviour of water would be similar to that of the test solutes. By that somewhat indirect method they calculated that $P_d = 5\cdot 4 \times 10^{-4}$ cm sec^{-1}. Of course, this indirect estimate of P_d needs to be corrected for unstirred layers because the solute permeabilities upon which it is based also suffer from that source of error. Subsequently Landis & Pappenheimer (1963) pointed out that the 'osmotic transient' method ignored the reflexion coefficients of the test solutes and their attempts to correct for that omission yielded $P_d = 2\cdot 8 \times 10^{-4}$ cm sec^{-1}. Thus, taking $P_d = 2\cdot 8 \times 10^{-4}$ cm sec^{-1} and $L_p = 2\cdot 5 \times 10^{-5}$ cm sec^{-1} atm^{-1} (Table 8.2) for the muscle capillaries and inserting them into equation 3.36, gives an equivalent pore radius of 41·6 Å. The validity of the conclusion, that the 'continuous' capillary wall is a porous structure, rests heavily on the reliability of the water permeabilities, particularly P_d. Moreover, it is hardly proper to use an indirect estimate for P_d for that sort of test. In the case of the vasa recta, Morgan & Berliner's (1968) more reliable data show that $(L_p RT/\bar{V}_w P_d)$ is about 21 and, as we have seen before, the ratio of the osmotic to the diffusional water permeability must significantly exceed 2 before we may conclude that the barrier is porous. The preceding calculation, therefore, does provide a clue that this fenestrated capillary behaves as a porous barrier, with the equivalent pore radius being about 17 Å. Because of the uncertainties about the accuracy of the water permeabilities, particularly P_d, the preceding argument constitutes a rather poor guide to the porosity of the capillary wall. Thus, one must turn to the solute permeability characteristics of the capillary wall to see if they yield any insights about pathways and mechanisms of transport.

Solute transport

Solute permeabilities

It is now established that capillary walls are relatively permeable to solutes in comparison to cell membranes. Pappenheimer and his collaborators were among the first to determine solute permeability coefficients for capillary walls. Pappenheimer *et al.* (1951) published a theory of solute transport in porous membranes (see Chapter 3) and used this approach to obtain estimates of solute permeabilities for the capillary walls in the perfused hindlimb of the cat. The basis of their experimental approach was an analysis of the so-called *osmotic-transient* generated by the addition of a quantity of a test substance to the arterial blood supplying the tissue. During the response to the increase in osmotic pressure there was a tendency for water to be drawn osmotically into the capillary lumen, and this was accompanied by an efflux of the test substance into the interstitial fluid. Pappenheimer and his colleagues adjusted the mean hydrostatic pressure in the capillaries during the response so that the osmotic withdrawal of fluid was prevented; they argued that the magnitude of the *osmotic-transient*, monitored by variations in the required capillary pressure, depends upon the substance and its concentration whereas its time constant is inversely related to the diffusion coefficient of the test solute. Furthermore, the net efflux of the test solute across the capillary walls during the *osmotic-transient* is given by the product of the solute concentration difference (between arterial and venous blood) and the blood flow; in fact, blood flow was maintained constant during the transient. Their experiments yielded the solute permeabilities from a knowledge of both the efflux rate and the effective difference in concentration of the test solutes;

the latter was deduced from the osmotic pressure determined during the transient. It is explicitly assumed in their treatment that the osmotic pressure $\Delta\pi$ exerted by the test solutes across the capillary wall is related directly to the mean difference in solute concentration Δc_s by the van't Hoff law, i.e. $\Delta\pi = RT\Delta c_s$. On the assumption that 100 g of skeletal muscle contained 7000 cm^2 of capillary surface, it was possible to obtain the solute permeability from these measurements (Pappenheimer, 1953). Table 8.3 shows the permeability of muscle capillaries to a variety of solutes; the third row shows the estimates which Pappenheimer obtained from the *osmotic-transient* data. It is evident that these solute permeabilities are much larger than the equivalent values for cell membranes. For example, the urea permeability of the membranes in the alga, *Nitella translucens*, has been found to be 2×10^{-5} and 7.4×10^{-7} cm sec^{-1} for the plasmalemma and tonoplast respectively (Dainty & Ginzburg, 1964b). On the other hand, the urea permeability of the human red cell is about 4×10^{-4} cm sec^{-1} (Sha'afi, Rich, Mikulecky & Solomon, 1970); however, as Pappenheimer (1953) has pointed out, the red cell is an exceptional case of a cell which is almost as permeable to solutes as the capillary wall. The salient features of the permeability data in Table 8.3 are, first, that the solute permeabilities are large and, secondly, that they seem to depend upon the size of the test solutes. Indeed, Pappenheimer *et al.* (1951) interpreted those data for the muscle capillaries in terms of a porous-barrier model and they estimated the *restricted pore area per unit path length*, $(A_p'/\Delta x)$, of the diffusion pathway for each solute. The latter parameter is given by equation 3.3 and $A_p' = (D_s'/D_s)A_p$, where A_p is the true pore area and (D_s'/D_s) is the ratio of the restricted diffusion coefficient to the free one (see page 79). Of course, if we are dealing with a barrier where the pore radii are much larger than the radii of the permeant solutes then there is no restriction and $A_p' = A_p$. For the cat's hindlimb Pappenheimer estimated that $(A_p'/\Delta x)$ for NaCl was 1.03×10^5 cm and taking $\Delta x = 1$ μm (i.e. thickness of capillary wall) he found that the restricted pore area was 10 cm^2. On the assumption that this approach could be applied to any molecule of known molecular radius he obtained a value of 12 cm^2 for water. Thus, for both NaCl and water molecules transversing the capillary wall the area of the diffusion pathways is about 10 cm^2 or less than 0.2% of the total surface area (7000 cm^2).

TABLE 8.3. Solute permeabilities of some capillary walls

	Permeability × 10⁵ (cm sec⁻¹)									
Tissue	Labelled water	Urea	Glycerol	Glucose	Sucrose	Raffinose	Inulin	Myoglobin	Haemoglobin	Reference
Vasa recta (rat)	200	47	—	—	—	—	—	—	—	Morgan & Berliner (1968)
Mesentery (rat)	121	63	—	—	—	—	—	—	6·6	Rasio (1970)
Skeletal muscle (cat)	—	—	—	48	—	6	—	10·4	0·00	Pappenheimer (1953)
Skeletal muscle (cat)	54*	26	—	9	5	4	0·5	0·04	—	Landis & Pappenheimer (1963)
Skeletal muscle (cat)	28*	14	—	6	4	3	0·3	0·1	0·001	Garlick (1970)
Skeletal muscle (cat)	—	—	—	—	2·1	—	0·26	—	—	Crone (1963a)
Skeletal muscle (dog)	—	—	—	—	0·74	—	0·26	—	—	Trap-Jensen & Lassen (1970)
Skeletal muscle (human)	—	2·9	—	1·3	0·9	0·5	0·09	—	—	Schafer & Johnson (1964)
Heart muscle (rabbit)	—	—	—	—	11	—	0·4	—	—	Vargas & Johnson (1967)
Heart muscle (rabbit)	—	—	—	10	6	—	0·5	—	—	Alvarez & Yudilevich (1969)
Heart muscle (dog)	—	3	1·5	1·0	0·8	—	0·27	—	—	Crone (1965)
Brain (dog)	—	0·44	0·21	0·16	0·00	—	0·00	—	—	

The above approach to solute permeation was an exceedingly useful starting point because it presented a sort of comprehensive account of solute transport and it suggested that the capillary wall behaved as a porous structure; furthermore it suggested that the aqueous channel has either an equivalent pore radius of 31 Å or slit width of 37 Å. At the time these estimates of the size of the porous channel were apparently very significant because they were in accord with the dimensions of some solutes, such as haemoglobin, which just failed to penetrate the capillary wall. However, the above analysis has been subjected to considerable modification for several reasons which will now be discussed.

An important assumption in the work of Pappenheimer *et al.* (1951) is that the mean concentration difference of the test solute across the capillary wall during the *osmotic-transient* can be estimated from the osmotic pressure by the van't Hoff relation. Because the capillary is permeable to the test solute, the effective osmotic pressure is less than the theoretical van't Hoff osmotic pressure and, in fact, it is given by $\sigma_s RT \Delta c_s$ where σ_s is the reflexion coefficient of the capillary wall for the solute s. The omission of the reflexion coefficient from the original theoretical analysis has been criticized by several workers (Ussing, 1953; Grim, 1953; Kedem & Katchalsky, 1958) and this source of error leads to an overestimate of solute permeability.

Landis & Pappenheimer (1963) have attempted to correct for the omission of the reflexion coefficient from the original analysis. Their approach, however, is an indirect one based on the assumption that the reflexion coefficient can be computed from a frictional model of solute and water transport in a porous barrier. Thus, their method of correction is rather unsatisfactory for two reasons: first, it does not rest on experimentally determined values of the reflexion coefficients and, secondly, the expression for the reflexion coefficient is open to criticism. In its favour can be claimed that it does give a rough quantitative guide of how much the solute permeabilities, and consequently the pore radius, need to be altered. They found, in fact, that their corrections yielded pore radii in the range 41–45 Å in contrast to the original value of about 31 Å. Landis & Pappenheimer (1963) also gave estimates of the solute permeabilities obtained in their revised treatment (see row 4, Table 8.3). Thus, failure to take account of the reflexion coefficient leads to an underestimate of the equivalent pore radius and an overestimate of the solute permeabilities. The magnitude of the appropriate corrections can be established only when the reflexion coefficients have been measured in the capillaries of mammalian skeletal muscle. At present no reliable experimental data are available for that preparation. Kedem & Katchalsky (1958) obtained estimates of the reflexion coefficients for glucose, sucrose and inulin from the data of Pappenheimer *et al.* (1951); however since their argument demands a knowledge of the hydrostatic and osmotic pressure differences across the capillary at the beginning of the *osmotic-transients* and since the original experiments were not designed to yield that information, their estimates of σ_s must be treated with caution. Some workers have tried to measure reflexion coefficients in other capillary preparations with the aim of assessing the porosity of the capillary walls. In particular, Vargas & Johnson (1964) estimated σ_s for urea, sucrose, raffinose and inulin from osmotic experiments on the excised rabbit heart perfused with different solutions. Using the expression for σ_s employed by several workers (Durbin, 1960), namely $\sigma_s = 1 - A_{sf}/A_{wf}$, they concluded that the equivalent pore radius was about 35 Å. In later work Vargas & Johnson (1967) obtained estimates for the solute permeabilities from the same experimental data (see row 9, Table 8.3). Several correction factors of an arbitrary nature were employed by Vargas & Johnson (1967) in their analysis of the experimental data and so their values for solute permeability must be subject to doubt. Pappenheimer (1970) has criticized the basic assumption of both Kedem & Katchalsky (1958) and Vargas & Johnson (1964) that the difference in concentration of the test solutes across the capillary wall can be estimated at the beginning (zero time) of the *osmotic transient* by their extrapolation procedures. Indeed, Pappenheimer claims that 'the mean concentration difference across the wall is in fact zero at zero time and is never more than a small fraction of the increment in arterial plasma except for large molecules' because of the very rapid rates of diffusion across the capillary wall. In order to be able to accept or reject Pappenheimer's argument we need to know not only the longitudinal gradients of solute in the capillary and in the adjacent interstitial fluid but also how these gradients are influenced by blood flow and the true solute permeability of the capillary wall. Of course, we could turn Pappenheimer's argument about the mean concentration difference on to his own *osmotic-*

transient data and ask how it is possible to measure an equivalent osmotic pressure transient for each solute if the efflux of solute is so rapid.

Undeniably it is technically difficult to get accurate estimates of solute permeabilities and reflexion coefficients for capillaries. One source of difficulty is the possible influence of the rate of blood flow on the measurements. In the capillaries of skeletal muscle Pappenheimer *et al.* (1951) found that the time course of the *osmotic-transient* was relatively independent of the rate of blood flow (see their Figs. 5 and 6). There are other instances, however, where the rate of blood flow is important—for example, in the exchange of labelled water.

Apart from the *osmotic-transient* approach there are other ways for measuring solute permeabilities. Crone (1963a), for example, has used the 'indicator diffusion' technique. This method consists in comparing the venous concentrations of a permeant (test) solute and of an impermeant (indicator) solute after both solutes have been injected simultaneously into the arterial blood supplying the tissue. Martin de Julian & Yudilevich (1964) have defined the extraction, E, of the test solute by

$$E = \lim_{t \to 0} \left[1 - \frac{c(t)}{C(t)} \right] \qquad 8.2$$

where $c(t)$ and $C(t)$ are the venous concentrations of the test and indicator solutes respectively. Provided that there is no recirculation of the blood during the experimental period relation 8.2 may be used to estimate E. For example, Alvarez & Yudilevich (1969) obtained E for an isolated perfused heart by measuring $c(t)$ and $C(t)$ in successive venous samples taken at 0·24 sec intervals after a rapid injection of labelled solutes. Their data permitted an accurate estimate of $c(t)/C(t)$ at zero time to be obtained. Knowing the extraction, E, for the test solute it is possible to calculate its permeability, P_s, from the following expression given by Renkin (1959) and Crone (1963a)

$$P_s A = -Q \ln (1-E) \qquad 8.3$$

where A is the capillary area in 1 g of tissue and Q is the rate of blood flow (cm^3 sec^{-1} per g of tissue).

Equation 8.3 was obtained in the following manner. Let us assume that c_a and c_v are the arterial and venous concentrations of the test solutes.

The quantities of the test solute entering and leaving the microcirculation in unit time are Qc_a and Qc_v respectively. Thus, the quantity of the solute which diffuses out of the capillaries is $Q(c_a - c_v)$ which may be equated with $P_s A \Delta c_s$, where Δc_s is the mean concentration difference across the capillary wall. According to Crone (1963a) the solute concentration profile along the capillary lumen may be taken as exponential, in which case the mean concentration in the capillary lumen is $(c_a - c_v)/(\ln(c_a/c_v))$. The latter term is equal to Δc_s, since the external concentration of test solute is probably negligible. Thus, we have

$$P_s A \Delta c_s = Q(c_a - c_v) \qquad 8.4$$

Substituting for Δc_s and re-arranging equation 8.4 gives

$$P_s A = Q \ln \left(\frac{c_a}{c_v} \right) \qquad 8.5$$

which is identical to equation 8.3 when it is realized that the extraction E is equivalent to $(1 - c_v/c_a)$.

Inspection of equation 8.3 shows that when $(P_s A/Q)$ is very large the extraction of the test solute will also be large. That condition means that in capillaries with high permeabilities to the test solute the efflux of the solute is limited by the rate of blood flow. In contrast, when $(P_s A/Q)$ is small the exchange of the solute is rate-limited by its diffusion through the capillary wall. In principle, the preceding theory offers us a method of estimating solute permeability provided E can be estimated accurately. In fact, expressions 8.2 and 8.3 relating P_s and Q have been widely used despite some doubts about their applicability, particularly to solute transport in capillary beds. Renkin (1959), for example, found that the clearance of radioactive potassium ions from blood into skeletal muscle was not related to blood flow as the theory predicted and he argued that the area of the capillary surface was dependent upon the perfusion rate. It might be contended, however, that no reliable extimate of P_s can be obtained in that case because the the exchange is entirely rate-controlled by blood flow. In a later paper Renkin (1967) demonstrated that the theory holds for flux measurements in an artificial kidney possessing an invariant area of membrane (Wolf, Remp, Kiley & Currie, 1961). Moreover, Alvarez & Yudilevich (1969) showed that there were no changes in the capillary area of the canine heart during their measurements of P_s by the 'indicator diffusion' method.

It is an important assumption in the derivation of the theory of 'indicator diffusion' that the concentration of labelled solutes on the outside of the capillary wall is zero. That assumption is probably not valid, especially when the extraction of labelled solute is large. Actually it has been contended (Vargas & Johnson, 1964; Chinard, Enns, Goresky & Nolan, 1965) that the 'indicator diffusion' method yields estimates for the exchange of solutes across the capillary wall which are rate-limited solely by blood flow. In order to examine that objection Alvarez & Yudilevich (1969) studied the transcapillary exchange of certain solutes in the isolated canine heart. They measured the fractional extraction at different rates of perfusion and they found that the analysis (equations 8.2 and 8.3) applied to the data except at relatively low rates of blood flow for the relatively permeant solutes. Under the latter circumstances, where $(P_s A/Q)$ is very large, P_s cannot be obtained by 'indicator diffusion' because Q, by itself, determines the rate of solute exchange. The values for the permeabilities which Alvarez & Yudilevich (1969) obtained are shown in row 10 of Table 8.3.

Vesicular transport

According to some workers, including Pappenheimer and his colleagues, small solutes penetrate the capillary wall via a small-pore system. The transport route occupies about 0·1% of the capillary surface and is comprised of numerous equivalent pores of about 40 Å in radius and 10,000 Å in length. In contrast to the rapid movement of small molecules across the capillary wall there is a relatively slow transport of large molecules, such as proteins. Grotte (1956) has suggested that those large solutes move through a thinly scattered system of pores with radii in the range 120–350 Å. Evidently this so-called large-pore system possibly occupies about 0·0001% of the capillary surface. Given that sort of density for such pores, it is hard to expect electron microscopical studies to offer convincing evidence for or against a large-pore system. At present, for example, there is no strong support for the particular view that a number of large extracellular spaces may be the source of the large-pore system.

Actually the large-pore system now seems to be a red herring, since Bruns & Palade (1968b) have presented a strong case for vesicular transport as the mechanism whereby certain large solutes cross the endothelium. They found that the vesicles in the endothelial cells could be labelled with ferritin injected intra venously. Since ferritin is quite a large molecule (diameter 110 Å it was not surprising that Bruns and Palade found no ferritin in the intercellular spaces. Their study demonstrated that its transport path across the endothelium is entirely a transcellular one Within the endothelial cells the labelling of the vesicles adjacent t the capillary lumen was found to be more intense than of thos vesicles on the opposite side of the endothelial cells and, further more, only a small fraction (10%) of the total number of vesicle was labelled with ferritin.

Several theoretical models of vesicular transport have bee published (Renkin, 1964; Shea & Karnovsky, 1966; Bruns & Palade, 1968b; Tomlin, 1969; Shea, Karnovsky & Bossert, 1970 In particular, the probabilistic model of Shea et al. (1970), whicl is based on the anatomical data of Bruns and Palade (1968b), pre dicts that only about 10% of the observed vesicles will cross th endothelial cell from the capillary lumen to the exterior and that th transit time is about one second. Earlier estimates (Renkin, 1964 Bruns & Palade, 1968b) of the transit time for vesicular transpor across an endothelial cell, say 0·3 μm thick, ranged from 300 t 24 sec and these are probably unreliable because they are based on the false assumption that all of the observed vesicles mus successfully complete a transport pathway across the endothelium Not only are the earlier estimates of transit time unreliable bu also they are larger than those expected from the known rates o transport of large molecules, such as dextrans, across the capillary wall (Renkin, 1964). Finally, the vesicular transport model of Shea et al. (1970) correctly predicts that after labelled molecules are introduced into the capillary lumen the majority of the labelled vesicles will be in the vicinity of the blood side of the endothelia cells even under conditions of steady-state transport of the labelled molecules across the endothelium. Moreover, the in homogeneity of the vesicular labelling will break down only if the cells are extremely thin.

Apart from the theoretical support for a vesicular mechanism of transport across the capillary endothelium, there is the addi tional favourable experimental evidence that the extent of 'molecular sieving' is very poor for molecules in the molecular weight range 100,000 to 400,000 (Grotte, 1956; Mayerson Wolfram, Shirley & Wasserman, 1960).

The route of solute and water transport
Continuous capillaries

There are several possible pathways for solute and water transport across continuous capillaries. Figure 8.5 depicts the alternative routes.

Let us concern ourselves first with the direct route (1) across the endothelial cells. If all of the water flow, whether it be driven by diffusion, filtration or osmosis, passes across the endothelial

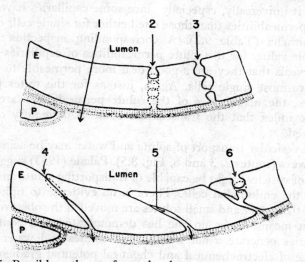

FIG. 8.5. Possible pathways for molecular transport across a continuous endothelium. E signifies endothelium with underlying basement membrane and P, part of a pericyte. **1**; Direct pathway across the endothelial cell, probably the main pathway for small molecules, such as gases and water, and possibly for lipid-soluble substances. **2**; Transport of vesicles which form at one cell surface and migrate towards the other. **3**; Vesicles emptying into each other. **4**; Passage along the junctions. The first part of this pathway appears closed by a 'tight' junction but the junction may not be completely impermeable. The *dotted lines* indicate that the passage might be temporarily forced open, or that it might be permanently open at certain points which have as yet escaped observation by electron microscopy. Recent work (Karnovsky, 1967), in fact, indicates that junctions may be 'gap' junctions or *macula occludentes*. **5**; Bypass of the junction by diffusion, i.e. a combination of diffusion across a thin layer of endothelial cytoplasm and channelling along part of the junctions. **6**; Bypass of the junction by vesicles which may carry materials to and from the junction, beyond the apparent 'tight' seal (Majno, 1965: Fig. 30).

cells then the hydraulic conductivity of the endothelial cell membranes lies in the range $3–60,000 \times 10^{-7}$ cm sec^{-1} atm^{-1} (Table 8.2). Some of these values are quite close to the quoted permeabilities for erythrocytes (see Table 5.2) and for some plant cells (see Table 5.4). Simply on the basis of the water permeabilities of capillaries the endothelial cells seem to be as permeable to water as some single animal and plant cells. While one has no reason, therefore, to doubt the transcellular route for water transport across some capillaries, neither does one have a good reason to accept it universally, especially since some capillaries have larger water permeabilities than those cited either for single cells or even for epithelia (Table 9.2). A corresponding inspection of the available values for the solute permeabilities of capillaries (Table 8.3) reveals that they are a good deal more permeable to solutes than are most single cells. Again, just as for the water permeabilities, the magnitudes of the solute permeabilities are not a reliable index that the transcellular route is obligatory or even preferred.

The vesicular transport of solute and water may be achieved in three ways (routes 2, 3 and 6, Fig. 8.5). Palade (1953) suggested a traffic of vesicles might be capable of transporting solute and water across the endothelial cells. There is no evidence to uphold the notion that water and small solutes are moved at the observed rates by that means. Since no one has demonstrated that continuous capillaries generate a net transport of solutes and water in the absence of electrochemical and chemical potential gradients, one must assume that vesicular transport generates a one-to-one exchange, if any, of some substances across the capillary wall. Consequently it is extremely difficult to envisage how vesicular transport of fluid might play an important role in filtration or osmosis, for example. Where vesicular transport does seem to be important is in the movement of large molecules from the blood to the interstitial space. A large host of molecules and small particles, including ferritin, gold, thorium dioxide and colloidal carbon, are observed within vesicles and probably transported outwards by that mechanism since such particles do not enter the intercellular spaces and are rarely seen as free particles in the cytoplasm of endothelial cells. It seems reasonable to expect that the vesicular mechanism can transport some large molecules in the opposite direction, that is from the interstitial space into blood, although such movement has not been examined extensively. Brandt (1962), for example, found that ferritin molecules injected into the skeletal muscle of the cat eventually occupied vesicles in the endothelium of the muscle capillaries.

Apart from the transport of large molecules across the capillary wall the vesicular movement of other substances, particularly water and small solutes, is probably insignificant. At least there is no compelling evidence in favour of that mechanism for small molecules. In fact, it is easier to accept that small molecules cross the wall either by the transcellular path alone or by the intercellular one alone (route 4), or by both together, since the water and solute permeabilities are relatively large.

The intercellular route has been the source of a great deal of speculation, especially because the work of Pappenheimer et al. (1951) suggests that the transport of water and small 'lipid-insoluble' molecules occurs in aqueous pores occupying a minute fraction of the capillary surface. Of course, it has been pointed out that the main weaknesses of their theoretical treatment are, first, that the so-called 'lipid-insoluble' solutes and water molecules are assumed to permeate the wall solely via aqueous channels and, secondly, that the effect of solvent drag on solute flow is not properly taken into account. The results of some recent work (Wright & Diamond, 1969a,b) on epithelial tissues has some bearing on the first weakness, since it has been demonstrated convincingly that the transmural movement of solutes involves a lipid route and an aqueous route both of which are perhaps associated with transport across the plasma membranes of the epithelial cells. Notwithstanding those objections to the pore treatment of capillary permeability, numerous workers have proposed that the small-pore system can be identified with the intercellular spaces. The attraction of the proposal lies not only in the similarity between the width (about 100–200 Å) of the intercellular spaces and the diameter (about 90 Å) of the equivalent pores but also in the fact that both occupy about 0·1% of the capillary surface for a path length of 1 μm. Until recently the general reluctance to accept intercellular transport was buttressed by the argument that this route was occluded by the 'tight' junction (see for example Bruns & Palade, 1968a). The postulated impermeability of the 'tight' junction has been questioned by Luft (1965) who suggested that the central region of the 'tight' junction may behave as a relatively permeable filter equivalent to a slit of 40 Å in width. Such a permeable slit is certainly compatible with Pappenheimer's conclusion that the equivalent slit width, as an alternative to the equivalent pore radius, was 37 Å. Both estimates are open to dispute, but Luft's views are interesting because they go against the established dictum that the 'tight' junctions are truly *tight*. Recent work on water and ion transport in epithelial tissues furnishes us with some evidence that the absolute impermeability of 'tight' junctions between epithelial cells cannot be taken for granted (Frömter & Diamond, 1972). In this context it is pertinent to note that Karnovsky (1971) claims that the junctions between endothelial cells are not invariably 'tight' since he has observed (Karnovsky, 1967) occasionally that junctions appear patent when gaps of about 40 Å are noted between adjacent cells. Furthermore, that observation is compatible with the finding that both horseradish peroxidase (molecular weight 40,000) and the smaller molecule cytochrome C (molecular weight 12,000) permeate from the capillary lumen into the intercellular spaces (Karnovsky, 1967, 1970, 1971). Given that the junctions, whether they be *zonulae* or *maculae occludentes*, represent the pore system envisaged by Pappenheimer and others then one meets a further objection, namely that the total area of those junctions is too small (Bruns & Palade, 1968a). According to Pappenheimer's model the total area of the equivalent aqueous channels ought to be about 0·1%. The latter area, however, was obtained on the basis of a path length (1 μm) equivalent to the width of an endothelial cell. If one considers alternatively that the 'tight' junction itself, but not the rest of the intercellular route, rate-controls solute and water movement then the equivalent path length becomes about 200 Å—the length of the intercellular junction. What was determined in Pappenheimer's experiments was $(A_p'/\Delta x)$, i.e. the restricted pore area divided by the path length. In his original calculations Δx was taken as 1 μm or 10,000 Å but, if we now take Δx as 200 Å, A_p' becomes 0·002% of the total capillary area. Another example of this can be found in the study of Alvarez & Yudilevich (1969) who estimated the area available for solute diffusion across capillaries in the canine heart. They too assumed that the length of the transport pathway was identical to the depth of the intercellular junction (100–400 Å). Their estimates of A_p' lie in the range 0·001–0·01% of the entire capillary surface and these values are compatible with independent estimates of the relative area (0·02%) of the junctions in those capillaries by Karnovsky (1967).

Of some importance to the question of possible aqueous channels in the capillary wall is the view of Pappenheimer *et al.* (1951) that molecules traversing that route experience some hindrance due to the fact that the channels are probably of molecular dimensions, say 80 Å in diameter. Actually the estimates of equivalent pore radius and the like are affected by the assumption that solute diffusion suffers some restriction. Crone (1963b), for example, has claimed that the size of equivalent pores, if they exist at all, in muscle capillaries should be larger than the values given by Pappenheimer and his colleagues. The essence of Crone's objection is that he was unable to observe any restriction to the diffusion of inulin and sucrose; that is, the ratio of the permeability coefficients for inulin and sucrose was similar to the corresponding ratio of the free diffusion coefficients for those solutes. Vargas & Johnson (1964), however, contended that the solute fluxes in Crone's experiments were rate-limited by blood flow rather than by the capillary permeability. Subsequently Alvarez & Yudilevich (1969) re-examined the question of restriction to diffusion in canine heart capillaries. They used the indicator-diffusion method (see equations 8.2, 8.3) and they obtained $P_s A$ for a number of solutes as a function of Q. For the solutes listed in Table 8.4 they observed that $P_s A$ was independent of Q above certain values of Q. The data in Table 8.4 show that the ratio

TABLE 8.4. Capillary permeability coefficients for canine heart

	$P_s \times 10^5$ (cm sec^{-1})	$D_s \times 10^5$ (cm^2 sec^{-1})	P_s/D_s (cm^{-1})
Inulin	0·27	0·18	1·5
Sucrose	0·8	0·55	1·5
Glucose	1·0	0·68	1·5
Glycerol	1·5	0·89	1·6
Urea	3·1	1·45	2·1

Modified from Alvarez & Yudilevich, 1969.

(P_s/D_s) is relatively constant or, in other words, that all of the test solutes permeated the capillary walls at rates which seem to be related to their respective diffusion coefficients in free solution.

Consequently Alvarez & Yudilevich (1969) estimated that the width of the transport pathway must be larger than 80 Å and they speculated that the characteristics of the intercellular junctions determined the permeability properties of the wall to small solutes and water. In support of that hypothesis they cited the fact that the brain capillaries have the lowest value for solute permeabilities (Table 8.3) whereas liver sinusoids are extremely permeable even to large molecules (Yudilevich, Renkin, Alvarez & Bravo, 1968). In the former type of capillary all of the endothelial cells are joined by genuine 'tight' junctions which do not allow horseradish peroxidase to penetrate into the intercellular spaces from the capillary lumen (Reese & Karnovsky, 1967). Such efficient junctions are not found in other continuous capillaries.

Thus, the key to understanding the water and solute permeability characteristics of continuous capillaries may be the degree of tightness of the junctions between the endothelial cells.

Beyond the 'tight' junctions the intercellular spaces, which are approximately 150 Å in width, probably represent an almost insignificant resistance to the movement of solute and water molecules which have entered them across the junctions or across the plasma membranes (routes 4 and 5, Fig. 8.5).

Fenestrated and discontinuous capillaries

In general, the fenestrated capillaries are more permeable to water than the continuous capillaries (see Table 8.2). Their permeability to large molecules is also relatively large. For example, Mayerson, Wolfram, Shirley & Wasserman (1960) found that the capillary bed of the intestinal mucosa was more permeable to dextrans than are the continuous capillaries of skeletal muscle. Such differences in the permeability properties of those two types of capillaries must surely result from differences in their structures. The morphology of the fenestrated capillaries has been described in detail by Clementi & Palade (1969) and from what we know of their structure it is probable that their permeability characteristics are intimately connected with the existence of fenestrations. Although Clementi & Palade (1969) presented ultrastructural evidence that the main route for the transport of large molecules was the fenestrations, it is perhaps not safe to assume that water and small solutes also rely uniquely on those transmural channels. Even for the movement of large molecules, the picture is complicated further by the observation by Clementi and Palade that fenestrations with diaphragms were relatively impermeable to ferritin but were permeable to the smaller molecule of horseradish peroxidase. One might speculate that the diaphragms contain a small-pore-system equivalent, but not identical, to that of the continuous capillaries. As yet the appearance of the diaphragms in electron micrographs of fenestrated capillaries does not give a reliable guide to their permeability characteristics for large molecules or, indeed, for small molecules, such as water.

The sinusoidal capillaries are highly permeable to large molecules. Grotte (1956), Mayerson *et al.* (1960) and relatively recently Yudilevich *et al.* (1968) have demonstrated that the liver sinusoids are the most permeable capillary bed examined so far. In fact, the total concentration of protein in hepatic lymph is almost equal to that of plasma. Of course, the permeable nature of the hepatic capillaries is what one would expect in capillary walls where the basement membrane is discontinuous and the endothelial cells are separated by gaps some 1000 to 10,000 Å in width. There can be no doubt that the latter channels constitute the major route for water and solute flow across the capillary walls in the liver and probably across other sinusoids too.

9

PERMEABILITY CHARACTERISTICS OF EPITHELIA

Passive water transport	319
Diffusional permeability to water	319
Hydraulic conductivity	322
Epithelia as composite barriers	329
Comparison between osmosis and filtration	333
Comparison between osmosis and diffusion	339
The mechanism of action of ADH	343
Passive transport of solute	346
Solute permeability	346
Reflexion coefficient	351
Electrokinetic phenomena in epithelia	356
Solvent drag	358
Summary	361
Ultrastructural route of water transport	362
Urinary bladder	362
Amphibian skin	365
Gall bladder	367
Renal collecting ducts	369
Effect of temperature on P_d and L_p	372
Rectification of osmotic water flow	378
Osmotic behaviour of the permeability barrier	379
Unstirred layers	383
Flow-induced deformations of the permeability barrier	385
Asymmetrical double-membrane model	386
Conclusion	389

EPITHELIA play an exceedingly important role in the maintenance of salt and water balance in animals and this is exemplified well by such tissues as the Malpighian tubules of insects or the proximal convoluted tubules of the mammalian kidney. For that reason alone certain epithelia have been studied extensively. In general, they are extremely suitable for permeability studies because they can be isolated conveniently in the form of sheets, tubes or sacs.

The main advantage of these preparations lies in the ease with which it is possible to control the composition of the bathing solutions and to measure solute and water fluxes.

Passive water transport
Diffusional permeability to water

Estimates of P_d have been obtained for a large number of epithelial tissues and Table 9.1 contains some representative values. An interesting feature in the table is that antidiuretic hormone (ADH) appears to create an increase in P_d; this effect will be discussed later in the light of the hormone's effect on the hydraulic

TABLE 9.1. Apparent values for P_d

Tissue	$P_d \times 10^5$ (cm sec^{-1})		Reference
	AHD	present	
Proximal tubule			
Rat	560	–	Persson & Ulfendahl (1970)
Distal tubule			
Rat	160	330	Persson (1970)
Descending loop of Henle			
Rat	119	95	Morgan & Berliner (1968)
Ascending loop of Henle			
Rat	50	39	Morgan & Berliner (1968)
Cortical collecting duct			
Rabbit	38	97	Grantham & Burg (1966)
Medullary collecting duct			
Rat	53	95	Morgan, Sakai & Berliner (1968)
Ependyma			
Goat	28	–	Heisey, Held & Pappenheimer (1962)
Rumen			
Sheep	18	–	Engelhardt & Nickel (1965)
Calf	11	–	Engelhardt & Nickel (1965)
Gall Bladder			
Dog	11	–	Grim & Smith (1957)
Fish	8·3	–	Diamond (1962c)
Rabbit	184*	–	Van Os & Slegers (1973)
Cloaca			
Fowl	8·3	–	Skadhauge (1967)
Intestine			
Dog colon	8·2	–	Grim (1962)
Dog ileum	2·2	–	Grim (1962)

TABLE 9.1—*continued*

Tissue	$P_d \times 10^5$ (cm sec^{-1})		Reference
		ADH present	
Dog jejunum	1·7	–	Grim (1962)
Dog duodenum	1·2	–	Grim (1962)
Human ileum	1·9	–	Soergel, Whalen & Harris (1968)
Human jejunum	1·5	–	Soergel, Whalen & Harris (1968)
Gastric mucosa			
Frog	4·9	–	Durbin et al. (1956)
Dog	3·8	–	Altamirano & Martinoya (1966)
Mouse	1·9	–	Öbrink (1956)
Urinary bladder			
Frog (*R. esculenta*)	10	–	Maetz (1968)
Toad (*B. marinus*)	9·5	17	Hays & Leaf (1962a)
Toad (*B. marinus*)	12†	110†	Hays & Franki (1970)
Dog	6·7	–	Johnson, Cavert, Lifson & Visscher (1951)
Skin			
Toad (*B. bufo*)	10	11	Koefoed-Johnsen & Ussing (1953)
Toad (*B. regularis*)	5·7	–	Maetz (1968)
Toad (*Xenopus laevis*)	3·0	–	Maetz (1968)
Frog (*R. temporaria*)	6·5	–	Dainty & House (1966b)
Frog (*R. temporaria*)	15†	–	Dainty & House (1966b)
Frog (*R. esculenta*)	5·5	–	Maetz (1968)
Gills			
Goldfish	3·4	–	Motais, Isaia, Rankin & Maetz (1969)
Eel (sea water)	1·8	–	Motais, Isaia, Rankin & Maetz (1969)
Eel (fresh water)	2·5	–	Motais, Isaia, Rankin & Maetz (1969)
Flounder (sea water)	1·7	–	Motais, Isaia, Rankin & Maetz (1969)
Flounder (fresh water)	2·7	–	Motais, Isaia, Rankin & Maetz (1969)
Perch (sea water)	1·0	–	Motais, Isaia, Rankin & Maetz (1969)
Brine shrimp (*Artemia salina*)	0·69	–	Smith (1969)

* Corrected for unstirred-layer effect.
† P_d determined under vigorous stirring conditions.

conductivity. The range of values of P_d is quite large and it will be remembered that cells also show that feature. Just how much of the variation in P_d values for epithelia is genuine is difficult to assess. A contributory factor is likely to be variations in the influence of unstirred layers on the kinetics of labelled water exchange. Dainty & House (1966b) studied the effects of stirring on the apparent P_d of the isolated frog skin. They found that increasing the agitation of the bathing media led to an increase in the magnitude of P_d for the skin (Fig. 9.1) and they attributed this

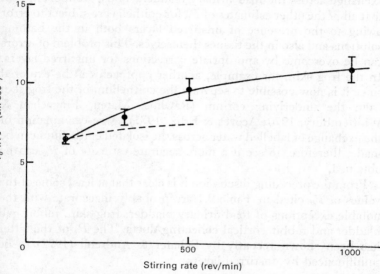

FIG. 9.1. The apparent diffusional permeability (P_d) of tritiated water in frog skin as a function of the stirring rate in the bathing media. Each point is the mean of eighteen measurements on eighteen skins, while the bars indicate ±S.E. The interrupted line is the theoretical relation between P_d and the stirring rate predicted on the basis of estimates of unstirred layer thickness from Dainty & House (1966a) (Dainty & House, 1966b: Fig. 3).

effect to a reduction in the magnitude of the unstirred layers at the surfaces of the skin; the interrupted line in Fig. 9.1 shows the increase in P_d which was predicted from independent estimates of the unstirred layers and their dependence on the stirring rate (Dainty & House, 1966a). These experiments demonstrate that the 'true' values for P_d must be significantly larger than the value obtained for the skin under normal stirring conditions. The correction, which must be applied to the data for the whole skin to obtain the 'true' P_d for the epithelial cells alone, must also include the effect of additional diffusion delays within the tissue. For example, on the serosal surface of the frog skin epithelium there is a connective tissue layer (corium) about 200 μm thick. After making allowance for such diffusion barriers in the skin, Dainty & House (1966b) obtained a corrected value of $15\cdot4 \times 10^{-5}$ cm sec^{-1} for P_d as opposed to a value of $6\cdot5 \times 10^{-5}$ cm sec^{-1} for the entire skin under poor stirring conditions. Rather similar results have been obtained by Hays & Franki (1970) in a study of labelled water exchange across the toad urinary bladder. Thus, we must expect that all of the other estimates of P_d for epithelia are subject to error owing to the presence of unstirred layers both in the bathing solutions and also in the tissues themselves. This problem of errors can be overcome by appropriate corrections for unstirred layers. In the frog skin, for example, another approach can be explored since it is now possible to separate the epithelium of the frog skin from the underlying corium (Rawlins, Mateu, Fragachan & Whittembury, 1970; Aceves & Erliz, 1971). A re-examination of the exchange of labelled water across the isolated epithelium can be made, therefore, to see if a more accurate estimate of P_d can be obtained.

From the preceding discussion it is clear that at least some of the values of P_d cited in Table 9.1 are probably inaccurate with the notable exceptions of toad urinary bladder, frog skin, rabbit gall bladder and rabbit cortical collecting ducts. The P_d of the latter has been shown recently by Schafer & Andreoli (1972) to be uninfluenced by unstirred layers.

Hydraulic conductivity

In contrast to the relatively few measurements of P_d for epithelia there are numerous estimates of L_p in a variety of epithelial tissues (Table 9.2). The techniques of measuring the net volume flux in those osmotic studies are principally either volumetric (e.g. Koefoed-Johnsen & Ussing, 1953) or gravimetric (e.g. Diamond, 1962c). Table 9.2 displays a wide range of magnitudes for L_p, namely $0\cdot1$–1800×10^{-7} cm sec^{-1} atm^{-1}, and indicates that ADH increases L_p even in the most permeable epithelium. The effect of ADH on L_p will be discussed later (see page 343). Inevitably there are some difficulties associated with comparing L_p values for different epithelia; for instance, the actual areas of the epithelial membranes are difficult to estimate accurately and

TABLE 9.2. Apparent values for L_p

Tissue	$L_p \times 10^7$ (cm sec^{-1} atm^{-1})		Reference
		ADH present	
Proximal tubule			
Rat	1640	1800	Ullrich, Rumrich & Fuchs (1964)
Rabbit	290–630	–	Burg & Grantham (1971)
Necturus	33	–	Bentzel, Davies, Scott, Zatzman & Solomon (1968)
Descending loop of Henle			
Rabbit	1710	–	Kokko (1970)
Rat	440	–	Morgan & Berliner (1968)
Ascending loop of Henle			
Rat	33	–	Morgan & Berliner (1968)
Distal tubule			
Rat	270	790	Ullrich *et al.* (1964)
Necturus	4·1	–	Maude, Shehadeh & Solomon (1966)
Cortical collecting duct			
Rabbit	49	160	Grantham & Burg (1966)
Medullary collecting duct			
Rat	31	220	Morgan & Berliner (1968)
Ependyma			
Goat	190	–	Heisey *et al.* (1962)
Rumen			
Goat	55*	–	Englehardt (1969)
Goat	1·6*	–	Englehardt (1969)
Cow	22†	–	Dobson, Sellers & Shaw (1970)
Sheep	3·7	–	Warner & Stacy (1972)
Cloaca			
Fowl	20	–	Skadhauge (1967)
Gastric mucosa			
Dog	490‡	–	Moody & Durbin (1969)
Dog	4·5‡	–	Moody & Durbin (1969)
Frog	8·1	–	Durbin *et al.* (1956)
Gall Bladder			
Dog	49	–	Grim & Smith (1957)
Fish (freshwater)	45	–	Diamond (1962*c*)
Rabbit	18	–	Diamond (1964*b*)
Intestine			
Eel (seawater)	89	–	Skadhauge (1969)
Eel (freshwater)	46	–	Skadhauge (1969)
Rat ileum & jejunum	77	–	Smyth & Wright (1966)
Human jejunum	35	–	Soergel *et al.* (1968)
Human ileum	17	–	Soergel *et al.* (1968)

TABLE 9.2—*continued*

Tissue	$L_p \times 10^7$ (cm sec^{-1} atm^{-1})	ADH present	Reference
Fish (sea water)	27	–	House & Green (1965)
Dog ileum	13	–	Visscher *et al.* (1944)
Choroid plexus			
Rabbit	28	–	Welch, Sadler & Gold (1966)
Corneal epithelium			
Rabbit	67§	–	Mishima & Hedbys (1967)
Rabbit	4§	–	Green & Green (1969)
Urinary bladder			
Dog	7·7	–	Johnson *et al.* (1951)
Frog (*R. esculenta*)	5·5	–	Maetz (1968)
Toad (*B. marinus*)	2·9	154	Hays & Leaf (1962a)
Turtle	3·1¶	–	Schilb & Brodsky (1970)
Skin			
Toad (*B. bufo*)	5·0	11	Koefoed-Johnsen & Ussing (1953)
Toad (*B. regularis*)	11	–	Maetz (1968)
Toad (*Xenopus laevis*)	2·0	–	Maetz (1968)
Frog (*R. pipiens*)	16	–	Franz & Van Bruggen (1967)
Frog (*R. esculenta*)	4·1	–	Maetz (1968)
Frog (*R. temporaria*)	3·9	7·1	House (1964a,b)
Lamprey	1·3	–	Bentley (1962)
Eel	0·06	–	Bentley (1962)
Gills			
Goldfish	1·6	–	Motais *et al.* (1969)
Eel (fresh water)	0·58	–	Motais *et al.* (1969)
Eel (*sea water*)	0·14	–	Motais *et al.* (1969)
Flounder (fresh water)	0·51	–	Motais *et al.* (1969)
Flounder (sea water)	0·10	–	Motais *et al.* (1969)
Perch (sea water)	0·08	–	Motais *et al.* (1969)
Brine shrimp (*Artemia salina*)	0·71	–	Smith (1969)

* The larger value of L_p was obtained by applying hydrostatic pressure to the luminal surface of epithelium: both values are expressed for the estimated actual area of the mucosal surface (Engelhardt, *personal communication*).

† Value is expressed for the estimated actual area of the mucosal surface (Dobson, *personal communication*).

‡ Again, the larger value of L_p was obtained by pressure applied to the luminal survace: both values are expressed for the estimated surface area of the mucosa (see Altamirano, 1969).

§ Large unexplained disparity.

¶ Obtained from original data by assuming area of bladder is 9 cm^2 (Brodsky, *personal communication*).

the measurements are performed at different temperatures (see *Effect of temperature on P_d and L_p*, page 372). Nevertheless, the extent of the range of L_p values cannot be explained away on that basis, and it must reflect genuine variation in the water permeability. Outside the spectrum of permeabilities cited in the table is the recent L_p value of 6200×10^{-7} cm sec^{-1} atm^{-1} estimated for the ciliary epithelium by Green & Pederson (1972a,b). This was obtained for pressure-driven flow whereas almost all of the other values (but see Moody & Durbin, 1969 and Engelhardt 1969) in Table 9.2 were determined for osmotic flow. These different methods for measuring L_p in epithelia do not give identical results and this will be discussed later (see *Comparison between osmosis and filtration*, page 333).

Epithelial tissues are about as permeable to water as the capillary walls, which have been studied, except of course for the exceedingly permeable fenestrated capillaries. In the case of capillary walls one is inclined to feel that some of the variation in L_p values found for different capillaries may stem from the degree of tightness of the junctions between neighbouring endothelial cells, particularly in the continuous capillaries. Epithelia possess genuine 'tight' junctions or *zonula occludentes* (Farquhar & Palade, 1963; Brightman & Reese, 1969) as opposed to the gap junctions or *macula occludentes* occasionally found in some continuous capillaries (Karnovsky, 1967, 1970). A notable exception to that general rule for epithelia is the ruminal epithelium where there are *macula occludentes* or gap junctions rather than true 'tight' junctions at the apical membranes of the cells lining the luminal surface (Henrikson, 1970; Henrikson & Stacy, 1971). Nevertheless, it is now evident that even the 'tight' junctions in epithelia cannot be regarded simply as impermeable seals. Such junctions may be permeable to water and small solutes, and consequently the intercellular pathway in epithelia might represent an important route for the transmural movement of some substances. That is clearly the case for ion transport across certain epithelia, including the cat sublingual gland (Lundberg, 1957), the frog skin (Ussing & Windhager, 1964), the proximal tubule of the newt kidney (Hoshi & Sakai, 1967), the rabbit gall bladder (Barry, Diamond & Wright, 1971) and *Necturus* gall bladder (Frömter & Diamond, 1972). In those tissues the conductance of the shunt pathway (intercellular route ?) is a significant fraction of the conductance of the entire cellular wall. In particular, Frömter & Diamond (1972) and Frömter (1972) demonstrated that the epithelial cell junctions are the sites of relatively high conductance pathways across the gall bladder by recording the voltage responses to transepithelial current passage with a microelectrode placed either above the cells or the junctions.

Unfortunately the corresponding evidence about the role of the intercellular pathway in transepithelial water movement is more difficult to obtain. An indirect way of examining this problem is to see if there is some relation between the 'leakiness' of epithelia to ions and their permeabilities to water. Frömter & Diamond (1972) tabulated certain physiological data which they considered might be correlated, perhaps loosely at least, with the tightness of epithelial cell junctions. In their analysis (Table 9.3) 'leaky' epithelia were differentiated from 'tight' epithelia on the basis of their electrical resistance, hydraulic conductivity and transport

TABLE 9.3. Physiological correlates of junctional tightness

Epithelium	Temperature (°C)	$L_p \times 10^7$ (cm sec^{-1} atm^{-1})	Resistance (Ω cm^2)	Transport P.D. (mV)	Ion transport rate (μMcm^{-2} hr^{-1})
		'LEAKY' JUNCTIONS			
Rat proximal tubule	37	1800	6	0	55
Rat jejunum	37	90	30	11	14
Rabbit gall bladder	37	18	28	0	13
Fish gall bladder	20	45	113	0	2·4
Frog choroid plexus	20	$\leqslant 20$*	73	0	1·6
Necturus proximal tubule	20	18	80	?	0·4
		'TIGHT' JUNCTIONS			
Rat submandibular salivary duct	37	?	400	71	12
Human submandibular salivary duct	37	?	365	65	>6
Frog gastric mucosa	20	9	500	30	4
Frog skin	20	4·5	2000	100	1·5
Toad urinary bladder	20	4	800	35	1·6
Turtle urinary bladder	20	3·6	650	70	4·5

* E. M. Wright (*personal communication*).
Modified from Frömter & Diamond, 1972.

potential. According to Frömter and Diamond's scheme the rabbit gall bladder is a 'leaky' epithelium while the frog skin is a 'tight' one. The so-called 'leaky' epithelia have relatively large water permeabilities, low electrical resistances and potentials; one might argue, therefore, that the passive transport of water and ions occurred perhaps predominantly through the intercellular spaces in those epithelia and due to this junctional shunting the potential generated by the active ion transport is never fully observed, if at all, across the whole epithelium. On the other hand, certain epithelia—the 'tight' ones—have relatively low water permeabilities, high electrical resistances and transport potentials; in those epithelia electrical shunting of the transport potential is quite poor and, perhaps, there is relatively little osmotic water flow through the intercellular spaces. To get a quantitative idea of the relative shunting efficiency of the intercellular pathways in both 'leaky' and 'tight' epithelia, one can compare the shunt conductance with the conductance of the epithelium itself. Frömter & Diamond (1972) found that the shunt conductance was about 95% of the total conductance of the (leaky) gall bladder whereas the data of Ussing & Windhager (1964) show that it represents only about 40% of the conductance of the (tight) frog skin.

Comparing the hydraulic conductivities of the 'leaky' and 'tight' epithelia it may seem surprising to find that the difference is not larger; for example, the rabbit gall bladder is apparently only about four or five times more permeable than the frog skin. Now it is thought highly likely that significant water transport through the intercellular spaces of an epithelium during an osmotic experiment will render an erroneous value for L_p; the reason for this will be discussed below. Indeed, this effect leads to an underestimate of L_p and consequently the values of L_p, particularly for the 'leaky' epithelia, probably ought to be considerably higher. The nature of this source of error is intimately connected with the possible existence of osmotic water flow through the intercellular spaces and the existence of unstirred layers.

The presence of unstirred layers, inevitably associated with epithelia, means that one can never equate the osmotic pressure difference between the bathing media with that across the epithelial cells alone. A number of workers (Dainty, 1963a; Brodsky & Schilb, 1965; Diamond, 1966a,b) have discussed the importance of unstirred layers in osmotic studies and it is agreed that the effective osmotic gradient is invariably less than the apparent osmotic gradient between the bathing media. Moreover, the discrepancy between the effective and the apparent osmotic gradients is particularly serious when the osmotic water flow is large. It is never the existence of this unstirred-layer phenomenon which is in question, but rather it is its quantitative effect on osmotic flow which is difficult to estimate. The difficulty arises because epithelia are generally considered, for the sake of theoretical simplicity, to be analogous to homogeneous plane sheets whereas the osmotic water flow may cross the tissue at selective sites, such as the intercellular spaces. In order to appreciate the nature of the problem let us look at the work of Diamond and his collaborators on the rabbit gall bladder.

Diamond (1966b) estimated from the time course of diffusion potentials arising from small concentration gradients of NaCl across the gall bladder that the thickness of the unstirred layer adjacent to the mucosal surface was 119 μm while that derived from streaming potential measurements was 108 μm. In contrast to these estimates, the epithelial cells are separated from the serosal solution by about 300 μm of connective tissue which significantly impedes the diffusion of solutes; for instance, Diamond estimated that the diffusion coefficient of sucrose in the serosa was only about 20% of its value in free solution. Assuming that the osmotic flow of water passes uniformly across the tissue Diamond estimated the effect of unstirred layers on the apparent value of L_p from the expression

$$C(\text{O}) = C_o \exp\left(-J_v \delta / D_s\right) \quad 9.1$$

where $C(\text{O})$ is the solute (sucrose) concentration at the membrane, C_o the concentration of sucrose in the bathing medium, J_v (cm³ cm⁻² sec⁻¹) is the net water flow, δ the thickness of unstirred layer and D_s is the diffusion coefficient of sucrose in free solution. From the known values of these parameters (Diamond, 1966a,b) the effects of the unstirred layers on L_p were assessed: the mucosal unstirred layer is responsible for a maximal reduction of about 2·5% in L_p; the thicker serosal unstirred region produces a maximal reduction of about 8·5% in L_p.

Recent experimental evidence about the pathways of osmotic water flow through the gall bladder create considerable doubt, however, about the quantitative aspects of the foregoing argument.

For instance, Tormey and Diamond (unpublished observations, cited by Wedner & Diamond, 1969) have concluded from an examination of the intercellular spaces that they represent an important pathway for osmotic water flow, and Smulders, Tormey & Wright (1972) have confirmed that this is so not only for water but also for solute transport too. It must be strongly emphasized here that, although most of the osmotic water flow may go through the intercellular spaces, there is no convincing evidence that it reaches these regions via the 'tight' junctions in the gall bladder (Wright, Smulders & Tormey, 1972). Given the situation that the bulk of the osmotic flow actually passes through the intercellular spaces then the linear velocity of water flow must be considerably larger than that employed above on the basis of the participation of the entire surface area. In view of this, Wedner & Diamond (1969) reassessed the significance of the unstirred layer effect on the apparent value of L_p and they concluded that there may be a substantial underestimate (probably > 25%) of L_p. In a relatively more detailed study of the significance of the intercellular transport route Wright *et al.* (1972) suggested that the gall bladder L_p may be at least ten times larger than the apparent value due to the rate-limiting influence of the unstirred-layer effect in such narrow channels. Consequently, it appears that osmotic flow through a sparse system of narrow unstirred layers such as the intercellular spaces could give rise to serious underestimates of L_p, especially for 'leaky' epithelia.

The variation in the apparent values for L_p in Table 9.2 could be interpreted as a consequence of differential 'leakiness' of the 'tight' junctions themselves but it is more likely that it is due to a genuine differential permeability of the epithelial cells which is perhaps accentuated by errors arising from unstirred layer effects. As yet we have no reliable guide about which effect will predominate in any given epithelium except a crude rule-of-thumb which says that the more permeable an epithelium is to water the more likely it is that the osmotic water flow permeates almost exclusively through the intercellular pathways. In addition, the more water-permeable an epithelium appears to be the more likely it is that its true L_p is substantially larger than its apparent value.

Epithelia as composite barriers

The values of L_p in Table 9.2 are quoted really as the effective water permeabilities of the epithelia as if they were homogeneous sheets but we must acknowledge that any given epithelium is a composite barrier. Actually we have dealt previously with part of its composite nature in the preceding discussion about the parallel routes of osmotic water transport; on the one hand, water moves across the epithelial cells and, on the other, between the cells. The transcellular route, however, is comprised of at least two barriers in series—the basal and apical membranes of the cells. Finally, one finds invariably a basement membrane and occasionally a connective tissue layer in series with both of the parallel transepithelial pathways. The contributions of both the basement membrane and any connective tissue layer to the effective L_p may not be entirely negligible even although neither of them retards significantly the movement of small solutes, for example. Aside from the relative contributions of the transcellular and intercellular pathways, therefore, the remaining question concerns the permeabilities of the apical and basal epithelial cell membranes in series. There have been very few attempts to measure the water permeabilities of the opposite faces of the epithelial cells.

MacRobbie & Ussing (1961) have examined the water permeabilities of the outer and inner boundaries of the frog skin epithelium. They mounted the isolated skin in a bath placed on a microscope stage; the thickness of the entire multi-layered epithelium was measured microscopically as the difference between a suitable reference point on the outer surface of the skin and a melanophore cell under that point below the epithelium. Because the skin did not alter in area under the conditions of their experiments, the thickness of the entire epithelium was taken as a measure of its volume. The total thickness of the epithelium was about 60 μm of which only about 20 μm participated in osmotic swelling or shrinkage. Figure 9.2 shows the changes which MacRobbie & Ussing (1961) observed in a skin bathed in sulphate Ringer solutions when the osmolarity of the inner medium was changed. Evidently part of the epithelium swelled when the inner solution was diluted and subsequently it shrank when the Ringer's solution was replaced. Apparently there is no net solute influx or efflux in the epithelial cells during the experiment, otherwise the reversibility of the volume changes would not have been so good. Using this technique, MacRobbie and Ussing estimated that the water permeabilities of the inner and outer boundaries of the epithelium were about 17×10^{-7} and 0.8×10^{-7} cm sec^{-1} atm^{-1} respectively, and so the inner membrane is much more permeable to water than the outer. The latter value is their estimate of the upper limit for the water permeability of the outer boundary because they found that the outer barrier was unresponsive to changes in the external osmolarity. An exception, however, to this was found when the inner solution contained ADH. Under those circumstances the water permeability of the outer barrier increased to values in the

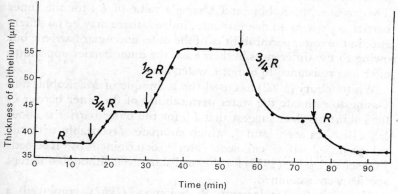

FIG. 9.2. Volume changes in the frog skin epithelium associated with changes in the osmotic pressure of the inner (serosal) bathing solution. The outer solution was distilled water and the inner solution was initially sulphate Ringer. The inner solution was changed to various dilutions ($\frac{3}{4}$R, $\frac{1}{2}$R) of sulphate Ringer (R) and the thickness of the epithelium was measured microscopically at certain times (●) (MacRobbie & Ussing, 1961: Fig. 2).

range $1.6-4.7 \times 10^{-7}$ cm sec^{-1} atm^{-1}. These values for the permeability of the outer barrier are somewhat less than the corresponding hydraulic conductivity for the entire skin. For instance, House (1964a,b) found that L_p for isolated frog skin was about 4×10^{-7} and 7×10^{-7} cm sec^{-1} atm^{-1} in the absence and presence of ADH respectively.

The identification of the outer and inner barriers of the frog skin epithelium is not an easy task. Probably the outer barrier corresponds to the outer membranes of the epithelial cells immediately below the cornified layer (Ussing & Windhager, 1964; Voute & Ussing, 1970). The identity of the inner barrier is less certain;

it may be composed of the inward-facing membranes of the multi-layered epithelial syncytium. The experiments of MacRobbie & Ussing reveal that less than 50% of the epithelium is osmotically active, so perhaps the inward-facing membranes of the epithelial cells have widely different permeabilities. Another feature of that epithelium is that the area of the outer barrier is much less than that of the inner barrier. Given that the inner barrier actually corresponds to the inner surface of the epithelium then its area is probably about sixty times that of the outer barrier (Smith, 1971). That means MacRobbie and Ussing's value of L_p for the inner barrier is perhaps an overestimate; indeed, there may be no difference in the water permeabilities of the outer and inner barriers, but owing to the differences in their areas the inner barrier appears to offer less resistance to osmotic water flow.

Whittembury (1962) has used the technique of MacRobbie and Ussing to estimate the water permeability of the outer barrier of toad skin; his data suggest that L_p for the outer barrier is about $1 \cdot 5 \times 10^{-6}$ cm sec^{-1} atm^{-1}, which compares favourably with the value of $3 \cdot 2 \times 10^{-6}$ cm sec^{-1} atm^{-1} determined by Koefoed-Johnsen & Ussing (1953) for the toad skin, even although different species were examined.

Alternatively Lindemann & Solomon (1962) employed a gravimetric method to measure the hydraulic conductivity of the luminal border of the mucosal cells of the rat small intestine. They obtained $L_p = 6 \times 10^{-6}$ cm sec^{-1} atm^{-1} which is quite close to the hydraulic conductivity of the entire tissue ($7 \cdot 7 \times 10^{-6}$ cm sec^{-1} atm^{-1}) obtained by Smyth & Wright (1966) from streaming potential measurements.

Thus, in the preceding cases there is strong evidence that L_p for the luminal or outer membrane is practically identical to that for the entire epithelium; those studies do not show unequivocally that epithelia have asymmetrical permeability characteristics for water. Good evidence for such asymmetry, however, has been obtained in the cortical collecting ducts of the rabbit kidney. Grantham & Burg (1966) found that, when the lumen of the isolated collecting duct was perfused with a dilute solution, there was net fluid absorption into the Ringers solution bathing the outer or peritubular surface; they noted that ADH increased the net fluid transport by increasing L_p for the duct epithelium. In a later study, Ganote, Grantham, Moses, Burg & Orloff (1968) monitored the epithelial volume by measuring microscopically the thickness of the duct wall. They noted that ADH induced cellular swelling during net osmotic water movement from the lumen. On the other hand, Grantham, Ganote, Burg & Orloff (1969) recorded cellular swelling even in the absence of ADH provided a suitable osmotic gradient was established across the peritubular surface. These experiments demonstrate that the rate-limiting barrier to osmotic flow is located at the luminal border of the cells and that the peritubular border is relatively permeable to water. Of course, we ought not to expect that all epithelia are asymmetrically permeable to water just like the collecting ducts.

In a number of epithelia it has been found that the relation between the net volume flux and the osmotic pressure gradient across the tissue is not linear. This phenomenon has been called rectification of volume flow or non-linear osmosis, and one of the possible explanations for that phenomenon rests on the assumption that epithelia behave like double-membrane systems with asymmetrical water permeabilities. Non-linear osmosis will be discussed later (see page 378).

Comparison between osmosis and filtration

Almost invariably the hydraulic conductivities of epithelia have been measured under circumstances where the driving force on water is an osmotic gradient rather than a pressure gradient. From studies on artificial membranes (Durbin, 1960; Mauro, 1957) one might expect that given differences in osmotic and hydrostatic pressure would be equally effective as driving forces for water transport. However, the effects of those gradients on epithelia are not equivalent. In fact, hydrostatic pressure gradients exert rather complex actions on fluid transport across epithelia. Broadly speaking, those actions fall into two classes depending on the direction of the pressure gradient: first, when the pressure on the mucosal surface of some tissues exceeds that on the serosal surface there is occasionally an enhancement of the fluid transport accompanying active salt transport; secondly, when the pressure gradient is in the reverse direction the hydraulic conductivity of the tissues increases. Discussion of the first type of effect on water transport will be deferred to the next chapter (see page 402). Throughout the following discussion of the effects of pressure on passive water flow the expressions *mucosal pressure* and *serosal pressure* will signify the excess pressure applied to the mucosal and serosal fluids respectively.

Probably the most extensive set of measurements of the effects of serosal (and mucosal) pressure on water transport is that of Hakim & Lifson (1969) on the *in vitro* canine intestine. Isolated sheets of small intestine (epithelium and lamina propria only) were bathed on both sides by identical Ringer solutions and a hydrostatic pressure difference was obtained by setting the surfaces of the bathing media at different heights. Figure 9.3 displays the relationship which they found between water transport and the hydrostatic pressure difference. The net water movement is expressed in cm³ hr⁻¹ and refers to an apparent surface area of 6·4 cm² while positive values of water flow refer to absorption (mucosa-to-serosa) and *vice versa* for negative values. In the absence of a pressure difference the water absorption proceeds at a rate of about 1 cm³ hr⁻¹; since Grim & Smith (1957) estimated that the mucosal surface area of the dog ileum is approximately 8 times larger than the apparent area, the net water flux is equivalent to 19 μl cm⁻² hr⁻¹, which is in good agreement with the corresponding value given by Visscher, Fetcher, Carr, Gregor, Bushey & Barker (1944).

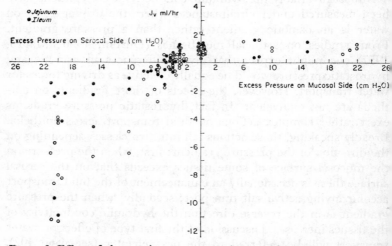

FIG.9 .3. Effect of changes in pressure on the rate of fluid transport across the canine intestine. Positive values of 'J_v', rate of fluid movement, signify transport from mucosa-to-serosa (Hakim & Lifson, 1969: Fig. 1).

Increments in the mucosal pressure on the canine intestine failed to increase the net water flow. On the other hand, small increments (2–26 cm H_2O) in the serosal pressure produce marked reductions in the rate of water absorption, with the effects being more pronounced in the jejunum than ileum. Elevations of the serosal pressure above 4 cm H_2O induced a net fluid secretion (serosa-to-mucosa) and Hakim and Lifson estimated that the 'mean value for the slope of a straight line fitted to the secretory points' was about 0·4 cm^3 hr^{-1} (cm $H_2O)^{-1}$; assuming that the actual surface area of the intestine is 51 cm^2 then the apparent L_p under those conditions is $2·2 \times 10^{-3}$ cm sec^{-1} atm^{-1}, which is about 10,000 times larger than that derived by osmotic methods (Table 9.2).

That such small serosal pressures can generate dramatic effects on water transport has been noted in other tissues. For instance, Wilson (1956) found that a serosal pressure of only 4 cm H_2O was enough to stop fluid absorption *in vitro* by the everted small intestine of the hamster, and quite small serosal pressures exert a similar kind of effect on the rabbit gall bladder. Dietschy (1964) and Tormey & Diamond (1967) found that serosal pressures of 5–10 cm H_2O applied to the everted gall bladder made the epithelium grossly leaky. In contrast, mucosal pressures of up to 20 cm H_2O produced no such effects in the fish gall bladder (Diamond, 1962c). Similar effects of hydrostatic pressure have been found in the isolated rumen of the goat. Engelhardt (1969) found that a hydrostatic pressure of 40 cm H_2O applied to the luminal surface led to a progressive decline in net water flow whereas the same pressure applied to the serosal surface produced a large net water flux which increased with time.

What gives rise to the increase in L_p when serosal pressure is applied to those various epithelia? In the case of the canine intestine Hakim & Lifson (1969) favoured the hypothesis that even quite small serosal pressures can disrupt the 'tight' junctions between the epithelial cells. Moreover, since these lateral junctions are found near to the luminal border of the cells perhaps serosal pressure forces fluid into the lateral intercellular spaces and the consequent distension of the spaces damages the intercellular junctions. Such a scheme could account for the failure of mucosal pressure to generate a corresponding increase in L_p since distension of the intercellular spaces would probably not result. Compatible with that view is the observation by Hakim and Lifson that certain solutes, such as Na^+, Cl^-, urea, glucose and inulin, experience solvent drag during the application of serosal pressures. Furthermore, Evans blue dye, ferritin and even erythrocytes were transported from the serosal surface into the mucosal fluid when the serosal pressure exceeded 10 cm H_2O. An alternative explanation put forward by Hakim and Lifson for such grossly leaky behaviour of the intestinal wall is that the gaps (left in the epithelium by the shedding of cells) remain open or are widened when serosal pressure is high. It is difficult, however, to reconcile that proposal with the failure of mucosal pressure to elicit an increase in L_p.

Thus, the general trend in epithelia is for fluid filtration in the serosa-to-mucosa direction to occur at a faster rate than one would expect from the L_p values determined in osmotic experiments. In all of these epithelia the relatively low resistance to fluid filtration is associated with intercellular spaces facing in the opposite direction to water flow and it possibly arises from disruption of the 'tight' junctions terminating these spaces. The canine gastric mucosa, however, is an exception to that rule because it exhibits a relatively low resistance to filtration occurring in the same direction as the intercellular spaces point.

Moody & Durbin (1969) observed that there was a profound difference in the resistances to water flow in the canine gastric mucosa when they were measured during the separate applications of a mucosal pressure and an equivalent osmotic pressure difference. They measured the net flow of water across a segment of secreting gastric mucosa (with essentially intact blood circulation) and they found that the hydraulic conductivity was 490×10^{-7} cm sec^{-1} atm^{-1} ($L_p{}^h$) for pressure-driven flow whereas it was $4 \cdot 5 \times 10^{-7}$ cm sec^{-1} atm^{-1} ($L_p{}^o$) for an osmotic gradient of NaCl. The difference between those measurements is extremely large and, indeed, the ratio of the 'hydrostatic permeability' to the 'osmotic permeability' is almost identical to the value reported by Vargas (1968a) for the squid axon (see page 166). Vargas suggested that the discrepancy arose in the axon because the axolemma possessed a small number of large pores in addition to a relatively large number of small pores. This is a plausible explanation for the axon but it is an even more plausible one for the gastric mucosa as Altamirano & Martinoya (1966) have deduced such a distribution of pore radii from a study of solute permeabilities. Again, as for the squid xon, there is really no adequate explanation for the discrepancy between $L_p{}^h$ and $L_p{}^o$. Despite the astounding quantitative similarity between those ratios in such different preparations there might still be different reasons for the discrepancies. For example, in the gastric mucosa there might be spurious data arising from changes in blood flow whereas in the *in vitro* experiments on the squid axon no such artifacts can occur. It is worth recording here that the goat's rumen shows a similar discrepancy between $L_p{}^h$ and $L_p{}^o$. Engelhardt (1969) has shown that when a small mucosal pressure (20 cm H_2O) is applied to the isolated rumen $L_p{}^h$ is about 5×10^{-7} cm sec^{-1} atm^{-1} whereas corresponding osmotic gradients yield an $L_p{}^o$ of $1·6 \times 10^{-7}$ cm sec^{-1} atm^{-1} (see Table 9.2). It should be emphasized that this disparity arises, as it does in the canine gastric mucosa, when $L_p{}^o$ is compared with $L_p{}^h$ derived from filtration experiments where mucosal rather than serosal pressures are employed.

Aside from the actual differences between the values of L_p obtained with osmotic and hydrostatic pressure gradients the canine gastric mucosa exhibits other interesting features. Moody and Durbin observed that the apparent value for L_p in the osmotic pressure experiments was dependent on the nature of the osmotic solute used (see Table 9.4). They interpreted these data on the

TABLE 9.4. Apparent L_p values in canine gastric mucosa; Effect of different osmotic solutes

Solute	$L_p \times 10^7$ (cm sec^{-1} atm^{-1})	$D_s \times 10^5$ (cm^2 sec^{-1})
Urea (13)	4·1	1·4
NaCl (12)	4·5	1·5
Sucrose; 1st period (14)	0·66	0·52
2nd and 3rd periods (6)	1·8	–

Values of L_p have been calculated from original data on the assumption that the actual area of the mucosa is 4·7 times the apparent area (Altamirano, 1969). Figures in parentheses are numbers of test periods. Solute diffusion coefficients, D_s, are free (dilute) solution values at 25°C.
Modified from Moody & Durbin, 1969.

secreting gastric mucosa in terms of the 'sweeping out' effect described by Rehm, Schlesinger & Dennis (1953) who suggested that acid secretion carries exogeneous solutes out of the gastric pits by exerting a solvent drag on them. Accordingly the concentration, $C(O)$, of solute at the base of the pits is given by an identical equation to 9.1 where C_o is the solute concentration in the luminal solution, $\mathcal{J}_v$ is the velocity (cm sec^{-1}) of the acid solution flowing out the pits, D_s has its usual meaning and δ is the depth of the pits. Moody and Durbin estimated from their data ($\mathcal{J}_v = 10^{-3}$ cm sec^{-1}, $\delta = 2 \times 10^{-2}$ cm) that for urea and sucrose $C(O)/C_o$ has the values 0·24 and 0·02 respectively; the corresponding value for NaCl must be about equal to that for urea because of the similarities in their free diffusion coefficients. On the basis of such a model one would expect, first, that the values of L_p determined in osmotic pressure experiments would be less than those for pressure-driven flow and, secondly, that the apparent value for L_p would be dependent upon D_s for the solute used in the osmotic experiments. Both of these expectations are realized. However, the magnitude of the actual discrepancy between L_p and $L_p{}^o$ is larger than predicted by the values $C(O)/C_o$ above. In fact, we might expect from the 'sweeping out' model that $(L_p{}^h/L_p{}^o)$ would fall in the range 4–50. Perhaps part of the answer is that the values of D_s are inappropriate and should be replaced by values representing some degree of restriction to diffusion somewhere within the pits.

It follows from the above analysis that $L_p{}^o$ for the secreting gastric mucosa ought to be less than that for the 'resting' mucosa. Moody and Durbin did not measure the latter value but they did refer to measurements of Altamirano & Martinoya (1966) on the resting gastric mucosa of the dog. Their value for $L_p{}^o$ was about 33×10^{-7} cm sec^{-1} atm^{-1}, significantly larger than that of Moody and Durbin for the secreting mucosa, and this supports the view that 'sweeping out' of solutes plays an important role in diminishing the effective osmotic pressure gradients across this tissue.

It might be contended that the reflexion coefficients for the osmotic solutes should be taken into account because low values for σ_s would lead to an underestimate of $L_p{}^o$. In the canine gastric mucosa, however, the permeability to small solutes, such as urea, is relatively low (Altamirano & Martinoya, 1966) and thus the reflexion coefficient for such molecules is probably close to unity. Furthermore, if the reflexion coefficients for the solutes used in the study of Moody & Durbin (1969) were less than unity the apparent $L_p{}^o$ measured with urea as the osmotic solute ought have been less than that for sucrose (see Table 9.4). This was not so.

The study of water flow in the canine gastric mucosa by Moody and Durbin is extremely significant because it raises valid doubts about measurements of L_p under osmotic conditions. Could it be that most values for L_p are underestimates because of 'sweeping out' effects? Of course, if epithelia behave as asymmetrical double-membrane systems then we might expect $L_p{}^h$ to exceed $L_p{}^o$ (see Chapter 2) but probably not by such a wide margin as found in the gastric mucosa or in the rumen. Finally, there is the possibility that mucosal pressure applied to these tissues makes the 'tight' junctions somewhat leaky just as serosal pressure does in other epithelia.

Comparison between osmosis and diffusion

A number of comparisons between P_d and the osmotic permeability, $L_p RT/\bar{V}_w$, have been made for certain epithelia. Almost invariably it has been found that the ratio $(L_p RT/\bar{V}_w P_d)$ exceeds unity and becomes even larger when ADH is present (Table 9.5). A notable exception is the gill of the seawater-adapted eel where the ratio lies between 0·86 and 0·17 for the temperature range 25–5°C (Motais & Isaia, 1972). In this case, however, L_p was determined by estimating the net water flux indirectly from the rates of urinary flow and drinking in the fish and by assuming that the salt concentration gradient across the gills exerted its full van't Hoff osmotic pressure difference. It is quite possible that the NaCl reflexion coefficient for the gill epithelium is significantly less than unity and consequently that L_p has been underestimated. Motais & Isaia (1972) also suggested that an alternative explanation for the underestimate of L_p is that a region of the gill is specialized for water absorption which is presumably linked to solute transport; thus, the estimated net osmotic water flux across the gill inferred by their indirect procedure is smaller than the actual water flux. Obviously the low ratio of the water permeabilities for the gill epithelium of seawater-adapted fish must be investigated further to understand why L_p is underestimated in this case.

Generally, comparisons between P_d and L_p are open to the same criticism that has been levelled at similar studies in single cells (Dainty, 1963a). In particular, we might consider that most of the estimates of P_d for epithelia are actually erroneous because of the

TABLE 9.5. Apparent values of $(L_p RT / \overline{V}_w P_d)$ for certain epithelia

Tissue	$\left[\dfrac{L_p RT}{\overline{V}_w P_d}\right]$	ADH present	Reference
Human jejunum	299	–	Soergel et al. (1968)
Human ileum	115	–	Soergel et al. (1968)
Goat ependyma	93	–	Heisey et al. (1962)
Dog ileum	81	–	Visscher et al. (1944)
Fish gall bladder	74	–	Diamond (1962c)
Dog gall bladder	60	–	Grim & Smith (1957)
Rat descending loop of Henle	46	–	Morgan & Berliner (1968)
Bird cloaca	36	–	Skadhauge (1967)
Toad skin (B. regularis)	27	–	Maetz (1968)
Dog gastric mucosa	27	–	Altamirano & Martinoya (1966)
Frog gastric mucosa	23	–	Durbin et al. (1956)
Rat proximal tubule	–	38	Persson & Ulfendahl (1970)
Rabbit cortical collecting duct	19	24	Grantham & Burg (1966)
Rat distal tubule	15	27	Persson (1970)
Brine shrimp gill	14	–	Smith (1969)
Frog skin (R. temporaria)	10	–	Maetz (1968)
Toad skin (Xenopus laevis)	9	–	Maetz (1968)
Rat medullary collecting duct	9	33	Morgan & Berliner (1968)
Rat ascending loop of Henle	8	–	Morgan & Berliner (1968)
Toad skin (B. bufo)	7	14	Koefoed-Johnson & Ussing (1953)
Frog urinary bladder (R. esculenta)	7	–	Maetz (1968)
Goldfish gill	6	–	Motais et al. (1969)
Toad urinary bladder (B. marinus)	6	110	Hays & Leaf (1962a)
Frog skin (R. temporaria)	5	–	Hevesy et al. (1935)
Toad urinary bladder (B. marinus)	4*	23*	Hays & Franki (1970)
Rabbit gall bladder	3·1†	–	Van Os & Slegers (1973)
Eel gill (fresh water)	3	–	Motais et al. (1969)
Flounder gill (fresh water)	2·6	–	Motais et al. (1969)
Frog skin (R. temporaria)	2*	–	Dainty & House (1966b)
Eel gill (sea water)	1·1	–	Motais et al. (1969)
Perch gill (sea water)	1·0	–	Motais et al. (1969)
Flounder gill (sea water)	0·8	–	Motais et al. (1969)

* P_d values obtained under vigorous stirring conditions.
† Corrected for unstirred-layer effect.

influence of unstirred layers. Many of the values for L_p which are currently quoted for epithelia may be erroneous too, since recent work on the rabbit gall bladder reveals that L_p is underestimated by an order of magnitude due to an unstirred layer effect (Smulders et al., 1972). The study of Moody & Durbin (1969) also raises the possibility that the values for L_p are underestimates. It is far too early to assess the overall significance of the possible errors in L_p but the suggested errors would cancel, if not exceed, the errors in the values of P_d. In view of the dubieties about the errors in P_d and L_p values one has no proper right to compare them. Nevertheless, I shall do so.

Dainty & House (1966b) obtained a value for P_d which had been corrected for the influence of unstirred layers. They found that their corrected value for $(L_p RT/\bar{V}_w P_d)$ was about 2. That value of the ratio is within the range which has been found for non-porous artificial membranes (Thau et al., 1967). Earlier estimates of this ratio for frog skin (Hevesy, Höfer & Krogh, 1935) were larger and fell in the range 3–5. However, the magnitude of the correction required for P_d has been confirmed by King (1969) in a study of the diffusion of deuterated and tritiated water across the isolated frog skin.

Later Hays & Franki (1970) performed a similar study on the urinary bladder of the toad and they obtained corrected ratios of about 6 and 23 for that epithelium in the absence and presence of ADH respectively. The original estimate of $(L_p RT/\bar{V}_w P_d)$ by Hays & Leaf (1962a) indicated that the equivalent pore radius in the toad bladder was about 8 Å in the absence of ADH and 40 Å in its presence. These dimensions, however, did not seem compatible with the failure of ADH to increase the permeability of the bladder to most small solutes, such as thiourea; a similar discrepancy was found in the frog skin. In order to surmount that difficulty of interpretation both Andersen & Ussing (1957) and Leaf & Hays (1962a) proposed that the outer or luminal surface of the epithelial cells is a composite barrier including a fine diffusion barrier in series with second porous membrane sensitive to ADH. The former barrier presents a restriction to the passage of small molecules whereas the latter barrier, because of its porous nature, is the rate-limiting one for osmotic flow of water.

Aside from frog skin and toad urinary bladder, there are several epithelia where the ratio $(L_p RT/\bar{V}_w P_d)$ is exceedingly large (see

Table 9.5). It is difficult to predict the extent of the corrections that might need to be made to these data on account of unstirred layers. Persson (1969) has argued from a theoretical standpoint that unstirred layers create an insignificant error in the P_d measurements in the renal tubules (Persson & Ulfendahl, 1970; Persson, 1970). No experimental check on his argument was made however, and if one accepts the data for rat renal tubules as correct then equivalent radii of the aqueous pathways in these tissues are about 25 Å and 20 Å for proximal and distal tubule respectively when ADH is present. These are large values and consequently the kidney tubules ought to be very permeable to small solutes. This is difficult to accept in the absence of experimental evidence and perhaps the renal tubules, like frog skin and toad urinary bladder, possess aqueous channels in association with a barrier restraining the movement of small solutes. Even after corrections for unstirred layers, Hays & Franki (1970) found that the equivalent pore radius in the toad bladder was 18 Å in the presence of ADH. That value is still incompatible with the transport rates of small solutes across the bladder. Indeed, Hays and Franki argued that to reduce the estimated pore radius to a size compatible with the permeability of the bladder to small solutes the value of $(L_p RT/\bar{V}_w P_d)$ must be significantly less than the apparent one.

Comparisons between the rates of diffusion of water molecules and of osmotic flow across a number of epithelia suggest that those tissues possess aqueous channels. In only two of those preparations have experiments been performed to determine the magnitude of the unstirred-layer effects on P_d, and when the appropriate corrections were made $(L_p RT/\bar{V}_w P_d)$ still exceeded unity. Now it seems likely that, when the unstirred layer effects on L_p are made, the value of that ratio will become larger or at least remain the same. The latter correction is an extremely important one since it has been assumed previously that unstirred layers do not interfere with the determination of L_p. However, in one epithelium at least there is evidence (Smulders *et al.*, 1972) that osmotic flow passes through the intercellular channels and since these occupy a small fraction of the total cross-sectional area of the epithelium the velocity of osmotic flow through these channels might be substantially larger than that expected for the entire tissue surface. Consequently a large underestimate of L_p may result. In contrast

DiBona, Civan & Leaf (1969a,b) presented ultrastructural evidence that osmotic water flow in the toad urinary bladder passes through the cells rather than between them, but their techniques could not exclude the possibility that water moves only through the apical 'tight' junctions and then into the cells across their lateral surfaces.

In addition to such complexities there is another problem of a more fundamental nature about comparisons between P_d and $_pRT/\bar{V}_w$. When such comparisons are made it is tacitly assumed that a single membrane rate-controls both diffusional and osmotic flows of water. The value of $(L_p RT/\bar{V}_w P_d)$ has a doubtful meaning if that is not the case. In fact, the work done on cellulose acetate membranes illustrates the problem of interpretation for those membranes consist of a thin dense 'skin' (about 0·2 μm thick) which rate-limits osmotic flow supported by a thick (100 μm) porous layer which rate-limits diffusion of water. For example, Nays (1968) found that the L_p values for the 'skin' and the porous layer were 1·3 and 32×10^{-7} cm sec^{-1} atm^{-1} whereas the corresponding P_d values were 30×10^{-4} and 3×10^{-4} cm sec^{-1} respectively. Thus, $(L_p RT/\bar{V}_w P_d)$ values for the dense 'skin' and thick porous layer are about 6 and 1300 respectively whereas the corresponding ratio for the entire membrane is about 47. Based on the false assumption that the cellulose acetate membrane behaves as a single rate-limiting barrier for osmosis and diffusion, one would use $(L_p RT/\bar{V}_w P_d) = 47$ to compute the pore radius as 25Å. In contrast one obtains a value of 8 Å for the dense 'skin' which is the actual rate limiting one for osmotic flow. This piece of work exemplifies the doubtful merit of estimating equivalent pore radii from measurements of L_p and P_d made on complex structures like epithelia.

It seems that we are still without an adequate picture of the nature of the barrier to osmotic water movement in epithelia. Whether the 'tight' junctions terminating the intercellular spaces or small pores in the cell membranes constitute the rate-limiting elements is a difficult problem to solve. Undoubtedly ADH increases the hydraulic conductivity of the osmotic barrier in a large number of epithelia and its mechanism of action may provide a clue to the nature of the barrier.

The mechanism of action of ADH

Antidiuretic hormone increases both the hydraulic conductivity and the diffusional permeability of several epithelia (Tables 9. and 9.2). Apparently the hormone exerts a more pronounce effect on L_p than on P_d. The largest increases in L_p were foun in the toad urinary bladder (Hays & Leaf, 1962a), and the medul lary collecting ducts of rats (Morgan & Berliner, 1968) where th water permeabilities rose to 40 and 7 times their original value respectively. In contrast, Hays and Leaf found ADH raised P for the bladder to about 1·5 times its original value. Koefoed Johnsen & Ussing (1953) encountered the same discrepanc between the effects of the neurohypophyseal hormone on L_p an P_d in the isolated frog skin and they proposed that the hormon increases the diameter of porous channels in the epithelial ce membranes. Their hypothesis was an attractive one because explained why ADH creates a bigger increase in L_p than in P_d the former is probably a function of r^4 and the latter of r^2, where was the radius of the aqueous channels.

Of course, as already mentioned, the simple porous-membran model of epithelia had to be modified to explain the fact th the water permeability of frog skin and toad bladder was increase by ADH even though the permeability to small solutes remaine unchanged. From that emerged the composite barrier mod of the cell membrane (Anderson & Ussing, 1957; Leaf & Hay 1962) comprised of a thin dense barrier in association with a thic porous one responsive to ADH. Later work by Lichtenstein Leaf (1965) seemed to confirm the validity of that model when the found that amphotericin B, an antibiotic, substantially increase the permeability of the bladder to labelled water molecule urea, thiourea and potassium and chloride ions but left L_p u affected. Thus, their work evidently suggested that amphoterici B selectively affects the thin dense barrier. Mendoza, Handler Orloff (1967), however, refuted that interpretation when th noted that amphotericin B did increase the hydraulic conductivi of the toad bladder. It is clear, therefore, that the experimen with this antibiotic do not demonstrate that the individu components of the proposed composite barrier can be alter selectively.

The main theory of ADH's action which we have discussed far is the pore-enlargement model stemming from Ussing's wor An alternative to that hypothesis was put forward by Ginetzins (1958) who suggested that the increase in L_p was due to a lowerin of the resistance to intercellular water flow rather than to transcellular flow. In particular, he attributed the increase in L_p of the collecting ducts to the presence of hyaluronidase which tended to break down the intercellular cement joining adjacent cells. According to this hypothesis the neurohypophyseal hormones stimulate epithelial cells to release hyaluronidase or a similar enzyme which depolymerizes the mucopolysaccharides of the intercellular cement. Indeed, such an enzyme is present in the urine from mammalian kidneys although it has a lower optimum pH than that of commercial hyaluronidase. This difference between the natural and the commercial enzyme preparations may explain why the latter did not increase L_p of the toad bladder (Leaf, 1960). Moreover, Ivanova & Natochin (1968) noted that hyaluronidase at a suitable pH produced an increase in L_p of the frog urinary bladder and that this effect, like that of ADH, was inhibited by elevated calcium concentrations. It might be contended that Ginetzinsky's mechanism is at variance with the observations on the frog skin (MacRobbie & Ussing, 1961) and on the collecting ducts (Ganote et al., 1968) that ADH induces cellular swelling when a dilute solution bathes the outer or luminal surface of the epithelium. Such swelling could result, however, from dilution of the fluid in the intercellular spaces if osmotic flow proceeds predominantly through the intercellular pathways as Ginetzinsky has proposed. However, there is good evidence, based on ultrastructural studies, against the notion that osmotic water flow proceeds exclusively through the intercellular spaces in the collecting duct; it has been summarized by Grantham (1971) and it will be discussed later (see *Renal collecting ducts*, page 369). Probably the most cogent piece of evidence diametrically opposed to the views of Ginetzinsky is that of Burg, Helman, Grantham & Orloff (1970) who showed not only that ADH increased the water permeability of the collecting duct cells but also that there is no solvent-drag effect exerted by the osmotic flow on the movement of small solutes, such as urea.

So far we have two rather divergent views of how ADH may increase the hydraulic conductivity of epithelial tissues. Recently Hays and his collaborators have done some work which tends to narrow the gap between those apparently dissimilar models. The significance of their experiments hinges upon the effect of ADH on P_d. In view of the increase in P_d which occurs when there is vigorous agitation of the bathing solutions, it is possible that the relatively small effect of ADH on P_d may be due to the masking effect of unstirred layers. Dainty & House (1966b) found that ADH did not increase the apparent P_d for frog skin bathed by strongly agitated media but they were not inclined to accept that result at face value because of the presence of anatomical unstirred regions in the skin. Hays & Franki (1970) looked at this question again in a study of water transport across the urinary bladder. They found that after unstirred layer corrections there was at least a ten-fold increase in P_d after the application of ADH in comparison to the earlier two-fold increase recorded by Hays & Leaf (1962a). As their unstirred-layer corrections for P_d did not take into account diffusion delays within the epithelium itself, such as the retarding influence of the cytoplasm, one might speculate that ADH exerts an even greater influence on P_d than that estimated. Consequently Hays & Franki (1970) and Hays (1971) have argued that ADH increases the number rather than the diameter of small aqueous channels in the luminal cell membrane. This model also explains how the luminal cell membrane maintains its selectivity towards small solutes even when ADH is present, since the proposed new aqueous channels are assumed to have the same physical and geometrical properties as their original counterparts. The hypotheses of Ginetzinsky and Hays are rather similar in the sense that both postulate the opening of new aqueous channels in response to ADH. Of course they differ on the site of action of ADH and this underlines the difficulty over the ultrastructural route of osmotic water transport across epithelia. At present the weight of evidence is set against Ginetzinsky's mechanism for the action of ADH on water permeability but clearly the intercellular flow of water in epithelia cannot be safely ignored, and it may well be that ADH decreases the resistance of that pathway, particularly in the regions of the 'tight' junctions about which we know so little.

The ultrastructural evidence about the route of osmotic water transport will be fully discussed later.

Passive transport of solute

Solute permeability

Some studies of the solute permeability characteristics of epithelia have been specifically designed to yield information about the nature of the transport pathways. For instance, there are the investigations of Höber & Höber (1937) on rat small intestine, or Durbin, Frank & Solomon (1956) on frog gastric mucosa and of Smulders & Wright (1971) on rabbit gall bladder.

In particular, Höber and Höber studied the rates of transport of various water-soluble substances across the intestine. They concluded that the tissue had a sieve-like character since all molecules larger than mannitol (molecular radius = 4 Å) did not permeate the intestinal wall. Their data indicated that this epithelium may be characterized by an equivalent pore radius of 4 Å.

This has been confirmed by Lindemann & Solomon (1962) who measured the reflexion coefficients of the luminal surface of the rat small intestine for various solutes, including mannitol.

On the other hand, Durbin et al. (1956) and Altamirano & Martinoya (1966) have applied the membrane-pore model of Renkin (1954) to studies of solute and water transport in the gastric mucosa of the frog and dog respectively. According to Renkin's analysis the effective pore radius of the membrane is given by equation 3.8, namely

$$\frac{A_s}{\Delta x} = \frac{A_p}{\Delta x}\left(1-\frac{a_s}{r}\right)^2\left[1-2\cdot104\left(\frac{a_s}{r}\right)+2\cdot09\left(\frac{a_s}{r}\right)^3-0\cdot95\left(\frac{a_s}{r}\right)^5\right]$$

where A_s and A_p are the areas available for solute diffusion and that of the postulated pores respectively in unit area of membrane. Both Durbin et al. (1956) and Altamirano & Martinoya (1966) found that the relation between $(A_s/\Delta x)$ and a_s required a theoretical fit based on two kinds of pore. Moreover, the latter workers noted that the sizes of the pore radii depended upon which solute was used to maintain isotonicity of the bathing solutions so that there would be no net volume flow during the permeability measurements. Their analysis showed that the gastric mucosa was functionally equivalent to a porous membrane possessing a large group (90% of total population) of small pores with radii of about 2·5 Å and the remainder with large radii of about 90–100 Å. Durbin et al. (1956) reached a similar conclusion for the frog gastric mucosa. On the other hand, in both studies the magnitude of the ratio $(L_p RT/\bar{V}_w P_d)$ was compatible with a single population of pores with radii of about 18 Å. Such porous channels, predicted on the basis of $(L_p RT/\bar{V}_w P_d)$ cannot be regarded as representative of any actual transport pathway and it is more likely that there is actually a continuous distribution of pore radii from about 90 Å down to 2 Å. In fact, Altamirano and Martinoya argued that because the majority of the pores have small radii (≤ 2 Å) an osmotic pressure gradient will drive water preferentially through the small pores whereas a hydrostatic pressure gradient will produce bulk flow through the wide pores. Consequently the observed magnitude of the hydraulic conductivity for the stomach ought to depend upon the nature of the driving force; in fact, such an effect has been observed in canine gastric mucosa (Moody & Durbin, 1969) and, incidentally, in single cells (Vargas, 1968a). Moreover, on the basis of small and large pores one might expect that hydrostatic pressure-driven flow would create a negligible ultrafiltration effect and this has been confirmed by Altamirano (1963).

The preceding analysis of epithelial transport has not been extensively applied probably because of its many limitations. For instance, this particular treatment of solute permeation ignores the alternative transport pathway through the so-called 'lipid-route' of the cell membrane; Wright & Diamond (1969b) have demonstrated the importance of the transport pathway through the lipid phase of epithelial cell membranes for a large number of substances and we shall consider this in the next section on the reflexion coefficients of epithelial tissues.

Recently Smulders & Wright (1971) have surveyed the permeability of the rabbit gall bladder to a series of nonelectrolytes. They measured the apparent permeabilities and the reflexion coefficients for sixteen solutes and Table 9.6 displays those data and the true permeabilities, obtained after appropriate corrections for unstirred layers. For inulin and sucrose the unstirred-layer correction, employed by Smulders and Wright, was negligible whereas at the other extreme it was of paramount importance for butanol and propanol, so much so that the permeation of those solutes is rate-controlled entirely by the unstirred layers. For the majority of the solutes lying between those extremes the unstirred layer corrections led invariably to 'true' solute permeabilities larger than the apparent values. For the sake of analysis Smulders and Wright adopted the view that the 'true' permeability of the gall bladder epithelium for a given solute ought to be related to its partition coefficient between water and lipid. As a measure of the latter parameter they employed the partition coefficient

K_s^{oil}, K_s^{ether} or $K_s^{isobutanol}$ for olive oil, ether or isobutanol respectively. Figure 9.4. shows their plot of P_m against $K_s^{isobutanol}$ and there is evidently a strong correlation between the permeability and the partition coefficient. These authors argued that if the model system, in this case isobutanol, mimicked the behaviour

TABLE 9.6. Apparent and true solute permeabilities of the rabbit gall bladder epithelium to various solutes

Compound	Apparent permeability $P \times 10^5$ (cm sec^{-1})	'True' permeability $P_m \times 10^5$	Reflexion coefficient
Inulin	0·06	0·06	—
Sucrose	0·38	0·4	1·00
Erythritol	1·2	1·4	0·95
Glycerol	1·5	1·8	0·92
1,4-Butanediol	2·6	3·4	0·74
Nicotimamide	2·7	3·8	0·65
1,2-Propanediol	2·9	3·7	0·63
Acetamide	4·7	7·0	0·50
Urea	5·5	8·9	0·48
1,6-Hexanediol	3·9	6·4	0·33
Antipyrine	3·9	7·6	0·28
1,7-Heptanediol	4·0	6·8	0·11
Caffeine	3·9	7·6	0·10
Ethyl acetoacetate	8·6	⩾10	0·06
n-Propanol	14·2	⩾10	0·01
n-Butanol	16·6	⩾10	0·00

Modified from Smulders & Wright, 1971.

of the biological membrane then the slope of the plot in Fig. 9.4 ought to be unity. In fact it is less than unity and they concluded that the rate-limiting barrier to nonelectrolyte penetration in the gall bladder is more hydrophilic than isobutanol. Furthermore they contended that this barrier appeared to be more hydrated than the membranes of single cells. Supporting that contention was their observations that the apparent activation energies for some of their test solutes were significantly lower than the corresponding values for single cells, such as the erythrocytes. Three solutes of the five compounds, whose apparent activation energies for permeation were measured, had activation energies in the range 5–7·5 kcal mole^{-1} which corresponds to the range found for their diffusion in aqueous solutions. The three solutes were urea, acetamide and sucrose. The co-ordinates for urea and acetamide in Fig. 9.4 indicate that these solutes have higher permeabilities than those predicted by their partition coefficients. Thus, both the activation energies and the permeabilities of the small polar molecules, urea and acetamide, favour the view that they permeate the gall bladder epithelium via an aqueous route presumably in

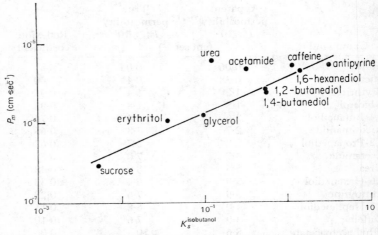

FIG. 9.4. The correlation between permeability coefficients and isobutanol partition coefficients. The permeabilities of the rabbit gall bladder were corrected for unstirred layers and expressed per unit area of the mucosal surface which is approximately 14 times larger than the serosal area (Smulders & Wright, 1971: Fig. 4).

parallel with the 'lipid route' rate-limiting the penetration of the other solutes. The only other exceptions to the general pattern of solute permeation in the gall bladder are sucrose and inulin for Smulders and Wright noted that the ratio of their permeabilities was practically identical to the ratio of their diffusion coefficients in bulk solution: moreover, as we have mentioned before, the activation energy for sucrose permeation suggests that it passes through an aqueous route in the epithelium. Smulders and Wright postulated that the aqueous route was probably due to the existence of some 'pores' somewhat larger than 12 Å in radius, which were distributed homogeneously over the epithelial surface. An alternative possibility is that the tissue was damaged, particularly at its edges where it is in contact with the experimental chambers; however, it was noted that the sucrose permeability was unaffected by varying the amount of possible edge damage relative to total surface area of the epithelium. Thus, in the gall bladder epithelium there is some evidence from solute permeability studies for aqueous 'pores' permitting the passage of large molecules and, of course, smaller molecules, too. Perhaps these 'pores' are present at the 'tight' junctions of the epithelial cells as Smulders and Wright have speculated. Other studies of the solute permeabilities of epithelia, particularly the gastric mucosa (Durbin *et al.*, 1956; Altamirano & Martinoya, 1966) and the small intestine (Loehry, Axon, Hilton, Hider & Creamer, 1970), have also pointed to the presence of a relatively small number of large 'pores'; again these may be located at the 'tight' junctions. In particular, the sparse system of large aqueous channels in the gall bladder represents a 'free-solution shunt', as Smulders and Wright refer to it, which lies in parallel with the 'lipid-route' across the epithelial cell membranes but according to their calculations it is not sufficiently effective to account for the transport rates of urea and acetamide. Such small polar solutes, therefore, probably pass through an aqueous route in the cell membranes. The latter pathway may be analagous to the small 'pore' system proposed for the gastric mucosa (Durbin *et al.*, 1956; Altamirano & Martinoya, 1966).

Reflexion coefficient

The most extensive measurements of reflexion coefficients for an epithelium have been made on the rabbit gall bladder (Wright & Diamond, 1969*a,b*). Their method for determining σ_s rested on two basic premises: first, the osmotic flow across the epithelium which is produced by a given concentration gradient of permeant molecules is less than that produced by a similar gradient of impermeant molecules; secondly, the rate of osmotic flow under these circumstances is directly proportional to the observed streaming potential.

The first premise actually embodies the definition of the reflexion coefficient where the difference in concentration, Δc_i, of the impermeant solute, i, exerts its full van't Hoff osmotic pressure $RT\Delta c_i$ ($\sigma_i = 1$) across the membrane and that for the permeant solute, s, is given by $\sigma_s RT\Delta c_s$. Of course, this approach rests on the assumption that the bathing solutions are perfectly stirred right up to the membrane.

The second premise is based on the observations (Diamond, 1962c; Pidot & Diamond, 1964; Dietschy, 1964) that the gall bladder exhibits a streaming potential in response to an osmotic gradient. Such streaming potentials have been found across artificial membranes too, and Schmid & Schwarz (1952) have shown that the true electrokinetic potential arising from bulk flow through a charged artificial membrane is accompanied by two other components of electrical potential. These particular components arise from local alterations in the ionic gradients not only in the unstirred layers adjacent to the membrane but also in the membrane itself. Such complexities are also experienced during streaming potential measurements in the gall bladder (Wedner & Diamond, 1969), in the alga *Nitella* (Tazawa & Nishizaki, 1956; Barry & Hope, 1969a,b) and in the squid axon (Vargas, 1968b). Despite those complications the measurement of streaming potentials in the rabbit gall bladder still provides an apparently rapid and reliable determination of the osmotic water flux, since Diamond (1966a) has demonstrated that the rate of osmotic water is directly proportional to the potential across the gall bladder wall in a variety of circumstances. Recent experiments by Wright *et al.* (1972) reveal, however, that streaming potentials in the rabbit gall bladder are correlated poorly with osmotic water flows when the direction of flow is from mucosa-to-serosa whereas in the opposite direction the correlation is good. The latter result is an important one since Wright & Diamond (1969a,b) employed similar osmotic gradients of their test solutes for the reflexion coefficient measurements; that is, the solutes were invariably added to the mucosal solution bathing the gall bladder.

Thus, for the gall bladder we might expect that the apparent streaming potential arising from a given gradient of the test solute divided by that produced by an identical gradient of an impermeant solute should yield the reflexion coefficient for the test solute. To confirm the validity of their approach, Wright and Diamond cite evidence that the reflexion coefficients determined by their electrical method in the rat small intestine (Smyth & Wright, 1966) agree with estimates obtained by a gravimetric procedure in the same tissue (Lindemann & Solomon, 1962). Table 9.7 shows those independent estimates of σ_s for a variety of solutes; the agreement between the separate sets of data is very

good except for ethylene glycol, which is relatively permeant.
Using the procedure described above, Wright and Diamond measured the reflexion coefficients of 206 non-electrolytes and they attempted to analyse the variation in the reflexion coefficients where Δx is the membrane thickness and K_s is the ratio (at

TABLE 9.7. *Reflexion coefficients of rat ileum for several solutes*

Solute	Reflexion coefficient	
	Streaming potential method (Smyth & Wright, 1966)	Gravimetric method (Lindemann & Solomon, 1962)
Sucrose	–	0·99
Lactose	0·97	–
Mannitol	–	0·99
Erythritol	0·87	0·93
Urea	0·82	0·81
Formamide	0·24	0·22
Ethylene glycol	0·17	0·27

from the standpoint that a given permeability, P_s, may be correlated with the lipid solubility of a solute, s, relative to its water solubility. According to that hypothesis P_s should be given by

$$P_s = \frac{K_s D_s}{\Delta x} \qquad 9.2$$

equilibrium) of the concentration of s in a lipid phase to its concentration in an adjacent water phase and D_s is the solute diffusion coefficient in the lipid phase. Provided that permeation is limited solely by diffusion in the lipid phase of the membrane, then σ_s is given by

$$\sigma_s = 1 - \frac{\omega_s \bar{V}_s}{L_p} = 1 - \frac{P_s \bar{V}_s}{RTL_p} \qquad 9.3$$

Thus, P_s and hence σ_s ought to be dependent upon K_s. In accord with equation 9.3 these workers noted that the reflexion coefficient dropped from unity to zero as K_s increased in magnitude and consequently they concluded that, in general, the penetration of non-electrolytes through the gall bladder was determined by their interactions with the lipid phase of the cell membranes. Apparently the permeability of the lipid phase for a given solute s was adequately reflected by either the partition coefficient, K_s^{oil}, of s in olive oil or the corresponding parameter K_s^{ether}. As the diffusion coefficient of s in liquids is approximately proportional to $1/\sqrt{M_s}$, where M_s is its molecular weight, Wright and Diamond also examined the dependence of σ_s on $K_s^{oil}/\sqrt{M_s}$, and also $K_s^{ether}/\sqrt{M_s}$. For instance, Fig. 9.5 shows the relation which they found between σ_s and $K_s^{ether}/\sqrt{M_s}$ and it supports the view that there is a 'main pattern' of non-electrolyte permeability. Within their so-called 'main pattern' solute permeability is correlated

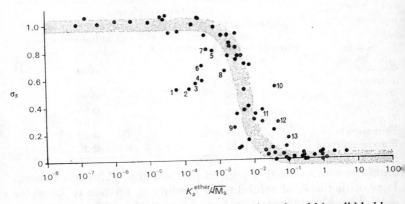

FIG. 9.5. Reflexion coefficients for non-electrolytes in rabbit gall bladder. The abscissa represents the ether:water partition coefficient divided by the square root of the molecular weight. Points referring to small solutes and branched solutes are numbered; small solutes, 1 = urea, 2 = methyl urea, 3 = formamide, 4 = acetamide, 5 = ethylene glycol, 6 = dimethyl urea, 7 = ethyl urea, 8 = propionamide, 9 = dimethyl formamide: branched solutes, 10 = pinacol, 11 = isovaleramide, 12 = 2-methyl-2,4-pentanediol, 13 = triacetin. The shaded line is drawn to indicate the general pattern of the other points and has no theoretical significance (Wright & Diamond, 1969b: Fig. 2).

Used by permission of The Royal Society

with the partition of solutes in the lipid phase of the cell membrane. Indeed, their study of the reflexion coefficient parallels the corresponding analysis of solute permeability coefficients in the alga *Chara ceratophylla* by Collander & Barlund (1933) and, of course, both pieces of work vindicate the conclusions of Overton (1896, 1899, 1902) who stressed the importance of the lipid component of the cell membrane in relation to its permeability characteristics.

Some solutes, however, do not obey the 'main pattern' of permeability. In Fig. 9.5 points numbered 1–9 signify solutes constituting the first type of deviation from the 'main pattern';

they are all small solutes with low lipid solubility which permeate more rapidly than expected, i.e. anomalously low values of σ_s. On the other hand, points 10–13 refer to solutes with highly branched structures which permeate more slowly than expected, i.e. anomalously high values of σ_s. The existence of the first kind of anomalous transport was considered by Wright and Diamond to indicate that those small polar solutes probably pass through an aqueous route associated with the polar groups of the membrane lipids. Such a transport pathway may also be the site where solvent drag or other coupling phenomena between water and small polar molecules might occur. This view has been questioned recently by Hays (1972a) on the grounds that some amides, such as urea, may cross the toad bladder predominantly by carrier-mediated diffusion, independent of osmotic flow. Finally these authors accounted for the abnormally low permeabilities of the non-electrolytes with highly branched structures by proposing that such molecules encounter relatively greater steric hindrance than straight-chain molecules do during passage through the orderly array of the cell membrane's lipid molecules.

This study of the reflexion coefficient casts serious doubt on a group of theoretical and experimental attempts to estimate equivalent pore radii for biological membranes, particularly where unwarranted assumptions have been made that the test solutes are 'lipid-insoluble'. In a subsequent review Diamond & Wright (1969) concluded that 'either the separate determination of σ and P and use of equation 2 [$\sigma_s = 1 - \omega_s \bar{V}_s/L_p - K_s^c f_{sw}^c/(f_{sw}^c + f_{sm}^c)$], or else the measurement of enough σ's or P's to reconstruct the main pattern in the particular cell under study, would be a desirable adjunct of pore-radius determinations'. As a postscript to their conclusion it is worth emphasizing that the expression (their equation 2) for the reflexion coefficient (equation 3.20) was obtained by Dainty & Ginzburg (1963) for a model of the cell membrane containing transport pathways through both the lipid phase and aqueous pores embedded in it.

Similar studies of the reflexion coefficients of epithelial membranes have been made on the frog choroid plexus (Wright & Prather, 1970), the toad duodenum (Tay & Findlay, 1972) and the goldfish gall bladder, guinea-pig intestine and the gall bladder and intestine of the bullfrog (Hingson & Diamond, 1972). Basically the results of these experiments are like those on the rabbit gall bladder.

Electrokinetic phenomena in epithelia

Both electro-osmosis and streaming potentials—electrokinetic phenomena arising from the frictional coupling of ions and water in membranes—have been observed *apparently* in a number of epithelia. Electro-osmosis has been recorded in frog skin (House 1964a), rat small intestine (Clarkson, 1967) and rabbit gall bladder (Wedner & Diamond, 1969) whereas streaming potentials have been noted in fish gall bladder (Diamond, 1962c), rabbit gall bladder (Pidot & Diamond, 1964; Dietschy, 1964; Diamond & Harrison, 1966; Wedner & Diamond, 1969), rat small intestine (Smyth & Wright, 1966) and frog choroid plexus (Wright & Prather, 1970).

Recently, however, it has been realized that the occurrence of such electrokinetic phenomena in epithelia, or in single cells for that matter, is complicated by the presence of transient changes in the local concentrations of ions in the neighbourhood of cell membranes (Barry & Hope, 1969a,b; Wedner & Diamond, 1969; Barry & Diamond, 1970). In streaming potential measurements the error due to unstirred layers shows itself as a 'polarization p.d.' whereas in electro-osmotic experiments it appears as a false volume flow resulting from a transport number effect. Even early work on electro-osmosis in artificial membranes (e.g. Stewart & Graydon, 1957) had revealed that during current passage local ionic concentration gradients develop in the adjacent unstirred layers. Now we have an explanation for their origin. Essentially such local concentrations occur because the transport numbers of the current-carrying ions may be different in the membrane, whether it be artificial or biological, from those in free solution (Barry & Hope, 1969a,b, see Chapter 3). The only epithelium in which that source of difficulty with electro-kinetic measurements has been explained quantitatively is the gall bladder. Wedner & Diamond (1969) observed the development of a 'polarization p.d.' across the rabbit gall bladder when current was passed across the tissue. The sign, magnitude and subsequent time course of decay of the 'polarization p.d.' were consistent with current-induced local concentrations of ions in the vicinity of the epithelial cell membranes. Indeed, the dissipation of such 'polarization p.d.s.' can be related to the diffusion of ions out of the unstirred layers not only into the solution but also back across the cell membranes; an additional source of dissipation is the net flow of water which is driven by the local osmotic gradient. Because that volume flow occurs in the same direction as the electro-osmotic flow under these circumstances it manifests itself as a spurious large component of an apparent electro-osmotic flow. Wedner and Diamond used the magnitude of the 'polarization p.d.s.' to estimate the size of the osmotic flows in their electro-osmosis experiments (see Table 9.8). Their data show that a substantial fraction, if not all, of the observed electro-osmotic flow in the gall bladder is not truly electro-osmotic in character. No similar analysis has been carried out on the other epithelia—rat small intestine and frog skin—which allegedly permit electro-osmosis. In the particular case of the frog skin the converse electrokinetic phenomenon, namely streaming potentials, has never been reported. Thus, it seems extremely doubtful that genuine electro-osmosis has actually been observed in skin, small intestine or gall bladder.

TABLE 9.8. Estimates of osmotic contribution to current-induced water flow across rabbit gall bladder

Solution bathing gall bladder	Current (mA)	Observed flow (μl hr^{-1})	'Polarization p.d.' (mV)	Calculated osmotic flow (μl hr^{-1})
KCl Ringer	3·4	81	8·2	51
	1·6	62	6·8	48
	1·5	22	3·5	24
	1·5	32	2·9	14
	1·0	21	1·1	20
Na$_2$SO$_4$ Ringer	1·0	39	0·16	33
	1·0	46	0·20	41
	1·0	41	0·30	60
	0·5	31	0·13	25
	1·0	41	0·14	27

The third column shows the measured rate of water flow caused by the applied current in the second column. Wedner & Diamond (1969) estimated the local osmotic contribution (fifth column) to the observed flow from the 'polarization p.d.' shown in the fourth column (see text). Modified from Wedner & Diamond, 1969.

Wedner and Diamond also noted that transient changes in the local ionic concentrations within the unstirred layers associated with the gall bladder were responsible for the generation of boundary diffusion potentials of the same sign as the apparent streaming potentials. Schmid & Schwarz (1952) have also established that a component of the streaming potentials observed across artificial membranes is a diffusion potential stemming from the unstirred layer effect. In order to estimate indirectly the true magnitude of the streaming potential Wedner and Diamond applied the Helmholtz–Onsager relation (equation 3.60) to the true (?) electro-osmotic volume flow evaluated from Table 9.8. It turned out that the true (?) streaming potential was probably about 15% or less of the apparent streaming potential. The results of their study do not detract from the merit of using the apparent streaming potential as an index of net water flow across the gall bladder, however, since those two parameters are related empirically to one another under certain circumstances—see Wright et al. (1972) for more details. Nevertheless, neither true streaming potentials nor genuine electro-osmosis may actually occur in the gall bladder, which is the most thoroughly studied epithelium from an electrokinetic standpoint. In the light of this it appears that the reliability of earlier electrokinetic studies on other epithelia must be questioned.

Solvent drag

Another way of looking at possible interactions between passive solute and water transport in epithelia is to examine the effect that osmotic water flow exerts on solute permeation. In 1944 Visscher et al. found that the ratio of unidirectional fluxes of labelled water across the canine intestine was not identical to the ratio of the water activities of the bathing media. From that observation they concluded that there was an active transport of water across the gut wall. It was contended by Ussing (1952) and, in greater detail, by Koefoed-Johnsen & Ussing (1953), however, that the data of Visscher et al. (1944) could be explained by postulating a solvent-drag effect. Such an interaction between water and labelled water molecules can arise if water penetrates through aqueous channels in the membrane. A consequence of Ussing's explanation is that there should be solvent drag on small solute molecules too; indeed, such an effect on the unidirectional fluxes of solute ought to be quantitatively related to the net water flux across the membrane. Andersen & Ussing (1957) derived an expression for the solute flux ratio in terms of the net volume flux on the basis that the solute traversed the membrane solely via the porous route. To test their theory Andersen and Ussing measured the unidirectional fluxes of heavy water and of the solutes thiourea and acetamide since they are poorly soluble in lipid and, hence, might be expected to pass through the porous route, if it exists at all. It is interesting to note here that the reflexion coefficient measurements of Wright & Diamond (1969a,b) revealed that thiourea belonged to the 'main pattern' of permeation (lipid route) whereas acetamide deviated from the 'main pattern' and probably permeated through an aqueous route in the gall bladder epithelium. Figure 9.6 shows

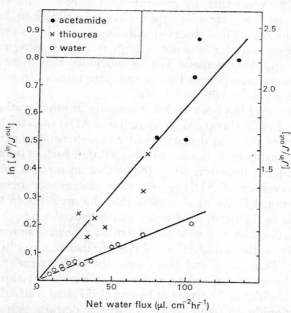

FIG. 9.6. The relation between the flux-ratio for solutes and water and the net water flux across toad skin. The graph includes only data from experiments on skins treated with neurohypophyseal hormone (Andersen & Ussing, 1957: Fig. 2).

the results of Andersen & Ussing's (1957) study of solvent drag in the isolated toad skin. They predicted that a graph of the logarithm of the ratio of the unidirectional fluxes (or apparent permeabilities) against the net osmotic water flux should be linear and Fig. 9.6 confirms that this was so, when antidiuretic hormone was present. Even in the absence of the hormone there was similar evidence for solvent drag but the data were not plotted for the sake of the figure's clarity. An additional prediction of their solvent-drag model was that the ratio of the slope of the line for each test solute to that for heavy water ought to be equal to (D_w/D_s), the theoretical value; both ratios were found equal to 2·8 in comparison with the theoretical values of 2·4 for acetamide and 2·7 for thiourea. Those solvent-drag experiments suggest the toad skin contains aqueous porous channels, and Andersen and Ussing estimated their equivalent radius (see Table 9.9). The pore radii obtained are significantly larger than the radii of the test solutes and consequently one might expect that the concentration of test solute molecules in the solvent-drag stream ought to have been similar to that in the bulk solution. The observed efficiency of the solvent-drag mechanism, however, was not as high as the foregoing argument suggests and Andersen and Ussing concluded that there must be an additional non-porous barrier to solute transfer in series with the porous barrier within the skin.

Solvent drag has been found *apparently* in other epithelia. For instance, Leaf & Hays (1962) reported that ADH increased the permeability of the toad urinary bladder to water and to some small solutes, notably urea, but not to other solutes, such as thiourea and chloride ions; moreover, they found that an osmotic water flow in the presence of ADH generated a discrepancy between the unidirectional fluxes of urea such that the urea flux in the same direction as the net flux of water was enhanced while that in the opposite direction was retarded. A similar solvent drag was exerted evidently on the unidirectional fluxes of labelled water in the toad bladder. Thus, Leaf and Hays were led to the same conclusion about the nature of solute and solvent movement through the urinary bladder as Andersen & Ussing (1957) had reached for skin. Another epithelial tissue which apparently permits osmotic water flow to exert a convective drag on small solutes and labelled water is the intestine of the dog (Lifson & Hakim, 1966; Lifson, Gruman & Levitt, 1968).

It must be stressed, of course, that the presence of unstirred layers can, in principle, give rise to spurious solvent-drag phenomenon (see page 113). Moreover, experimental confirmation of this unstirred layer effect in artificial lipid membranes has been obtained by Andreoli, Schafer & Troutman (1971) (see page 141). and recently in toad urinary bladder by Hays (1972b). Thus, it is crucial to have quantitative studies of the unstirred layer effect in solvent-drag experiments in other epithelia.

Summary

The passive transport of solutes and water across epithelia has been examined by a variety of methods. Measurements of water permeabilities, solute permeabilities and reflexion coefficients as well as studies of electrokinetic phenomena and solvent drag all help to characterize the transport pathways across the epithelium. Undeniably all of those approaches are open to both theoretical or experimental objections of one sort or another. Nevertheless, taken together they offer evidence that some epithelia behave at least partially like porous barriers; of course, the porous-membrane model is too crude to explain all of the facets of solute and water transport. It is not surprising that those different ways of exploring transport phenomena do not offer a

TABLE 9.9. Comparison of values for equivalent pore radius obtained by different methods

Tissue	Equivalent pore radius (Å)			Reference
	Solute permeation	Reflexion coefficient	$\left(\dfrac{L_p RT}{\bar{V}_w P_d}\right)$	
Rat intestine	4·0	–	–	Höber & Höber (1937)
	–	4·0	–	Lindemann & Solomon (1962)
	–	–	36*	Curran & Solomon (1957)
Toad skin	–	4·5	–	Whittembury (1962)
	–	–	9·5†	Koefoed-Johnsen & Ussing (1953)
	–	(Solvent drag; 6–20 Å)	–	Koefoed-Johnsen & Ussing (1953)
Dog gastric mucosa	2·5 (88% of population)	–	18	Altamirano & Martinoya (1966)
	90 (12%)			
Goat ependyma	8·2	–	36†	Heisey et al. (1962)

* This value was obtained by comparing the rate of mannitol permeation with that of the net water flux, i.e. analogous to an estimate from $L_p RT/\bar{V}_w P_d$ value.

† These estimates have been calculated from the original L_p and P_d values by equation 3.36.

coherent picture of the magnitude of the aqueous channels. Table 9.9 summarizes some of the equivalent pore-radius determinations for certain epithelia and the rationale behind the choice of those particular epithelia is that in every case at least two independent estimates are available for the size of the postulated aqueous channels. The general lack of agreement between the results of the different methods seems to underline the dubiety involved in applying certain theoretical approaches to transport phenomena in biological membranes. Such reservations, however, about estimates of equivalent pore radii in epithelial tissues should definitely not discourage that avenue of research, for there are undoubtedly transport routes in epithelia where interaction between some solutes and water can occur, and the approximate size of such channels is an aid to other transport studies.

Ultrastructural route of water transport

The results of certain physiological studies of osmotic water transport across epithelia can be interpreted in terms of two views which are diametrically opposed. On the one hand, we can postulate that the mucosal or outer cell membranes of some epithelia represent the rate-limiting barrier to water flow which is almost exclusively transcellular. On the other hand, the water flow could occur predominantly through the intercellular channels and, in fact, might be rate-limited by the 'tight' junctions. The weight of experimental evidence is insufficient to force us to abandon completely either of those models. It seems at present that neither transcellular nor intercellular water flow can be completely ignored. Several workers have attempted to resolve the dichotomy by looking at the ultra structure of certain epithelia and those studies have provided valuable evidence which will now be considered.

Urinary bladder

Under the influence of neurohypophyseal hormones, the urinary bladder of amphibians becomes relatively permeable to water. The apical plasma membranes of the epithelial cells probably rate-limit osmotic water flow in this tissue; evidence for this is that the presence of dilute media on the mucosal surface of the bladder does not lead to cellular swelling although swelling does occur after AHD is added (Peachey & Rasmussen, 1961; Carasso, Favard & Valérien, 1962). However, Carasso *et al.* (1962) also drew attention to the possible importance of the intercellular spaces in the toad bladder as a transport route for water (see also Pak Poy & Bentley, 1960). The cellular swelling induced by ADH in the presence of an osmotic gradient apparently concurs with the observation (Hays & Leaf, 1962*a*) that neurohypophyseal hormones decrease the inulin space (or extracellular space) of the bladder probably as a result of cellular swelling. Natochin, Janáček & Rybova (1965), however, refuted evidence based on the inulin space in the presence of osmotic water flow; in fact, they argued that under those conditions the size of the inulin space depended inversely upon the size of the net water flux.

Although the foregoing evidence implies that the mucosal border of the epithelial cells in the urinary bladder is the permeability barrier to osmotic water movement the argument is complicated by the presence of four types of the cell (granular, mitochondria—rich, goblet and basal) in the mucosal epithelium (Choi, 1963); only the first three types contact the mucosal surface of the bladder (DiBona, Civan & Leaf, 1969*a*). In a later study DiBona, Civan & Leaf (1969*b*) concluded that the volume of the granular cells was selectively increased when an osmotic influx of water occurred in the presence of ADH even although all of the cell types were capable of swelling when the serosal medium was diluted. DiBona *et al.* (1969*b*) attached no importance to the possible transport route between the cells, although they did note that ADH caused an enlargement of the intercellular spaces in both the presence and absence of osmotic water transfer. According to Carasso, Favard, Bourguet & Jard (1966), however, neurohypophyseal hormones cause distension of the intercellular spaces only in the presence of osmotic water flow.

Apart from the conventional view that the osmotic water flow across the urinary bladder is a truly transcellular one cannot ignore the possible role of the intercellular spaces. Indeed some evidence indicates that such spaces serve as part of the route for water transport (Pak Poy & Bentley, 1960; Carasso *et al.*, 1962; Carasso *et al.*, 1966; Grantham, Cuppage & Fanestil, 1971). The approach in those studies was to establish a relation between the degree of distension of the intercellular spaces and the rate of net fluid transport. However, in the bladder the apparent width of such spaces may not be a reliable index of water transport since DiBona & Civan (1969, 1970) have demonstrated that distended spaces occur in bladders exhibiting no net water transport and that the stretch exerted on the tissue influences the width of the spaces. To surmount that difficulty Wade & Discala (1971) employed horseradish peroxidase as a marker in the intercellular spaces. Although this substance cannot penetrate the spaces from the mucosal surface it can do so when it is placed on the serosal surface of the bladder (Mazur, Holtzman, Schwartz & Walter, 1971). After labelling the spaces in that way with horseradish peroxidase Wade and Discala found that either establishing an osmotic gradient by diluting the mucosal fluid or applying ADH did not alter the distribution of the enzyme; however, when ADH was applied in the presence of an osmotic gradient they found that horseradish peroxidase was rapidly removed from the intercellular spaces, presumably by solvent drag. Thus, their work vindicates the notion that osmotic flow proceeds at least partially through the intercellular spaces but it does not settle the key question about how water enters the spaces. Wade and Discala hold the opinion that osmotic water flow is probably rate-limited by the luminal border of the cells and that water enters the intercellular spaces across the lateral plasma membranes of the epithelial cells. To support their view they cite the facts that ADH causes cellular swelling during osmotic flow and that it does not apparently affect the structure of the 'tight' junctions terminating the intercellular transport route.

DiBona (1972) has recently reported that raising the osmotic pressure of the mucosal solution bathing toad urinary bladders enlarges regions of the 'tight' junctions due to the formation of fluid 'blisters'; this structural change in the 'tight' junctions might be associated with the large (65%) drop in electrical resistance of the bladders which DiBona observed. On the other hand, raising the serosal osmotic pressure did not apparently alter the structure of the 'tight' junction nor did it decrease the resistance; in fact, there was a slight increase of 11% in the transmural resistance. DiBona concluded that the accessibility of the 'tight' junctions to water and presumably also solute molecules means that these 'junctions probably limit—rather than prevent—passage of water and ions between cells'.

It is clear that we must await more experiments before we can really decide how tight the 'tight' junctions really are in the urinary bladder.

Amphibian skin

Although there is evidence that the main resistance to osmotic water transport in this multi-layered epithelium is located near to the outer surface of the skin (MacRobbie & Ussing, 1961), water may move through the intercellular spaces too, rather than strictly by a transcellular pathway alone. Ussing (1965) referred to unpublished observations of Zadunaisky that the application of hydrostatic pressures of 80–90 cm H_2O on the inner surface of the frog skin produced fluid blisters between the corium and the epithelium; this confirms older reports of such a phenomenon (Reid, 1890). Later Voute & Ussing (1970) examined the effect of hydrostatic pressure on the skin's structure and they found that the intercellular spaces expanded reversibly when the internal hydrostatic pressure was raised. Since both the corium and the basement membrane probably offer relatively little resistance to pressure-driven water flow, it is easy to visualize how such a driving force can generate an expansion of the intercellular spaces provided that these channels are closed by 'tight' junctions at the level of the first cell layer below the stratum corneum. Incidentally, Voute and Ussing found that in the absence of a pressure gradient there was a correlation between the volume of the intercellular spaces and the rate of active sodium transport by the skin. It seems that distension of the intercellular spaces, as a consequence of active solute and passive water transport, is a general feature of most transporting epithelia and it will be fully discussed in the final chapter.

Actually similar effects of hydrostatic pressure have been observed in the goat's rumen—an epithelium rather similar in structure to the frog skin. Engelhardt (1969) found that a hydrostatic pressure of 40 cm H_2O applied to the luminal surface led to a progressive decline in the net water flow whereas the same pressure applied to the serosal surface produced a large net water flow which increased with time. The latter experimental conditions were accompanied by the formation of fluid blisters below the stratum corneum (Engelhardt & Schwartz, 1968). Engelhardt (1969, 1970) has proposed that an increase in luminal pressure reduces the dimensions of the superficial intercellular spaces and consequently the resistance to water flow increases while pressure on the serosal surface generates the opposite effects.

Aside from the pressure-driven flow experiments we now have evidence from other experiments involving osmotic gradients that the main channel for water movement, under certain circumstances, is the intercellular route. The principle of this experimental procedure is to place barium and sulphate ions respectively in the outer and inner solutions which bathe the skin, and subsequently the tissue is examined for sites of formation of barium sulphate crystals (Ussing, 1970). Such locations of precipitation, therefore, will mark the common transport pathways for both of these ions. Provided no osmotic gradient existed across the tissue, Ussing found practically no precipitation except in a small number of cases where some occurred in the intercellular spaces. However, when there was an osmotic water flow from the inner to the other surface, precipitate was observed universally in the intercellular spaces. Those observations have led Ussing (1970, 1971) to speculate that when there is no osmotic gradient the skin is relatively impermeable to water because the 'tight' junctions between the outermost layers of epithelial cells are quite 'tight' but when there is an osmotic water efflux the junctions become relatively permeable. Unfortunately we do not know enough about the properties of 'tight' junctions to be able to accept or reject Ussing's model, but at least we know that his hypothesis is consistent with the results of other experiments on frog skin. For instance, earlier work by Ussing & Andersen (1956) had shown that when the osmolarity of the outer solution exceeded that of the inner one the electrical resistance of the skin dropped precipitously; it was confirmed later (Ussing & Windhager, 1964; Ussing, 1966) that the establishment of an osmotic gradient, rather than simply an elevation of external osmotic pressure, was the key factor in eliciting the drop in resistance. It seems that this decrement can be attributed to an increase in the conductance of the shunt pathway through the skin and it is tempting to identify the shunt pathway with the 'tight' junctions and their associated intercellular spaces.

Apart from the increase in the skin's conductance that is induced by such osmotic gradients, a number of test molecules, such as sucrose, experience an anomalous inward driving force (Ussing, 1966; Franz & Van Bruggen, 1967). The force on these molecules is anomalous in the sense that it goes against the net osmotic water flow and yet is both passive in character and rather non-specific as regards the solutes that it transports. It is interesting to note that a similar phenomenon has been observed in artificial membranes too (e.g. Franz, Galey & Van Bruggen, 1968; Galey & Van Bruggen, 1970). In particular, Franz *et al.* (1968) attributed the net flux of the test solute molecules against the osmotic gradient to a 'solute-drag' effect between them and the solute molecules employed to create the osmotic gradient (see Chapter 4, page 143). Other workers, notably Ussing (1966, 1971) and Ussing & Johansen (1969) have advanced the idea that there is 'anomalous solvent drag' on the test solutes due to the heteroporous nature of the membrane. This mechanism has been discussed in detail in Chapter 4. With particular reference to the amphibian skin Ussing (1971) has suggested that osmotic water efflux possibly makes the 'tight' junctions more permeable and that there is a concomitant circulation of water flow such that osmotic efflux is partially counter-balanced by a water influx through the leaky 'tight' junctions. It is this water influx which generates the 'anomalous solvent drag' on the solute molecules and causes them to flow in across the skin by the permeable intercellular route. A detailed discussion of anomalous solvent drag in heteroporous membrane has been published by Patlak & Rapoport (1971) (see pages 143–144).

Gall bladder

A large number of structural studies have demonstrated that the distension of the intercellular spaces in the gall bladder epithelium may well be related to its reabsorptive function (Yamada, 1955; Hayward, 1962*a,b*; Johnson, McMinn & Birchenough, 1962; Evett, Higgins & Brown, 1964; Kaye, Wheeler, Whitlock & Lane, 1966; Tormey & Diamond, 1967). In particular, both Kaye *et al.* (1966) and Tormey & Diamond (1967) discovered that fluid reabsorption in the absence of osmotic gradients was associated with distension not only of the intercellular spaces but also of the subepithelial capillaries. Several experimental conditions, such as omission of sodium from the bathing media, produced a drastic reduction in the width of the intercellular spaces. Moreover, both sets of workers found that an adverse osmotic gradient, which significantly diminished the net water flow, also led to the collapse of the intercellular spaces. All of this structural evidence invites the speculation that the intercellular spaces constitute an important route for water transport in the gall bladder. We would envisage, for example, that water may pass across the apical plasma membranes into the cells and thence across the lateral membranes into the intercellular spaces or, alternatively, it may enter the intercellular spaces through the 'tight' junctions.

Recently attempts (Smulders *et al.*, 1972; Wright *et al.*, 1972) have been made to delineate the ultrastructural route of water flow by analysing the changes in the permeability and structure of the gall bladder exposed to different osmotic gradients. Smulders *et al.* (1972), for example, found that increasing the osmolarity of the mucosal solution caused a reduction in the width of the intercellular spaces and an accompanying decrement in the permeabilities of certain solutes; on the contrary, increasing the osmolarity of the serosal solution caused a dilatation of the spaces but no change in the solute permeabilities. Smulders *et al.* (1972) concluded that the salient feature of the osmotically induced structural changes was the collapse of the intercellular spaces. Indeed, they argued that as the spaces collapse solute diffusion within them rate-limits the transmural solute flux and consequently that this effect accounts for the decrease in the solute permeabilities. Wright *et al.* (1972) made a corresponding study of osmotic water flows and they found that the L_p for mucosa-to-serosa water transport was four times larger than for that in the opposite direction. Again, these workers argued that the dilatation of the intercellular spaces, that occurred when serosal osmotic pressure was raised, could account for the larger value of L_p under those conditions. They considered too the possible significance of the 'tight' junctions as transport routes but discarded them on the grounds that their hydraulic conductivity was probably not large enough to account for the overall L_p of the gall bladder. Of course, their argument involved the arbitrary, but reasonable, choice of an equivalent pore radius (12 Å) for the 'tight' junction; it will be remembered that Smulders & Wright (1971) had previously speculated on the basis of solute permeability measurements that 'tight' junctions might behave as an aqueous transport route with equivalent pores of 12 Å. It could be contended that the 'tight' junctions are considerably more porous (when there is an osmotic water flow from the mucosal to the serosal solution) than is suggested by the solute permeability measurements of Smulders & Wright (1971); however, Wright *et al.* (1972) showed that doubling the pore radius was still not sufficient to account for the observed water permeability. Unfortunately their arguments like all theoretical ones, are subject to some uncertainties and, therefore, it remains possible that a significant fraction of osmotic water flow passes through the 'tight' junctions. Furthermore, the fact that the intercellular spaces dilate when water flows from the mucosal to the serosal surface of the tissue indicates that there is a significant water permeability barrier between the ends of the spaces and the serosal surface, as Smulders *et al.* (1972) suggest.

The work of Smulders *et al.* (1972) and Wright *et al.* (1972) suggests that solute and water molecules pass through the intercellular spaces during their transit across the gall bladder epithelium and that probably water enters these spaces predominantly across the lateral surfaces of the cells and not the 'tight' junctions. Nevertheless, we must await further experiments designed to tell us quantitatively just how 'tight' these junctions are to water.

Renal collecting ducts

Burg, Grantham, Abramow & Orloff (1966) have devised a technique for *in vitro* experiments on isolated perfused renal segments. In their procedure, lengths (1–5 mm) of collecting tubules are dissected from the rabbit's renal cortex so that measurements of water and solute fluxes across the tubule wall can be made (e.g. Grantham & Burg, 1966; Grantham & Orloff, 1968). Using such preparations, Burg, Helman, Grantham & Orloff (1970) found that ADH increased the permeability of the tubular cells to water but not some solutes. It is very important in such studies to know whether water moves across two plasma membranes in series or alternatively if it passes through extracellular channels between the cells. This problem has been tackled in a number of ways.

As described earlier, the volume of the duct epithelium has been recorded microscopically by Ganote *et al.* (1968) and Grantham *et al.* (1969), who found that ADH induced no change in cellular volume in the absence of osmotic gradients; but that it did so when hypotonic fluid bathed the luminal surface. Their data indicate that the rate-limiting barrier to osmotic flow is located at the luminal border of the tubular cells, and apparently that there is negligible osmotic flow through the intercellular junctions. During those experiments on the collecting duct it was noted that the intercellular spaces became distended when ADH was applied to ducts whose lumens were perfused with hypotonic solutions. Electron micrographs of the collecting-duct epithelium reveal that there are two types of epithelial cell present and that all of the cells are separated by very narrow intercellular spaces. In contrast, electron micrographs of collecting ducts fixed 30 min after the addition of ADH showed cellular swelling and dilatation of the intercellular spaces owing to the enhanced net water flux from the lumen to the peritubular surface. It is worth emphasizing that the widening of the intercellular channels is not an obligatory concomitant of osmotic swelling of the cells since such swelling can be generated by diluting the serosal medium and under these conditions the intercellular spaces are still collapsed (Grantham *et al.*, 1969). In contrast to certain other epithelia, where the size of the intercellular spaces have been correlated roughly with the rates of active salt transfer and passive water flow, the spaces between the tubular cells in the collecting ducts are enlarged solely by osmotic water flow from the lumen to the blood side. Furthermore, the morphological evidence is in concordance with the general notion that the intercellular channels are separated from the luminal medium by apical 'tight' junctions.

The possibility that osmotic water flow occurs solely between the tubular cells has been considered by Grantham (1971) who estimated the width of the necessary channel through the apical junctional complex. According to Grantham's calculations the width of such a perforation would need to be at least 42 Å. The structural studies, however, have not revealed such channels in the apical 'tight' junction, although it must be admitted that the structural examination of such junctions is probably in its infancy. Certainly a more powerful argument against intercellular osmotic flow is the notable absence of a solvent-drag effect on small solutes, such as urea (Burg *et al.*, 1970). It appears rather that osmotic water transport proceeds through the apical plasma membranes and subsequently through the lateral plasma membrane into the intercellular spaces opening to the peritubular surface. In accord with that view, depicted schematically in Fig. 9.7, Grantham *et al.*, (1969) have concluded that the lateral plasma membranes are relatively permeable to water since they observed that the epithelium swelled rapidly when the external (serosal) medium was diluted. When the tubules were replaced in isotonic Ringer solution, however, the epithelial cells shrank while the extracellular spaces became distended. Subsequently the width of the intercellular spaces returned to its normal limits within about 200 sec. The fact that it took such a long time for

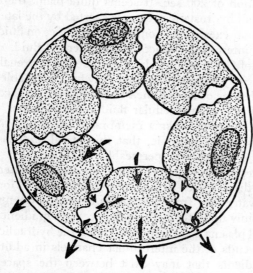

FIG. 9.7. Proposed pathways of osmotically induced net water movement in collecting tubules treated with vasopressin. The arrows indicate that the osmotic flow must first cross the luminal border of the cells and then may reach the peritubular fluid by crossing the opposite face of the cells or via the lateral membranes and intercellular spaces (Grantham et al., 1969: Fig. 12).

fluid to leave the intercellular spaces implies that the junctional complex at the apex of each space offers a significant hydraulic resistance to water flow. Other evidence in their paper upholds the concept that during shrinkage of the tubular cells water leaves the cell relatively rapidly not only across the lateral plasma membranes but also across the serosal surface of the epithelium. Grantham (1971) has estimated from his measurements of the mechanical properties of the membranes of the collecting duct that the hydrostatic pressure exerted on the fluid in the enlarged intercellular spaces might be about 10 cm H_2O. In contrast to the possible difference in osmotic pressure between the intercellular fluid and the serosal fluid such a hydrostatic pressure seems exceedingly small. However, Grantham et al. (1969) showed that if the luminal pressure exceeded the serosal pressure by only 5 cm H_2O the emptying time of the intercellular spaces in the shrinkage experiments was markedly reduced to about 10–30 sec as opposed to the normal duration of 200 sec. It seems quite plain, therefore, that relatively small hydrostatic pressures exerted by the lateral plasma membranes may exert a rate-controlling influence on fluid exchange between the intercellular spaces and the serosal fluid.

The work of Grantham and his colleagues on the renal collecting ducts is especially illuminating since they have been able to analyse the water relations of this epithelium into its separate components that is, cellular and intercellular fluid compartments. They have shown that the apical plasma membrane rate-limits the osmotic water flow across the tubule, that the lateral and basal plasma membranes are relatively permeable to water, that the intercellular spaces are an important route for osmotic water flow and, finally, that these spaces behave as compartments. Moreover, the kinetics of water flow from that series of compartments seems to depend possibly upon the hydrostatic pressure gradient developed by the lateral plasma membranes and upon the hydraulic resistance of the basal ends of the intercellular channels in addition to any osmotic gradients that may exist between the spaces and the serosal fluid.

Effect of temperature on P_d and L_p

In view of the striking effect which temperature has on the water permeabilities of cells (see Table 5.6), it is surprising to find that comparatively little attention has been paid to this phenomenon in epithelial transport.

Hays & Leaf (1962b) studied the temperature dependence of both P_d and L_p of the toad urinary bladder. They found that the apparent activation energy, E_a, for diffusion of labelled water across the tissue was 9·8 kcal mole^{-1} in the absence of ADH and 4·1 kcal mole^{-1} when the hormone was present. Hays and Leaf argued that this pronounced drop in E_a for labelled water diffusion was caused by an expansion of aqueous channels in the rate-limiting barrier to water transfer. In contrast, they found that E_a for the diffusional movement of urea was 4·1 and 3·9 kcal mole^{-1} in the absence and presence of ADH respectively. The similarity of the two latter values seems remarkable when one considers that ADH increased the urea permeability of the urinary bladder about tenfold. In order to explain this Hays and Leaf postulated that urea permeated through the urinary bladder *via* pathways which contain water with the same properties as bulk water. On the other hand, they postulated that labelled water passed through aqueous channels containing 'organized water' principally associated with the channel's walls. In terms of their model ADH was assumed to increase the diameter of the channels and consequently to reduce the rate-limiting effect of the 'organized water' in them. Their model is not considered fully here because we now know that it is based on unreliable estimates of P_d as a result of large errors from unstirred-layer effects. For example, Hays & Franki (1970) have found, after corrections for unstirred layers, that P_d for the toad bladder treated with ADH is about 11×10^{-4} cm sec^{-1} as opposed to $1 \cdot 7 \times 10^{-4}$ obtained by Hays & Leaf (1962*a*). This discrepancy plainly calls for a re-examination of the effect of temperature on P_d in the toad bladder and Hays, Franki & Soberman (1971) have carried out such a study. When the appropriate corrections for diffusion in the unstirred layers situated in the solutions and the tissue had been made they found that E_a for P_d in the epithelial layer was 11·7 kcal mole^{-1} in the absence of ADH and 10·6 kcal mole^{-1} in its presence. These data are clearly at variance with those of Hays & Leaf (1962*b*) and the nature of the discrepancy confirms the contention that the permeation of labelled water across the toad bladder in the presence of ADH is rate-limited by diffusion across unstirred layers. Both Hays and his co-workers have argued that the identity between the new values for E_a in the presence and absence of hormone signifies that water molecules participate in the same degree of hydrogen-bonding during transit through the epithelial cell membranes. Furthermore, they visualize that ADH probably brings about the opening of new aqueous channels with identical geometrical and physical characteristics to those in the unstimulated bladder. If this is not the case it becomes difficult to explain how ADH can accomplish an increase in P_d without altering the apparent activation energy for water diffusion across the bladder. In order to account for the hormone's failure to increase the permeability to small solutes Hays *et al.* (1971) have attributed the bladder's selectivity to solutes to the small size of their postulated aqueous channels. Indeed, they envisage that such aqueous channels are close to the size of the water molecule itself. Of course, the proposition that the urinary bladder exerts selectivity to the penetration of some solutes and not others rests on solute permeabilities which have not been corrected for unstirred-layer effects. That particular objection is not likely to be a serious one for most of the solutes studied earlier (Leaf & Hays, 1962) since their apparent permeabilities (about 10^{-6} cm sec^{-1}) are lower than one might expect even for quite thick unstirred layers.

In contrast to the relatively high values of E_a for the diffusion of labelled water across the bladder, Hays & Leaf (1962b) found that the corresponding value for E_a for osmotic water flow in the presence of ADH was 4·6 kcal mole^{-1}. The latter value is quite close to that for viscous flow of water in bulk liquid. (Unfortunately E_a was not measured in the absence of ADH.) It is hard to reconcile their value of E_a for osmotic flow with the current picture of water diffusion across the bladder (Hays et al., 1971) that is, if the rate-limiting barrier contains aqueous channels of the same dimensions as the individual water molecules it is somewhat surprising that the apparent activation energy for osmotic flow is so close to that for bulk viscosity. Probably the most satifactory way of resolving this difficulty is to re-examine the effects of temperature on L_p after unstirred-layer corrections have been made. Such corrections are necessary and may have profound significance in this context, since recent work on another epithelium—rabbit gall bladder (Wright et al., 1972)—has indicated that the apparent L_p is a gross underestimate of the 'true' L_p.

It is regrettable that there are so few studies of the temperature dependence of passive water transport across epithelia because this approach can tell us something about the interactions which water molecules experience during their transit in the tissue. Apart from the toad bladder the only other epithelia which have been investigated in this context are apparently the frog skin, the human skin, fish gills, the rabbit gall bladder and the rabbit ciliary epithelium.

Grigera & Cereijido (1971) obtained P_d values for the outer barrier of frog skin by measuring the uptake of labelled water across the skin's outer surface. Assuming, as they did, that the outer unstirred layer is about 50 μm they concluded that the 'true' permeability of the outer membrane was probably underestimated in their experiments by only a few per cent. Figure 9.8 shows their data for the dependence of P_d upon the absolute temperature. The relation is non-linear and they considered that it could be represented by three straight lines corresponding to apparent activation energies of 4·3, 8·5 and 16·7 kcal mole^{-1}. Their interpretation is that 'the state of water or the state of the

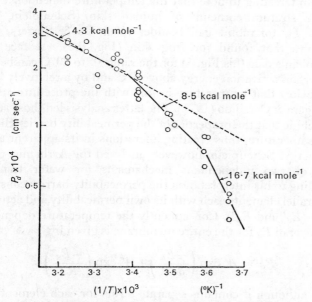

FIG. 9.8. Logarithmic plot of the diffusional permeability to water (P_d) against $1/T$ for frog skin. Each point represents a single determination. The straight lines for a given set of observations were fitted by the method of least squares. The interrupted line corresponds to self-diffusion of water molecules (Grigera & Cereijido, 1971: Fig. 2).

membrane (or both) is not the same throughout the range of temperature studied'. Above 25°C the activation energy for P_d is similar to that for the self-diffusion coefficient of water (4·6 kcal mole^{-1}, Wang et al., 1953), whereas below that temperature the relatively large activation energies signify that water permeation occurs through 'non-liquid water' in the outer barrier of the skin according to Grigera and Cereijido. Of course, it is quite possible that such large activation energies are characteristic of a lipid route that water molecules must traverse; after all, large activation energies have been recorded both for water diffusion through artificial phospholipid membranes (Price & Thompson, 1969) and for solute permeation through an epithelium where the solubility in lipid, or rather something analogous to lipid, of some solutes seems to be strongly related to their permeabilities (Smulders & Wright, 1971).

It is interesting to note that the temperature dependence of P_d for the stratum corneum of human skin (Scheuplein, 1966) and of L_p for rabbit gall bladder (Van Os & Slegers, 1973) resembles that found for frog skin (Fig. 9.8). At face value Scheuplein's data (his Fig. 5) for the range 0° to 20°C can be fitted by a certain activation energy above 6°C and by a relatively higher value below that temperature in line with the procedure adopted by Grigera & Cereijido (1971). The latter analysis led those authors to conclude that the properties of the permeability barrier changed with temperature, thus creating alterations in its apparent activation energy. Scheuplein, however, analysed the Arrhenius plot in terms of two independent mechanisms for water transport. According to his interpretation the permeability barrier consists of two parallel elements each with its own permeability and activation energy (E_a' and E_a''). Consequently the temperature dependence of the overall P_d for the entire membrane is given by

$$P_d = A' \exp\left(\frac{-E_a'}{RT}\right) + A'' \exp\left(\frac{-E_a''}{RT}\right) \qquad 9.4$$

which, although it contains separate terms for each element, is of the same form as equation 5.15 for self-diffusion of water. Scheuplein employed a graphical solution and he obtained values of 3×10^{-3} and $2 \cdot 8 \times 10^{-7}$ cm sec^{-1} for A' and A'' and 6·0 and 19·7 kcal mole^{-1} for E_a' and E_a'' which he attributed respectively to 'pore and bulk diffusion' of water molecules. The former aqueous pathway occupies about 0·001% of the membrane's area and permits self-diffusion of water molecules in rather the same way that certain porous artificial membranes with similar low activation energies do (Northrop & Anson, 1929; Stokes, 1950). According to Scheuplein one possible site for the occurrence of 'pore diffusion' in human skin was occasional dilatation of the intercellular spaces. The 'bulk diffusion' pathway, on the other hand, possessed a large activation energy and he concluded that this route contained water with a more organized structure than that of its bulk liquid. As mentioned earlier, such high values for E_a could be alternatively attributed, however, to the kinetics of water transport through a lipid barrier in the skin. Possibly a similar conclusion applies to osmotic flow across rabbit gall bladder (cf. Van Os & Slegers, 1973).

In addition to the estimates of activation energy for water diffusion across frog skin there are corresponding values for osmotic flow across toad skin *in vivo*. Dicker & Elliott (1967) found that E_a for osmotic water entry was 12 kcal mole^{-1} and that ADH caused a reduction to 8 kcal mole^{-1}. Thus, if osmotic water flow goes through a porous route in the toad skin the fluid in such channels does not have the same properties as bulk water, otherwise E_a would be identical to that for η_w, namely 4·6 kcal mole^{-1} (Wang *et al.*, 1953). Of course, it could be argued that the large value for E_a arises because the rate-limiting step for osmotic water flow is diffusion across, say, the outer membranes of the epithelial cells and that ADH creates a porous route where part of the water flux moves by bulk flow with an activation energy of 4·6 kcal mole^{-1}. Indeed, one can see a parallel between the influence of ADH on E_a in the toad skin and comparative studies of the temperature dependence of the water permeabilities of cellulose acetate membranes with different degrees of porosity (Gary-Bobo & Solomon, 1971) (see page 128).

Apparently the only other reported value for E_a for hydraulic flow across an epithelium is that of 5·9 kcal mole^{-1} given by Green & Pederson (1972*b*) for rabbit ciliary epithelium. This value seems appropriately close to what one might expect for such a permeable ($L_p = 6 \times 10^{-4}$ cm sec^{-1} atm^{-1}) epithelium.

High values in the range 11·8–19·9 kcal mole^{-1} have been obtained for E_a for water diffusion across the gills of the eel (Motais & Isaia, 1972). Since these measurements were performed *in vivo* we cannot simply conclude that such activation energies pertain to the permeability properties of the epithelium itself. We must take account of the different patterns of blood flow through the gills at different temperatures. For instance, Motais & Isaia (1972) have pointed out that at high temperatures relatively more blood might flow through the lamellae which are presumably highly permeable to water whereas at low temperatures the blood flow may be partially shunted through the less permeable filaments. Unfortunately this argument is quite speculative, particularly because there is no quantitative evidence whatever about the relative water permeabilities of the filaments and the lamellae of the fish gill.

Rectification of osmotic water flow

So far we have tacitly assumed that the rate of osmotic water flow across epithelia is directly proportional to the osmotic gradient. Direct proportionality has been observed in some cases. For instance, Diamond (1962c) observed a linear relation when small osmotic gradients (≤ 30 m-osm) were established across the fish gall bladder and Warner & Stacy (1972) have noted a corresponding linear plot for the sheep rumen exposed to relatively large osmotic gradients. Nevertheless, certain epithelia exhibit non-linear relations between the net water flux and the osmotic gradient, especially when the gradients are large. The list of epithelia showing such phenomena is extensive and includes the urinary bladders of the toad (Bentley, 1961) and the turtle (Brodsky & Schilb, 1965), the frog skin (House, 1964a, 1965; Franz & Van Bruggen, 1967), and fowl cloaca (Skadhauge, 1967), the rabbit gall bladder (Diamond, 1966a; Wright et al., 1972), the frog intestine (Loeschke, Bentzel & Csaky, 1970), the proximal tubule of *Necturus* (Bentzel, Parsa & Hare, 1969) and the insect rectum (Phillips, 1970; see also Phillips & Beaumont, 1971). This non-linear osmotic behaviour falls into classes. In the first type the apparent resistance to osmotic water flow increases with the magnitude of the osmotic gradient whereas in the second type there is a genuine rectification of osmotic flow inasmuch as the apparent resistance to water transport in one direction is less than in the opposite direction. The difference between these classes will not be examined at length in the following discussion for I prefer to discuss the possible mechanisms of non-linear osmosis as generally as possible.

It should be emphasized, of course, that such non-linear osmosis is a passive phenomenon; that is, it is completely independent of the active solute and concomitant water transport that normally occur in epithelia. Nevertheless, most studies of non-linear osmosis have been deliberately performed in the absence of active salt transport to avoid possible spurious effects.

This phenomenon might arise from several sources. Although these will be discussed separately it must be stressed that possibly several effects may act conjointly to yield non-linear osmosis in a given epithelium.

Osmotic behaviour of the permeability barrier

Until recently one of the most plausible explanations for non-linear osmosis was the hypothesis that the water permeability of the membrane varied with its degree of hydration. Studies of artificial membranes have shown that their hydraulic conductivities are correlated with their water content so that the larger the water content the higher the value of L_p (see Table 4.3). This has prompted the suggestion that the apparent rectification of volume flow occurring in single animal and plant cells results from variations in the hydration of their cell membranes. One can envisage that not only the entire cell but also its plasma membrane swells or shrinks in response to changes in the external osmolarity. The membrane is considered to be more water permeable in the swollen state and less permeable in the shrunken state. The alternative view is that the cell membrane does in fact rectify the volume flow by some intrinsic mechanism as yet not understood. The mechanism of rectification of water flow has been discussed for the algae, *Chara corallina* and *Nitella translucens*, by Dainty & Ginzburg (1964a) and for the erythrocyte by Rich *et al.* (1968), Farmer & Macey (1970) and Blum & Foster (1970). This topic has been dealt with in Chapter 6 (see page 236).

The possible osmotic behaviour of the permeability barrier in epithelia has been put forward for the toad urinary bladder (Earley, Sidel & Orloff, 1962) and the rabbit gall bladder (Diamond, 1966a).

Let us consider Diamond's work on the gall bladder, for it illustrates clearly the general principles necessary for the study of non-linear osmosis. Figure 9.9 depicts the results of an experiment where different sucrose solutions were placed in the lumen of the bladder and Ringer solution bathed its serosal surface. The gall bladder evidently rectifies the water flow in such a way that osmotic water flow from mucosa to serosa experiences apparently less resistance than water flow in the opposite direction. However, the interpretation of such data is not quite so easy. In fact, Diamond concluded from other osmotic experiments that his results could be explained by postulating that the L_p of the gall bladder was dependent on the average osmotic pressure of the solutions bathing the tissue. Inspection of Fig. 9.9, for example, reveals that as the luminal sucrose concentration decreases the apparent

L_p increases in just the sort of manner one might expect if the permeability barrier in the epithelium behaved like an osmometer. In particular Diamond envisaged that aqueous channels in the epithelial cell membrane themselves behaved as osmometers and that when these channels shrank in the presence of hypertonic

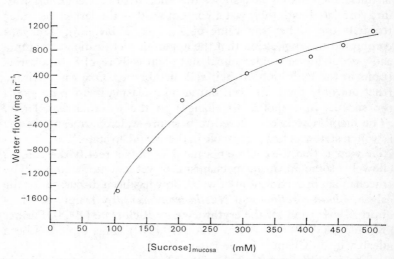

FIG. 9.9. Rate of water flow measured gravimetrically (experimental points) as a function of osmotic gradient across rabbit gall bladder. The smooth curve is the water flow as reconstructed from streaming potential measurements. The serosal solution was NaCl Ringer's solution throughout, and the mucosal solution was sucrose Ringer's solution. Osmotic gradients were produced by varying the mucosal sucrose concentration, plotted on the abscissa (sucrose Ringer's solution is isotonic when it contains 257 mM sucrose). Positive values of water flow on the ordinate indicate flow from serosa-to-mucosa and negative values flow in the opposite direction (Diamond, 1966a: Fig. 1).

solutions they offered a larger resistance to water flow. Unfortunately the evidence upholding Diamond's model of non-linear osmosis was based on his method of determining water flow indirectly from streaming potentials. Recently Wright *et al.* (1972) have shown that this method is unreliable for monitoring osmotic water flows from mucosa to serosa. In fact, Wright and his coworkers demonstrated by direct measurements of water fluxes that there was a genuine asymmetry of osmotic flow across the gall bladder. They noted that for a given osmotic gradient of 300 mM sucrose the water flux from mucosa to serosa was about three times larger than that in the opposite direction. This finding is diametrically opposed to the view that the water permeability barrier itself behaves as an osmometer (cf. Diamond, 1966a) unless the barrier is an asymmetrical structure composed of two or more layers with different osmotic properties (Hartley, 1948). In the gall bladder, therefore, there is a genuine rectification of water flow and its characteristics are similar to those of certain other epithelia, such as the frog intestine (Loeschke et al., 1970) and the toad urinary bladder (Bentley, 1961). In these tissues osmotic water movement from mucosa to serosa meets a lower resistance than that in the opposite direction.

Although the preceding discussion seems to suggest that there is a unique pattern of osmotic rectification, this is misleading. Admittedly the epithelia that have been examined do permit osmotic flow to pass more easily into the blood stream or the serosal fluid than in the opposite direction but in many cases insufficient information is at hand and consequently alternative mechanisms cannot be excluded. For instance, Machin (1969) found that the isolated toad skin rectified osmotic water flow in the same way as the intestine, gall bladder and urinary bladder. (Incidentally this is diametrically opposed to observations of Huf (1955) on the frog's skin, but Huf's experiments have been criticized because very dilute serosal solutions (10% Ringer) were used.) In Machin's experiments the osmotic influx was obtained by bathing the outer and inner surfaces with distilled water and Ringer solution respectively, whereas osmotic effluxes occurred from the serosal Ringer solution into different hypertonic solution of sucrose on the outside. Machin considered that the non-linear osmosis was probably related to the varying water content of the skin under those different osmotic conditions; that is, the efflux measurements were accompanied by a reduction in the water content owing to the high ambient osmotic pressure whereas the converse held for the osmotic influx. Thus, we cannot differentiate on the basis of this study between genuine rectification of water flow and the possible dependence of the hydraulic conductivity on osmotic pressure. The picture seems to become more confusing when we take further experimental data into account. For instance Fig. 9.10a shows the osmotic water fluxes across the isolated frog skin exposed to different concentrations of sucrose added to the normal Ringer solutions (House, 1964a). In one set of experiments (○) there is an additional net water influx accompanying active salt transport; this water flux is evident as the intercept on

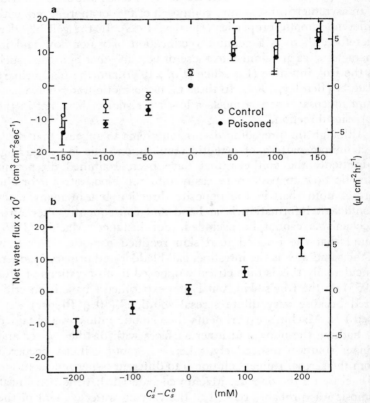

FIG. 9.10. Net water flow as a function of osmotic gradient across frog skin. In **a** each point is the mean of 10 measurements obtained in 5 experiments using a pair of skins from the same frog for each experiment, one serving as a control (○) and one treated with 1 mM KCN (●). The bars indicate ±S.D. The osmotic gradients were obtained by bathing one surface of the skin with NaCl Ringer and the other with a similar Ringer containing sucrose. In **b** each point is the mean of 30 measurements on 10 skins bathed in sulphate Ringer to which sucrose is added in different concentrations. In both **a** and **b** the abscissa is the difference in the sucrose concentration, C_s, across the skin (House, 1964a: Fig. 2. Dainty & House, 1966b: Fig. 2).

the ordinate axis when no sucrose gradient is present. In the other set of measurements (●) active salt transport and its associated water flow have been abolished by cyanide. A statistical analysis of these data revealed that there was a genuine non-linearity in the relations between the flux and the osmotic gradient (House, 1965), with the apparent L_p decreasing as the osmotic gradient increases, irrespective of its direction. These results seem to favour the view that the skin's hydraulic conductivity depends on the osmotic pressure. However, when similar experiments (Dainty & House, 1966b) were performed on skins bathed in sulphate Ringer solutions plus different sucrose concentrations, the osmotic relations were linear (Fig. 9.10b). In the frog skin non-linear osmosis occurs when permeant NaCl is present in the bathing solutions but not when impermeant Na_2SO_4 is used as its replacement. This discrepancy means that L_p does not depend on osmotic pressure and we have no adequate explanation as yet for non-linear osmosis in the frog skin.

In summary, it seems that no epithelium which exhibits non-linear osmosis does so because its hydraulic conductivity decreases as the external osmotic pressure increases.

Unstirred layers

The influence of unstirred layers on measurements of L_p has been discussed in Chapter 4. Briefly we can recapitulate by emphasizing that unstirred layers of solution and (or) unstirred regions in the tissue reduce the size of the osmotic gradient across the epithelium itself below that of the apparent gradient between the bathing fluids. Owing to this effect the apparent L_p will be less than the 'true' L_p for the epithelium alone. The salient feature of this discrepancy is that it is a function of the net volume flux and, therefore, the larger the osmotic flow the more will the apparent L_p deviate from the 'true' L_p. In other words, some epithelia may exhibit non-linear osmosis due to perturbations in the effective osmotic gradients. This raises the question: can the rectification of osmotic flow observed in several epithelia be due to unstirred layers?

When Diamond (1966a,b) examined the role of unstirred layers in non-linear osmosis in the gall bladder he found that the mucosal unstirred layer was responsible for a maximal reduction of about 2·5% in the apparent L_p whereas the serosal unstirred layer gave a reduction of 8·5%. The disparity between these values is due to the differences in the thicknesses of the unstirred layers on each side of the tissue. Recalling our discussion of the unstirred layer effect on L_p in Chapter 4, it will be remembered that osmotic water flow across a membrane tends to raise the solute concentration at the membrane on one side and lower it on the other side. Inspection of equations 4.12 and 4.13 shows that the factor expressing the deviation between the solute concentrations at the membrane and those in the bulk solutions is given by $\exp(\mathcal{J}_v\delta/D_s)$, where $\mathcal{J}_v$ is the net volume flux, δ is the thickness of the unstirred layer and D_s is the solute diffusion coefficient in the unstirred layer. Now in a typical epithelium the mucosal unstirred layer is generally much smaller than the serosal one, particularly because the latter contains an additional connective tissue layer which impedes solute diffusion. To illustrate the argument let us consider an epithelium where the mucosal and serosal unstirred layers give respectively 10% and 20% deviations in the solute concentrations at the surfaces of the epithelial cells. If we perform osmotic experiments on this tissue it will rectify the water flow. We can see why this is so from the following example. Let us assume that the mucosal solution contains 200 mM of a given solute and the serosal solution contains 400 mM of the same solute, then the solute concentrations at the mucosal and serosal surfaces of the cells will be 220 mM and 320 mM. The actual osmotic gradient driving water from mucosa to serosa is, therefore, 100 mM instead of the nominal value of 200 mM. When the mucosal and serosal solutions are now interchanged the new surface concentrations will be 360 mM and 240 mM. The actual osmotic gradient driving water from serosa to mucosa is now 120 mM. Consequently the osmotic water flux from serosa to mucosa will be larger than that in the opposite direction for the same apparent osmotic gradient. This example illustrates, in principle, how different unstirred layer thicknesses at each side of an epithelium could generate rectification of osmotic water flow. The crucial point about this argument, however, is that the predicted direction of rectification is opposite to that actually observed in epithelia, such as the intestine, urinary bladder and gall bladder. Consequently the unstirred layer effect must be discarded as a mechanism for rectification in those epithelia.

Flow-induced deformations of the permeability barrier

Another source of non-linear osmosis might be that the pathways for osmotic flow suffer some deformation related to the rate of water movement. This is akin to a valve-like behaviour of the water permeability barrier and Brodsky & Schilb (1965) have contended that such an effect is partially responsible for rectification of osmotic water flow in the turtle urinary bladder. Provided that osmotic water flow is governed by Poiseuille's law, one would expect that the hydraulic conductivity of the epithelium will be a function of the dimensions of the aqueous channels through which water flow occurs. Brodsky and Schilb pointed out that in the turtle bladder L_p decreases with increasing osmotic gradients and that this might reflect changes in the dimensions of water-filled pores. In support of this argument they cited evidence about the changes in the water content of the bladders which occurred in response to the osmotic gradients employed; however, it must be admitted that the connexion between changes in the bladder's water content and possible alterations in the dimensions of a relatively sparse system of aqueous channels is tenuous.

Recently this proposed mechanism of rectification has been reconsidered for the gall bladder by Wright *et al.* (1972). These authors argued that, since the width of the lateral intercellular spaces depends on the direction of osmotic flow (Smulders *et al.*, 1972), these channels may rate-limit osmotic flow particularly in the serosa-to-mucosa direction. Their calculations showed that, if osmotic flow proceeds through the intercellular spaces, the apparent L_p of the epithelium should decrease when the spaces are collapsed. As they point out, there are many uncertainties about such calculations but in principle this mechanism can account for the observed rectification of osmotic water flow. It is probably highly relevant that other rectifying epithelia, namely *Necturus* proximal tubule (Bentzel *et al.*, 1969), frog intestine (Loeschke *et al.*, 1970) and toad urinary bladder (DiBona, 1972) also exhibit similar changes in the dimensions of the intercellular spaces during osmotic experiments. Perhaps the most significant, and as yet unexplained, feature of these structural studies is the fact that during osmotic flow from the mucosa to serosa the spaces become distended. This must mean that a significant hydraulic resistance exists between the open ends of the spaces and the serosal surface of the tissue. Smulders *et al.* (1972), for example, consider that the subserosal layer of collagen fibres might operate as such a barrier but they also cite Grantham's unpublished observations that the basement membrane of renal tubules has an 'appreciable hydraulic resistance'.

It seems that structural studies of certain epithelia demonstrate that osmotic water flow meets an additional permeability barrier situated between the cellular layer and the serosal surface. If this is so, then an alternative model of rectification must be considered. This hypothesis has been called the asymmetrical double-membrane model and its relevance to osmotic flow across epithelia will now be discussed.

Asymmetrical double-membrane model

Theoretical accounts (Kedem & Katchalsky, 1963c; Patlak *et al.*, 1963) of the transport characteristics of a series array of two dissimilar membranes a and b separated by an aqueous layer have been discussed in Chapter 2. Consider such a double-membrane system bathed by identical solutions of a permeant solute s. Provided that both membranes have identical reflexion coefficients for s then the overall hydraulic conductivity, L_p, of the system is given by

$$\frac{1}{L_p} = \frac{1}{L_p{}^a} + \frac{1}{L_p{}^b}$$

where $L_p{}^a$ and $L_p{}^b$ are the respective hydraulic conductivities of the membranes a and b. If the corresponding reflexion coefficients, $\sigma_s{}^a$ and $\sigma_s{}^b$, are not equal, then L_p is given by equation 2.125, namely

$$\frac{1}{L_p} = \frac{1}{L_p{}^a} + \frac{1}{L_p{}^b} + c_s{}^* \frac{(\sigma_s{}^a - \sigma_s{}^b)^2}{\omega_s{}^a + \omega_s{}^b}$$

See *Membranes in series* (Chapter 2) for an account of the derivation of this equation. It indicates that under those circumstances the apparent hydraulic conductivity will invariably be less than $L_p{}^a L_p{}^b/(L_p{}^a + L_p{}^b)$, the effective hydraulic conductivity of an asymmetrical double-membrane without a central compartment. In fact, the additional term in the equation for L_p exists because the net volume flux, $\mathcal{J}_v$, across the double-membrane system will either produce depletion or accumulation of s in the central compartment depending on the direction of flow. This is a solvent-drag effect and in this scheme it exerts an influence on the solute concentration of the central aqueous layer. Such an effect on the steady-state distribution of permeant solutes across a single membrane was predicted by Nims (1961) and confirmed experimentally for a Millipore filter by Nims & Thurber (1961). In the case of the double-membrane model the dependence of c_s^* (the average concentration of s for membranes a and b) on J_v produces a non-linear relationship between volume flow and the driving force, whether it is a hydrostatic pressure gradient or an osmotic gradient created by an impermeant solute (see equation 2.126). The effect gives the asymmetrical double-membrane system the power to rectify volume flow. For the sake of simplicity consider that only a difference of hydrostatic pressure, Δp, exists across the system then the relation between J_v and Δp, i.e. equation 2.127, can be reduced to

$$\alpha J_v - \beta J_v^2 = \Delta p \qquad 9.5$$

where α and β replace the rather cumbersome coefficients of the original equations; α is always positive but the sign of β depends upon that of $(\sigma_s^a - \sigma_s^b)^3$. In the first instance, when $\sigma_s^a > \sigma_s^b$, β is positive and a plot of the general relationship shows that the volume efflux will be rectified (see Fig. 2.4a). On the other hand, when $\sigma_s^a < \sigma_s^b$, β is negative and the corresponding graphical relation is depicted in Fig. 2.4b where the volume influx is rectified. Although the preceding argument is developed for hydrostatic pressure gradients the same results hold when there are equivalent osmotic pressure gradients, due to the presence of different concentrations of impermeant solutes. In order to see how this theoretical scheme can be applied to rectification of osmotic flow in epithelia let us consider, for example, the gall bladder, where the most extensive studies have been made.

Figure 9.11 shows schematically the changes that occur in the gall bladder epithelium, particularly the intercellular spaces, during osmotic experiments (Smulders *et al.*, 1972). Let us suppose that these spaces constitute the central compartment of the double-membrane model described above. Then the epithelial cell membranes or possibly even the 'tight' junctions can be taken as the outer (or mucosal) membrane a of the model and the basement membrane or the subserosal layer as the inner (or serosal) membrane b. Although we have no quantitative estimates of the water and solute permeabilities of membrane b it is quite likely, indeed highly probable, that it is substantially more permeable than membrane a. Thus, according to this double-membrane model of the gall bladder there should be rectification of the osmotic efflux, i.e. in the direction serosa-to-mucosa, just as is observed in practice (Wright et al., 1972). Moreover, this model also seems to explain the changes in the distension of the intercellular spaces under different osmotic conditions.

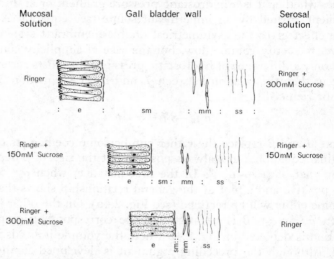

FIG. 9.11. Diagrammatic representation of changes in structural geometry of the rabbit gall bladder wall during osmotic experiments. Approximate dimensions based on the data of Smulders et al. (1972). The thicknesses of the component layers are shown by e (epithelium), sm (submucosa), mm (muscularis) and ss (subserosa). The salient changes are (i) collapse of the intercellular spaces and (ii) pronounced decrease in sm when osmotic flow occurs in the serosa-to-mucosa direction.

For such a double-membrane system separating two solutions of different osmotic pressure, namely π_s^o and π_s^i, it can be shown that there will be a difference in hydrostatic pressure, Δp, between the central compartment and the bathing solutions. Δp is given by

$$\Delta p = \frac{L_p^o \sigma_s^o (\pi_s^o - \pi_s^c) - L_p^i \sigma_s^i (\pi_s^c - \pi_s^i)}{L_p^o + L_p^i} \qquad 9.6$$

where π_s^c is the osmotic pressure of the fluid in the central compartment. The derivation of this relation is similar to that of Ogilvie et al. (1963) (see page 150). When there is an osmotic inflow across the system $(\pi_s^o - \pi_s^c)$ will be negative whereas $(\pi_s^c - \pi_s^i)$ will be negligible because membrane b is so permeable. Equation 9.6 predicts, therefore, that Δp will be negative during osmotic flow from outside (mucosa) to inside (serosa). Such a pressure gradient could be responsible for the observed dilatation of the intercellular spaces during osmotic flow from mucosa to serosa. On the other hand, when there is an osmotic efflux across the system $(\pi_s^o - \pi_s^c)$ will become positive while $(\pi_s^c - \pi_s^i)$ will be again negligible. According to equation 9.6 Δp will be positive and this could account for the collapse of the intercellular spaces during osmotic flow from serosa to mucosa.

It is significant that the asymmetrical double-membrane model of epithelia can explain qualitatively not only the observed rectification of osmotic flow but also the degree of distension of the intercellular spaces—the designated central compartment. It is necessary, however, to obtain quantitative experimental support for this scheme before it can be accepted.

Conclusion

Non-linear osmosis has been observed in a number of epithelia. Almost invariably this phenomenon exhibits itself in the form of rectification of water movement so that the apparent resistance to osmotic flow is larger in the serosa-to-mucosa direction than in the opposite one. Of the mechanisms that have been put forward to explain rectification only the double-membrane model (Kedem & Katchalsky, 1963c; Patlak et al., 1963) and the theoretical argument (Wright et al., 1972) about the role of the intercellular spaces seem to be acceptable. It is not certain which, if any, of these hypotheses is the correct one principally because the experimental data on this phenomenon is so meagre.

10

ACTIVE SALT AND WATER TRANSPORT

Osmosis	391
Plant roots	393
Lizard's cloaca and cow's rumen	396
Role of hydrostatic pressure gradients	398
Gastric mucosa	400
Mammalian intestine	402
Proximal tubule of mammalian kidney	404
Conclusion	406
Electro-osmosis	407
Thermo-osmosis	408
Pinocytosis	410
Active water transport	413
Energy requirements	414
Examples of active water transport?	418
Verdict	427
Asymmetrical double-membrane model	431
Rabbit gall bladder	436
Turtle urinary bladder	440
Local osmosis	442
Experimental evidence	443
Mechanism of local osmosis	450
'Standing-gradient osmotic flow'	452
Conclusion	470

MY final chapter deals with the ability of biological systems to transport fluid in an active sense. This kind of fluid transfer is exemplified especially well in epithelia, such as the kidney proximal tubule and the gastric mucosa which can absorb and secrete fluids of different ionic composition. Here one meets the net flow of water under conditions where no external driving force for water may exist. Such fluid transport has been the subject of an extensive series of experiments and from these have stemmed many speculative theories some of which involve coupling between water and active solute transport. In many instances, however, it has been suggested that water itself is actively transported. It is tempting to expect that all epithelia share a unique mechanism of fluid transport but it is more likely that several alternative mechanisms have evolved to achieve the same end. Recently some investigators have sought to explain the power of fluid absorption or secretion by epithelia in relation to the ultrastructural routes of water and solute movements across them. Such studies have revealed the complexity of this general problem and, in particular, the inadequacies of several earlier theories of fluid absorption and secretion.

In order to simplify the discussion of this topic I have decided to deal mainly with mechanisms rather than with individual tissues.

Osmosis

Excluded from the following discussion are the obvious cases of osmotic water reabsorption in the toad urinary bladder and the mammalian renal collecting ducts where osmotic gradients are established by virtue of prior active salt reabsorption at a different site. Instead, we shall discuss whether active salt transport can create instantaneously significant osmotic gradients across tissues. For instance, micro-perfusion studies of single proximal tubules in the rat kidney have revealed that water reabsorption is dependent upon the luminal sodium concentration and on the rate of sodium reabsorption (e.g. Giebisch, Klose, Malnic, Sullivan & Windhager, 1964). One might expect that salt reabsorption in the proximal tubule ought to reduce the luminal osmolarity while it increases that of the blood in the peritubular capillaries. However, as the experiments of Giebisch *et al.* (1964) and others have demonstrated, net fluid absorption continues even although there is actually no osmotic pressure difference between blood and lumen. This has been noted widely, in a large list of epithelia, including intestine, frog urinary bladder, frog skin, dog gastric mucosa, rabbit gall bladder and the Malpighian tubules, rectum and salivary glands of insects. Moreover, water transport can occur against osmotic gradients experimentally established between the bulk solutions bathing epithelia. In fact, a large number of epithelia actually absorb or secrete isotonic fluids, so that the mechanism of coupling between active salt and water transport is not intimately linked with external osmotic gradients because they do not generally exist nor are they generated by the fluid absorption or secretion. The failure of active solute transport to create osmotic gradients between the fluids bathing epithelia is readily understood when one considers the magnitudes of the solute fluxes and the fluid compartments involved. Let us now examine critically the osmotic model in a particular case, say the absorption of salt and water by the isolated urinary bladder of the turtle (Brodsky & Schilb, 1965).

Consider the turtle urinary bladder (area A cm^2) immersed in a large volume of Ringer solution. When the lumen (volume v_0 cm^3) is filled with Ringer solution the bladder reabsorbs salt and water from the lumen over a period of several hours. Let us assume that the net flux of salt, $\mathcal{J}_s$ (mole cm^{-2} sec^{-1}) cannot create a significant alteration in the osmolarity of the large serosal volume but that it does lead to a progressive dilution of the mucosal solution. Then the salt concentration, C_t (mole cm^{-3}), in the luminal solution will be a function of time t given by

$$C_t = \frac{m_0 - \mathcal{J}_s A t}{v_t} \qquad 10.1$$

where m_0 (mole) is the total amount of salt in the lumen at zero time and v_t is the inner (luminal) volume at t. Under those conditions there will be a net flux of water, ϕ_w (cm^3 cm^{-2} sec^{-1}), given by

$$\phi_w = L_p RT(C_o - C_t) \qquad 10.2$$

where C_o (mole cm^{-3}) is the salt concentration in the serosal solution throughout the experiment and also in the luminal solution at zero time. Substituting for C_t (equation 10.1) into equation 10.2 gives

$$\phi_w = L_p RT \left[\frac{-m_0 + \mathcal{J}_s A t + v_t C_o}{v_t} \right] \qquad 10.3$$

This equation predicts that ϕ_w is zero at $t = 0$ and that it increases subsequently with time. Both of these predictions are at variance with the experimental facts (Brodsky & Schilb, 1965). Moreover, we can take the analysis a stage further by replacing m_0 with $v_0 C_o$. This step gives

$$\phi_w = L_p RT \left[\frac{\mathcal{J}_s A t + (v_t - v_0) C_o}{v_t} \right] \qquad 10.4$$

Since the observed water flux does not produce a significant decrease in the luminal volume we may equate v_0 and v_t to yield

$$\phi_w = \frac{L_p RT \mathcal{J}_s At}{v_0} \qquad 10.5$$

Equation 10.5 provides a guide to the osmotic water flow generated under those circumstances. Taking $L_p = 3\cdot6 \times 10^{-7}$ cm sec^{-1} atm^{-1} $RT = 2\cdot5 \times 10^4$ atm cm^3 mole^{-1}, $\mathcal{J}_s = 0\cdot55 \times 10^{-9}$ mole cm^{-2} sec^{-1} (or approximately 2 μmole cm^{-2} hr^{-1}), $A = 9$ cm^2, $v_0 = 3$ cm^3 and $t = 3\cdot6 \times 10^3$ sec (one hour) we find that the osmotic water flow is about 54×10^{-9} cm^3 cm^{-2} sec^{-1} (or $0\cdot2$ μl. cm^{-2} hr^{-1}), which is less than 5% of the observed water flow. Thus, a coupling between active salt movement and water flow based solely on classical osmosis must be discarded in this tissue. The preceding argument (House, 1968) has been applied also to the water absorption achieved by isolated frog skin, and again the conclusion was also that the predicted osmotic flow was about 5%, or less, of the observed water flux. Even for epithelia, like fish gall bladder (Diamond, 1962c), which are relatively permeable to water, the osmotic mechanism does not hold.

Plant roots

At present there is a common view that the formation of a relatively concentrated exudate at the basal ends of excised plant roots involves osmotic water flow (Arisz, Helder & Van Nie, 1951; Arisz, 1956; House & Findlay, 1966a). It has been proposed that some salts, such as KCl or KNO$_3$, are actively transported into small confined regions—the xylem vessels—within the root. By the line of argument in the preceding section one might expect that the salt concentrations in the xylem vessels ought to be highly dependent on the net salt flux. In other words, solute is being transferred from a large compartment, namely the external solution, to a relatively minute one, namely the xylary fluid. On such grounds one might expect the accompanying water flow to be driven by an osmotic gradient created by active salt transport. This view concurs with the observations that the salt concentration in the fluid exudate elaborated at the root's basal end is invariably higher than that of the external medium and that the rate of fluid exudation is loosely dependent on the salt concentration of the exudate itself. House & Findlay (1966a) furnished quantitative support for this plausible osmotic model of water transport in plant roots. They measured the concentration, C_s^x, of KCl in the fluid exudates from isolated maize roots placed in different KCl solutions and also the fluid absorption rates in these roots. The latter net water fluxes, ϕ_w, were expressed as the rate of fluid exudation (μl hr^{-1}) from the root's basal end divided by the root's external surface area. On the basis that there is a steady-state equivalence between the rate of net salt transfer into the xylem vessels and its rate of exit in the exudation stream, House and Findlay estimated the net flux of KCl from the expression $\phi_w C_s^x$. Their experiments demonstrated a linear relationship between the net water flux and the difference in salt concentration between the exudate and the bathing medium, C_s^0 (Fig. 10.1). A statistical treatment of

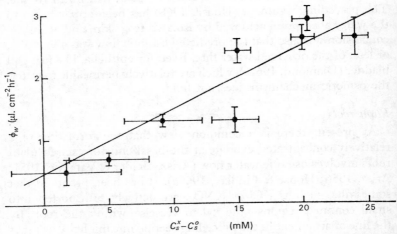

FIG. 10.1. The relation between the exudation rate, ϕ_w, from isolated maize roots and the KCl concentration difference between exudate and medium, $(C_s^x - C_s^0)$. Each point is the mean value of at least 14 measurements and the vertical and horizontal bars indicate $\pm$S.E. (House & Findlay, 1966a: Fig. 4).

their data showed that ϕ_w can be expressed by the relation

$$\phi_w = L_p RT(C_s^x - C_s^0) + \phi_w^0 \qquad 10.6$$

and, thus, the slope of the regression line in Fig. 10.1 gave the equivalent hydraulic conductivity of the root surface; the L_p value obtained was 5.7×10^{-7} cm sec^{-1} atm^{-1} but this assumes that the reflexion coefficient for KCl is unity and is, therefore, possibly an underestimate. Estimating L_p in terms of the root's surface area is merely a convenient way of expressing the data. Actually we have no right to assume that the water permeability barrier is located at the root's surface; indeed, there is some evidence (House & Findlay, 1966c) that it exists in the vicinity of the xylem vessels. Given that the osmotic barrier exists somewhere *within* the root, its actual hydraulic conductivity ought to be significantly higher than that quoted above. The intercept, ϕ_w^0, on the ordinate axis was found to be significantly different from zero and House and Findlay enumerated some possible explanations for this net water flux which was apparently unrelated to the osmotic gradient. First, it might be a genuine water flow accompanying active salt transport in an analogous way to that found in animal epithelia bathed in isotonic solutions. Secondly, it might be an osmotic water flow arising from the additional presence of a small concentration of unidentified solute. Finally, these authors stressed that ϕ_w^0 might arise spuriously because the magnitudes of the active salt flux and (or) L_p were dependent upon position along the root's length; consequently, although water transport into the xylem vessels at any given point along the root may be truly an osmotic flow, a spurious 'non-osmotic' component may appear because average parameters (e.g. L_p) are being employed. House & Findlay (1966b) presented further evidence (based on a linear correlation between ϕ_w and L_p) for osmotic water flow in isolated maize roots.

It is interesting to examine whether or not the obvious dependence of the exudate salt concentration on the external salt concentration is compatible with osmotic water transport. In order to do that one needs to know quantitatively how the net salt flux, $\mathcal{J}_s$, depends on C_s^0: House & Findlay (1966a) obtained $\mathcal{J}_s$ from the experimental values of ϕ_w and C_s^x and the relation, $\mathcal{J}_s = \phi_w C_s^x$, mentioned previously. Substituting $\phi_w = \mathcal{J}_s/C_s^x$ into equation 10.6 one obtains

$$\mathcal{J}_s = L_p RT (C_s^x)^2 + (\phi_w^0 - L_p RT C_s^0) C_s^x \qquad 10.7$$

Solving this equation for C_s^x gives

$$C_s^x = \frac{C_s^0 - \dfrac{\phi_w^0}{L_p RT} + \sqrt{\left(C_s^0 - \dfrac{\phi_w^0}{L_p RT}\right)^2 + 4\dfrac{\mathcal{J}_s}{L_p RT}}}{2} \qquad 10.8$$

The alternative solution is inadmissible since it gives a negative value for C_s^x. Before equation 10.8 can be used to predict C_s^x at any value of C_s^0 it is necessary to measure L_p, ϕ_w^0 and the dependence of $\mathcal{J}_s$ on C_s^0. After the appropriate measurements (House & Findlay, 1966*a,b,c*) had been made, it was found that the values of C_s^x predicted by equation 10.8 agreed well with the experimental ones over a wide range of C_s^0, thus supporting this osmotic model for fluid exudation.

Whether or not a small component of the fluid exudation from isolated roots is driven by some mechanism other than osmosis has not been resolved experimentally, but this problem will be taken up again under the heading *Active water transport* (see page 425). Anderson, Aikman & Meiri (1970) have suggested that water is absorbed into a local osmotic compartment within the root. This compartment *may* be the xylem vessels, and according to their model it functions as a standing-gradient osmotic flow system similar to that proposed originally for water reabsorption by the gall bladder (Diamond & Bossert, 1967; see page 452). Applying this model to the root means that no ϕ_w^0 is required provided that the hydraulic conductivity is permitted to vary along the root's length in a manner *qualitatively* similar to that observed by House & Findlay (1966*b*). As yet, however, such a local osmotic compartment remains to be identified convincingly in the plant root.

Lizard's cloaca and cow's rumen

Murrish & Schmidt-Nielsen (1970) investigated the possibility that water reabsorption in the lizard's cloaca is active. Normally the urine entering the cloaca is dehydrated until a urinary pellet containing about 46% water is formed. To determine the 'force' that withdraws fluid from the cloacal lumen Murrish and Schmidt-Nielsen placed a wick into the empty cloaca and recorded the suction pressure, $-p$, required to prevent fluid absorption. The average values of p for normal and dehydrated lizards were 207 and 255 mm H_2O respectively. These values were close to the colloid osmotic pressures (215 and 267 mm H_2O) of plasma in those animals. The authors also found that when the wick was soaked with 5% bovine albumin in Ringer solution p was only about 10 mm H_2O. Finally, they noted that the suction 'force' developed by an isolated sample of urine rose to about 250 mm H_2O as it dried to its normal water content of 46%. (This force is presumably generated by capillarity between the particles of the dried urinary pellet.) On the basis of these results Murrish and

Schmidt-Nielsen argued that the colloid osmotic pressure was the significant driving force responsible for the dehydration of the cloacal contents. If this is so, it constitutes the unique case where water retrieval by an epithelium is powered by the osmotic pressure gradient due to plasma proteins.

It should be emphasized, however, that the size of the driving force reported by these authors is extremely small in relation to the total osmotic pressure of plasma. Expressed in more conventional units of pressure it only amounts to 0·025 atm which is equivalent to the osmotic gradient generated by about 1 mM of impermeant solute. In fact, it seems quite probable that the colloid osmotic pressure itself is dwarfed by the local osmotic gradients established by active salt reabsorption across the cloacal wall. It would therefore be interesting to examine cloacal reabsorption in the lizard by employing other methods (cf. Skadhauge, 1967) for studying salt and water transport across such epithelia.

Possibly a more subtle and convincing case of osmotic water flow masquerading as an apparent active transport of water is found in the isolated ventral sac of the cow's rumen. In this preparation Dobson, Sellers & Shaw (1970) measured the net water flux as a function of the osmotic gradient in a series of experiments (Fig. 10.2). Inspection of their data indicates that there was no significant net water transfer across the rumen wall in the absence of an osmotic gradient provided the solutions in the rumen contained NaCl and KCl and were bubbled with nitrogen. However, when the luminal solution contained fatty acids and was bubbled with carbon dioxide, net water transport occurred apparently in the absence of osmotic gradients. This is an interesting result because the latter experimental conditions within the rumen are quite close to those of normal rumen contents, particularly with respect to acidity and concentrations of sodium, fatty acids and carbon dioxide. Also of interest is the fact that the rate of apparent active water transport is about 15 ml min^{-1}; taking the mucosal area in those experiments as 10^5 cm^2 (Dobson, *personal communication*) the net water flux can be calculated to be about 9 μl cm^{-2} hr^{-1} which is similar to that recorded in other epithelia (see Table 10.1 later). Thus, the experiments of Dobson *et al.* (1970) demonstrate that the cow's rumen can be induced by the presence of carbon dioxide in the lumen to transport water against an osmotic gradient. Recent work has shown that the apparent active transport of water is probably driven osmotically. It turns out that the passage of carbon dioxide from lumen to blood which occurs in these experiments leads to a significant rise in the osmotic pressure of plasma. When this rise in the plasma osmotic

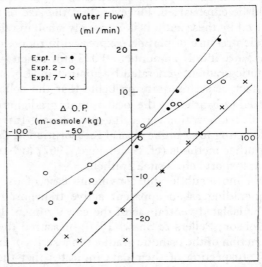

FIG. 10.2. Relation between rate of appearance of water in rumen and osmotic pressure difference across rumen epithelium. Δ O.P. = osmotic pressure of ruminal fluid − osmotic pressure of plasma. Expt. 1; NaCl + KCl: N_2. Expt. 2; NaCl + KCl + Na phosphate: N_2. Expt. 7; Na fatty acid + KCl: CO_2 (Dobson et al., 1970: Fig. 4).

pressure is taken into account (Dobson, Harrop & Phillipson 1972) it is found to be sufficient to generate the observed water flow.

Role of hydrostatic pressure gradients

There is convincing evidence against the concept that hydrostatic pressure is the driving force for fluid absorption or secretion in a number of epithelia. For instance, Heisey et al. (1962) found that the secretory rate of the goat's choroid plexus was not influenced by pressure gradients of up to 30 cm H_2O. Notwithstanding the existence of small pressure gradients across certain epithelia it has been concluded that such gradients are ineffectual because the hydraulic conductivities of the epithelia are too low to permit significant water flow. For instance, the pressure exerted on the lumen of the gall bladder may be 10–15 cm H$_2$O above that on the serosal surface (McMaster & Elman, 1926). Diamond (1962c) demonstrated, however, that elevating the luminal hydrostatic pressure in the fish gall bladder by as much as 20 cm H$_2$O did not affect the rate of fluid reabsorption. In accord with that result, Dietschy (1964) found that the everted gall bladder of the rabbit can transport fluid against a hydrostatic pressure exerted on its serosal surface. The insignificance of fluid filtration in the gall bladder can be explained on the basis of its value for L_p. For rabbit gall bladder Diamond (1964b) found that $L_p = 2 \times 10^{-6}$ cm sec^{-1} atm^{-1} and, hence, a hydrostatic pressure of 20 cm H$_2$O (or about 2×10^{-2} atm) would produce a relatively small water flow of about 0·14 μl cm^{-2} hr^{-1} in comparison with the normal fluid reabsorption rate of 50 μl cm^{-2} hr^{-1}. Apparently the values of L_p for other epithelia generally support the same view. Nevertheless, the preceding argument relies partially on the plausible assumption that the value for L_p derived from osmotic experiments is an accurate estimate of the hydraulic conductivity. Evidently that premise must now be questioned since recent work on rabbit gall bladder by Wright et al. (1972) indicates that unstirred-layer effects cause L_p to be underestimated possibly by more than ten-fold. Moreover, Moody & Durbin (1969) have demonstrated for the canine gastric mucosa that the resistance to pressure-driven flow is substantially lower than the apparent resistance to osmotic flow (see page 336 et seq.). Thus, one is forced to the conclusion that small hydrostatic pressures may not be as inconsequential as some investigators have forecast.

Indeed, quite small hydrostatic pressure gradients do exert profound effects on fluid movements across some epithelia. For instance, the formation of the aqueous humour of the eye relies partially on ultrafiltration since Davson (1953) has shown that raffinose and p-aminohippuric acid enter the aqueous humour. It is extremely unlikely that these molecules are 'pumped' along with the primary secretion from the ciliary body and, furthermore, since their rates of penetration are equal it is possible that they are filtered through a system of relatively large pores. (In this connexion it is interesting to note that the capillaries of the rabbit ciliary body contain fenestrations of about 300–400 Å in diameter (Pappas & Tennyson, 1962) although they are less porous than those diameters suggest because colloidal gold particles of 100 Å in diameter do not pass across them.) Other evidence supporting the notion that ultrafiltration is partially responsible for the formation of aqueous humour in the rabbit is the finding of Cole (1960) that even in the presence of 2,4-dinitrophenol or fluoracetamide the rate of formation is still about 30% of the normal rate. Recently Green & Pederson (1972*a*) have estimated that ultrafiltration represents an even higher proportion (about 80%) of the secretory flux across the rabbit ciliary epithelium.

Let us now turn to some other specific examples of epithelia where hydrostatic pressure gradients may exert an influence on their secretory or absorptive function.

Gastric mucosa

The secretion produced by cells other than the parietal cells of the gastric mucosa has been called the 'alkaline component' and Altamirano (1963) has studied the elaboration of the 'alkaline component' in the canine gastric mucosa in order to elucidate its secretory mechanism. In accord with earlier reports he found that intra-arterial injections of Acetylcholine (ACh) induced an alkaline-secretion; moreover, he noted that small doses of ACh produced an acid secretion. On the basis of certain criteria he established that the gastric secretions produced by large doses of ACh did not contain a significant amount of acid from the parietal cells and consequently he considered that they were 'pure' alkaline secretions. Altamirano found that the concentrations of certain solutes (Na^+, H^+, Ca^{++}, HCO_3^-, inorganic phosphate, ammonia and urea) were typical of an ultrafiltrate of plasma; furthermore, the secretion contained mucoproteins probably from the mucous cells and possibly a small amount of plasma proteins. The belief that the alkaline secretion from the canine gastric mucosa is formed by ultrafiltration was substantiated by Altamirano in two ways. Both of his tests concern the hypothesis that the hydrostatic pressure inside the gastric capillaries and small arterioles is large enough to generate the observed secretory rate. First, he applied a counter pressure to the lumen so that no fluid transfer occurred across the blood vessels enclosed within his experimental chamber. In the absence of ACh the mean counter pressure required was 15·2 mm Hg whereas in the presence of ACh the counter pressure had to be raised to 66·1 mm Hg. Taking into account the colloid osmotic pressure of the plasma proteins Altamirano concluded that the hydrostatic pressure in the capillaries and arterioles of the gastric mucosa attained a maximum of 90 mm Hg. Thus, pressure in the blood vessels of the canine gastric mucosa may reach values similar to those found in the afferent arterioles of the mammalian kidney. For his second method of estimating the capillary pressure Altamirano measured the secretory rate at certain counter pressures and subsequently he obtained the particular counter pressure required to halt secretion completely by extrapolation of the data to zero volume flow. The results of the second technique were in accord with those of the first.

Altamirano's experiments show quite clearly that the mechanism of alkaline secretion in the canine gastric mucosa is ultrafiltration. It is of considerable interest, therefore, to examine the effects of hydrostatic pressure on the secretion of acid by the gastric mucosa. Davenport & Fisher (1940) suggested that the water accompanying acid secretion was driven by the hydrostatic pressure difference between the blood vessels and the lumen of the stomach. Later Davies & Terner (1949) repudiated that opinion by demonstrating that the isolated frog gastric mucosa continues to secrete HCl and water even in the absence of its blood supply. In particular, Davies and Terner found that small applied pressures (5 cm H_2O) failed to affect the rate of fluid transfer. If the hydraulic conductivity of the frog gastric mucosa is taken as 8×10^{-7} cm sec^{-1} atm^{-1} (Durbin et al., 1956) then the pressures applied in the experiments of Davies & Terner (1949) ought to have produced small net water flows (about 0·02 μl cm^{-2} hr^{-1}) below their limits of detection. The role of blood pressure, therefore, as a possible driving force for water flow across the *in vivo* gastric mucosa cannot be dismissed completely yet. Certainly in the frog gastric mucosa fluid secretion is not coupled tightly, if at all, to acid secretion (Durbin et al., 1956; Villegas & Sananes, 1968). However, the work of Moody & Durbin (1969) indicates that a luminal pressure of about 90 cm H_2O is sufficient to halt both acid and water secretion in the canine gastric mucosa. At present the nature of this inhibition is not understood; for example, it might be mediated by restriction of the local blood supply.

In the canine gastric mucosa relatively small hydrostatic pressures exert marked effects on the ability of the tissue to elaborate both alkaline and acid secretions. It seems that the alkaline secretion is achieved almost exclusively by ultrafiltration but unfortunately we are not able to affirm exactly what role blood pressure plays as a driving force for the water flow accompanying acid secretion in the gastric mucosa.

Mammalian intestine

The effects of elevating intraluminal pressure, or mucosal pressure as it is occasionally called, on net fluid transfer across the mammalian intestine have been examined by a number of workers. The point of those studies was to assess the effect which the normal mucosal pressure exerts on fluid absorption. Abbott, Hartline, Hervey, Ingelfinger, Rawson & Zetzel (1943), for instance, found that the mucosal pressure in the human small intestine is usually about 8–10 cm H_2O above that of the serosal pressure and that this pressure difference may attain values of up to 50 cm H_2O for brief periods. On the basis of the values of L_p for mammalian intestine (see Table 9.2) one would expect that such pressure gradients could not influence the normal rate of fluid absorption. Nevertheless, the mucosal pressure exercises some control over water transport.

Wells (1931) reduced the intraluminal pressures in the canine intestine and found that small decrements (8–26 cm H_2O) below atmosphere pressure were sufficient to stop absorption of water completely. These data are in discord with the values predicted from the hydraulic conductivity since the expected filtration flux (2×10^{-7} cm^3 cm^{-2} sec^{-1}) given by $L_p \Delta p$ is considerably less than the usual fluid absorption flux. To my knowledge there is no satisfactory explanation for the discrepancy. Several effects could be responsible. First, the value of L_p derived from osmotic flow determinations may be a serious underestimate of the true L_p (cf. canine gastric mucosa; Moody & Durbin, 1969; rabbit gall bladder; Wright *et al.*, 1972); or secondly, alterations in blood flow may occur and give rise indirectly to changes in fluid absorption. Thirdly, the degree of distension of the tissue may affect in some way the active rate of salt transport to which fluid absorption is coupled.

The effects of increasing the mucosal pressure on the rat intestine have been described by several workers (Fisher, 1955; Smyth & Taylor, 1957; Lee, 1963). In particular, Hakim &

Lifson (1969) have compared the data on rat intestine by adopting a normalization procedure (see Fig. 10.3). Normalizing the water transport rates was achieved by expressing the rate of net fluid transport at a given mucosal pressure as a percentage of the rate at a mucosal pressure of 20 cm H₂O. According to Fig. 10.3 fluid absorption in the rat intestine shows a pronounced dependence

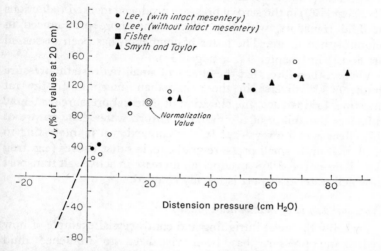

FIG. 10.3. Composite results for the relationship between fluid transport rate, 'J_v', and pressure for rat *in vitro* preparations. Observations of the various series have been normalized by setting J_v equal to 100 when the distension pressure is 20 cm H₂O. The data are taken from Lee (1963), Fisher (1955) and Smith & Taylor (1957). The interrupted line represents probable extrapolation according to the results of Wilson (1956) for hamster intestine (Hakim & Lifson, 1969: Fig. 5).

upon the mucosal (or 'distension') pressure. At face value these data seem to depict a possible link between the mucosal pressure and the hydraulic conductivity but that is not the case. Smyth & Taylor (1957) reported that a similar range of mucosal pressures did not alter the net water flux across the rat intestine when either glucose was absent from the bathing solutions or when phlorrhizin was present. Under both of those conditions water movement is entirely passive and, hence, an increase in mucosal pressure does not actually increase L_p. In the canine intestine Hakim, Lester & Lifson (1963) also found that elevating mucosal pressure did not increase L_p. The interpretation of the mucosal pressure effects on the rat intestine, therefore, must involve some direct or indirect link between the mechanism of water absorption or, rather, active solute absorption and the pressure gradient. In contrast to the results of the rat intestine, increments in the mucosal pressure on the canine intestine failed to increase water absorption (Hakim & Lifson, 1969) whereas small increments (2–26 cm H_2O) in the serosal pressure produced marked reductions in fluid transport, with this effect being more pronounced in jejunum than ileum. The latter phenomenon has been discussed previously in Chapter 9 (see page 335).

We see, therefore, that the effects of small hydrostatic pressure gradients are diverse in the mammalian intestine. In the rat intestine, for instance, the elevation of mucosal pressure somehow enhances the ability of the tissue to absorb water. The nature of this effect is still a mystery! In this connexion it is interesting to recall that quite small pressure gradients in other tissues (e.g. frog skin, Nutbourne, 1968) produce an increase in active salt transport to which water transport is usually coupled.

Proximal tubule of mammalian kidney

Probably the most intriguing and controversial example of how hydrostatic pressure has been visualized to influence fluid absorption is to be seen in the kidney proximal tubule.

The concept that alterations in the filtered load of certain substances, notably sodium ions (due to changes in glomerular filtration rate), were exactly matched by appropriate adjustments in the tubular reabsorption rate stems from many sources (e.g. Walker, Bott, Oliver & MacDowell, 1941). This phenomenon was termed 'glomerulo-tubular balance' by Smith (1951) who discussed its possible sources. Previously Richards & Schmidt (1924) had demonstrated that only some of the glomeruli in the frog kidney are active in the renal circulation at any given time. It might be contended, therefore, that glomerulo-tubular balance in the mammalian kidney was caused simply by adjustments in the numbers of active glomeruli. Nevertheless, Smith (1951) marshalled considerable evidence against this proposal for chicken, dog, cat, rabbit, rat, sheep and human kidneys. Alternatively he favoured the notion that the distal nephron was capable of changing its rate of reabsorption to meet the demands of the alterations in glomerular filtration rate. Recently, however, the general emphasis has shifted to the proximal tubule as a possible site for glomerulo-tubular balance and perhaps this shift of opinion may be eventually redressed in Smith's favour.

The study of glomerulo-tubular balance in the mammalian kidney, has been advanced considerably by the advent of micropuncture methods, and the degree to which proximal reabsorption of sodium ions and water balance the glomerular filtration rate has been the subject of many investigations. At first sight the results of those studies seem confusing. For instance, both Lassiter, Mylle & Gottschalk (1964) and Malnic, Klose & Giebisch (1966) showed that the rate of reabsorption of sodium in the proximal tubules of the rat was not in constant proportion to the glomerular filtration rate when the rat was loaded with hypertonic saline. In this case the glomerular filtration rate increases considerably (Lassiter *et al.*, 1964) but the ability of the proximal tubules to reabsorb sodium decreases. Moreover, there is evidence that glomerulo-tubular balance is absent in rats (Cortney, Mylle, Lassiter & Gottschalk (1965) and dogs (Dirks, Cirksena & Berliner, 1965) loaded with isotonic salines. In short, under diuretic conditions induced by perfusion of either isotonic or hypertonic salines the proximal reabsorption of sodium and water is reduced, probably due to the enhanced salt backflux that occurs from capillaries to the tubular lumen through the intercellular spaces (Boulpaep, 1972). On the other hand, in the non-diuretic condition the proximal tubule is capable of reabsorbing a constant fraction of the filtered load over quite a wide range of glomerular filtration rates (Giebisch & Windhager, 1964; Gertz, Mangos, Braun & Pagel, 1965).

Several hypotheses have been put forward to account for glomerulo-tubular balance in mammalian proximal tubules under conditions where there are spontaneous variations in glomerular filtration rate. In particular, considerable attention has been paid to the model of Gertz *et al.* (1965) who suggested that the hydrostatic pressure inside the proximal tubules contributed indirectly to the regulation of salt and water reabsorption. According to their model, the resistance of the thin limb of Henle's loop to fluid flow is sufficient to ensure normally that the rate of volume flow out of the proximal tubule is about one-third of the glomerular filtration rate. If an increase in the latter occurs, the intraluminal pressure in the proximal tubule will tend to rise and produce an expansion of the tubule. Furthermore, these authors postulated that the rate of reabsorption of salt and water is directly proportional to the intratubular volume and, hence, to the square of the tubule radius, r. According to this model glomerulo-tubular balance was achieved because an increase in glomerular filtration rate produced an appropriate distension of the tubule and *vice versa*. The bulk of the evidence supporting Gertz's model has been attacked on technical grounds (see Orloff & Burg, 1971). The specific objections will not be enumerated here except to say that the crucial evidence diametrically opposed to Gertz's model is that of Burg & Orloff (1968) and Morgan & Berliner (1969). The former authors used the isolated perfused tubules of rabbit kidney whereas the latter perfused the rat proximal tubules *in vivo*. Both investigations revealed that tubular distension is not associated with an increase in the reabsorption rate and, thus, the conclusion that glomerulo-tubular balance is not an inherent feature of the proximal tubule is inevitable. Now one can see that this work runs with the general tide of opinion that a humoral mechanism is involved in glomerulo-tubular balance. A humoral mechanism is also compatible with an alternative hypothesis (Leyssac, 1963) that an alteration in proximal reabsorption is the primary event which in turn, changes the intraluminal pressure and consequently the glomerular filtration rate. At present the identity of the proposed hormone remains elusive and, indeed, around the whole question of glomerulo-tubular balance confusion seems to flourish.

Conclusion

Evidently filtration, as such, does not play a salient part in the normal absorption or secretion of fluid by epithelia, although the elaborations of the 'alkaline component' in the canine gastric mucosa and of the aqueous humour in the rabbit are probable exceptions to the rule. In fact, the hydrostatic pressure gradient which normally exist across epithelial tissues are, at face value insufficiently large to be the chief driving force for primary water transport. Despite this, small changes in such pressure gradient may generate alterations in the active salt and water absorption in some epithelia. Unfortunately we simply do not know enough about this facet of salt and water transport. Moreover, quite small pressure gradients can produce marked increases in the water and solute permeabilities of some epithelia, such as intestine, gall bladder and rumen. What we do know about the responses of epithelia to hydrostatic pressure gradients suggests that they exert influences which are completely unexpected on the basis of equivalent osmotic pressures. Therefore, we cannot simply equate hydrostatic pressures with their equivalent osmotic pressures even from a phenomenological standpoint. This disparity may mean that seemingly inconsequential hydrostatic pressure gradients within epithelial tissues could exert an influence on active salt and water transport.

Electro-osmosis

A great deal of speculation has revolved around the suggestion that the coupling mechanism between active ion transport and the accompanying water flow is electro-osmosis. Normally epithelia are transporting ions and water under open-circuit conditions where the net flow of ionic currents is zero. If one believes that the observed water is electro-osmotic in origin one needs to postulate from the outset that the individual ionic currents occur in pathways which permit quite high degrees of frictional interaction between ions and water. For example, in an epithelium such as the kidney proximal tubule, which reabsorbs an isotonic solution of NaCl, this mechanism demands that each ion must carry with it about 150 to 200 water molecules during transit. This is an extremely large number of water molecules to be dragged electro-osmotically, but even so it is still the minimum number from a hypothetical standpoint. Probably we should expect on the basis of electrokinetic measurements in the gall bladder (Diamond & Harrison, 1966; Wedner & Diamond, 1969) that the aqueous channels would be negatively charged and in this case only the counterions, sodium say, might participate in electro-osmotic drag. Thus, if sodium ions alone exert an electro-osmotic drag on water, the required ratio of water molecules per ion rises to 300 or 400. There is no evidence whatever for such high electro-osmotic permeabilities in epithelial tissues, or in artificial membranes for that matter (cf. values of β, page 141). Indeed, the existence of electro-osmosis in epithelia has been apparently established only in one case—rabbit gall bladder (Wedner & Diamond, 1969)—and even in that instance there is still some dubiety (see page 357). From the data of Wedner and Diamond it follows that the electro-osmotic coupling in the gall bladder may achieve a flow of about 20 water molecules or less per ion transported. Therefore, our sole example of an epithelium that exhibits electro-osmosis cannot achieve an electro-osmotic coupling which exceeds 10% of that observed during active salt and water reabsorption. The likelihood that electro-osmosis plays an important role in fluid transport in other epithelia seems equally remote.

The existence of electro-osmosis has been established convincingly only in the large plant cells—*Nitella* and *Chara*. One might well ask what physiological significance, if any, does electro-osmosis have in these cells. Bennet-Clark (1959), for instance, has suggested that an electro-osmotic water movement accompanying active ion transport into the vacuole of a plant cell would tend to raise the turgor pressure of the cell above that value expected on the basis of equilibrium between the vacuolar fluid and the external medium. Dainty (1963a) has treated that proposal quantitatively. Assuming that the number of water molecules dragged electro-osmotically attained a plausible maximum of 55 mole Faraday^{-1} Dainty calculated that the excess hydrostatic pressure generated by that effect would be about 5×10^{-3} atm. This is negligible compared to the actual turgor pressure (about 8 atm.) of plant cells. Moreover, his value for the the number of water molecules carried electro-osmotically by each ion was deliberately chosen to obtain in theory the maximal effect. We know now that the most reliable value (38 mole Faraday^{-1}: Barry & Hope 1969b) of this electro-osmotic ratio for plant cells is quite close to Dainty's estimated maximum. Thus, electro-osmosis contributes a negligible component to the turgor pressure of plant cells.

Thermo-osmosis

There is no satisfactory evidence for or against thermo-osmosis in cells or tissues because practically no attention has been paid to this mechanism. The only report alleging the existence of thermo-osmosis, of which I am aware, is that of Vecli & Bianchi (1966) for isolated frog skin. Those workers claimed that a difference in temperature of a 'few hundredths of a degree centigrade' was established across skin sacs placed in an aerated bathing solution. Unfortunately they are not explicit about the direction of the temperature gradient but presumably aeration reduced the temperature of the bathing (serosal) solution as a result of water evaporation into the air bubbles. Vecli and Bianchi concluded that thermo-osmosis was responsible for a component of the observed water absorption and that the remaining fraction of the net water flux was driven by an osmotic gradient created by th active salt transfer. On the basis of the argument presented earlier in the section on *Osmosis*, House (1968) showed that the latter osmotic component of water absorption is probably trivial and represents only about 5% of the observed net flux of approximately 1 μl cm^{-2} hr^{-1}. According to Vecli and Bianchi the net water flux (0·7 μl cm^{-2} hr^{-1}) which occurred at the beginning of their experiments and consequently could not be an osmotic flow, must be attributed solely to thermo-osmosis. Nevertheless, they observed that a net water flow occurred across skin sacs placed in a non-aerated bathing solution—conditions under which, according to their hypothesis, no temperature gradient ought to develop. The temperature gradient was not monitored under those conditions, however. It is therefore hard to accept their conclusion that net water transport does not occur in the absence of thermal and osmotic gradients, particularly because the existence of such gradients has not been established conclusively in the first place.

In the absence of experimental data on thermo-osmosis in biological systems, and particularly in epithelia, its potential significance as a device for moving water must remain entirely in the realm of speculation. Nevertheless, the extent of such speculation can be constricted by the following considerations.

Probably the most difficult hurdles to be jumped before thermo-osmosis can be accepted are questions about the size, direction and mechanism of such gradients. We cannot really answer any of those questions yet. If one ignores them and instead speculates that heat may be released asymmetrically within, say, an epithelium then other problems arise. For example, can such thermal gradients be maintained in the face of dissipation by heat conduction across the tissue? Both Spanner (1954) and Katchalsky & Curran (1965) have concluded from their analysis of thermo-osmosis that the answer to that question is probably no. It seems unwise to duplicate or amplify their arguments since we have practically no experimental information to discuss. Instead let us focus attention on certain evidence which is out of step with the view that thermo-osmosis is responsible for fluid absorption. In nearly all epithelia involved in salt and fluid transport both the direction and rate of fluid transport are somehow related obligatorily to the corresponding characteristics of active solute transport (e.g. see Fig. 10.4). When active salt transport is halted, net water transport also vanishes provided, of course, that there is no other driving force for water movement. The observed relationship *per se* between salt and water transport in epithelia does not rule out the possibility that water, rather than salt, is the substance which is undergoing 'active transport'. Set against that concept, however, are two experimental facts. First, active ion transport is noted under circumstances where no net transport of water occurs and, secondly, although there is some evidence for solvent-drag effects on salt transfer, such as streaming potential measurements in some epithelia, the latter measurements are now regarded as somewhat dubious. Even if they are taken as genuine they represent an efficiency of coupling between water and ionic movements which is too low to be applicable here.

In spite of the foregoing arguments against the possible importance of thermo-osmosis we cannot discard it completely for there are cases where fluid transport is apparently not coupled closely, if at all, to active solute transport. In particular, the most outstanding cases are the integument and rectum of insects where water, apparently by itself, can be absorbed against huge differences of chemical potential. Schmidt-Nielsen (1969) has considered the hypothesis that insects may employ temperature gradients between themselves and the atmosphere as a mechanism for water uptake. He rejected such a general mechanism on the grounds that the necessary differences in temperature lie in the range 2–14 centigrade degrees. Admittedly such thermal gradients are far beyond the values that might be realized; however, they were estimated on the basis that water vapour is simply condensed at the interface between the insect's integument and the atmosphere. Until we know more about the state of water molecules at such liquid-gas interfaces and what forces dictate the movement of water molecules across such interfaces we cannot completely disregard thermo-osmosis. It is conceivable that quite small temperature gradients could exert marked effects on water uptake at that kind of boundary, but again this is mere speculation!

Pinocytosis

Undoubtedly pinocytosis occurs in some protozoa and in the endothelial cells of continuous capillaries. In the latter case the pinocytotic vesicles offer a transport route for the passage of large molecules across the capillary wall (see Chapter 8). In epithelia too there is some evidence for pinocytosis; for instance, in the mammalian intestine the epithelial cells are capable of engulfing latex spheres, 1000 Å in diameter (Sanders & Ashworth, 1961), and dye particles (Barrnett, 1959) within vesicles. Inevitably the question arises about the possible role of such a mechanism in the active transfer of solutes and water across epithelia. In this context Hogben (1960) has referred to pinocytosis as the 'last refuge of the intellectually bankrupt' although he conceded that it was probably responsible for the 'intestinal absorption of intact protein' in some instances. However, pinocytosis should not be dismissed so easily as that.

Both Grim (1963) and Frederiksen & Leyssac (1969) have claimed that fluid reabsorption in the gall bladder is due to the transfer of small volumes of isotonic salt solution across the epithelial cells. Grim's hypothesis is based on the concept that the reabsorption of salt is entirely passive and that the salt movement is influenced by solvent drag. In contrast to Grim's analysis of the steady-state fluxes of salt and water across the gall bladder Frederiksen & Leyssac (1969) investigated the transient changes in water and salt transfer following sudden changes in the bathing media on both sides of the tissue. During the immediate onset of such transients the net water flux remained constant while the net flux of salt changed abruptly. From this and other observations Frederiksen and Leyssac concluded that fluid transport was the primary event and that salt was a secondary consequence of it. The solute fluxes during the transients, however, were not measured; actually they were obtained indirectly from the net water fluxes on the erroneous assumption that the tonicity of the transported fluid changed instantaneously to its new value when the media were altered (cf. Diamond, 1964b). The interpretation of their transient experiments, therefore, is complex and their conclusions are probably not justified. These authors proposed that the mechanism of water transport was a 'mechanical volume pump' presumably more like peristalsis than pinocytosis. A similar mechanism has been invoked by Leyssac (1966) to account for water reabsorption in the proximal tubular cells of the mammalian kidney but here too the evidence for this mechanism is not compelling.

While some experimental data on fluid transport in epithelia may be compatible with pinocytotic transfer, other results are in conflict with it. In a large number of epithelia the net water flux is proportional to the net solute flux and it disappears when active solute movement is abolished. Moreover, active solute transport can continue even when net water transfer is halted. Finally, there is the outstanding selectivity of the mechanism responsible for active solute transport; for example, pinocytosis or other sorts of 'solution pump' cannot account for the ability of the rabbit gall bladder to transport almost exclusively only sodium and chloride ions (Diamond, 1964*a,b*).

The physiological evidence for pinocytotic transfer of salt and water in epithelia is poor. On the other side of the coin we have the structural evidence to consider. Parsons (1963) estimated that each epithelial cell in the mammalian small intestine would need to produce 1000 vesicles (diameter 600 Å) per second just to account for the absorption of the digestive juices. If that sort of argument is applied to other epithelia similar rates of vesicular formation are required. In particular, one can calculate the number of vesicles that ought to be present in a typical epithelial cell at any instant. In the mammalian small intestine, for example, the rate of fluid absorption is about 20 µl cm^{-2} hr^{-1} which can be expressed as a velocity, namely 6×10^{-6} cm sec^{-1}. Let us assume that the vesicles are formed at the mucosal surface and travel a distance of say 20 µm at that speed to the other side of each cell. Their transit time, therefore, must be about 300 seconds and during that period vesicles will have been formed at the minimal rate given above, i.e. 1000 per sec. Thus, at any instant we would expect to find about 300,000 vesicles in each cell. This number is probably an underestimate because the calculation assumes that every vesicle formed at the mucosal surface will discharge its contents at the serosal surface of the cell; by analogy with theories of vesicular transport in endothelial cells it seems more likely that only a small proportion of the total number of vesicles formed will make a complete transit of the cell. If that is so, the transit time required is probably considerably smaller than 300 sec and the number of vesicles required is probably considerably larger than 300,000. Epithelial cells do not possess such large numbers of vesicles. For instance, both Kaye *et al.* (1966) and Tormey & Diamond (1967) reported that the rabbit gall bladder epithelium contained relatively few vesicles and certainly not the number that might be expected if they were involved in fluid transport.

Pappas & Smelser (1958) suggested that the vesicles that they observed in the cells of the ciliary body were probably involved in the mechanism of fluid secretion across this epithelium. Both Tormey (1963) and Missoten (1964) have questioned the existence of such vesicles in the ciliary epithelium and, in particular, Tormey concluded that the vesicles are illusory because they result from sections of an extensive network of cytoplasmic tubules.

Another site where vesicular transport of fluid has been invoked is the contractile vacuole of amoeba. Schmidt-Nielsen & Schrauger (1963) claimed that the enlargement of the contractile vacuole was generated by the fusion of small vesicles, and the structural study of Mercer (1959) is in accord with such a mechanism. Nevertheless, the nature of the involvement, if any, of vesicular transport is open to debate. In this connexion, it is interesting to note that Riddick (1968) occasionally recorded abrupt increases in the size of the contractile vacuoles. However, such discontinuities in the growth of contractile vacuoles are probably due to the fusion of small vacuoles with the main contractile vacuole rather than to spontaneous fusion of exceedingly large numbers of vesicles. The growth of the contractile vacuole will be discussed later in the following section on active water transport.

Active water transport

The notion that epithelial tissues, or single cells for that matter, may be able to actively transport water has been surrounded by controversy for decades. Several workers have got themselves into hot water over this vexed question (e.g. see Robinson, 1965). Probably only physiologists working on the water relations of insects have some right to retain a belief in active water transport but there is dissension even in that camp. Recently Ramsay (1971) has put the case rather elegantly for believing that active water transport occurs in the insect rectum and he sums up his case for the defence of active water transport in the following way:

'The principle seems to have gained acceptance that the accused is presumed to be guilty until he can be proved to be innocent—the onus of proof is placed upon those who support the view that water is actively transported. And the course of justice is likely to be further perverted by the system of appeals. If it is demonstrated with full thermodynamic rigour that water is actively transported across the

rectal wall the prosecution will then require it to be demonstrated that water is actively transported across the rectal epithelium. If this is established then they will submit that water might move passively in consequence of solute movements within some limited region of the epithelial cell—and so on until the case is transferred from the jurisdiction of thermodynamics to the jurisdiction of quantum mechanics.'

One feels that the ranks of the prosecution ought to demand the crucial evidence for active water transport, namely proof that there is a flow of water across the epithelium in the absence of any other flow of material, except that due to the metabolic reaction driving the water pump. Let us now review the case for active water transport even although that key witness for the defence may not appear.

Energy requirements

Probably the most general criticism levelled at active water transport as a possible mechanism of fluid transfer is that its energy requirements may be excessively large. For instance, Pitts (1968) claims that the 'formation of concentrated urine by active reabsorption of water would be metabolically wasteful and that the active reabsorption of ions and osmotic equilibration of water would be less demanding'. That may be so! However, the argument with which he upholds that view is unsatisfactory. Pitts contends that active reabsorption of a volume, say 1 cm³, of isotonic saline due to active sodium transport involves about 10^{20} combinations (and subsequent dissociations) of sodium ions with a carrier whereas it would require 300 times that number of combinations of water molecules with a carrier if water alone were transported actively. Surely the argument should include also the energetics of combination and dissociation of sodium and water with their carriers, if they exist at all, during transit in the membrane. We simply do not have that sort of information at hand and, therefore, we cannot rule out active water transport quite so easily.

One model of active water transport in epithelia which has been criticized by Brodsky, Rehm, Dennis & Miller (1955) solely on energetic grounds, is the 'osmotic diffusion pump' of Franck & Mayer (1947). In that model it was proposed that polymerization of some small solutes into large molecules occurred at one end of the cell; subsequently water accompanied the large molecules as they diffused across to the other side of the cell where they were split into the original substances. Accordingly the rate of water transport across the cell should be related to the rate of diffusion of the 'carrier' molecules. There are several reasons why this is an unacceptable model of water transport; in particular, Brodsky *et al.* (1955) computed from the change of free energy for the diffusion process across the cell that the energy requirement for this mechanism is about 1000 times larger than the maximal metabolic rate of any cell.

We may not conclude, however, from the foregoing specific example that active water transport is too expensive from an energetic standpoint. It is preferable to establish within as general a framework as possible the work performed during water transport across a membrane. Consider a membrane separating two solutions of an impermeant solute i. The work may be divided into two terms. First, there is the energy expended reversibly in moving water against its difference of chemical potential, $\Delta\mu_w$; the work done by a mole of water is given by

$$\Delta\mu_w = \bar{V}_w \Delta p - \bar{V}_w RT \Delta c_i \qquad 10.9$$

where Δp and Δc_i are the differences in the hydrostatic pressure and in the concentration of i across the membrane. In addition to that form of work there is the irreversible expenditure of energy occurring during the passage of water molecules through the membrane. Thus, net transport of water, even in the absence of a chemical potential gradient for water, involves the performance of work and in order to estimate the magnitude of that work one must postulate some mechanism of water transfer. Heinz & Patlak (1960) have obtained the following expression for the work expended during the active transport of one mole of uncharged particles across a membrane possessing a carrier system

$$RT \ln \left(\frac{\mathcal{J}^{\mathrm{in}}}{\mathcal{J}^{\mathrm{out}}} \right) \qquad 10.10$$

where $\mathcal{J}^{\mathrm{in}}$ and $\mathcal{J}^{\mathrm{out}}$ are the unidirectional fluxes of the particle; Patlak (1961) has confirmed theoretically that the expression holds for the active transport of charged particles too. A similar expression has been used by other workers (e.g. Zerahn, 1956;

Martin & Diamond, 1966) to examine the energetics of active ion transport in epithelia. In order to estimate the work done during active transport of water on the above basis one must obtain the unidirectional fluxes of water in the absence of osmotic and hydrostatic pressure gradients. Such measurements have been made in a number of epithelia but it is worth stressing that almost all of them need to be corrected for unstirred layers and consequently cannot be used here. However, the data on the toad urinary bladder are reliable and they are used in the following rough calculation. Hays & Leaf (1962a) found that in the presence of ADH there was a net flux of water (3 µl cm^{-2} hr^{-1}) from the mucosal to the serosal surface of the bladder when no osmotic or hydrostatic pressure gradients existed. In these experiments the unidirectional fluxes of water were about 400 µl cm^{-2} hr^{-1} and these measurements exhibit a feature common to most, if not all, epithelia, namely that the rate of fluid absorption or secretion is always small compared to the unidirectional water fluxes. The disparity is actually larger than that reported above because Hays & Franki (1970) have found that after corrections for unstirred layers the unidirectional fluxes attain values of about 4000 µl cm^{-2} hr^{-1}. If the reabsorption of fluid is driven by a carrier-mediated water pump, then the work done per mole of water transported is

$$RT \ln \left(\frac{4003}{4000}\right) = 580 \times 0.001 \text{ calories}$$

at 20°C, where the approximation, $\ln(1+x) = x$ when x is small, has been used. Thus the active transport of 1 mole, or 18 cm^3, of water requires about 0.6 calories. How does this compare with the energy required for active ion transport? Leaf, Anderson & Page (1962) found that the unidirectional sodium fluxes were 2.1 and 0.6 µ mole cm^{-2} hr^{-1} in the absence of ADH. If the net sodium flux in their experiments is active the work done per mole of sodium transported is

$$RT \ln \left(\frac{2 \cdot 1}{0 \cdot 6}\right) = 580 \times 1 \cdot 25 \text{ calories}$$

For the toad bladder, therefore, the energy requirement for active water transport is about 10^{-4} cals cm^{-2} hr^{-1} whereas that for active sodium transport is about 10^{-3} cals cm^{-2} hr^{-1}. In this case, therefore, the reabsorption of salt solution is probably more expensive in terms of energy when it is driven solely by active ion transfer and passive water flow than *vice versa*. Incidentally, it should be noted that the isolated toad bladder can supply the necessary energy for active sodium transport (Leaf, Page & Anderson, 1959). Unfortunately it is not possible to extend this argument validly to other epithelia since the ratio of unidirectional water fluxes is probably overestimated for most of them.

Another way of tackling the question of energetics is to look at the size of the adverse chemical potential gradient which will halt fluid absorption of secretion. For example, the mammalian intestine, in common with a large number of epithelia, can transport fluid against an osmotic gradient. At one time this property of epithelia was considered as evidence for active water transport but that criterion has now been discarded by almost all physiologists (see later). If we assume for the moment that water transport is active and that it can be counterbalanced by an appropriate adverse osmotic gradient, then this difference in the chemical potential for water is a measure of the minimal energy required by the postulated active water pump. For instance, in the rabbit gall bladder a variety of *in vitro* and *in vivo* techniques (Wheeler, 1963; Dietschy, 1964; Diamond, 1964b; Whitlock & Wheeler, 1964) have demonstrated that an adverse osmotic gradient of about 80 m-osm or 2 atm is required to abolish fluid reabsorption. This is equivalent to a difference in the chemical potential for water of about 0·9 cal per mole at 37°C. In their study of the energetics of active ion transport in the rabbit gall bladder Martin & Diamond (1966) considered that the reabsorption of one mole of NaCl required 508 calories. Since the gall bladder reabsorbs an isotonic fluid the reabsorption of one mole of NaCl must be accompanied by about 6·6 litres or 366 moles of water. Therefore, if the reabsorption process were driven by an active water transport and passive salt flow, the minimal energy required for the reabsorption of one mole of NaCl would be $366 \times 0·9 = 330$ calories. This estimate, like that for active salt reabsorption, is considerably less than the total energy (4060 calories) which can be supplied by this tissue during the time required for the reabsorption of one mole of salt.

The preceding theoretical discussion about active water transport reveals that a water pump should not be ruled our purely on energetic grounds. Needless to say this argument does not prove its existence nor signify its likelihood.

Examples of active water transport?

Since the beginning of this century it has been asserted by some workers that active water transport exists in a number of cells and tissues of both plants and animals. To illustrate the pitfalls and difficulties which surround the question of active water transport and also perhaps to find a genuine example of it we shall examine some specific claims.

Mammalian intestine. In an admirable set of experiments Reid (1892b, 1901) found that the isolated rabbit ileum transports fluid from its mucosal to its serosal surface even when it is bathed on both sides by identical Ringer's solutions. Reid concluded that fluid absorption was not driven by osmotic or hydrostatic pressure gradients but rather that it stemmed from secretory activity of the epithelial cells. In 1938 Ingraham, Peters & Visscher proposed a 'fluid-circuit' theory in which they envisaged an active transport of water across the mammalian intestine from the lumen to plasma and a smaller fluid stream in the opposite direction. According to their model the net salt absorption from the lumen can be attributed to solvent-drag effects in the two opposing streams, with a higher salt concentration in the active water stream than that in the other. Although solvent drag occurs in the movement of small uncharged solutes across the intestine of the rat (Fisher, 1955; Fullerton & Parsons, 1956) and dog (Hakim & Lifson, 1964), it probably exerts no effect on sodium ion movement in the rat intestine (Curran & Solomon, 1957; Curran, 1960; Green, Seshadri & Matty, 1962). Other evidence for the fluid-circuit theory was derived from the relative sizes of the unidirectional fluxes of labelled water (Visscher *et al.*, 1944) across the canine intestine but it was shown later that the disparity between the unidirectional fluxes could be attributed to passive water transport through an aqueous route in the tissue (Koefoed-Johnsen & Ussing, 1953).

Net fluid transfer across the intestine persists even when the luminal fluid is made hypertonic. This has been observed in the rat colon (Parsons & Paterson, 1965), rat jejunum (Parsons & Wingate, 1961) and dog ileum (Vaughan, 1960; Hakim, Lester &

Lifson, 1963). The adverse osmotic gradients required to cancel net fluid transfer across those tissues are large; in particular, Hakim et al. (1963) found that fluid absorption ceased only when the adverse osmotic pressure reached about 4·5 atm (or 200 m-osm) whereas Vaughan (1960) reported that the required value was about double that. Despite these observations that water can be transported against a chemical potential gradient, such fluid movement may still be entirely passive even although it is apparently dependent on glucose (e.g. Fisher, 1955; Barry, Matthews & Smyth, 1961; Hakim et al., 1963) and is abolished in the presence of metabolic inhibitors (Parsons, Smyth & Taylor, 1958; Smyth & Taylor, 1957). The reason for the reluctance in accepting those findings as valid proof of active water transport is that they fail to satisfy the crucial requirement that water is transported across the intestinal wall in the absence of net solute movement. In fact, there is a close relation between the rates of net salt and water transport across the intestine. For instance, Clarkson & Rothstein (1960) demonstrated that the net water flux across the rat small intestine was linearly related to the net flux of solute (chiefly NaCl) over a wide range of transport rates (Fig. 10.4). Moreover, evidence for active solute transport in the intestine of the rat and other mammals is overwhelming (see Schultz & Curran, 1968). It seems from the tight coupling between water and active salt transport, therefore, that we need not invoke active water transport in the mammalian intestine. That view is upheld by the data (typical of mammalian intestine) presented in Fig. 10.4 which illustrates the salient point that there is no net water transport in the absence of active salt transport.

Thus, in the mammalian intestine and in a large number of epithelia which exhibit similar relations between water and salt transport, we should look for the mechanism of coupling between active salt transfer and water movement since there is nothing to suggest unequivocally that water transport itself is active.

Analysis of fluid transport in the mammalian intestine has revealed that neither dependence of water transport upon metabolic energy nor transport of water against its chemical potential gradient constitutes adequate proof of active water transport.

Insect rectum. The ability of some insects to absorb water from the atmosphere into their haemolymph is outstanding.

For instance, Edney (1966, 1967) has shown that after a period of dehydration the desert cockroach absorbs water from the ambient air until its water balance is restored completely. In fact, some insects can reduce the relative humidity of a given air space well below the value which would be in equilibrium with their body fluids. The site of water uptake has been difficult to

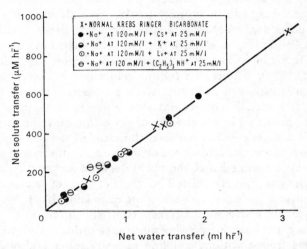

FIG. 10.4. Relation between water and solute fluxes across small intestine of rat *in vitro*. The main solute transported is NaCl and different rates of solute transport were obtained by partial replacement of Na by other cations. The slope of the line relating water flow and solute transport indicates that the transported fluid is approximately isotonic with the bathing solutions (Clarkson & Rothstein, 1960: Fig. 2).

establish; efforts to ascertain whether or not the tracheal system is involved are plagued by dubiety because elimination of the respiratory system leads naturally to anoxia. The body surface or integument is probably the site of water absorption in some species (Beament, 1964, 1965) but recently Noble-Nesbitt (1970) has shown that the ability of the firebrat to absorb water from the atmosphere is stopped by blocking its anus; this also applies to the mealworm. Thus, the site of water absorption in the firebrat and in the mealworm, is the rectum. Here one is dealing with the absorption of water from a gaseous compartment, i.e. the rectal lumen, to a fluid compartment, i.e. the haemolymph. It is possible in some cases that the water uptake could be achieved by the establishment of a local osmotic gradient within the rectal wall (see *Local osmosis*). However, the lower limit to which the relative humidity can be reduced by rectal absorption is exceedingly low. The firebrat apparently holds the world record because it can reduce the relative humidity of an air space to 45% (Beament, Noble-Nesbitt & Watson, 1964), which is considerably less than the relative humidity in equilibrium with its haemolymph or even with saturated potassium chloride or saturated sodium chloride for that matter! Thus, water absorption in insects can proceed against huge differences in the chemical potential for water. Beament (1965) has estimated that some insects can absorb water from the atmosphere against gradients in the range 50–300 atm; in fact, water uptake in the firebrat can be achieved against a huge gradient of about 2000 atm.

Fluid absorption in the isolated rectum of the locust has been studied by Phillips (1964*a–c*). Phillips ligated the hindgut and subsequently altered the luminal contents by injecting and withdrawing fluid through the anus. He concluded that sodium, potassium and chloride ions are all actively reabsorbed from the rectal lumen. The net transport of water, however, is not closely coupled to solute movement for Philips found that water absorption is at a maximum in dehydrated locusts whereas ion absorption is reduced. Further evidence for the independence of water and solute absorption in the locust rectum was obtained when Phillips introduced into the rectal lumen pure solutions of xylose which may be considered practically impermeant. Under those conditions water absorption continued. Stobbart (1969) has confirmed that there is an *in vivo* absorption of water from pure sugar solutions placed in the rectum of the desert locust and similar results have been found in the blowfly (Phillips, 1961, 1969), and in the cockroach (Wall, 1967). Again, just as in the epithelia of vertebrates water can be moved across the insect rectum against adverse osmotic gradients; in the locust rectum the osmotic gradients required to abolish water absorption lie in the range 300–1000 m-osm (or 7–22 atm) considerably larger than the corresponding values (40–400 m-osm) for vertebrate epithelia. Compatible with the view that such water absorption may be active is the finding that it is abolished by certain metabolic inhibitors (see Table 3, Phillips, 1970). Nevertheless, the existence of active water transport is not *established* by such observations (cf. mammalian intestine). The nub of the problem in the insect rectum is whether or not water moves in the complete absence of solute transfer. Phillips (1964a) showed that after the introduction of pure sugar solution into the rectal lumen the rectal fluid never contained more than 3 mM NaCl. Consequently there must be a relatively efficient absorption of all ions diffusing passively from the haemolymph into the rectal lumen. In other words, there must be recycling of ions in the rectal wall; of course, the net movement of salt under these circumstances is exceedingly small and the absorbed fluid is hypotonic to the haemolymph. In confirmation of the view that water and solute movements across the rectal wall are not truly independent, Phillips (1971) has shown recently that there is actually some coupling between water and active ion transport in the locust rectum. Can we explain rectal water absorption in the insect in terms of some coupling mechanism which arises from the passive and active transport of ions within the rectal wall? This point will be discussed later (see page 462).

Rabbit blastocyst. The experiments of Tuft & Böving (1970) showed that water moves into the blastocoel against its chemical potential gradient. As in the insect rectum, one meets in the rabbit blastocyst a net transport of hypotonic fluid. Tuft and Böving argued that water uptake into the blastocoel is active although their only grounds for this conclusion was that water moves against its chemical potential gradient. Clearly this is not an adequate criterion for active water transport. Much more experimental work needs to be done not only on water transport but also solute transport in the rabbit blastocyst before we can begin to understand the mechanism of water uptake in this tissue. It is cited here because it is one of the few preparations which can apparently transport hypotonic fluid.

Contractile vacuole. Micropuncture techniques have been used to examine the contents of contractile vacuoles in *Amoeba proteus* and the giant amoeba, *Pelomyxa carolinesis* (*Chaos chaos*). Schmidt-Nielsen & Schrauger (1963) found that the osmolarity of the cytoplasm in *Amoeba proteus* was 101 m-osm whereas that of the contractile vacuole fluid was only 32 m-osm. In a similar sort of study on *Pelomyxa carolinesis* Riddick (1968) found values of 117 and 51 m-osm for cytoplasm and vacuole respectively; he also noted that the cytoplasmic concentrations of sodium and potassium were 0·6 and 31 mM while the corresponding values for vacuolar fluid were 19·9 and 4·6 mM. Thus, in addition to the hypotonicity of the vacuolar fluid one also has to account for its different ionic composition from that of cytoplasm.

In this study Riddick found that contractile vacuoles could be isolated and maintained in a solution of similar ionic composition to that of cytoplasm. Since the vacuoles did not shrink in this medium it implies that their volumes are actively regulated in some way or that they are very impermeable to water. The latter possibility does not seem likely since the addition of NaCl or sucrose to the bathing solution caused the isolated vacuoles to shrink 'instantaneously'. On the other hand, Riddick noted that increasing the external osmotic pressure by raising the normal levels of ions (particularly potassium) produced a rather slow shrinkage of the isolated vacuoles. He concluded from the latter experiments that the water permeability of the vacuole was low, but the work of Hopkins (1946) indicated that isolated vacuoles are quite water permeable. Hence, it seems probable that the slow osmotic withdrawal of water probably occurred because a permeant solute was employed (i.e. $\sigma_{KCl} < 1$).

It is true that the growth of contractile vacuoles in the cytoplasm of certain amoebae could be due to active water transport in association with appropriate passive ion transport. Schmidt-Nielsen and Schrauger suggested, however, that vacuolar enlargement resulted from the coalescence of vesicles containing fluid with the same composition as that of cytoplasm. According to their view the hypotonicity of the vacuolar fluid was produced by subsequent reabsorption of a hypertonic fluid. However, Riddick favoured the view that cytoplasmic vesicles are formed originally with isotonic contents and that subsequently sodium ions are actively secreted into the vesicle while potassium ions are actively extruded. Provided that the vesicular membrane has a low permeability to water and the rate of potassium efflux is greater than sodium influx, then a hypotonic vesicle will be formed by this process. Finally, these hypotonic vesicles fuse together to form and subsequently enlarge the vacuole. The alternative view of Schmidt-Nielsen and Schrauger about the hypotonicity of the vacuolar fluid was refuted by Riddick on the grounds that the vacuolar osmolarity was relatively constant and unrelated to vacuolar size. However, their hypothesis is not out of step with that feature of vacuolar enlargement and this is confirmed by the following argument.

The total amount, Q, of solute in the vacuole is given by $Q = C^v V$, where C^v is the total concentration of solute and V is the volume of the vacuole. Since C^v is observed to be constant, then we have

$$\frac{dQ}{dt} = C^v \frac{dV}{dt} \qquad 10.11$$

Riddick's experiments showed that the radius, r, of the vacuole increases linearly with time. Hence, equation 10.11 can be re-written as

$$\frac{dQ}{dt} = C^v 4\pi r^2 \frac{dr}{dt} \qquad 10.12$$

because $V = \frac{4}{3}\pi r^3$. Let the surface area of the vacuole be A which is equal to $4\pi r^2$. According to the model of Schmidt-Nielsen and Schrauger, solute is added to the vacuole in isotonic proportions with that of the cytoplasmic solute concentration, C^c, i.e. at a rate equal $C^c(dV/dt)$. In contrast, solute is removed at a rate given by $J_s A$, where J_s is the net flux of solute per unit area in unit time. Solute absorption from the vacuole is unaccompanied by water flow. Thus, the model of Schmidt-Nielsen and Schrauger indicates that

$$\frac{dQ}{dt} = C^c \frac{dV}{dt} - J_s A \qquad 10.13$$

Equation 10.13 can be re-written in the following form

$$\frac{dQ}{dt} = C^c 4\pi r^2 \frac{dr}{dt} - J_s 4\pi r^2 \qquad 10.14$$

This model is compatible with the condition that C^v remains constant when equation 10.12 is identical to equation 10.14, viz.

$$C^v \frac{dr}{dt} = C^c \frac{dr}{dt} - J_s \qquad 10.15$$

Equation 10.15 contains the constant terms C^v, C^c and (dr/dt) and, therefore, it can be satisfied only when J_s is a constant given by $(C^c - C^v)(dr/dt)$. The condition that J_s is constant is quite a reasonable one: moreover, the value for J_s (66 p mole cm^{-2} sec^{-1} or 0·24 μ mole cm^{-2} hr^{-1}) calculated by the above expression from Riddick's data lies within the range of active solute transport rates normally observed in animal tissues.

The formation of the hypotonic fluid in the contractile vacuole could be explained on the basis of an isotonic transport of fluid into the vacuole and an opposing reabsorption of solute. Sodium ions are certainly accumulated in the vacuole against a concentration gradient whereas the vacuolar potassium concentration is kept relatively low. Davson (1970) has pointed out that the equilibrium potentials of the vacuolar fluid with respect to the cytoplasm are about -90 mV and $+50$ mV for sodium and potassium respectively, but that the potential of the vacuole is about $+15$ mV with respect to the cytoplasm (Prusch & Dunham, 1967). That suggests that sodium is actively pumped into the vacuole while potassium is actively extruded, although the flux ratios for these ions are really required before a definite conclusion about active ion transport can be drawn. This rather vague picture of ion transport in the contractile vacuole could be welded on to the model of Schmidt-Nielsen & Schrauger (1963) by postulating that there is an isotonic fluid secretion, powered by active sodium transport, and an opposing hypertonic fluid absorption from the vacuole, powered by active potassium transport. Clearly, we need a thorough study of ion and water transport across the vacuolar membrane before we can build up a good picture of how the hypotonic vacuolar fluid is elaborated.

An alternative description of the formation of the vacuolar fluid is to postulate that there is active fluid secretion and passive ion transport into the vacuole. Riddick observed that the vacuole's radius increased linearly with time and that $(dr/dt) = 0.6$ μm min^{-1}. Given that the vacuole is a sphere whose area can increase appropriately as it grows, then the net fluid flux across its membrane is $(dV/dt)/4\pi r^2 = (dr/dt)$. Thus, the net fluid flux into the vacuole is 0.6 μm min^{-1} or 10^{-6} cm^3 cm^{-2} sec^{-1} or 3.6 μl cm^{-2} hr^{-1}, and this rate of water transport is similar to that reported for numerous epithelia (see Table 10.1). An active water pump and, in particular, vesicular fluid transport into the vacuole could account for that influx of water—we certainly cannot disregard active water transfer as a mechanism. Nevertheless, in this matter I am prejudiced and believe that the elaboration and composition of the vacuolar fluid depends ultimately on active ion transport and concomitant passive water movement.

Plant roots. The ability of excised plant roots placed in dilute solutions to elaborate a relatively concentrated fluid from their basal ends has been discussed in the section on *Osmosis*. Although the rate of fluid exudation is determined predominantly by the osmotic pressure difference between the exudate and the bathing solution a certain small component of this fluid movement continues *apparently* in the absence of an osmotic gradient. Currently there is no satisfactory explanation, backed up by experimental work, for this disparity between the water flow and its driving force. Moreover, in the isolated plant root the external osmotic pressure, which is required to stop fluid exudation, exceeds that of the exudate collected immediately before its abolition. Of course, the osmotic pressure of the exudate formed in the absorbing apical region of the root may exceed that of the exudate emerging from the basal regions. Such a longitudinal concentration gradient within the exudation pathway in the root has not been demonstrated. Instead, some workers have been content to conclude that the fluid exudation is driven partially by an active mechanism rather than entirely by osmosis. This proposal has met some strong criticism, but interest in active water transport in plant roots has been revived recently by the work of Ginzburg & Ginzburg (1970*a,b*, 1971). In a particular set of experiments they employed isolated maize roots from which the steles had been removed and they called this type of preparation a cortical 'sleeve' (Ginzburg & Ginzburg, 1971). Fig. 10.5 depicts a schematic cross-section of a primary maize root in which the cortex is delineated from the inner stele by the endodermis. In common with other workers, Ginzburg and Ginzburg found that the stele could be extracted from the root after the cortex and endodermis had been disrupted by bending the root; The prepared 'sleeves' consisted of the epidermis, cortex and the outer regions of the broken endodermis and they could be cut to desired lengths for perfusion experiments. When they perfused the 'sleeves' with different solutions they found that the relation between net water transport and the osmotic gradient was linear; however, extrapolation of the linear plots to a zero osmotic gradient displayed intercepts significantly different from zero. Their data concur with similar studies on intact maize roots (e.g. see Fig. 10.1) and those results can be expressed quantitatively by equation 10.6, where $\phi_w{}^o$ denotes the size of the intercept at zero osmotic gradient. The value they obtained for $\phi_w{}^o$ was close to that obtained by House & Findlay (1966*a*) for intact maize roots; however, a non-zero value of the intercept, $\phi_w{}^o$, is not convincing evidence for active water transport. The net solute fluxes across the cortical 'sleeves' were only about 1% of those recorded in intact roots whereas the net water fluxes were roughly the same. Ginzburg and Ginzburg showed that $\phi_w{}^o$ was independent of the net solute flux and that it was reduced but not abolished by either potassium cyanide or 2,4-dinitrophenol. These observations, particularly their experiments with the metabolic inhibitors, are rather puzzling and they suggest that the observed $\phi_w{}^o$ contains a relatively large and spurious component. Although their evidence suggests, at

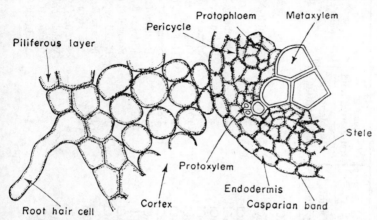

FIG. 10.5. Transverse section of a plant root, such as maize, showing the arrangement of different cells in the cortex and the stele (Sutcliffe, 1962: Fig. 3b).

face value, that the plant root, or rather the cortical and epidermal cells, can actively transport water, one of the authors in a later theoretical paper (Ginzburg, 1971) rejects active water transport in favour of 'a mechanistic model which does not necessitate a "water pump" to explain the ... water flow in root cortex'. His model is rather similar to the asymmetrical double-membrane to be discussed in the next section.

Verdict

In some epithelia, such as the mammalian intestine, where fluid absorption was considered previously to be active, it is now taken to be passive and coupled to active solute transport. The nature of this coupling will be discussed later. Within this group of epithelia we find that the absorbate or secretion is either isotonic or hypertonic to plasma (Table 10.1). Outside this group we find

TABLE 10.1. Characteristics of certain fluid secretions and absorbates

Tissue	Net fluid flux (μl cm^{-2} hr^{-1})	Chief solute transported	Osmolarity of transported fluid / Osmolarity of bathing fluid	Reference
Trout urinary bladder	6.7	NaCl	5.6	Lahlou & Fossat (1971)
Marine bird { Petrel	?	NaCl	5	Schmidt-Nielsen (1960)
nasal gland { Cormorant	?		2	
Marine fish intestine	8.1	NaCl	4.5	House & Green (1965)
Fowl cloaca	8.4	NaCl	4	Skadhauge (1967)
Goose nasal gland	?	NaCl	3	Hokin (1967)
Necturus distal tubule	4.3	NaCl	2.5	Maude et al. (1966)
Dog ileum	17	NaCl	2.2	Visscher et al. (1944)
Frog skin	1.1	NaCl	2.1	Huf et al. (1951)
Human ileum	3.0	NaCl	1.8	Soergel et al. (1968)
Turtle urinary bladder	5.2	NaCl	1.6	Brodsky & Schilb (1965)
Frog gastric mucosa	12	HCl	1.3	Villegas & Sananes (1968)
Human jejunum	6.9	NaCl	1.2	Soergel et al. (1968)
Cat sweat gland	?	NaCl	1.1	Brusilow & Gordes (1964)
Dog gastric mucosa	75*	HCl	1.1 / 0.97	Moody & Durbin (1969): Thull & Rehm (1956): Altamirano et al. (1969)
Rabbit choroid plexus	60†	NaCl	1.03	Welch et al. (1966): Welch (1967)
Insect (*Calliphora*) salivary gland	1.2‡	KCl	1.02	Oschman & Berridge (1970)
Insect (*Rhodnius*) (Malpighian tubule)	198 (diuretic hormone present)	KCl	1.02	Maddrell (1969)

Tissue		Solute	Ref.
Rabbit corneal endothelium		NaHCO$_3$	(...)
Cat pancreas	?	NaHCO$_3$	Case et al. (1969)
Dogfish rectal gland	?	NaCl	Burger & Hess (1960)
Fish gall bladder	15	NaCl	Diamond (1962a,b,c)
Guinea pig gall bladder	23	NaCl	Diamond (1964a)
Rabbit gall bladder	50	NaCl	Diamond (1964a)
Dog gall bladder	66	NaCl	Grim & Smith (1957); Ravdin et al. (1932); Gilman & Cowgill (1933)
Rat ileum	27	NaCl	Curran & Solomon (1957)
Rabbit ciliary body	9·6	NaCl	Cole (1962)
Rat submaxillary gland	?	NaCl	Young & Schögel (1966)
Rat proximal tubule	2·2	NaCl	Giebisch et al. (1964)
Rabbit blastocyst (8 days post coitum)	7·6	NaCl	Daniel (1964); Tuft & Böving (1970); Smith (1970)
Insect (Dixippus) Malpighian tubule	0·4	KCl	Ramsay (1954)
Amoeba contractile vacuole	3·6	?	Riddick (1968)
Frog gastric mucosa (sulphate Ringer solutions)	12	HCl	Villegas & Sananes (1968)
Insect (Schistocerca) rectum	17	KCl	Phillips (1964a)

Values in the numeric column read 1·0, 1·0¶, 1·0, 1·0, 1·0, 1·0, 1·0, 1·0, 1·0, 1·0, 0·96, 0·92, 0·44, 0·09, 0·01‖

* Computed from secretory rates given by Moody & Durbin (1969) on the basis that the actual area of the mucosa is 4·7 times the apparent area (see Altamirano, 1969).

† Computed from data of Welch et al. (1966).

‡ M. Berridge (*personal communication*).

§ S. Hodson (*personal communication*).

¶ Concentration of NaCl in secretion was approximately twice that in plasma.

‖ Obtained from experiments where the lumen of rectum was bathed with sugar solutions (Phillips, 1964a); the author estimated that the net solute flux was not greater than 0·05 μmole per hour whereas the net water flux was 17 μl per hour.

other systems—insect rectum, frog gastric mucosa, rabbit blastocyst, the contractile vacuole of amoebae and the plant root—where the transported fluid is hypotonic. Our present knowledge of those systems suggests that water transfer *may* possibly be active; however, the experimental data on which that conclusion is based is sparse indeed for the blastocyst, the contractile vacuole and the plant root. The experiments of Phillips and others on the insect rectum, however, offer more information which upholds the opposite conclusion that water transport in the insect rectum is probably passive and driven in some way by the movements of ions within the rectal wall.

In all of the biological systems which transport hypotonic, isotonic or hypertonic fluids there is no pressing need to invoke active water transport. Of course, the state of our knowledge in many instances does not exclude the possibility of active water movement but the weight of it certainly favours passive water transfer coupled to active ion transport.

Finally, there is possibly the most compelling case for active water transport—uptake of water from the atmosphere by insects. Here, there is absorption of water molecules alone against an extremely steep gradient of chemical potential. The site of this water uptake is the insect's integument or more probably its rectum (Noble-Nesbitt, 1970). In this instance one cannot assert that water is driven osmotically into the haemolymph from the atmosphere unless one is prepared to accept the implication that the body fluids are saturated salt solutions or much worse! (Dr. J. M. Diamond has suggested that such an intense osmotic gradient *might* be established with a substance, e.g. glycerol, infinitely soluble in water.) Despite this overwhelming evidence for active water transport in some insects I remain unconvinced principally because water is undergoing a phase transition at the surface of the insect and we simply do not know enough about it. For example, the heat of transfer for water moving across the insect's external surface may be extremely high and consequently one might speculate that thermo-osmosis could make an important contribution to the driving force for water uptake.

At present any decision that one makes about the existence of active water transport is fraught with difficulties chiefly because of lack of experimental data. It is notable also that the number of so-called firmly established cases of active water transport has steadily declined over the last decade or so. This is largely as a result of an improved criterion for active water transport. Ramsay (1971) seems to be calling for a 'not guilty' or a 'guilty' verdict, or in other words, acceptance or rebuttal of active water transport. I would prefer an alternative verdict—'not proven'; by this I mean to indicate that there is insufficient evidence for active transport of water as a phenomenon but there is no theoretical argument or experimental fact that excludes its possible existence somewhere in the animal or plant kingdom.

Asymmetrical double-membrane model

The concensus of experimental evidence and theoretical views about fluid transport in epithelia directs us towards a final aim—the resolution of the mechanism by which passive water transport is coupled to active solute movement. When the osmolarity of the transported fluids is examined it is found that the majority of tissues transfer either isotonic or hypertonic fluids (Table 10.1). Relatively few tissues seem capable of achieving hypotonic fluid transport and, in fact, such hypotonic fluids may be formed by the primary transport of an isotonic fluid at one site and the subsequent removal from this precursor solution of a hypertonic fluid at another site. This point will be discussed later. For the moment we have to explain how isotonic or hypertonic fluids are transported. The problem has been rephrased by some physiologists, notably Diamond, in the following way. How does the active transport of one molecule of solute achieve the concomitant transfer of about 300 or 400 molecules?

Curran & Solomon (1957) proposed that intestinal water absorption is a passive process resulting from active salt transport. In order to explain this remarkable coupling in the rat ileum and other epithelia Curran (1960) presented a model consisting of two different membranes in series. His model is illustrated in Fig. 10.6 where two identical solutions, A and C, are separated by a 'membrane' composed of a central compartment, B, between two different membranes. It is assumed that active solute transport occurs across the first membrane, 1 (thin barrier with small pores), from region A to B. The second membrane, 2, is a relatively thick one containing large pores. Curran suggested that in the *in vitro* preparation of rat intestine the first membrane might represent one of the plasma membranes of the epithelial cells whereas the second membrane might be the serosa which exhibits little hindrance to solute and water movement.

According to Curran's hypothesis, active solute transport into region B will create an osmotic pressure gradient driving water from A to B. Since membrane 2 is exceedingly permeable to solutes, i.e. its reflexion coefficient for the actively transported solute is practically zero, the accumulation of solute in B will produce

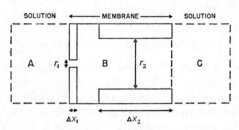

FIG. 10.6. Schematic model system for water transport. A and C represent external solutions. The thickness and width of the pores in the individual permeability barriers are denoted by Δx and r respectively (Curran, 1960: Fig. 5).

little water movement from C to B. Due to osmotic water flow from A to B there will be a tendency for the hydrostatic pressure in B to rise and move water from B to C. Thus, the overall effect of osmotic water flow across membrane 1 and filtration across membrane 2 is that there is a net flow of water from A to C.

Curran argued that the operation of such a model was compatible with other characteristics of water flow in the mammalian intestine. For instance a suitable hydrostatic pressure applied at C (serosal surface) will halt net water flow whereas an identical pressure applied at A (mucosal surface) will not accelerate net water flow because the membrane 1 has a lower L_p than that of membrane 2. However, as we have seen before, these characteristics are observed in the canine intestine (Hakim & Lifson, 1969) but are not duplicated exactly in the rat intestine (Smyth & Taylor, 1957) where the application of small mucosal pressures accelerates fluid absorption by some unexplained mechanism.

It is important to note that the intestine's ability to absorb water against an osmotic gradient is also compatible with Curran's model. Water will continue to flow from A to C even when the solute concentration at A exceeds that in C provided that active solute transport proceeds fast enough across membrane 1 to raise the solute concentration at B above that at A.

In principle, the double-membrane model can account for the observed coupling between active ion transport and passive water not only in the intestine but also in other epithelia. Moreover, it offers a plausible mechanistic view of water absorption and consequently we ought to examine it in more detail.

Both Curran & McIntosh (1962) and Ogilvie *et al.* (1963) have successfully tested Curran's series-membrane model in an artificial system (see *Asymmetrical double-membrane system*, Chapter 4). The salient point of their study was that a net volume flow could be elicited across the system even when both external solutions were identical. These authors stressed that the so-called double-membrane effect demanded the existence of a positive hydrostatic pressure in the central compartment and different values for the reflexion coefficients of the individual membranes. The latter condition is equivalent to saying that the overall 'membrane' must exhibit asymmetrical permeability characteristics.

The double-membrane model has been treated in detail by Patlak *et al.* (1963) and Kedem & Katchalsky (1963c). Both sets of workers have derived equations which describe the flows of solute and water across a system composed of two membranes arranged in series. Both treatments are similar except that Patlak *et al.* (1963) have included active solute transport in their theoretical scheme. Consequently their conclusions seem more directly applicable to our current discussion of active salt and water transport in epithelia than those of Kedem & Katchalsky (1963c).

Patlak *et al.* (1963) considered a series-membrane composed of two homogeneous membranes arranged as before with a central region m and they embodied certain assumptions in their analysis. The relatively important ones were that all solutions contained a single neutral solute, which underwent active transport across the first membrane (flux j), and that there was no physical interactions between the active solute transport and the passive fluxes of solvent and solute. Provided that the solutions in the outer, o, and inner, i, compartments are identical ($c^o = c^i = c$) and at the same hydrostatic pressure ($p^o = p^i = p$) and that unstirred layers can be neglected, these authors were able to obtain a simple expression for the osmolarity ($\mathcal{J}_s/\mathcal{J}_v$) of the transported fluid. The principal results of their paper are the following.

When active solute transport exists, i.e. $j \neq 0$, then

$$\frac{j}{\mathcal{J}_v(\sigma^o - \sigma^i)} > c > 0 \qquad 10.16$$

For the case where active solute transport occurs from outside to inside, i.e. $j > 0$, the inequality 10.16 indicates that the direction of the volume flow will also be positive only if $\sigma^o > \sigma^i$. Those conditions are similar to the preceding model of Curran. Moreover, the analysis of Patlak *et al.* (1963) shows that under these circumstances

$$\frac{\mathcal{J}_s}{\mathcal{J}_v} > c(1-\sigma^i) > 0 \qquad 10.17$$

In other words, the osmolarity of the transported fluid is hypertonic when c is small. If $\sigma^i = 0$, then the transported fluid is invariably hypertonic to the bathing concentration c; however, when $\sigma^i \neq 0$ there occurs a value of c above which the transported fluid becomes hypotonic. That is, the model predicts that active solute transport generates a concomitant flow of solution which may be hypertonic, isotonic or hypotonic to the bathing media when the second membrane is more permeable to the solute than the first. Under these circumstances the concentration of solute in the central compartment m always exceeds that of the bathing solutions. That is, solute tends to accumulate in m by virtue of the active transport system and subsequently it diffuses preferentially from m into the inner compartment. Moreover the hydrostatic pressure of m also exceeds that on both sides of m and this pressure gradient forces water preferentially across the second membrane.

For the case where active solute transport occurs from right to left, i.e. $j < 0$, the inequality 10.16 indicates that the direction of volume flow will also be negative only if $\sigma^o > \sigma^i$. Under these circumstances, however, there will be solute depletion in m, i.e. $c_m < c$, and also the hydrostatic pressure in m will be lower than that on both sides of m. Again it is found that the transported fluid may be hypertonic, isotonic or hypotonic for those conditions.

A feature of the double-membrane model, which has some interesting repercussions (see later) is the obligatory existence of a hydrostatic pressure gradient between the central compartment and the external compartments. Both House (1964a) and Schilb & Brodsky (1970), however, have published modified versions of the double-membrane model which require no such pressure gradients. Consider the usual double-membrane system where the compartments o and i are filled with solutions of a certain solute n, say NaCl, while the central compartment m is filled with a different solute k, say KCl. If the first membrane is permeable to n but not to k and *vice versa* for the second membrane, then the volume flows across each membrane can be written as follows

$$J_v^o = L_p^o RT(c_k - \sigma_n^o c_n) \qquad 10.18$$

$$J_v^i = L_p^i RT(c_n - \sigma_k^i c_k) \qquad 10.19$$

Under steady-state conditions J_v^i and J_v^o can be equated to yield

$$c_k = \frac{c_n(\sigma_n^o L_p^o + L_p^i)}{L_p^o + L^i \sigma_k^i} \qquad 10.20$$

Since both J_v^o and J_v^i are equal to the net volume flow, J_v, across the system, we can substitute equation 10.20 into equation 10.18 to give J_v, viz.

$$J_v = \frac{L_p^o L_p^i RT c_n(1 - \sigma_n^o \sigma_k^i)}{L_p^o + L_p^i \sigma_k^i} \qquad 10.21$$

Under conditions where no hydrostatic pressure gradient exists, therefore, there will be a net volume flow provided that either σ_k^i or σ_n^o is less than unity. In order to maintain this net volume flow at a steady value both c_k and c_n must remain constant with time.

Of course, this model also predicts that water can flow against an external osmotic gradient set up between compartments o and i. Let us suppose a certain concentration c_i, of an impermeant molecule is present in compartment o. Now J_v^o is given by

$$J_v^o = L_p^o RT(c_k - \sigma_n^o c_n - c_i) \qquad 10.22$$

Under steady-state conditions J_v^o is equal to J_v^i, given by equation 10.19 and hence the net volume flow can be found as before to yield

$$J_v = \frac{L_p^o L_p^i RT(c_n - \sigma_n^o \sigma_k^i c_n - \sigma_k^i c_i)}{L_p^o + L_p^i \sigma_k^i} \qquad 10.23$$

and net volume flow will be abolished when $c_i = c_n(1 - \sigma_n^o \sigma_k^i)/\sigma_k^i$.

House (1964a) invoked the preceding model to explain water absorption by the isolated frog skin and he suggested that the individual membranes were the outer and inward-facing plasma membranes of the epithelial cells. There is some evidence that the 'outer barrier' of the frog skin is permeable to NaCl but not to KCl and that the 'inner barrier' of the epithelial cells is permeable to KCl but not to NaCl. Unfortunately both

σ_{KCl} and σ_{NaCl} have not been measured in the frog skin and so the model can be regarded only as a plausible explanation.

Schilb & Brodsky (1970) have presented a similar double-membrane model for the turtle urinary bladder where the interior of the mucosal cells is identified with the central compartment and their evidence for the model will be discussed later.

The double-membrane model of Patlak *et al.* (1963) is an exceedingly interesting one since it describes clearly the properties of a series-membrane which can generate net volume flow. It is, however, a difficult job to examine the applicability of the model to epithelial water transport since the individual membrane require to be identified and their transport coefficients (L_p, ω_s, σ_s) must be measured. This has not been done in any epithelium as far as I can gather. Despite that, the model has attracted many workers to draw parallels between it and certain transporting epithelia.

Rabbit gall bladder

Whitlock & Wheeler (1964) employed a volumetric technique, devised earlier by Wheeler (1963), to study the osmolarity of the transported fluid in the rabbit gall bladder (Fig. 10.7). With such an apparatus Whitlock and Wheeler recorded net fluid transport by periodically aspirating the serosal solution into the compartment C connected to a vertical burette. The height of the fluid volume in the burette gave the volume of the serosal solution to an accuracy of ± 10 μl. They also determined unidirectional ion fluxes and the electric potential across the isolated gall bladder. Using these techniques to record the net fluxes of solute and fluid, these workers found that the osmolarity of the absorbate was identical to that of the bathing media when hypertonic, isotonic and and hypotonic solutions of NaCl were tested. Table 10.2 shows those data and also the corresponding results when impermeant sucrose was added either to each bathing solution or to both bathing solutions. In contrast to their findings with NaCl Ringers the presence of sucrose, particularly in the serosal solution, caused the transported fluid to become significantly hypertonic to the bathing media. Whitlock and Wheeler sought to explain those results in terms of the double-membrane model.

These authors assumed that the reflexion coefficient of the first membrane for the transported solute (NaCl) was less than unity

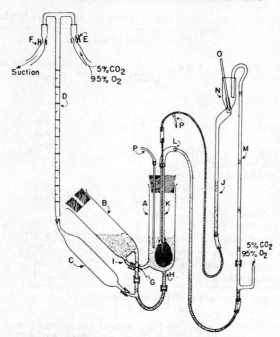

FIG. 10.7. A volumetric method for measuring water transport across the gall bladder wall. The volume of the external solution is periodically recorded by aspirating it into the glass bulb C and measuring its height in the 2 ml burette D. A, main chamber. B, serosal chamber for mixing and sampling. E, gas inflow clamp. F, suction clamp. G, line for adjustment of meniscus during volume measurement. H, I, constrictions in outflow and return tubes to fix fluid levels during volume measurement. J, inflow reservoir for mucosal fluid. K, cannula with attached gall bladder. L, mucosal return tubing with soft rubber connexion through which glass inflow tubing is inserted. M, glass tubing through which mucosal fluid is pumped back to inflow reservoir by gas mixture. N, bubble trap and sampling chamber. O, rubber-sheathed glass rod with which outflow from N can be stoppered for sampling. P, agar bridges (Wheeler, 1963: Fig. 1).

but greater than zero, that the reflexion coefficient of the second membrane was zero (i.e. non-selective to solute and solvent passage) and, finally, that both membranes were highly permeable to water. In their original paper they did not identify the hypothetical membranes but in a later publication (Kaye *et al.*, 1966) they postulated that the first membrane was the mucosal or apical

TABLE 10.2. Solute and water reabsorption by rabbit gall bladder

Experimental conditions	Bathing solution				Absorbate osmolarity (m-osm)
	Na	Cl	Sucrose	Osmolarity (m-osm)	
	(mM)				
1. Identical NaCl Ringer solutions					
a Hypotonic	97	82	–	200	192
b Isotonic	143	128	–	284	290
c Hypertonic	208	193	–	397	384
2. Sucrose substituted in both solutions					
a Isotonic	97	82	85	284	373
b Hypertonic	143	128	111	394	508
3. Sucrose substituted in one solution					
a { Mucosal solution	143	128	–	284 }	380
{ Serosal solution	97	82	85	284 }	
b { Mucosal solution	97	82	85	284 }	335
{ Serosal solution	143	128	–	284 }	

Modified from Whitlock & Wheeler, 1964

plasma membrane whereas the second membrane was the basal region of the lateral intercellular spaces (Fig. 10.8).

In this scheme the central compartment m is identified with the lateral intercellular spaces. Indeed, Kaye *et al.* (1966) considered that the distension of those spaces in gall bladders transporting fluid at a maximal rate was indirect evidence for their intimate involvement in the transport mechanism. A similar conclusion about the role of the intercellular spaces was also reached independently at the same time by Diamond & Tormey (1966). Although the latter authors envisaged a different mechanism of fluid transport (see *Local osmosis* later).

Whitlock and Wheeler found that it was possible to choose arbitrary, but reasonable, values for the coefficients (σ_s, P_s and L_p) for each membrane so that the mathematical equations describing the double-membrane model gave appropriate values for the absorbate osmolarities. For instance, the reflexion coefficient of the first membrane for NaCl was taken as 0·9 and that for the second membrane as zero. Clearly, the nature of their arbitrary assumptions, unsupported as they are by experimental evidence, weakens the argument for the double-membrane model. On the

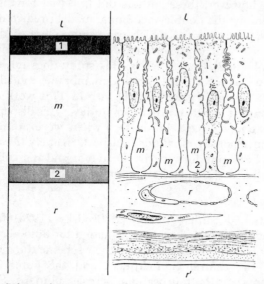

FIG. 10.8. Schematic representation of the Curran (1960) double-membrane model (left) and of the wall of the rabbit gall bladder (right). The Curran model is composed of three compartments in series, separated by two barriers (membranes). Barrier 1 (which separates compartments l and m) is assumed to be a semipermeable membrane in the model system and in actual systems is also presumed to be the site of active solute transport in the direction l to m. Barrier 2 is defined by Curran as a nonselective barrier which merely retards diffusion of solutes between compartments m and r. This model can account for net transport of water (coupled solute–solvent transport: see text).

On the right, a diagrammatic representation of the rabbit gall bladder wall is appropriately labelled to show the analogy between its structure and the Curran model.

The intercellular space fulfils all the requirements of the middle compartment. *In vivo*, the epithelial basal complex, consisting of the narrow channel, the basal lamina of the epithelium, the small amount of lamina propria, and the basal lamina and capillary endothelium, is the second barrier and the vascular compartment is the r compartment. *In vitro*, the serosal bath is made the third compartment (r') by virtue of the experimental design. Hence the second barrier must include the entire thickness of the wall of the gall bladder beneath the epithelium (Kaye et al., 1966: Fig. 33).

credit side, however, these authors did find that for their chosen set of conditions the double-membrane model predicted values of the absorbate osmolarities which were quite close to the experimental values both in the presence and absence of impermeant solute (Table 10.2). Nevertheless, this study does not offer solid evidence in favour of the double-membrane model of fluid absorption although it is compatible with it. This work does yield data which are at variance with other studies. This disparity will be discussed later (see page 446 *et seq.*) when we examine another proposed mechanism of fluid transport, namely *local osmosis* which has been put forward by Diamond and his colleagues to explain fluid absorption and secretion in epithelia.

Turtle urinary bladder

As mentioned earlier, Schilb & Brodsky (1970) favour the double-membrane model as an explanation for fluid absorption in the urinary bladder of the turtle. In their scheme the mucosal cell constitutes the central compartment and the outer and inner compartments are the mucosal and serosal solutions. Schilb and Brodsky assumed that sodium ions are readily permeant at the mucosal membrane but not at the serosal membrane and that the serosal membrane contains a 'pump' which extrudes sodium ions from the cell into the serosal fluid. In their model potassium ions are relatively more permeant at the serosal cell membrane than at the mucosal membrane. Their model can explain how water is absorbed from the mucosal solution into the serosal solution in the same way as House's (1964a) model accounts for the water absorption by the frog skin. Ultimately the model relies on the maintenance of the ionic gradients between the cellular fluid and the external media and such gradients for potassium and sodium ions are assumed to be maintained by active transport mechanisms.

Schilb and Brodsky examined experimentally the consequences of inhibition of sodium transport on the transport of fluid by the turtle bladder. On the basis of the double-membrane model they predicted that inhibition of active solute (sodium chloride) transport would 'cause a transient increase in the transmural flow of water', that during the period of accelerated water flow 'mucosal cells would accumulate sodium and retain potassium' and, finally 'that the mucosal cells would not swell sufficiently to prevent an increase in cellular osmolarity during the period of sodium accumulation'. In fact they found that acetyl-β-methylcholine (Mecholyl) produced a transient acceleration of net water transport across the turtle bladder even although it partially inhibits active sodium transport in this tissue (Schilb, 1969). Mecholyl increased the rate of water transport by about three-fold during a period of 30 minutes after its application to the serosal surface of the bladder whereas Schilb (1969) has shown that Mecholyl reduced the rate of active sodium transport by about 25% during that period. Schilb and Brodsky showed that the effect of Mecholyl on water absorption was genuine and that it could not be attributed solely to changes in the fluid content of the bladder wall (i.e. spurious water fluxes) or to an increase in the hydraulic conductivity of the tissue. They found, moreover, that partial inhibition of the active salt transport led to accumulation of sodium, but not potassium ions, in the mucosal cells, and they interpreted the transient acceleration of net water transport induced by Mecholyl in terms of their double-membrane model. Indeed, they argued that the increase in the cellular concentration of sodium ions produced an increase in the effective osmotic pressure of the cellular compartment. Furthermore, they assumed that across the mucosal and serosal membranes the effective osmotic gradient could be expressed as $\bar{\sigma}_1 \Delta \pi_1$ and $\bar{\sigma}_2 \Delta \pi_2$ respectively, where $\bar{\sigma}_1$, and $\bar{\sigma}_2$ are some 'average' reflexion coefficients and $\Delta \pi_1$, and $\Delta \pi_2$ are the differences in osmotic pressure across the respective membranes during the transient acceleration of water flow. According to their reasoning, the increase in the cellular concentration of sodium ions, induced by Mecholyl, ought to create a larger driving force on water across the mucosal membrane than the opposing force across the serosal membrane provided $\bar{\sigma}_1 > \bar{\sigma}_2$. Hence, net water transport ought to be initially accelerated in the mucosal to serosal direction. The effect of this enhanced water absorption through the mucosal cells would eventually tend to reduce the osmotic pressure of the cellular fluid. It is difficult to accept, however, that the small increase (10 mM) in the cellular sodium concentration could increase the net water flux as much as three-fold. Schilb and Brodsky also tested the effect of ouabain, another potent inhibitor of active sodium transport in the turtle bladder (Solinger, Gonzalez, Shamoo, Wyssbrod & Brodsky, 1968) and it also produced a transient acceleration of net fluid transport in the bladder.

The study of Schilb and Brodsky on the turtle bladder is an interesting one since it shows that the apparent stoichiometric relation between net water transport and active salt flow, which has been assumed to be an indisputable feature of all epithelial transport, can break down in some circumstances. Although these data do not prove the validity of the double-membrane model, they seem to be in conflict with other models, notably local osmosis. Nevertheless, Schilb and Brodsky's interpretation ought to be supported by a proper quantitative analysis of the double-membrane model under these transient conditions and by other experiments designed to substantiate their tacit assumption that no transient leakage of solutes, other than sodium and potassium, occurs from the epithelial cells.

Apparently there have been no analogous transient studies of fluid and salt transport across epithelia in the presence of inhibitors. Both Capraro & Marro (1963) and House (1964a), however, have shown that fluid absorption by the isolated frog skin persists for a longer period after the application of 2,4-dinitrophenol and ouabain than does active sodium transport. Of course, the interpretation of such experiments is usually difficult because the inhibitor of active ion transport possibly produces other effects such as changes in L_p and alterations in the balance of electrolytes and water in the tissue.

Obviously the application of the double-membrane model rests on certain plausible assumptions, but one of these assumptions is certainly invalid; that is, it is assumed that the central compartment is well stirred. Indeed, not only is this assumption false but also it now seems that lack of stirring in the central compartment, identified by Kaye *et al.* (1966) as the lateral intercellular spaces of the gall bladder, is of crucial importance to the operation of an alternative account of salt and water coupling. This other mechanism is *local osmosis* and it has steadily gained ground over the double-membrane model.

Local osmosis

This mechanism of fluid transport resembles classical osmosis because the water movement is driven by osmotic gradients established by the active transport of solute. The osmotic gradients, however, in this model are envisaged to exist within the tissue itself, possibly in a confined domain in the neighbourhood of the epithelial cells. Of course, if one inspects the double-membrane model of Curran (1960)—see Fig. 10.6—it becomes evident that the mechanism for moving water across the first membrane is indeed local osmosis. Actually the local-osmosis mechanism may be considered to be a special case of the double-membrane model. In particular, if the second membrane is entirely permeable to both water and solutes then both models become indistinguishable.

Both of these models for fluid transport can account for the coupling between active solute transport and passive water flow. Both can account for the observation that water can be moved against osmotic gradients. Finally, both can account for the elaboration of epithelial secretions and absorbates which are either isotonic or hypertonic. Nevertheless, the different theories are in conflict on the question of hypotonic fluid transport. The double-membrane model, in principle, permits hypotonic fluid transport to occur. On the other hand, if hypotonic fluid transport is to be achieved solely by local osmosis then at least two sites of salt and water transport in the tissue must be postulated. At the first site an isotonic secretion may be produced and at the second hypertonic fluid must be removed from the primary secretion to ensure that the fluid which is finally elaborated is hypotonic.

Experimental evidence

As Diamond (1964b) has pointed out, one can discriminate between local osmosis and other theories of fluid transport by studying the osmolarity of the transported fluid as a function of the osmolarity of the bathing solution. If local osmosis occurs then the transported fluid should be isotonic to the bathing solution even when it is varied over quite wide limits. On the other side of the coin, it is not easy to predict, say, from the double-membrane model what will happen to the transported osmolarity under those circumstances since the permeability coefficients (σ_s, L_p and P_s) of the individual membranes may depend upon the osmolarity of the bathing solutions. In principle, the model of Patlak *et al.* (1963) indicates that the osmolarity of the transported fluid will be hypertonic to dilute bathing solutions, isotonic to a certain value of the bathing osmolarity and, finally, hypotonic to relatively concentrated bathing solutions. Another possible mechanism of salt and water coupling put forward by Diamond (1962c) is *co-diffusion*. Diamond noted that the coupling between the passive transport of salt and passive water movement was so efficient in the fish gall bladder that a salt concentration gradient across the tissue produced a flow of isotonic salt solution. Moreover, he speculated that the coupling between active salt transport and passive water flow was somehow similar to that type of co-diffusion in order to account for the active reabsorption of isotonic fluid which normally occurred. A co-diffusion mechanism, therefore, would be expected to yield a reabsorbate with a unique osmolarity irrespective of the bathing osmolarity provided, of course, that the permeability characteristics of the epithelium were not altered by the experimental conditions.

In order to measure the osmolarity of the reabsorbate in the rabbit gall bladder Diamond (1964b) employed a 'unilateral' preparation. He placed a gall bladder filled with solution in a sealed Goetz tube containing moist oxygen (Fig. 10.9); in this preparation, therefore, there is no serosal (outside) solution, and this technique is similar to that used earlier by Smyth & Taylor (1957) to study the fluid adsorbed by the rat intestine. Under these con-

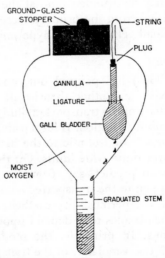

FIG. 10.9. 'Unilateral' gall bladder preparation. The rabbit gall bladder is cannulated and suspended in a Goetz tube filled with oxygen saturated with water vapour. Droplets of the reabsorbed fluid drop off the bladder's serosal surface into the graduated stem and subsequently the ionic composition of this fluid may be determined (Diamond, 1964b: Fig. 1).

ditions the 'unilateral' preparation continued to reabsorb salt and water and, according to Diamond, 'pure absorbate' was elaborated by the preparation and dripped off the serosal surface of the tissue to be collected in the graduated stem of the Goetz tube. Thus, the reabsorbed fluid from the 'unilateral' preparation could be collected and subsequently analysed in contrast to the rather more indirect method employed by Whitlock & Wheeler (1964) who measured not only fluid movement but also unidirectional solute fluxes (see Fig. 10.7).

Diamond (1964b) analysed the ionic composition and osmolarity of the transported fluid over a wide range (68–578 mM) of luminal osmolarity. The latter was varied by changing the NaCl concentration of the mucosal solution or by adding certain amounts of impermeant solutes—sucrose and raffinose. In particular, when the concentration of the main solute (NaCl) was varied Diamond noted a striking identity between the total concentration of cations ([Na+K] in the transported fluid and that in the mucosal

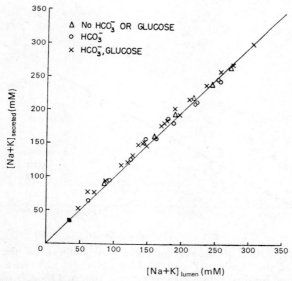

FIG. 10.10. The total concentration of sodium and potassium ions in the fluid secreted by 'unilateral' gall bladders when the luminal [NaCl] was varied. Abscissa, [Na+K] in the luminal bathing solution while the ordinate shows the corresponding value in the secreted (i.e. reabsorbed) fluid. The straight line is the line of 45° slope (Diamond, 1964b: Fig. 3).

solution (Fig. 10.10). This identity persisted even when the sodium concentration was decreased to 26·6 mM or increased to 294·5 mM. Diamond also found that the chloride concentration in the reabsorbed fluid was invariably similar to that in the mucosal solution in those experiments. The results of Diamond's study are in general agreement with those of Whitlock & Wheeler (1964) who used a chamber technique rather than a 'unilateral' gall-bladder preparation. Nevertheless, there is some disparity between the experimental data obtained by those different techniques since Whitlock and Wheeler found that the reabsorbed fluid became significantly hypertonic to the bathing media when sucrose was present in the solutions, particularly the serosal solution. On the other hand, Diamond's 'unilateral' preparation elaborated a fluid which was still isotonic to the mucosal fluid when it contained either sucrose or raffinose. Figure 10.11, for instance, shows the effects of the addition of sucrose (+) or raffinose (△) to the mucosal solution on the secreted fluid. According to Diamond's

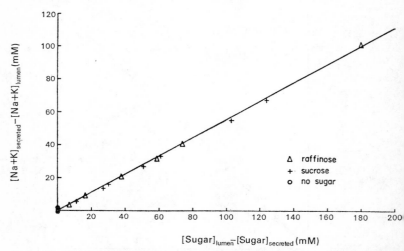

FIG. 10.11. Effect of addition of sucrose or raffinose to the luminal solution of the 'unilateral' gall bladder upon the difference between secreted and luminal cation concentrations. Luminal [NaCl] was always 100 to 125 mM, [sugar]$_{secreted}$ was very small, and [Na] was much larger than [K] in both luminal and secreted fluids. Thus, the main variation is in [Na]$_{secreted}$ as a function of [sugar]$_{lumen}$. The straight line, fitted by least squares, has a slope of 0·55 and an intercept on the ordinate of −0.3 mM (Diamond, 1964b: Fig. 6).

data, when the lumen contained an impermeant sugar the transported fluid had a higher total concentration of cation than the mucosal solution and it contained practically no sugar. In fact, the transported fluid contained approximately an extra 1 mM NaCl for every 2 mM of sugar added to the mucosal solution. In contrast to the data of Whitlock and Wheeler, therefore, Diamond's 'unilateral' preparation reabsorbed fluid which was isotonic to the luminal solution, whether or not impermeant sugar was present, over a large range of osmolarities. Figure 10.12 depicts Diamond's results expressed in terms of the osmolarities of both the transported and luminal solutions. What is the origin of the

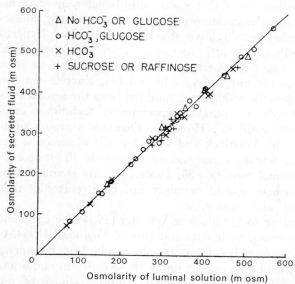

FIG. 10.12. Osmolarity of the secreted fluid as a function of the osmolarity of the luminal solution bathing 'unilateral' gall bladders. The straight line is the line of 45° slope (Diamond, 1964b: Fig. 8).

discrepancy between the results of Diamond (1964b) and Whitlock & Wheeler (1964) when impermeant sugar is present in the solutions bathing the gall bladder? Several 'explanations' are at hand.

According to Diamond (1968), for example, the experimental technique of Whitlock and Wheeler lends itself to 'diffusional artifacts'. To illustrate Diamond's point let us look at a specific set of circumstances. For instance, Diamond considered the case in

Whitlock and Wheeler's experiments where the gall bladder was bathed on both sides by solutions containing NaCl Ringers plus sucrose (see experimental conditions 2b, Table 10.2, page 438). For simplicity let us assume that both bathing solutions contain 100 mM NaCl and 200 mM sucrose. Diamond argued that in this instance the epithelium would elaborate a primary isotonic absorbate consisting of 200 mM NaCl at its serosal surface, whereupon diffusion of NaCl would occur down its concentration gradient towards the serosal solution. Furthermore, Diamond contended that the osmotic pressure of the fluid adjacent to the serosal membranes of the epithelial cells would drop slightly because of salt diffusion out of that region, and hence some decrease in the rate of water absorption would occur. Consequently 'this secondary modification would leave the transported fluid hypertonic'. If Diamond's argument is correct then it explains why Whitlock & Wheeler found a hypertonic fluid transport when an impermeant sugar was present and why he did not, since in the 'unilateral' preparation the transported fluid becomes the serosal solution. In other words Diamond's experimental technique exhibits no 'diffusional artifacts'. However, Diamond's argument does not explain why Whitlock and Wheeler still observed a hypertonic absorbate when sucrose was added only to the mucosal solution (experimental condition 3b, Table 10.2). By Diamond's reasoning the absorbate should certainly not be hypertonic under these conditions and yet it is.

According to Whitlock & Wheeler (1964) the apparent discrepancy between their data and that of Diamond (1964b) is not a true discrepancy because it actually results from the operation of a double-membrane system under those different conditions. They predicted from their somewhat arbitrary choice of membrane parameters that when sucrose is present in the mucosal solution the osmolarity of the transported fluid should be hypertonic by about 12% in their preparation, but only hypertonic by 3% in Diamond's 'unilateral' preparation. Since the precision of Diamond's experimental technique was almost but not quite good enough to distinguish between that level of hypertonicity and isotonicity itself, the 'explanation' of Whitlock and Wheeler remains untested.

There has been some criticism of the 'unilateral' preparation by Marro & Germagnoli (1966), who contended that the absorbate's osmolarity will be recorded invariably as isotonic to that of the mucosal solution because of the fairly rapid osmotic equilibration which occurs in the small volume of the serosal solution. According to their calculations, osmotic equilibration of the serosal and mucosal fluids is attained within about five minutes and thereafter the serosal solution, or rather the reabsorbed fluid, remains isotonic with the mucosal solution. Their argument depends on the small size of the fluid volume adjacent to the serosal surface of the epithelial cells and they assume that for a gall bladder area of 10 cm^2 the volume of serosal solution will be 0·1 cm^3. Their estimate seems too low to me, since the thickness of the serosa in the gall bladder is about 250 μm (Tormey & Diamond, 1967) and the serosal fluid adhering to the external surface of the 'unilateral' preparation might be, say, some 50 μm thick. This would mean that the total volume of the serosal solution in Diamond's experiments was about 0·3 cm^3. Consequently, if that value is the appropriate one for the serosal volume, the theoretical argument of Marro and Germagnoli indicates that osmotic equilibration of the serosal and mucosal fluids takes considerably longer than five minutes. Nevertheless, the point which they make does raise some doubt about the error involved in the determination of the absorbate's osmolarity by the 'unilateral' preparation and this doubt can be resolved adequately only by experiments and probably not by further theoretical arguments.

The preceding discussion about the osmolarity of the transported fluid and its dependence on the osmolarity of the bathing media displays some of the experimental results and difficulties encountered in attempts to devise a test for the local-osmosis model.

Apparently the only major point scored by the double-membrane model over the local-osmosis hypothesis is the observation of Whitlock and Wheeler that the absorbate becomes hypertonic when impermeant sucrose is present in the bathing solutions. Diamond, however, considers that such evidence is compatible with local osmosis when the effects of secondary transport processes are taken into account. Also of some importance to this discussion of the relative merits of these two theories of water transport is the observation that the lateral spaces of the gall bladder are distended during water reabsorption but collapsed in its absence (e.g. Tormey & Diamond, 1967). The double-membrane model of the gall bladder (Fig. 10.8) identifies these spaces as the central compartment and, thus, we would expect them to be at a slightly higher pressure than the bathing solutions and consequently distended. In the local-osmosis model, on the other hand, fluid transport meets no hydraulic resistance between the epithelial cells and the serosal surface of the tissue and the spaces need not be distended.

The characteristics of isotonic fluid secretion in at least four other epithelia are similar to that shown in Fig. 10.12 for rabbit gall bladder. Case, Harper & Scratcherd (1968) studied the secretion of ions and water in a perfused preparation of the isolated pancreas of the cat. They exposed the gland to a wide range of external osmotic pressure; the osmolarity of the pancreatic juice was identical to that of the perfusate whatever its chosen value. Maddrell (1969) has also found an isotonic relationship between the osmolarity of the fluid secretion from *Rhodnius* Malpighian tubules and that of the bathing fluid. Similar results have been noted for fluid absorption in the rat jejunum (Lee, 1968) and apparently for gastric secretion in the frog (Makhlouf, 1971).

Mechanism of local osmosis

This hypothesis assumes that there are local osmotic gradients in the vicinity of the epithelial cells. Where do such gradients become established?

Obvious sites for local concentrations of actively transported solute are the unstirred regions on both sides of the epithelial cell layer or layers. Such unstirred layers are associated with every epithelium; for instance, the epithelial cells of the intestine are bounded on their mucosal surface by unstirred layers of fluid and on their serosal surface by unstirred regions of extracellular fluid between smooth muscle cells and connective tissue. These are basically flat unstirred layers lying parallel to the epithelial tissue and facilitating osmotic equilibration of the transported fluid with the bathing media. Dainty & House (1966*a*) have considered whether such unstirred regions can account for the observed coupling of active salt and passive water transport. They assumed that water was driven by the local osmotic gradient across the tissue and their expression for J_v, the net water flux, was given by where

$$J_v = \frac{J_s L_p RT(\delta^o + \delta^i)}{D_s + c_s(\delta^o + \delta^i)L_p RT} \qquad 10.24$$

J_s is the net (active) solute flux, L_p is the hydraulic conductivity of the whole epithelium, c_s is the solute concentration in the bathing solutions, D_s the solute diffusion coefficient in the unstirred layers and δ^o and δ^i are the thicknesses of the outer and inner unstirred layers. When Dainty and House applied this expression to the relevant data for frog skin they noted the predicted water flux was significantly less than the observed values. For example, taking $J_s = 10^{-6}$ mole cm^{-2} hr^{-1} (Ussing & Zerahn, 1951), $L_p = 4 \times 10^{-7}$ cm sec^{-1} atm^{-1} (House, 1964a), $D_s = 4 \times 10^{-6}$ cm^2 sec^{-1} (Hoshiko, Lindley & Edwards, (1964), $c_s = 2 \times 10^{-4}$ mole cm^{-3}, $\delta^o = 5 \times 10^{-3}$ cm and $\delta^i = 2 \times 10^{-2}$ cm (Dainty & House, 1966a) one finds that $J_v = 0 \cdot 06$ μl. cm^{-2} hr^{-1}, somewhat less than 5% of the flux first measured by Reid (1892a). Applying the same argument to other epithelia yields essentially the same answer, namely that diffusion delays in flat unstirred layers are apparently too small to create sufficiently large local osmotic gradients. To use the preceding expression, however, one needs a reliable value for the hydraulic conductivity. Since the discussion of L_p values in Chapter 9 has shown that most of them are probably underestimated, particularly for the relatively permeable epithelia such as the proximal tubule or the gall bladder, we cannot conclude quite so readily that unstirred layers are insignificant in this connexion. For instance, the relevant data for the rabbit gall bladder indicates that J_v is about 6 μl cm^{-2} hr^{-1} (according to equation 10.24) whereas the observed reabsorption flux is 50 μl cm^{-2} hr^{-1} (see Table 10.1). Recently Wright *et al.* (1972) have concluded, however, that the L_p value for the gall bladder may be underestimated by as much as ten-fold and consequently the water flux predicted by equation 10.24 becomes practically identical to the observed flux.

Apart from the flat unstirred layers associated with epithelial tissues, other alternative sites for the establishment of local osmotic gradients have been proposed. For instance, active accumulation of solute in cytoplasmic vesicles might inevitably lead to a local osmotic water intake and subsequently net fluid transport would result from the emptying of these vesicles at the basal membranes of the epithelial cells. There is no compelling structural evidence to support that type of local osmosis since cytoplasmic vesicles are not observed in the necessary numbers. Other workers have directed their attention alternatively to the ultrastructural geometry of the epithelium and from their work has emerged an attractive picture of local osmosis which will now be described.

'Standing-gradient osmotic flow'

The question of the site where local osmosis occurs is a central one in the analysis of this mechanism. Diamond and his collaborators have resolved this difficulty by proposing that the local osmotic coupling is achieved by a 'standing-gradient osmotic flow' system (Diamond & Bossert, 1967) within the lateral intercellular spaces of the gall bladder. Prior to the theory of Diamond and Bossert, numerous workers (Yamada, 1955; Hayward, 1962a,b; Johnson, McMinn & Birchenough, 1962; Evett, Higgins & Brown, 1964; Kaye et al., 1966; Diamond & Tormey, 1966) reported that the distension of the intercellular spaces in gall bladder might be associated with its reabsorptive function. Diamond and Bossert, however, expressed that notion in a quantitative theory of active salt and passive water movement in the gall bladder and, as we shall see, their theory has potentially a wide application to all other epithelia in the transport of solute and water.

The upper part of Fig. 10.13 depicts their idealized geometry of the lateral intercellular spaces in the gall bladder. Essentially each intercellular space can be considered to be equivalent to a cylindrical channel although it actually has the shape of a thin cylindrical ring. At one end it is closed by a 'tight' junction and its other end opens at the serosal surface of the epithelial cells; Diamond and Tormey referred to this type of space as a 'forward channel' since it faced in the same direction as that of the reabsorbed fluid. Some epithelia, however, possess intercellular spaces which face in the opposite direction to that of fluid transport. Those spaces are the so-called 'backwards channels' treated theoretically by Diamond & Bossert (1968)—see lower part of Fig. 10.13. They occur, for example, in the choroid plexus of vertebrates, avian salt glands, insect Malpighian tubules and in the rectal glands of elasmobranchs. The theoretical analysis of 'backward channels' is exactly similar to that of the 'forward channels' but it does lead to some different predictions which will be discussed later.

Diamond & Bossert (1967, 1968) assumed that no solute or water traversed the 'tight' junction. Later work (Smulders et al., 1972) has shown that the passage of solutes and water through these junctions is probably not negligible, but at the moment we shall

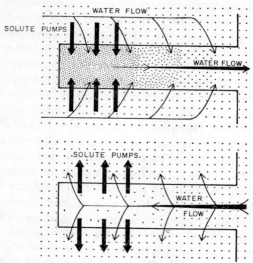

FIG. 10.13. Comparison of 'forwards' and 'backwards' operation of a standing-gradient osmotic flow system, which consists of a long narrow channel closed at one end (e.g. a basal infolding, lateral intercellular space, etc.). The density of dots indicates the solute concentration. Forwards operation (*top*); solute is actively transported into the channel across its walls, making the channel fluid hypertonic. As solute diffuses down its concentration gradient towards the open mouth, more and more water enters the channel across its walls due to the osmotic gradient. In the steady state a standing osmotic gradient will be maintained in the channel by active solute transport, with the osmolarity decreasing progressively from the closed end to the open end: and a fluid of fixed osmolarity (isotonic or hypertonic, depending upon the values of such parameters as radius, length, and water permeability) will constantly emerge from the mouth. Backwards operation (*bottom*); solute is actively transported out of the channel across its walls, making the channel fluid hypotonic. As solute diffuses down its concentration gradient towards the closed end, more and more water leaves the channel across its walls owing to the osmotic gradient. In the steady state a standing osmotic gradient will be maintained in the channel by active solute transport, with the osmolarity decreasing progressively from the open end to the closed end: and a fluid of fixed osmolarity (isotonic or hypertonic, depending upon the parameters of the system) will constantly enter the channel mouth and be secreted across its walls. Solute pumps are depicted only at the bottom of the channels for illustrative purposes but may have different distributions along the channel (Diamond & Bossert, 1968: Fig. 1).

ignore it and assume that the 'tight' junction represents an effective blind end. According to Diamond and Bossert's model, salt is actively pumped across the lateral plasma membranes into the lateral spaces which become hyperosmotic to the cells. The subsequent movement of the actively transported solute towards the serosal surface of the gall bladder *in vitro* or the serosal blood *in vivo* is achieved by convection and diffusion along the intercellular spaces. Our previous discussion of the kinetics of salt movement in flat unstirred layers indicated that it would be possibly too fast to allow adequate osmotic equilibration. However, in this case, where the exit of actively transported solute is confined to the long and narrow intercellular channels for a relatively longer time, it will suffer relatively more dilution owing to the osmotic entry of water across the lateral plasma membranes. Thus, the intercellular spaces offer a mechanism for local osmotic equilibration of the transported salt solution by a relatively prolonged constrainment of the actively transported solute within the neighbourhood of the epithelial cells. The quantitative theory of this model is built on the assumptions that solute is actively transported continuously into the 'forward channel' and that it moves along the channel not only by diffusion but also by a convective stream generated by the osmotic entry of water into the channel all along its length. Under these circumstances the theory of Diamond & Bossert (1967) predicts that in the steady state there will be a 'standing gradient' of solute concentration within the channel; the maintenance of the concentration gradient relies ultimately, of course, on the maintained presence of active salt transport. The independent variables of their theory are the channel radius, channel length, solute diffusion coefficient, the rate of solute transport, the hydraulic conductivity of the epithelium, the osmolarity of the bathing solution outside the channel and, finally, the distance x from the closed end of the 'forward channel'. Knowing the range of probable values for those variables in a number of epithelia, Diamond and Bossert were in a position to evaluate the dependent variables, namely the osmolarity, $C(x)$, and the linear velocity, $v(x)$, of the fluid in the channel at x. Figures 10.14 and 10.15 show profiles of concentration and velocity obtained from the numerical solutions of their differential equations for four sets of chosen values of the independent variables. With regard to $C(x)$, Fig. 10.14 reveals that the solute concentration invariably decreases from the blind end of the channel to the open end but the steepness of its profile depends on the chosen value of the channel's parameters. For instance, the only difference between

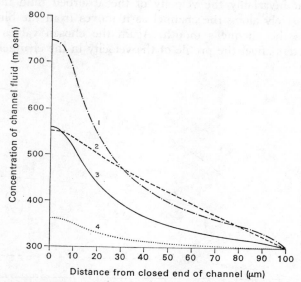

FIG. 10.14. Examples of solute concentration profiles in a standing-gradient osmotic flow system. The solute concentration in the channel in the steady state (ordinate) is plotted against x, the linear distance from the closed end of the channel (abscissa). The length, L, of the channel and its radius, r, were held fixed at 100 and 0·05 μm respectively. The solute concentration at the mouth of the channel was taken as 300 mM and the solute diffusion coefficient in the channel as 10^{-5} cm^2 sec^{-1}. The rate of active solute transport was held at 10^{-8} (curve 1), 5×10^{-9} (curve 3), or 10^{-9} (curves 2 and 4) mole cm^{-2} sec^{-1} for $0 < x < 10$ μm and at zero for $x > 10$ μm. The water permeability was held at 9×10^{-7} (curves 1, 3 and 4) or 0.45×10^{-7} (curve 2) cm sec^{-1} atm^{-1}. The corresponding calculated concentrations of the emergent fluid were 342 (curve 1), 803 (curve 2), 318 (curve 3) and 304 (curve 4) mM (Diamond & Bossert, 1967: Fig. 3).

the channel parameters yielding curves 1 and 4 is that the solute transport rate for the former is ten times that chosen for the latter curve. The apparent precision of the predicted concentration profiles is perhaps a little misleading, and it ought to be stressed that that they are presented only to illustrate how the characteristics of the channels are expected to affect the local osmotic equilibration which occurs in them. A similar opinion applies to the predicted velocity profiles displayed in Fig. 10.15. In this diagram we see that invariably the velocity of the absorbed fluid increased progressively along the channel as it moves from the blind end towards the channel's mouth. Again the chosen values of the parameters affect the profile of the velocity in the channel. What

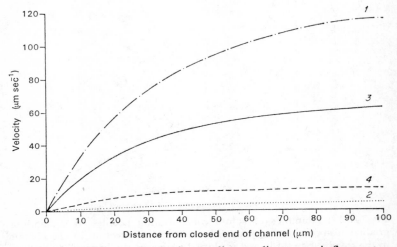

FIG. 10.15. Velocity profiles in the standing-gradient osmotic flow system corresponding to the solute concentration profiles in Fig. 10.14. The linear velocity of water flow in the channel in the steady state (ordinate) is plotted against x, the linear distance from the closed end of the channel (abscissa). Symbols and values as in Fig. 10.14 (Diamond & Bossert, 1967: Fig. 4).

we are eager to obtain from this, or any other model for that matter, is the concentration of the transported fluid. Diamond and Bossert emphasized that the concentration of the fluid emerging from the channel's mouth is not invariably identical to the concentration of the solute $C(L)$, at the end of the channel, i.e. $x = L$. In fact, $C(L)$ was unavoidably equated to the solute concentration, C_0, in the bathing solution as a boundary condition. One must expect, therefore, that the emergent concentration will exceed $C(L)$ since in general (dC/dx) at $x = L$ will not be zero and so there will be not only a convective flow but also a diffusive flux of solute at the channel's orifice. Only when (dC/dx) is zero at $x = L$ will the emergent concentration be equal to $C(L)$. Inspection of Fig. 10.14 indicates that the latter condition apparently holds only for curve 4.

In order to estimate the concentration of the fluid transported through the channel, Diamond and Bossert used the expression:

$$2\pi r \frac{\int_{x=0}^{x=L} N(x)\,dx}{\pi r^2 v(L)}$$

where $N(x)$ denotes the rate of active solute transport into the channel at x and r is the channel radius. The numerator gives the effective rate at which solute is delivered into the whole channel and in the steady state this must be equal to the rate at which solute emerges from the channel. Their expression, therefore, gives the emergent osmolarity because the denominator is the rate at which fluid moves out of the channel's end. These authors used that relation to explore the dependence of the emergent osmolarity upon certain factors, such as the radius and length of the channel and the hydraulic conductivity of the epithelium.

Figure 10.16, for example, displays the dependence of the osmolarity of the emergent fluid upon the length of the channel. Thus, their theory predicts that for relatively short channels (< 100 μm) the emergent osmolarity is substantially hypertonic to the osmolarity of the bathing fluid whereas for channels longer than 100 μm the transported fluid is virtually isotonic. This confirms the qualitative view that prolonging the diffusion pathlength will allow almost complete osmotic equilibration to occur.

On the other hand, Fig. 10.17 depicts the effects of increasing the channel's radius on the emergent osmolarity. In performing the calculations required for the data in Fig. 10.17, Diamond and Bossert assumed that L_p and the total amount of solute delivered into the channel were both constant. The implications of those assumptions are that a given narrow channel is less permeable to water than a relatively wide one and, moreover, that the net flux of solute is greater in the narrow than in the wide channels. Since the area of the channel depends on the radius whereas the channel's volume depends on the square of the radius, increasing the radius of the channel will lead to the transport of the same amount of solute into a relatively larger volume. Consequently, we might expect that the net water flux into the channel will be less than that required for a full osmotic equilibration of the channel's fluid with the bathing medium. Figure 10.17 illustrates that for relatively large radii (> 0·1 μm) full osmotic equilibration is not attained but that for narrow channels osmotic water entry into the smaller volumes is adequate to produce an isotonic reabsorbate.

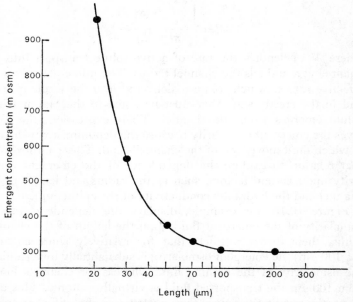

FIG. 10.16. The effect of varying channel length upon the osmolarity of fluid produced by a standing-gradient osmotic flow system. The parameters of the system other than length were held fixed at the following values; water permeability = 9×10^{-7} cm sec^{-1} atm^{-1}: radius = 0·05 μm: solute diffusion coefficient = 10^{-5} cm^2 sec^{-1}: external solute concentration = 300 mM: active solute flux = 10^{-9} mole cm^{-2} sec^{-1} for $0 < x < 10$ μm (and equal to zero for $x > 10$ μm). The abscissa is channel length plotted on a logarithmic scale (Diamond & Bossert, 1967: Fig. 5).

Needless to say the 'standing-gradient' theory of Diamond & Bossert (1967, 1968) is not free from some form of criticism or other. What theory is not?

In particular, recent experimental work has shown that the flow of solute and water through the 'tight' junctions on the rabbit gall bladder cannot be ignored and, therefore, the intercellular spaces or the 'forward channels' are not terminated by a completely impermeant seal.

Secondly, the theory has ignored the cellular compartment by assuming that solute concentration in it remains constant. It seems highly unlikely that there will not be some alteration in the cellular

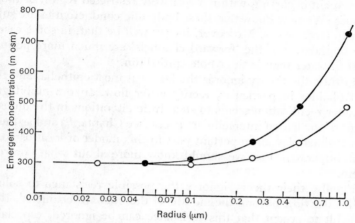

FIG. 10.17. The effect of varying channel radius upon the osmolarity of fluid produced by a standing-gradient osmotic flow system. Length was held fixed by 100 μm, external solute concentration at 300 mM and solute diffusion coefficient at 10^{-5} cm^2 sec^{-1}. The water permeability was taken as 9×10^{-7} (●—●) or at 18×10^{-7} (○—○) cm sec^{-1} atm^{-1}. The active solute transport rate per unit channel length was held fixed at $\pi \times 10^{-14}$ mole per cm, independent of variations of channel radius, by assuming that active solute flux was $\pi \times 10^{-14}/2\pi r$ mole cm^{-2} sec^{-1}. For a channel length larger than 10 μm it was assumed that the active solute flux was zero. The abscissa is the channel radius plotted on a logarithmic scale (Diamond & Bossert, 1967: Fig. 6).

concentrations of solutes, particularly when one remembers that in the vicinity of the closed end of the channel the fluid transported may be markedly hypertonic. Lindemann & Pring (1969) have paid some attention to that theoretical point and they conclude that solute concentration gradients probably do exist within the cells.

Thirdly, the theory offers a fruitful picture of how solute and water movements across the boundaries of 'forward channels' are coupled by local osmosis to yield hypertonic or isotonic fluids, but it says nothing at all about the mechanism governing salt and water transport across the other boundary of the epithelium, namely the mucosal or apical plasma membrane. The latter question is a difficult one and presumably its resolution may be tied up with solute concentration gradients within the epithelial cells.

Fourthly, the theory employs the hydraulic conductivity values obtained for entire epithelial tissues to construct a theory of water movement occurring within a relatively restricted region of those tissues. We now know that these hydraulic conductivities are subject to large errors. Moreover, it may well be that, in some cases, the boundaries of the 'forward channels' are much more permeable to water than is the whole epithelium.

Fifthly, the theory ignores the fact that most epithelia, and the gall bladder in particular, rectify water flow. In the rabbit gall bladder rectification seems to stem from alterations in the dimensions of the lateral intercellular spaces (see Chapter 9) and as these dimensions play an important part in this model of local osmosis it is imperative that we should know more about what controls them.

Finally, their theory ignores the possible resistance to solute and water flow beyond the mouths of the 'forward channels'. It is difficult to accept that this resistance can be ignored, especially when one considers Grantham's observation that the distended intercellular spaces in the collecting duct epithelium take some 200 sec to collapse as a result of passive solute and water transport out of them (Grantham *et al.*, 1969). When this barrier, albeit a quite permeable one, is taken into account there will be a true marriage between Curran's double-membrane model and Diamond's 'standing-gradient' hypothesis.

Even if those few objections could be answered and embodied in a modified theory of 'standing-gradient' osmotic flow it seems unlikely to me at least that anything new would emerge. The great success of the 'standing-gradient' model lies not in the precision of its predictions but in the quantitative support that it gives to a view which is intuitively correct.

The next step is to see how widely the 'standing-gradient' model may be applied to salt and water transport in general. Such an examination has several avenues. First of all what experimental evidence can be mustered in favour of the model? Secondly, do the ultrastructural details of all tissues that transport fluid match the requirements of the theory? Finally, as we have seen in Table 10.1, some tissues transport neither hypertonic nor isotonic fluids but hypotonic fluids: does this present a serious challenge to the theory?

Although the 'standing-gradient' model was originally proposed by Diamond and Bossert to explain fluid transport in animal epithelia it may well have an important bearing on salt and water transport in plants. For instance, Anderson *et al.* (1970) have suggested that this mechanism accounts for water absorption into the isolated plant root, where the xylem vessels are *perhaps* the equivalent 'forward channels' in this absorbing organ.

Recent work on certain epithelia containing so-called 'forward channels' has produced evidence for local osmosis and, in particular, for the 'standing-gradient' model.

Gall bladder. In an effort to check whether or not the rabbit gall bladder contained hypertonic fluid within the lateral intercellular spaces, Machen & Diamond (1969) employed an indirect electrical method. The basis of their technique was the fact that a diffusion potential occurs across the gall bladder when it separates two solutions of different NaCl concentration. Actually the dilute solution becomes electrically positive with respect to the concentrated solution because this epithelium is preferentially permeable to cations. Machen and Diamond argued that if the intercellular spaces contained an elevated NaCl concentration, as the 'standing-gradient' model predicted for maximal fluid transport, then that salt concentration gradient ought to generate a diffusion potential. The electrical sign of the potential should be such that the bladder's lumen is positive with respect to the intercellular spaces. Since the gall bladder possesses an electrically neutral NaCl 'pump', Machen and Diamond found it possible to estimate the magnitude of the secondary diffusion potential without any interference from possible changes in potential stemming from alterations in active pumping rates. They estimated from the magnitude of the observed diffusion potential that the intercellular spaces were hypertonic to the bathing solution by about 20 m-osm. Actually one might have expected the hypertonicity in the channels to have exceeded that value on the grounds that the normal absorption of fluid by the gall bladder proceeds until the mucosal fluid is about 80 m-osm hypertonic to the serosal fluid. However, it must be emphasized that the method of Machen & Diamond (1969) offers only a guide to the size of the actual local osmotic gradient which exists in the tissue. We must await confirmation of their results in the gall bladder by some direct methods. Clearly it will not be an easy technical feat to sample the salt concentration in the narrow (<1 μm) intercellular spaces.

Insect rectal pads. The general organization of the excretory system in the insect has been described by Wigglesworth (1932); Fig. 10.18 illustrates the system diagrammatically. It consists

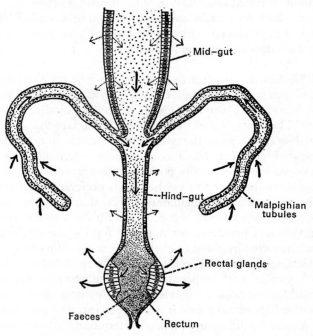

FIG. 10.18. Diagrammatic representation of the insect excretory system and the course of water circulation in alimentary and excretory systems. Arrows signify direction of water movements. Fluid is probably both secreted and absorbed in the mid-gut. It is absorbed in the hind-gut, especially in the rectum and returned to the gut by the Malpighian tubules. During antidiuresis water is absorbed from the rectal lumen against an increasing osmotic gradient as the concentrated faecal pellet is formed (Wigglesworth, 1932: Fig. 1).

of Malpighian tubules which produce an isotonic fluid containing predominantly KCl and some other constituents of haemolymph.

Thereafter the secreted fluid enters the colon and subsequently the rectum where reabsorption of solutes and water occurs. During antidiuresis about 90% of the fluid secreted by the Malpighian tubules is reabsorbed by the rectum to leave a faecal mass which is about three- or four-fold hypertonic to the haemolymph.

The insect rectum has thickenings termed rectal pads or papillae which are responsible for the reabsorption of water and solutes. Although the arrangement of the epithelial cells in the rectal pads is complex, the individual cells occupy a single layer and they are basically similar to other epithelial cells responsible for salt and water transport. Like the gall bladder epithelium, the columnar cells in the rectal pad have long narrow intercellular channels which open in the direction of fluid absorption (Fig. 10.19a). In contrast to the gall bladder, the intercellular spaces in the rectal pad are very narrow ($0.02\ \mu$m) and they form a ramifying network of channels which are apparently isolated from the rectal lumen and from the sinus by separate desmosomes. Nevertheless, these channels join with larger intercellular spaces in the apical region of the cells and the latter spaces are continuous with the sinus. Oschman & Wall (1969) suggest that the route of the absorbed fluid is via the cells, the intercellular spaces and thence into the sinus as shown in Fig. 10.19b. Their experiments indicate that the absorbed fluid passes through one-way 'valves' in the muscle layer into the haemolymph.

Since the rectal pad contain such large intercellular spaces (IS) it offers an almost unique opportunity for testing the 'standing-gradient' model of Diamond and Bossert. Wall, Oschman & Schmidt-Nielsen (1970) found that they could obtain minute fluid samples for analysis from these intercellular spaces. The significant result of their analysis was that the fluid collected from the large intercellular spaces was indeed hypertonic to the fluid in the lumen (over a wide range of luminal concentrations) by about 130 m-osm on average. Their measurements represent the first direct vindication of the 'standing-gradient' model. An interesting feature of the insect rectum, however, is that it transports hypotonic fluid (Phillips, 1964*a–c*) which is apparently at variance with the local osmosis concept. For example, Wall & Oschman (1969) demonstrated that the absorbate collected from the sinus is hypotonic to the luminal fluid when the insect is in a state of antidiuresis whereas the later measurements of Wall *et al.* (1970) showed that

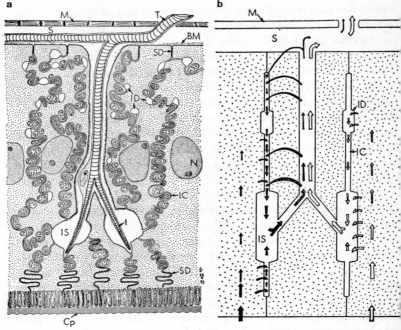

FIG. 10.19. **a.** The general organization of the insect rectal pad as revealed by electron microscopy. The muscle layer (M) facing the haemolymph is shown at the top and the cuticle (Cp) is shown at the bottom. Tracheae penetrate through the muscle layer and into the subepithelial sinus (S). Tracheae also penetrate into the pads following indentations of the basal surface that are lined with basement membrane (BM). The tracheae branch to send fine tracheoles (t) between the cells and into the large spaces of the intercellular sinus (IS). The lateral membranes of the pad cells (nuclei shown at N) are separated by both narrow intercellular channels (IC) and larger intercellular dilations (ID). These are apparently sealed from the apical and basal surfaces by separate desmosomes (SD). Mitochondria are closely associated with both apical and lateral plasma membranes.

b. An interpretation of the physical significance of the ultrastructure summarized in **a**. The open arrows on the right illustrate the supposed flow of water across the rectal pads, while the movement of solute molecules is shown at left with solid arrows. It is assumed that most of the water enters the intercellular sinuses because of the osmotic gradient created by pumping solute into them. Both solutes and water flow towards the subepithelial sinus in the spaces surrounding the penetrating tracheae. The membranes of these spaces are lined with basement membrane and are relatively impermeable to water. Solutes are taken up the precursor fluid in the intercellular spaces is actually hypertonic. One is forced to the conclusion that the primary absorbate suffers a secondary reabsorption (presumably of a very hypertonic fluid) which leaves the final absorbate significantly hypotonic. Wall (1971) draws an obvious parallel between such events occurring within the rectal epithelium itself and a broadly similar series of events in the salivary glands of some vertebrates where a primary isotonic secretion in the acini is finally rendered hypotonic by subsequent reabsorption of hypertonic fluid in the ducts.

Pancreas. Pancreatic juice collected under both *in vivo* and *in vitro* conditions is isotonic to plasma. In particular, Case *et al.* (1968) showed that the isolated pancreas of the cat secreted a fluid which was isotonic to the bathing fluid over a wide range of osmolarity (Fig. 10.20). To explain this phenomenon one might suggest

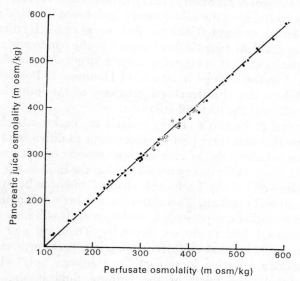

FIG .10.20. The relationship of perfusate osmolality and the osmolality of pancreatic juice from the isolated pancreas of the cat. The straight line is the line of 45° slope (Case *et al.*, 1968: Fig. 7).

---

from these spaces and perhaps also from the subepithelial sinus and returned to the narrow intercellular channels to be recycled through the system (Oschman & Wall, 1969: Figs. 33 & 34).

that there is a *local osmotic gradient* established somewhere within the gland. Recently Swanson & Solomon (1970) have presented evidence based on micropuncture studies on the rabbit pancreatic fluid which confirms that there is a 'standing-gradient' in the smallest extralobular ducts. The fluid in those ducts became about 20–40 m-osm hypertonic to the bathing fluid when pancreatic secretion was induced, but was isotonic during the spontaneous secretion of pancreatic juice. Swanson and Solomon concluded that the primary secretion of ions and water occurs in the intralobular ducts. Mangos & McSherry (1971), however, have shown that the sodium concentration in the acinar fluid significantly exceeds that of the rabbit's plasma, so possibly in the rabbit pancreas the 'standing-gradient' may originate in the acinar lumen. The existence of such 'standing-gradients' in pancreatic ducts is perhaps not a general feature of all species since the micropuncture work of Mangos & McSherry (1971) on rat pancreas reveals that the primary secretion in the acinar lumen has almost identical sodium and chloride concentrations to that in plasma. If there is a 'standing-gradient osmotic flow' system in the pancreatic ducts of some species then it is operating over a larger domain than that envisaged in the original treatment of Diamond & Bossert (1967, 1968), where the ultrastructural geometry of the individual epithelial cells was the focus of attention.

Diamond & Bossert's (1968) analysis of 'backwards channels' demonstrated that their performance ought to differ from that of the 'forward channels' in at least three aspects of salt and water coupling. First, their theory predicted that the fluid in the vicinity of the blind end of the 'backwards channel' ought to be hypotonic to the external medium. This claim has not been substantiated so far in any epithelium possessing such spaces; it will be a remarkable technical feat to do so. Secondly, Diamond and Bossert stressed the point that epithelia containing 'backwards channels' can generate effective local osmotic gradients only when the predominant constituent of the bathing fluid is the actively transported solute. This condition prompted them to assert that such epithelia are to be found transporting principally NaCl and water out of the plasma into other compartments. As examples of this one can cite the choroid plexus of vertebrates and the nasal salt gland of birds, both of which contain 'backwards channels' which can be identified with the basal infoldings of the cell membranes. The Malpighian tubules of insects, however, also contain similar 'backwards channels' (Berridge & Oschman, 1969) but they actively transport KCl rather than NaCl which is the principal constituent of the haemolymph (Ramsay, 1953; Maddrell, 1971). The third prediction about the 'backwards channels' of Diamond and Bossert's model is that there ought to be a considerable 'sweeping-in' effect on solutes not actively transported. That is, such solutes will tend to accumulate in the blind end of the channels, and consequently one might expect an enhanced diffusion of these solutes across the epithelium. Thus, in epithelia with 'backwards channels' small solutes ought to accompany the actively transported solute in minor proportions. The secretory behaviour of the insect Malpighian tubules (Ramsay, 1958) and the rabbit choroid plexus (Davson, 1967), especially with respect to the transport of small molecules, agrees with that prediction.

A common conclusion from the theories of both 'forward' and 'backwards channels' is that the osmolarity of the transported fluid should be virtually independent of transport rate. To a certain extent this prediction is at variance with an intuitive feeling held by many physiologists that the osmolarity of hypertonic absorbates and secretions ought to decrease as the transport rate falls and, hence, the time for osmotic equilibration rises. Nevertheless, in some epithelia, notably the avian salt gland (Schmidt-Nielsen, 1960) and the crocodile distal tubule (Davis & Schmidt-Nielsen, 1967) the rate of secretion and reabsorption of such hypertonic fluids does not influence their osmolarities. In a sense this is an indirect confirmation that the 'standing-gradient' model of local osmosis is basically sound. In some cases, such as the salivary glands of some species, where the transport rate does affect the osmolarity of the emergent fluid it has been shown that there is a secondary reabsorption of hypertonic fluid from the primary isotonic secretion and that this process exerts relatively larger effects at slower secretory rates.

If the 'standing-gradient' model is to be accepted as a universal explanation for coupling between active salt and water movement in all tissues, then probably one of the first prerequisites is that their ultrastructural geometries must favour local osmotic gradients. Do all epithelial cells have associated with them 'forwards' or 'backwards' channels? It should be pointed out, of course, that no matter what geometrical feature is identified as, say, a 'forwards' channel the adjacent regions to it must function as 'backwards' channels (see Maddrell, 1971). For instance, in the gall bladder epithelium the lateral intercellular spaces are 'forwards' channels but the epithelial cells themselves must be 'backwards' channels by Diamond and Bossert's own definition. Lateral intercellular spaces perhaps function as 'forwards' channels in other epithelia such as intestine, skin, urinary bladder and the kidney proximal tubules of some species. Other features of cellular geometry may function as 'forwards' channels; for example, we have the bile canaliculi in liver, the intracellular canaliculi in gastric parietal cells and the extensive infoldings of the basal membranes of epithelial cells in salt glands, ciliary body, salivary gland ducts, choroid plexus and kidney proximal and distal tubule. Another possible example is the brush border of microvilli observed on the mucosal surface of the gall bladder, intestine, proximal tubule and also the Malpighian tubules of insects. According to the 'standing-gradient' model, short channels (< 100 μm) ought to yield hypertonic absorbates or secretions, yet the epithelia such as the mammalian proximal tubule or the insect Malpighian tubule, which have relatively short (< 5 μm) basal infoldings and microvilli, still produce isotonic reabsorbates and secretions. Moreover, the rabbit's corneal endothelium is capable of absorbing fluid from the corneal stroma (Maurice, 1972) at a respectable rate of 6 μl cm^{-2} hr^{-1} (cf. Table 10.1) even although it seems to possess no convincing candidates for either 'forwards' or 'backwards' channels; the intercellular spaces in the endothelium are narrow (< 0.02 μm) and short (2 μm), the basal infoldings of the cell membranes are relatively sparse and there are no microvilli (Hodson, 1968). It is clearly not good enough to speculate about the possible transport behaviour of certain geometrical features of epithelial cells since it is impossible to decide whether or not such structural geometries as we observe, say, in the proximal tubule, are compatible with the 'standing-gradient' model. We need to know the water permeability and the transport rates across such boundaries before we can begin to apply Diamond and Bossert's theory widely to all epithelia.

In apparent conflict with the local osmosis model are the observations that the frog skin (Kirschner, Maxwell & Fleming, 1960), the frog gastric mucosa (Villegas & Sananes, 1968) and the insect rectum (Phillips, 1964a,b,c) exhibit some independence between the rate of active solute transport and the net fluid flux. The frog skin apparently absorbs a hypertonic salt solution from the external medium, but the degree of hypertonicity increases if the external medium is diluted (Huf, Parrish & Weatherford, 1951; Steinbach, 1967). Both the frog gastric mucosa and the insect rectum can transport hypotonic fluids, and this is a serious challenge to the 'standing-gradient' model. Undoubtedly the most widely studied example of a tissue that can transport hypotonic fluid is the insect rectum. In this case hypotonic fluid transport can be explained by postulating that the primary absorbate is hypertonic or isotonic and that its osmolarity is reduced well below that of the bathing fluids by subsequent reabsorption of a significantly more hypertonic solution. The preceding hypothesis is in good agreement with the experimental evidence of Wall & Oschman (1969) and Wall et al., 1970).

Whether or not other preparations elaborate hypotonic fluids by a similar two-step process is an open question. For example, the blastocoel of the rabbit blastocyst contains a hypotonic fluid and unfortunately we do not know how the composition of this fluid is controlled. At present, because of the lack of experimental data, the list of possible 'explanations' is large; there *might* be 'active transport' of water (Tuft & Böving, 1970) or a double-membrane system might be in operation (Enders, 1971) or, for that matter, there might be an isotonic secretion powered by active ion transport into the blastocoel cavity and an opposing absorption of a hypertonic fluid.

The problem of the formation of the vacuolar fluid in the contractile vacuoles of amoeba is rather similar to that of the blastocoel formation. In the case of the vacuole, however, this fluid cavity is lined by a plasma membrane and not by a layer of cells. This prompts several questions. Where does the local osmotic gradient exist in the vicinity of the contractile vacuole? How many processes are involved in the formation of the vacuolar fluid? Can the vacuolar membrane achieve isotonic secretion and simultaneous hypertonic reabsorption of fluid? It is ironic that such questions can be answered for some epithelia but not for a relatively simpler system—the contractile vacuole.

Conclusion

It is customary, though probably unwise in this case, to draw some conclusions. Without much doubt I feel that active salt and water coupling is achieved by local osmosis in a variety of epithelia and, in particular, that the 'standing-gradient' model of Diamond and Bossert is the best physical picture we have of that process. It is undeniably a successful theory, founded on both elegant experimental and theoretical work, but probably we should be a little cautious about accepting it as the only mechanism evolved by nature to achieve fluid transport in epithelia and other tissues. For example, there is evidence that in some tissues osmosis and filtration can play an important role in fluid transport. Nevertheless, the salient feature of these different mechanisms of fluid secretion and absorption is that water is moved passively. There is no compelling evidence for active water transport as yet.

REFERENCES

Abelson, H. T. & Smith, G. H. (1970). Nuclear pores; the pore-annulus relationship in thin sections. *Journal of Ultrastructure Research*, **30**, 558–588.

Abbott, W. O., Hartline, H. K., Hervey, J. P., Ingelfinger, F. J., Rawson, A. J. & Zetzel, L. (1943). Intubation studies of the human small intestine. XXI. A method for measuring intra-luminal pressures and its application to the digestive tract. *Journal of Clinical Investigation*, **22**, 225–234.

Aceves, J. & Erlij, D. (1971). Sodium transport across the isolated epithelium of the frog skin. *Journal of Physiology*. **212**, 195–210.

Adair, G. S. (1929). The thermodynamic analysis of the observed osmotic pressures of protein salts in solutions of finite concentration. *Proceedings of the Royal Society of London* A. **126**, 16–24.

Adolph, E. F. (1967). Ontogeny of volume regulations in embryonic extracellular fluids. *Quarterly Review of Biology*, **42**, 1–39.

Afzelius, B. A. (1955). The ultrastructure of the nuclear membrane of the sea urchin oocyte as studied with the electron microscope. *Experimental Cell Research*, **8**, 147–158.

Agutter, P. S. (1972). The isolation of the envelopes of rat liver nuclei. *Biochimica et Biophysica Acta*, **225**, 397–401.

Altamirano, M. (1963). Alkaline gastric secretion produced by intraarterial acetylcholine. *Journal of Physiology*, **168**, 787–803.

Altamirano, M. (1969). Action of concentrated solutions of non-electrolytes on the dog gastric mucosa. *American Journal of Physiology*, **216**, 33–40.

Altamirano, M., Izaguirre, E. & Milgram, E. (1969). Osmotic concentration of the gastric juice of dogs. *Journal of Physiology*, **202**, 283–296.

Altamirano, M. & Martinoya, C. (1966). The permeability of the gastric mucosa of dog. *Journal of Physiology*, **184**, 771–790.

Alvarez, O. A. & Yudilevich, D. L. (1969). Heart capillary permeability to lipid-insoluble molecules. *Journal of Physiology*, **202**, 45–58.

Andersen, B. & Ussing, H. H. (1957). Solvent drag on non-electrolytes during osmotic flow through isolated toad skin and its response to antidiuretic hormone. *Acta Physiologica Scandiavica*, **39**, 228–239.

Anderson, N. G. (1953). Studies on isolated cell components. VI. The effects of nucleases and proteases on rat liver nuclei. *Experimental Cell Research*, **5**, 361–374.

REFERENCES

Anderson, W. P., Aikman, D. P. & Meiri, A. (1970). Excised root exudation—a standing-gradient osmotic flow. *Proceedings of the Royal Society of London*, B. **174**, 445–458.

Andersson-Cedergren, E. (1959). Ultrastructure of motor end plate and sarcoplasmic components of mouse skeletal muscle fiber as revealed by three dimensional reconstructions from serial sections. *Journal of Ultrastructure Research*, Supplement **1**, 1–191.

Andreoli, T. E., Dennis, V. W. & Weigl, A. M. (1969). The effect of amphotericin B on the water and nonelectrolyte permeability of thin lipid membranes. *Journal of General Physiology*, **53**, 133–156.

Andreoli, T. E., Schafer, J. A. & Troutman, S. L. (1971). Coupling of solute and solvent flows in porous lipid bilayer membranes. *Journal of General Physiology*, **57**, 479–493.

Andreoli, T. E. & Troutman, S. L. (1971). An analysis of unstirred layers in series with 'tight' and 'porous' lipid bilayer membranes. *Journal of General Physiology*, **57**, 464–478.

Applebloom, J. W. T., Brodsky, W. A., Tuttle, W. S. & Diamond, I. (1958). The freezing point depression of mammalian tissues after sudden heating in boiling distilled water. *Journal of General Physiology*, **41**, 1153–1169.

Arisz, W. H. (1956). Significance of the symplasm theory for transport in the root. *Protoplasma*, **46**, 5–62.

Arisz, W. H., Helder, R. J. & Van Nie, R. (1951). Analysis of the exudation process in tomato plants. *Journal of Experimental Botany*, **2**, 257–297.

Austin, G., Sato, M. & Longuet-Higgins, H. C. (1966). Water permeability in *Aplysia* neuronal membrane. *American Biophysical Society Abstracts*, 130.

Auty, R. P. & Cole, R. H. (1952). Dielectric properties of ice and solid D_2O. *Journal of Chemical Physics*, **20**, 1309–1314.

Backmann, E. L. & Runstrom, J. (1909). Der osmotische Druck bei der Entwicklung von *Rana temporaria*. *Biochemische Zeitschrift*, **22**, 290–298.

Backmann, E. L. & Runstrom, J. (1912). Der osmotische Druck wahrend der Embryonalentwicklung von *Rana temporaria*. *Pflügers Archiv fur die gesamte Physiologie des Menchen und der Tiere*, **144**, 287–345.

Baker, P. F., Hodgkin, A. L. & Meves, H. (1964). The effect of diluting the internal solution on the electrical properties of a perfused giant axon. *Journal of Physiology*, **170**, 541–560.

Baldwin, H. H. & Taube, H. (1960). Flow adaptation on the isotopic dilution method for the study of ionic hydration. *Journal of Chemical Physics*, **33**, 206–210.

Bangham, A. D. & Bangham, D. R. (1968). Very long-range structuring of liquids, including water, at solid surfaces. *Nature*, **219**, 1151–1152.

Barker, J. N. (1961). Rat amniotic and yolk fluids. *Federation Proceedings*, **20**, 418.

Barlow, W. (1883). Probable nature of the internal symmetry of crystals. *Nature*, **29**, 186–188.

Barnes, P., Cherry, I., Finney, J. L. & Petersen, S. (1971). Polywater and polypollutants. *Nature*, **230**, 31–33.

Barrnett, R. J. (1959). The demonstration with the electron microscope of the end-products of histochemical reactions in relation to the fine structure of cells. *Experimental Cell Research*, Supplement **7**, 65–89.

Barry, B. A., Matthews, J. & Smyth, D. H. (1961). Transfer of glucose and fluid by different parts of the small intestine of the rat. *Journal of Physiology*, **157**, 279–288.

Barry, P. H. (1970a). Volume flows and pressure changes during an action potential in cells of *Chara australis*. I. Experimental results. *Journal of Membrane Biology*, **3**, 313–334.

Barry, P. H. (1970b). Volume flows and pressure changes during an action potential in cells of *Chara australis*. II. Theoretical considerations. *Journal of Membrane Biology*, **3**, 335–371.

Barry, P. H. & Diamond, J. M. (1970). Junction potentials, electrode standard potentials and other problems in interpreting electrical properties of membranes. *Journal of Membrane Biology*, **3**, 93–122.

Barry, P. H., Diamond, J. M. & Wright, E. M. (1971). The mechanism of cation permeation in rabbit gallbladder. Dilution potentials and biionic potentials. *Journal of Membrane Biology*, **4**, 358–394.

Barry, P. H. & Hope, A. B. (1969a). Electroosmosis in membranes: effects of unstirred layers and transport numbers I. Theory. *Biophysical Journal*, **9**, 700–728.

Barry, P. H. & Hope, A. B. (1969b). Electroosmosis in membranes: effects of unstirred layers and transport numbers II. Experimental. *Biophysical Journal*, **9**, 729–757.

Barton, T. C. & Brown, D. A. J. (1964). Water permeability of the fetal erythrocyte. *Journal of General Physiology*, **47**, 839–849.

Bascom, W. D., Brooks, E. J. & Worthington, B. N. (1970). Evidence that polywater is a colloidal silicate sol. *Nature*, **228**, 1290–1293.

Battin, W. T. (1959). The osmotic properties of nuclei isolated from amphibian oocytes. *Experimental Cell Research*, **17**, 59–75.

Beament, J. W. L. (1964). The active transport and passive movement of water in insects. *Advances in Insect Physiology*, **2**, 67–129.

Beament, J. W. L. (1965). The active transport of water: Evidence, models and mechanisms. *Symposia of the Society for Experimental Biology*, **19**, 273–298.

Beament, J. W. L., Noble-Nesbitt, J. & Watson, J. A. L. (1964). The waterproofing mechanism of arthropods. III. Cuticular permeability in the firebrat, *Thermobia domestica* (Packard). *Journal of Experimental Biology*, **41**, 323–330.

Beck, L. V. & Shapiro, H. (1936). Permeability of germinal vesicle of the starfish egg to water. *Proceedings of the Society for Experimental Biology and Medicine*, **34**, 170–172.

Beck, R. E. & Schultz, J. S. (1970). Hindered diffusion in microporous membranes with known pore geometry. *Science*, **170**, 1302–1305.

Bennett, H. S., Luft, J. H. & Hampton, J. C. (1959). Morphological classifications of vertebrate blood capillaries. *American Journal of Physiology*, **196**, 381–390.
Bennet-Clark, T. A. (1959). Water relations of cells. In *Plant Physiology*, Vol. II (ed. Steward, F. C.), pp. 105–191. New York; Academic Press.
Bentley, P. J. (1961). Directional differences in the permeability to water of the isolated urinary bladder of the toad, *Bufo marinus*. *Journal of Endocrinology*, **22**, 95–100.
Bentley, P. J. (1962). Permeability of the skin of the cyclostome *Lampetra fluviatilis* to water and electrolytes. *Comparative Biochemistry and Physiology*, **6**, 95–97.
Bentzel, C. J., Davies, M., Scott, W. N., Zatzman, M. & Solomon, A. K. (1968). Osmotic volume flow in the proximal tubule of *Necturus* kidney. *Journal of General Physiology*, **51**, 517–533.
Bentzel, C. J., Parsa, B. & Hare, D. K. (1969). Osmotic flow across proximal tubule of *Necturus*; correlation of physiologic and anatomic studies. *American Journal of Physiology*, **217**, 570–580.
Bentzel, C. J. & Solomon, A. K. (1967). Osmotic properties of mitochondria. *Journal of General Physiology*, **50**, 1547–1563.
Berendsen, H. J. C. (1962). Nuclear magnetic resonance study of collagen hydration. *Journal of Chemical Physics*, **36**, 3297–3305.
Berendsen, H. J. C. & Migchelsen, C. (1965). Hydration structure of fibrous macromolecules. *Annals of the New York Academy of Sciences*, **125**, 365–379.
Berendsen, H. J. C. & Migchelsen, C. (1966). Hydration structure of collagen and influence of salts. *Federation Proceedings*, **25**, 998–1002.
Bergmann, R. (1921). Beiträge zur Altersbestimmung von Kalbstöfen der schwartzbunten Niederungarasse. *Archiv für wissenschaltliche und praktische Tierheilkunde*, **47**, 293–315.
Bernal, J. D. (1964). The structure of liquids. *Proceedings of the Royal Society of London* A. **280**, 299–322.
Bernal, J. D. (1965). The structure of water and its biological implications. *Symposia of the Society for Experimental Biology*, **19**, 17–32.
Bernal, J. D. & Fowler, R. H. (1933). A theory of water and ionic solution, with particular reference to hydrogen and hydroxyl ions, *Journal of Chemical Physics*, **1**, 515–548.
Berntsson, K., Haglund, B. & Løvtrup, S. (1964). Water permeation at different tonicities in the amphibian egg. *Journal of Experimental Zoology*, **155**, 317–324.
Berntsson, K. E., Haglund, B. & Løvtrup, S. (1965). Osmoregulation in the amphibian egg. The influence of calcium. *Journal of Cellular and Comparative Physiology*, **65**, 101–112.
Berridge, M. J. & Oschman, J. L. (1969). A structural basis for fluid secretion by Malpighian tubules. *Tissue and Cell*, **1**, 247–272.
Birks, R. I. & Davey, D. F. (1969). Osmotic responses demonstrating the

extracellular character of the sacroplasmic reticulum. *Journal of Physiology*, **202**, 171–188.
Birks, R. I. & Davey, D. F. (1972). An analysis of volume changes in the T-tubes of frog skeletal muscle exposed to sucrose. *Journal of Physiology*, **222**, 95–111.
Bjerrum, N. (1952). Structure and properties of ice. *Science*, **115**, 385–390.
Blinks, J. R. (1965). Influence of osmotic strength on cross-section and volume of isolated single muscle fibres. *Journal of Physiology*, **177**, 42–57.
Blinks, L. R. & Airth, R. L. (1957). Electro-osmosis in *Nitella*. *Journal of General Physiology*, **41**, 383–396.
Blum, R. M. & Foster, R. E. (1970). The water permeability of erythrocytes. *Biochimica et Biophysica Acta*, **203**, 410–423.
Bogucki, M. (1930). Recherches sur la permeabilité des membranes et sur la pression osmotique des oeufs des salmonides. *Protoplasma*, **9**, 345–369.
Boulpaep, E. L. (1972). Permeability changes of the proximal tubule of *Necturus* during saline loading. *American Journal of Physiology*, **222**, 517–531.
Boyle, P. J. & Conway, E. J. (1941). Potassium accumulation in muscle and associated changes. *Journal of Physiology*, **100**, 1–63.
Bozler, E. (1959). Osmotic effects and diffusion of non-electrolytes in muscle. *American Journal of Physiology*, **197**, 505–510.
Bozler, E. (1961a). Distribution of non-electrolytes in muscle. *American Journal of Physiology*, **200**, 651–655.
Bozler, E. (1961b). Electrolytes and osmotic balance of muscle in solutions of non-electrolytes. *American Journal of Physiology*, **200**, 656–657.
Bozler, E. (1962). Osmotic phenomena in smooth muscle. *American Journal of Physiology*, **203**, 201–205.
Bozler, E. (1965). Osmotic properties of amphibian muscles. *Journal of General Physiology*, **49**, 37–45.
Brading, A. F. (1970). Osmotic phenomena in smooth muscle. In *Smooth Muscle*, (ed. Bülbring, E., Brading, A. F., Jones, A. W., Tomita, T.), pp. 166–196. London; Edward Arnold.
Brading, A. F. & Setekleiv, J. (1968). The effect of hypo- and hypertonic solutions on volume and ion distribution of smooth muscle of guinea-pig taenia coli. *Journal of Physiology*, **195**, 107–118.
Bradley, R. S. (1936). Polymer adsorbed films. Part I. The adsorption of argon on salt crystals at low temperatures, and the determination of surface fields. *Journal of the Chemical Society (Transactions)*, 1467–1474.
Brambell, F. W. R. (1954). Transport of proteins across the fetal membranes. *Cold Spring Harbor Symposia on Quantitative Biology*, **19**, 71–81.
Brandt, P. W. (1958). A study of the mechanism of pinocytosis. *Experimental Cell Research*, **15**, 300–313.

Brandt, P. W. (1962). A study of pinocytosis in muscle capillaries. *Anatomical Record*, **142**, 219.

Brandt, P. W. & Pappas, G. D. (1960). An electron microscopic study of pinocytosis in ameba. I. The surface attachment phase. *Journal of Biophysical and Biochemical Cytology*, **8**, 675–687.

Bratton, C. B., Hopkins, A. L. & Weinberg, J. W. (1965). Nuclear magnetic resonance studies of living muscle. *Science*, **147**, 738–739.

Briggs, G. E. (1967). Electro-osmosis in *Nitella*. *Proceedings of the Royal Society of London B*, **168**, 22–26.

Brightman, M. W. & Palay, S. L. (1963). The fine structure of the ependyma in the brain of the rat. *Journal of Cell Biology*, **19**, 415–439.

Brightman, M. W. & Reese, T. S. (1969). Junctions between intimately apposed cell membranes in the vertebrate brain. *Journal of Cell Biology*, **40**, 648–677.

Brodsky, W. A., Rehm, W. S., Dennis, W. H. & Miller, D. G. (1955). Thermodynamic analysis of the intracellular osmotic gradient hypothesis of active water transport. *Science*, **121**, 302–303.

Brodsky, W. A. & Schilb, T. P. (1965). Osmotic properties of isolated turtle bladder. *American Journal of Physiology*, **208**, 46–57.

Brown, E. & Landis, E. M. (1947). Effect of local cooling on fluid movements, effective osmotic pressure and capillary permeability in the frog's mesentery. *American Journal of Physiology*, **149**, 302–315.

Bruns, R. R. & Palade, G. E. (1968a). Studies on blood capillaries. I. General organization of blood capillaries in muscle. *Journal of Cell Biology*, **37**, 244–276.

Bruns, R. R. & Palade, G. E. (1968b). Studies on blood capillaries. II. Transport of ferritin molecules across the wall of muscle capillaries. *Journal of Cell Biology*, **37**, 277–299.

Brusilow, S. W. & Gordes, E. H. (1964). Solute and water secretion in sweat. *Journal of Clinical Investigation*, **43**, 477–484.

Buckley, K. A., Conway, E. J. & Ryan, H. C. (1958). Concerning the determination of total intracellular concentrations by the cryoscopic method. *Journal of Physiology*, **143**, 236–245.

Buijs, K. & Choppin, G. R. (1963). Near-infrared studies of the structure of water. I. Pure water. *Journal of Chemical Physics*, **39**, 2035–2041.

Bunch, W. & Edwards, C. (1969). The permeation of non-electrolytes through the single barnacle muscle cell. *Journal of Physiology*, **202**, 683–698.

Bunch, W. H. & Kallsen, G. (1969). Rate of intracellular diffusion as measured in barnacle muscle. *Science*, **164**, 1178–1179.

Burg, M. B. & Grantham, J. J. (1971). Ion movements in renal tubules. In *Membranes and Ion Transport*, Vol. 3 (ed. Bittar, E. E.), pp. 49–77. London; Wiley-Interscience.

Burg, M. B., Grantham, J. J., Abramow, M. & Orloff, J. (1966). Preparation and study of fragments of single rabbit nephrons. *American Journal of Physiology*, **210**, 1293–1298.

Burg, M. B., Helman, S. E., Grantham, J. J. & Orloff, J. (1970). Effect of vasopressin on the permeability of isolated rabbit cortical collecting tubules to urea, acetamide and thiourea. In *Urea and the Kidney*. pp. 193–199. Amsterdam; Excerpta Medica International Congress Ser. No. 195.

Burg, M. B. & Orloff, J. (1968). Control of fluid absorption in the reanl proximal tubule. *Journal of Clinical Investigation*. **47**, 2016–2024.

Burger, J. W. & Hess, W. N. (1960). Function of the rectal gland in the spiny dogfish. *Science*, **131**, 670–671.

Campbell, E. S., Gelernter, G., Heinen, H. & Moorti, V. R. G. (1967). Interpretation of the energy of hydrogen bonding; permanent multipole contribution to the energy of ice as a function of the arrangement of hydrogens. *Journal of Chemical Physics*, **46**, 2690–2707.

Capraro, V. & Marro, F. (1963). The net passage of water through the isolated skin of *Rana esculenta* in the absence of apparent osmotic gradient. *Archives Italiennes de Biologie*, **101**, 161–173.

Carasso, N., Favard, P., Bourguet, J. & Jard, S. (1966). Role du flux net d'eau dans les modifications ultrastructurales de la vessie de grenouille stimulée par l'ocytocine. *Journal de Microscopie*, **5**, 519–522.

Carasso, N., Favard, P. & Valérien, J. (1962). Variations des ultra-structures dans les cellules épithéliales de la vessie du crapaud après stimulation par l'hormone neurohypophysaire. *Journal de Microscopie*, **1**, 143–158.

Case, R. M., Harper, A. A. & Scratcherd, T. (1968). Water and electrolyte secretion by the perfused pancreas of the cat. *Journal of Physiology*, **196**, 133–149.

Cass, A. & Finkelstein, A. (1967). Water permeability of thin lipid membranes. *Journal of General Physiology*, **50**, 1765–1784.

Chapman, G. & McLauchlan, K. A. (1967). Oriented water in the sciatic nerve of rabbit. *Nature*, **215**, 391–392.

Chapman-Andresen, C. (1962). Studies on pinocytosis in amoebae. *Compte rendu des travaux du Laboratoire de Carlsberg*, **33**, 73–264.

Chapman-Andresen, C. & Dick, D. A. T. (1961). Volume changes in the amoeba *Chaos chaos* L. *Compte rendu des travaux du Laboratoire de Carlsberg*, **32**, 265–289.

Chapman-Andresen, C. & Holter, H. (1964). Differential uptake of protein and glucose by pinocytosis in *Amoeba proteus*. *Compte rendu des travaux du Laboratoire de Carlsberg*, **34**, 211–226.

Chappell, J. B. & Crofts, A. R. (1965). Calcium ion accumulation and volume changes of isolated liver mitochondria. Calcium ion-induced swelling. *Biochemical Journal*, **95**, 378–386.

Chinard, F. P., Enns, T., Goresky, C. A. & Nolan, M. F. (1965). Renal transit times and distribution volumes of T-1824, creatinine and water. *American Journal of Physiology*, **209**, 243–252.

Choi, J. K. (1963). The fine structure of the urinary bladder of the toad, *Bufo marinus*. *Journal of Cell Biology*, **16**, 53–72.

Churney, L. (1942). The osmotic properties of the nucleus. *Biological Bulletin of the Marine Biological Laboratory, Woods Hole, U.S.A.*, **82**, 52–67.

Clarkson, T. W. (1967). The transport of salt and water across isolated rat ileum. Evidence for at least two distinct pathways. *Journal of General Physiology*, **50**, 695–727.

Clarkson, T. W. & Rothstein, A. (1960). Transport of monovalent ions by the isolated small intestine of the rat. *American Journal of Physiology*, **199**, 898–906.

Claussen, W. F. (1951a). Suggested structures of water in inert gas hydrates. *Journal of Chemical Physics*, **19**, 259–260.

Claussen, W. F. (1951b). Erratum; suggested structures of water in inert gas hydrates. *Journal of Chemical Physics*, **19**, 662.

Claussen, W. F. (1951c). A second water structure for inert gas hydrates. *Journal of Chemical Physics*, **19**, 1425–1426.

Clementi, F. & Palade, G. E. (1969). Intestinal capillaries. I. Permeability to peroxidase and ferritin. *Journal of Cell Biology*, **41**, 33–58.

Cole, D. F. (1960). Effects of some metabolic inhibitors upon the formation of the aqueous humour in rabbits. *British Journal of Ophthalmology*, **44**, 739–750.

Cole, D. F. (1962). Transport across the isolated ciliary body of ox and rabbit. *British Journal of Ophthalmology*, **46**, 577–591.

Cole, K. S. (1932). Surface forces of the *Arbacia* egg. *Journal of Cellular and Comparative Physiology*, **1**, 1–9.

Collander, R. (1949). The permeability of plant protoplasts to small molecules. *Physiologia Plantarum*, **2**, 300–311.

Collander, R. (1950). The permeability of *Nitella* cells to rapidly penetrating non-electrolytes. *Physiologia Plantarum*, **3**, 45–57.

Collander, R. (1954). The permeability of *Nitella* cells to non-electrolytes. *Physiologia Plantarum*, **7**, 420–445.

Collander, R. & Bärlund, H. (1933). Permeabilitätsstudien an *Chara ceratophylla*. *Acta Botanica Fennica*, **11**, 1–114.

Collie, C. H., Hasted, J. B. & Ritson, D. M. (1948). Dielectric properties of water and heavy water. *Proceedings of the Physical Society*, **60**, 145–160.

Connick, R. E. & Poulson, R. E. (1959). Effect of paramagnetic ions on the nuclear magnetic resonance of O^{17} in water and the rate of elimination of water molecules from the first coordination sphere of cations. *Journal of Chemical Physics*, **30**, 759–761.

Conway, E. J. & Geoghegan, H. (1955). Molecular concentration of kidney cortex slices. *Journal of Physiology*, **130**, 438–445.

Conway, E. J., Geoghegan, H. & McCormack, J. I. (1955). Autolytic changes at zero centigrade in ground mammalian tissues. *Journal of Physiology*, **130**, 427–437.

Conway, E. J. & McCormack, J. I. (1953). The total intracellular concentration of mammalian tissues compared with that of the extracellular fluid. *Journal of Physiology*, **120**, 1–14.

Cook, J. S. (1967). Nonsolvent water in human erythrocytes. *Journal of General Physiology*, **50**, 1311–1325.

Cope, F. W. (1965). Nuclear magnetic resonance evidence for complexing of sodium ions in muscle. *Proceedings of the National Academy of Sciences, U.S.A.*, **54**, 225–227.

Cope, F. W. (1967a). NMR evidence for complexing of Na^+ in muscle, kidney, and brain, and by Actomyosin. The relation of cellular complexing of Na^+ to water structure and to transport kinetics. *Journal of General Physiology*, **50**, 1353–1375.

Cope, F. W. (1967b). A theory of cell hydration governed by adsorption of water on cell proteins rather than by osmotic pressure. *Bulletin of Mathematical Biophysics*, **29**, 583–596.

Cope, F. W. (1969). Nuclear magnetic resonance evidence using D_2O for structured water in muscle and brain. *Biophysical Journal*, **9**, 303–319.

Cortney, M. A., Mylle, M., Lassiter, W. E. & Gottschalk, C. (1965). Renal tubular transport of water, solute and PAH in rats loaded with isotonic saline. *American Journal of Physiology*, **209**, 1199–1205.

Coulter, N. A. (1958). Filtration coefficient of the capillaries of the brain. *American Journal of Physiology*, **195**, 459–464.

Crank, J. (1956). *The Mathematics of Diffusion*. 1st edition, p. 144, Oxford; Clarendon Press.

Crawford, J. D. & McCance, R. A. (1960). Sodium transport by the chorioallantoic membrane of the pig. *Journal of Physiology*, **151**, 458–471.

Crone, C. (1963a). The permeability of capillaries in various organs as determined by use of the 'indicator diffusion' method. *Acta Physiologica Scandinavica*, **58**, 292–305.

Crone, C. (1963b), Does 'restricted diffusion' occur in muscle capillaries? *Proceedings of the Society for Experimental Biology and Medicine*, **112**, 453–455.

Crone, C. (1965). The permeability of brain capillaries to non-electrolytes. *Acta Physiologica Scandinavica*, **64**, 407–417.

Cross, M. H. (1971). Rabbit blastocoele perfusion technique. *Nature*, **232**, 635–637.

Cross, M. H. & Brinster, R. L. (1969). Trans membrane potential of the rabbit blastocyst trophoblast. *Experimental Cell Research*, **58**, 125–127.

Curran, P. F. (1960). Na, Cl, and water transport by rat ileum *in vitro*. *Journal of General Physiology*, **43**, 1137–1148.

Curran, P. F. & McIntosh, J. R. (1962). A model system for biological water transport. *Nature*, **193**, 347–348.

Curran, P. F. & Solomon, A. K. (1957). Ion and water fluxes in the ileum of rats. *Journal of General Physiology*, **41**, 143–168.

Curran, P. F., Taylor, A. E. & Solomon, A. K. (1967). Tracer diffusion and unidirectional fluxes. *Biophysical Journal*, **7**, 879–901.

Curtis, H. J. & Cole, K. S. (1942). Membrane resting and action potentials from the squid giant axon. *Journal of Cellular and Comparative Physiology*, **19**, 135–144.

Dainty, J. (1963a). Water relations of plant cells. *Advances in Botanical Research*, **1**, 279–326.

Dainty, J. (1963b). The polar permeability of plant cell membranes to water. *Protoplasma*, **57**, 220–228.

Dainty, J. (1965). Osmotic flow. *Symposia of the Society for Experimental Biology*, **19**, 75–85.

Dainty, J., Croghan, P. C. & Fensom, D. S. (1963). Electro-osmosis with some applications to plant physiology. *Canadian Journal of Botany*, **41**, 953–966.

Dainty, J. & Ginzburg, B. Z. (1963). Irreversible thermodynamics and frictional models of membrane processes, with particular reference to the cell membrane. *Journal of Theoretical Biology*, **5**, 256–265.

Dainty, J. & Ginzburg, B. Z. (1964a). The measurement of hydraulic conductivity (osmotic permeability to water) of internodal characean cells by means of transcellular osmosis. *Biochimica et Biophysica Acta*, **79**, 102–111.

Dainty, J. & Ginzburg, B. Z. (1964b). The permeability of the cell membranes of *Nitella translucens* to urea, and the effect of high concentrations of sucrose on this permeability. *Biochimica et Biophysica Acta*, **79**, 112–121.

Dainty, J. & Ginzburg, B. Z. (1964c). The permeability of the protoplasts of *Chara australis* and *Nitella translucens* to methanol, ethanol and isopropanol. *Biochimica et Biophysica Acta*, **79**, 122–128.

Dainty, J. & Ginzburg, B. Z. (1964d). The reflection coefficient of plant cell membranes for certain solutes. *Biochimica et Biophysica Acta*, **79**, 129–137.

Dainty, J. & Hope, A. B. (1959). The water permeability of cells of *Chara australis*, R. Br. *Australian Journal of Biological Sciences*, **12**, 136–145.

Dainty, J. & House, C. R. (1966a). 'Unstirred layers' in frog skin. *Journal of Physiology*, **182**, 66–78.

Dainty, J. & House, C. R. (1966b). An examination of the evidence for membrane pores in frog skin. *Journal of Physiology*, **185**, 172–184.

Danford, M. D. & Levy, H. A. (1962). The structure of water at room temperature. *Journal of the American Chemical Society*, **84**, 3965–3966.

Daniel, J. C. (1964). Early growth of rabbit trophoblast. *American Naturalist*, **98**, 85–98.

Danielli, J. F. & Davson, H. (1935). A contribution to the theory of permeability of thin films. *Journal of Cellular and Comparative Physiology*, **5**, 495–508.

Davenport, H. W. & Fisher, R. B. (1940). The mechanism of the secretion of acid by the gastric mucosa. *American Journal of Physiology*, **131**, 165–175.

Davies, H. G. (1961). Structure in nucleated erythrocytes. *Journal of Biophysical and Biochemical Cytoplogy*, **9**, 671–687.

Davies, J. & Routh, J. I. (1957). Composition of the foetal fluids of the rabbit. *Journal of Embryology and Experimental Morphology*, **5**, 32–39.

Davies, R. E. & Terner, C. (1949). The effects of applied pressure on secretion by isolated amphibian gastric mucosa. *Biochemical Journal*, **44**, 374–384.

Davis, L. E. & Schmidt-Nielsen, B. (1967). Ultrastructure of the crocodile kidney (*Crocodylus acutus*) with special reference to electrolyte and fluid transport. *Journal of Morphology*, **121**, 255–276.

Davson, H. (1953). The penetration of large water-soluble molecules into the aqueous humour. *Journal of Physiology*, **122**, 10P.

Davson, H. (1967). *Physiology of the Cerebrospinal Fluid*. p. 41. London; Churchill.

Davson, H. (1970). *A Textbook of General Physiology*, Vol. 1. p. 1000. London; Churchill.

Dawson, I. M., Hossack, J. & Wyburn, G. M. (1955). Observations on the Nissl's substance, cytoplasmic filaments and the nuclear membrane of spinal ganglion cells. *Proceedings of the Royal Society of London* B, **144**, 132–142.

De Groot, S. R. (1952). *Thermodynamics of Irreversible Processes*. Amsterdam; North-Holland.

De Luque, O. & Hunter, F. R. (1959). Osmotic studies of amphibian eggs. I. Preliminary survey of volume changes. *Biological Bulletin of the Marine Biological Laboratory, Woods Hole, U.S.A.*, **117**, 458–467.

Denbigh, K. G. & Raumann, G. (1950). Thermo-osmosis of gases through a membrane, *Nature*, **165**, 199–200.

Dengel, O. & Riehl, N. (1963). Diffusion von Protonen (Tritonen) in Eiskristallen. *Physik der Kondensierten Materie*, **1**, 191–196.

Derjaguin, B. V. (1965). Recent research on the properties of water in thin films and in microcapillaries. *Symposia of the Society for Experimental Biology*, **19**, 55–60.

Derjaguin, B. V. (1966). Effect of lyophile surfaces on the properties of boundary liquid films. *Discussions of the Faraday Society*, **42**, 109–119.

Derjaguin, B. V., Churaev, N. V., Fedyakin, N. N., Talaev, M. V. & Ershova, I. G. (1967). The modified state of water and other liquids. *Bulletin of the Academy of Sciences, U.S.S.R., Division of Chemical Science*, **10**, 2095–2102.

Despic, A. & Hills, G. J. (1956). Electro-osmosis in charged membranes. The determination of primary solvation numbers. *Discussions of the Faraday Society*, **21**, 150–162.

DeVries, A. L., Komatsu, S. K. & Feeney, R. E. (1970). Chemical and physical properties of freezing-point depressing glycoproteins from Antarctic fishes. *Journal of Biological Chemistry*, **245**, 2901–2908.

DeVries, A. L. & Wohlschlag, D. E. (1969). Freezing resistance in some Antarctic fishes. *Science*, **163**, 1073–1075.

Diamond, J. M. (1962a). The reabsorptive function of the gall-bladder. *Journal of Physiology*, **161**, 442–473.

Diamond, J. M. (1962b). The mechanism of solute transport by the gallbladder. *Journal of Physiology*, **161**, 474–502.

Diamond, J. M. (1962c). The mechanism of water transport by the gallbladder. *Journal of Physiology*, **161**, 503–527.

Diamond, J. M. (1964a). Transport of salt and water in rabbit and guineapig gall bladder. *Journal of General Physiology*, **48**, 1–14.

Diamond, J. M. (1964b). The mechanism of isotonic water transport. *Journal of General Physiology*, **48**, 15–42.

Diamond, J. M. (1966a). Non-linear osmosis. *Journal of Physiology*, **183**, 58–82.

Diamond, J. M. (1966b). A rapid method for determining voltage-concentration relations across membranes. *Journal of Physiology*, **183**, 83–100.

Diamond, J. M. (1968). Transport mechanisms in the gall bladder. *Handbook of Physiology*. (*Alimentary Canal*), Chapter 115, 2451–2482. Washington D.C.: American Physiological Society.

Diamond, J. M. & Bossert, W. H. (1967). Standing-gradient osmotic flow. A mechanism for coupling of water and solute transport in epithelia. *Journal of General Physiology*, **50**, 2061–2083.

Diamond, J. M. & Bossert, W. H. (1968). Functional consequences of ultrastructural geometry in 'backwards' fluid-transporting epithelia. *Journal of Cell Biology*, **37**, 694–702.

Diamond, J. M. & Harrison, S. C. (1966). The effect of membrane fixed charges upon diffusion potentials and streaming potentials. *Journal of Physiology*, **183**, 37–57.

Diamond, J. M. & Solomon, A. K. (1959). Intracellular potassium compartments in *Nitella axillaris*. *Journal of General Physiology*, **42**, 1105–1121.

Diamond, J. M. & Tormey, J. M. (1966). Role of long extracellular channels in fluid transport across epithelia. *Nature*, **210**, 817–820.

Diamond, J. M. & Wright, E. M. (1969). Biological membranes; The physical basis of ion and nonelectrolyte selectivity. *Annual Review of Physiology*, **31**, 581–646.

Dianzani, M. U. (1953). On the osmotic behaviour of mitochondria. *Biochimica et Biophysica Acta*, **11**, 353–367.

DiBona, D. R. (1972). Passive intercellular pathway in amphibian epithelia. *Nature*, **238**, 179–181.

DiBona, D. R. & Civan, M. M. (1969). Toad urinary bladder; intercellular spaces. *Science*, **165**, 503–504.

DiBona, D. R. & Civan, M. M. (1970). The effect of smooth muscle on the intercellular spaces in toad urinary bladder. *Journal of Cell Biology*, **46**, 235–244.

DiBona, D. R., Civan, M. M. & Leaf, A. (1969a). The anatomic site of the trans-epithelial permeability barriers of toad bladder. *Journal of Cell Biology*, **40**, 1–7.

DiBona, D. R., Civan, M. M. & Leaf, A. (1969b). The cellular specificity of the effect of vasopressin on toad urinary bladder. *Journal of Membrane Biology*, **1**, 79–91.

Dick, D. A. T. (1959a). Osmotic properties of living cells. *International Review of Cytology*, **8**, 387–448.

Dick, D. A. T. (1959b). The permeation of water into chick heart fibroblasts in tissue culture. *Proceedings of the Royal Society of London* B, **150**, 43–52.

Dick, D. A. T. (1959c). The rate of diffusion of water in the protoplasm of living cells. *Experimental Cell Research*, **17**, 5–12.

Dick, D. A. T. (1964). The permeability coefficient of water in the cell membrane and the diffusion coefficient in the cell interior. *Journal of Theoretical Biology*, **7**, 504–532.

Dick, D. A. T. (1966). *Cell Water*. London: Butterworth.

Dick, D. A. T. (1970). Water movements in cells. In *Membranes and Ion Transport*, Vol. 3 (ed. Bittar, E. E.), pp. 211–250. New York; Wiley-Interscience.

Dick, E. G., Dick, D. A. T. & Bradbury, S. (1970). The effect of surface microvilli on the water permeability of single toad oocytes. *Journal of Cell Science*, **6**, 451–476.

Dick, D. A. T. & Lea, E. J. A. (1964). Na fluxes in single toad oocytes with special reference to the effect of external and internal Na concentration on Na efflux. *Journal of Physiology*, **174**, 55–90.

Dick, D. A. T. & Lowenstein, L. M. (1958). Osmotic equilibria in human erythrocytes studied by immersion refractometry. *Proceedings of the Royal Society of London*. B. **148**, 241–256.

Dicker, S. E. & Elliott, A. B. (1967). Water uptake by *Bufo melanostictus*, as affected by osmotic gradients, vasopressin and temperature. *Journal of Physiology*, **190**, 359–370.

Dickerson, J. W. T. & McCance, R. A. (1957). The composition and origin of the allantoic fluid in the rabbit. *Journal of Embryology and Experimental Morphology*, **5**, 40–42.

Dietschy, J. M. (1964). Water and solute movement across the wall of the everted rabbit gall bladder. *Gastroenterology*, **47**, 395–408.

DiPolo, R., Sha'afi, R. I. & Solomon, A. K. (1970). Transport parameters in a porous cellulose acetate membrane. *Journal of General Physiology*, **55**, 63–76.

Dirks, J. H., Cirksena, W. J. & Berliner, R. W. (1965). The effect of saline infusion on sodium reabsorption by the proximal tubule of the dog. *Journal of Clinical Investigation*, **44**, 1160–1170.

Dobson, A., Harrop, C. J. F. & Phillipson, A. T. (1972). Osmotic effects of carbon dioxide absorption from the rumen. *Federation Proceedings*, **31**, 260.

Dobson, A., Sellers, A. F. & Shaw, G. T. (1970). Absorption of water from isolated ventral sac of rumen of the cow. *Journal of Applied Physiology*, **28**, 100–104.

Dobson, A., Sellers, A. F. & Thorlacius, S. O. (1971). Limitation of diffusion by blood flow through bovine ruminal epithelium. *American Journal of Physiology*, **220**, 1337–1343.

Donahue, S. & Pappas, G. D. (1961). The fine structure of capillaries in the cerebral cortex of the rat at various stages of development. *American Journal of Anatomy*, **108**, 331–348.

Durbin, R. P. (1960). Osmotic flow of water across permeable cellulose membranes. *Journal of General Physiology*, **44**, 315–326.

Durbin, R. P., Frank, H. & Solomon, A. K. (1956). Water flow through frog gastric mucosa. *Journal of General Physiology*, **39**, 535–551.

Dydyńska, M. & Wilkie, D. R. (1963). The osmotic properties of striated muscle fibres in hypertonic solutions. *Journal of Physiology*, **169**, 312–329.

Earley, L. E., Sidel, V. W. & Orloff, J. J. (1962). Factors influencing permeability of a vasopressin-sensitive membrane. *Federation Proceedings*, **21**, 145.

Edelman, I. S. (1961). Transport through biological membranes. *Annual Review of Physiology*, **23**, 37–70.

Edney, E. G. (1966). Absorption of water vapour from unsaturated air by *Arenivago* sp. (*Polyphagidae*, Dictyoptera). *Comparative Biochemistry and Physiology*, **19**, 387–408.

Edney, E. B. (1967). Water balance in desert arthropods. *Science*, **156**, 1059–1066.

Eigen, M. (1963). Fast elementary steps in chemical reaction mechanisms. *Pure and Applied Chemistry*, **6**, 97–115.

Eigen, M. (1964). Proton transfer, acid-base catalysis, and enzymatic hydrolysis. *Angewandte Chemie*. International Edition in English, **3**, 1–19.

Eigen, M. & De Maeyer, L. (1958). Self-dissociation and protonic charge transport in water and ice. *Proceedings of the Royal Society of London*, A **247**, 505–533.

Eisenberg, D. & Kauzmann, W. (1969). *The Structure and Properties of Water*, 1st edn. p. 213. London; Oxford University Press.

Elias, H-G. (1961*a*). Osmose an permeablem Membranen. I. Der Lösungsmittel-Einfluss auf den Stavermann-Effekt. *Zeitschrift für Physikalische Chemie*, **28**, 303–321.

Elias, H-G. (1961*b*). Osmose an permeablen Membranen. II. Anomale Osmosen an ionenfreien Membranen. *Zeitschrift für Physikalische Chemie*, **28**, 322–331.

Enders, A. C. (1971). The fine structure of the blastocyst. In *The Biology of the Blastocyst* (ed. Blandau, R. J.), pp. 71–94. Chicago; University of Chicago Press.

Engelhardt, W. v. (1969). Der Wasserdurchtritt durch die Pansenschleimhaut Osmotische, hydrostatische, hämodynamische und humorale Einflüsse. *Zentralblatt für Veterinärmedizin*, A**16**, 597–625.

Engelhardt, W. v. (1970). Movement of water across the rumen epithelium. In *Physiology of Digestion and Metabolism in the Ruminant* (ed. Phillipson, A. T.) pp. 132–146. Newcastle; Oriel Press.

Engelhardt, W. v. & Nickel, W. (1965). Die Permeabilität der Pansenwand für Harnstoff. Antipyrin und Wasser. *Pflügers Archiv für die gesamte Physiologie des Menschen und der Tiere*, **286**, 57–75.

Engelhardt, W. v. & Schwartz, R. (1968). Der Nettowasserfluss durch die isolierte cutane Schleimhaut des Pansens bei hydrostatischen Druckgradienten. *Pflügers Archiv für die gesamte Physiologie des Menschen und der Tiere*, **300**, R36–R37.

Ernst, J. (1963). *Biophysics of the Striated Muscle*, 2nd edn., Budapest; Akademiaia Publishing Co.

Essig, A. (1966). Isotope interaction and 'abnormal' flux ratios—a frictional model. *Journal of Theoretical Biology*, **13**, 63–71.

Eucken, A. (1946). Zur Kenutnis der Konstitution des Wassers. *Nachrichten von der Akademie der Wissenschaften in Göttingen* (Mathematisch-Physikalische Klasse). **1**, 38–48.

Everitt, C. T. & Haydon, D. A. (1969). Influence of diffusion layers during osmotic flow across bimolecular lipid membranes. *Journal of Theoretical Biology*, **22**, 9–19.

Everitt, C. T., Redwood, W. R. & Haydon, D. A. (1969). Problem of boundary layers in the exchange diffusion of water across bimolecular lipid membranes. *Journal of Theoretical Biology*, **22**, 20–32.

Evett, R. D., Higgins, J. A. & Brown, A. L. (1964). The fine structure of normal mucosa in human gall bladder. *Gastroenterology*, **47**, 49–60.

Falk, M. & Ford, T. A. (1966). Infra-red spectrum and structure of liquid water. *Canadian Journal of Chemistry*, **44**, 1699–1707.

Farmer, R. E. L. & Macey, R. I. (1970). Perturbation of red cell volume; rectification of osmotic flow. *Biochimica et Biophysica Acta*, **196**, 53–65.

Farquhar, M. G. & Palade, G. E. (1963). Junctional complexes in various epithelia. *Journal of Cell Biology*, **17**, 375–412.

Faxen, H. (1923). Die Bewegung einer starren Kugel längs der Achse eines mit zäher Flüssigkeit gefüllten Rohres. *Arkiv för Matematik, Astronomi Och Fysik*, **17**, No. 27, 1–28.

Fedyakin, N. N. (1962). The motion of liquids in microcapillaries. *Russian Journal of Physical Chemistry*, **36**, 776–780.

Feldherr, C. M. (1964). Binding within the nuclear annuli and its possible effect on nucleocytoplasmic exchanges. *Journal of Cell Biology*, **20**, 188–192.

Feldherr, C. M. (1969). A comparative study of nucleocytoplasmic interactions. *Journal of Cell Biology*, **42**, 841–845.

Fensom, D. S. (1966). Action potentials and associated water flows in living *Nitella*. *Canadian Journal of Botany*, **44**, 1432–1437.

Fensom, D. S. & Dainty, J. (1963). Electro-osmosis in *Nitella*. *Canadian Journal of Botany*, **41**, 685–691.

Fensom, D. S., Ursino, D. J. & Nelson, C. D. (1967). Determination of relative pore size in living membranes of *Nitella* using the techniques of electro-osmosis and radioactive tracers. *Canadian Journal of Botany*, **45**, 1267–1275.

Fensom, D. S. & Wanless, I. R. (1967). Further studies of electro-osmosis in *Nitella* in relation to pores in membranes. *Journal of Experimental Botany*, **18**, 563–577.

Fenstermacher, J. D. & Johnson, J. A. (1966). Filtration and reflection coefficients of the rabbit blood-brain barrier. *American Journal of Physiology*, **211**, 341–346.

Fernandez, H. L., Burton, P. R. & Samson, F. E. (1971). Axoplasmic transport in the crayfish nerve cord. The role of fibrillar constituents of neurons. *Journal of Cell Biology*, **51**, 176–192.

Fernández-Móran, H. (1959). Fine structure of biological lamellar systems. *Review of Modern Physics*, **31**, 319–330.

Fernández-Móran, H. (1962). Cell-membrane ultrastructure. Low-temperature electron microscopy and x-ray diffraction studies of lipoprotein components in lamellar systems. *Circulation*, **26**, 1039–1065.

Ferry, J. D. (1936). Statistical evaluation of sieve constants in ultrafiltration. *Journal of General Physiology*, **20**, 95–104.

Fettiplace, R., Haydon, D. A. & Knowles, C. D. (1972). The action of lysine vasopressin on artificial lipid layers. *Journal of Physiology*, **221**, 18–20P.

Fiat, D. & Connick, R. E. (1968). Oxygen-17 magnetic resonance studies of ion solvation. The hydration of Aluminium (III) and Gallium (III) ions. *Journal of the American Chemical Society*, **90**, 608–615.

Finean, J. B. (1957). The role of water in the structure of peripheral nerve myelin. *Journal of Biophysical and Biochemical Cytology*, **3**, 95–102.

Finean, J. B. (1969). Biophysical contributions to membrane structure. *Quarterly Reviews of Biophysics*, **2**, 1–23.

Finean, J. B., Coleman, R., Green, W. A. & Limbrick, A. R. (1966). Low-angle X-ray diffraction and electron microscope studies of isolated cell membranes. *Journal of Cell Science*, **1**, 287–296.

Finean, J. B. & Millington, P. F. (1957). Effects of ionic strength of immersion medium on the structure of peripheral nerve myelin. *Journal of Biophysical and Biochemical Cytology*, **3**, 89–94.

Fisher, R. B. (1955). The absorption of water and some small solute molecules from the isolated small intestine of the rat. *Journal of Physiology*, **130**, 655–664.

Forslind, E. (1952). A theory of water. *Acta Polytechnica*, **115**, 9–43.

Franck, J. & Mayer, J. E. (1947). An osmotic diffusion pump. *Archives of Biochemistry*, **14**, 297–313.

Frank, H. S. (1958). Covalency in the hydrogen bond and the properties of water and ice. *Proceedings of the Royal Society of London* A. **247**, 481–492.

Frank, H. S. & Evans, M. W. (1945). Free volume and entropy in condensed systems. III. Entropy in binary liquid mixtures: partial molal entropy in dilute solutions: structure and thermodynamics in aqueous electrolytes. *Journal of Chemical Physics*, **13**, 507–532.

Frank, H. S. & Quist, A. S. (1961). Pauling's model and the thermodynamic properties of water. *Journal of Chemical Physics*, **34**, 604–611.

Frank, H. S. & Wen, W. Y. (1957). Structural aspects of ion-solvent interaction in aqueous solutions; a suggested picture of water structure. *Discussions of the Faraday Society*, **24**, 133–140.

Franks, F. & Good, W. (1966). Mechanism of viscous flow of water. *Nature*, **210**, 85–86.

Franz, T. J., Galey, W. R. & Van Bruggen, J. T. (1968). Further observations on asymmetrical solute movement across membranes. *Journal of General Physiology*, **51**, 1–12.

Franz, T. J. & Van Bruggen, J. T. (1967). Hyperosmolarity and the net transport of nonelectrolytes in frog skin. *Journal of General Physiology*, **50**, 933–949.

Frazini-Armstrong, C. & Porter, K. R. (1964). Sarcolemmal invaginations constituting the T-system in fish muscle fibers. *Journal of Cell Biology*, **22**, 675–696.

Frederiksen, O. & Leyssac, P. P. (1969). Transcellular transport of isosmotic volumes by the rabbit gall bladder *in vitro*. *Journal of Physiology*, **201**, 201–224.

Freeman, A. R., Reuben, J. P., Brandt, P. W. & Grundfest, H. (1966). Osmometrically determined characteristics of the cell membrane of squid and lobster giant axons. *Journal of General Physiology*, **50**, 423–445.

Frey-Wyssling, A. (1946). Sur Wasserpermeabilität des Protoplasmas. *Experientia*, **2**, 132–137.

Frömter, E. (1972). The route of passive ion movement through the epithelium of *Necturus* gallbladder. *Journal of Membrane Biology*, **8**, 259–301.

Frömter, E. & Diamond, J. (1972). Route of passive ion permeation in epithelia. *Nature New Biology*, **235**, 9–13.

Fugelli, K. (1967). Regulation of cell volume in flounder (*Pleuronectes flesus*) erythrocytes accompanying a decrease in plasma osmolarity. *Comparative Biochemistry and Physiology*, **22**, 253–260.

Fullerton, P. M. & Parsons, D. S. (1956). The absorption of sugars and water from rat intestine *in vivo*. *Quarterly Journal of Experimental Physiology*, **41**, 387–397.

Gaffey, C. T. & Mullins, L. J. (1958). Ion fluxes during the action potential in *Chara*. *Journal of Physiology*, **144**, 502-524.

Gainer, H. (1968). Osmotically inactive volume of whole frog sartorius muscle; a reappraisal. *Bioscience*, **18**, 702–704.

Gainer, H. & Grundfest, H. (1968). Permeability of alkali metal cations in lobster muscle. A comparison of electrophysiological and osmometric analyses. *Journal of General Physiology*, **51**, 399–425.

Galey, W. R. & Van Bruggen, T. J. (1970). The coupling of solute fluxes in membranes. *Journal of General Physiology*, **55**, 220–242.

Gall, J. G. (1964). Electron microscopy of the nuclear envelope. *Protoplasmatologia. Handbuch der Protoplasmaforschung*, **5**, 4–25.

Ganote, C. E., Grantham, J. J., Moses, H. L., Burg, M. B. & Orloff, J. (1968). Ultrastructural studies of vasopressin effect on isolated perfused renal collecting tubules of the rabbit. *Journal of Cell Biology*, **36**, 355–367.

Garlick, D. G. (1970). Factors affecting the transport of extracellular molecules in skeletal muscle. In *Capillary Permeability. The Transfer of Molecules and Ions between Capillary Blood and Tissue* (ed. Crone, C. & Lassen, N. A.) Proceedings of the Alfred Benzon Symposium II, pp. 228–238. Copenhagen; Munksgaard.

Gary-Bobo, C. M., DiPolo, R. & Solomon, A. K. (1969). Role of hydrogen-bonding in nonelectrolyte diffusion through dense artificial membranes. *Journal of General Physiology*, **54**, 369–382.

Gary-Bobo, C. M. & Solomon, A. K. (1971). Effect of geometrical and chemical constraints on water flux across artificial membranes. *Journal of General Physiology*, **57**, 610–622.

Gelfan, S. (1928). The electrical conductivity of protoplasm. *Protoplasma*, **4**, 192–200.

George, J. H. B. & Courant, R. A. (1967). Conductance and water transfer in a leached cation-exchange membrane. *Journal of Physical Chemistry*, **71**, 246–249.

Gertz, K. H., Mangos, J. A., Braun, G. & Pagel, H. D. (1965). On the glomerular tubular balance in the rat kidney. *Pflügers Archiv für die gesamte Physiologie des Menschen und der Tiere*, **285**, 360–372.

Giebisch, G., Klose, R. M., Malnic, G., Sullivan, W. J. & Windhager, E. E. (1964). Sodium movement across single perfused proximal tubules of rat kidneys. *Journal of General Physiology*, **47**, 1175–1194.

Giebisch, G. & Windhager, E. E. (1964). Renal tubular transfer of sodium, chloride and potassium. *American Journal of Medicine*, **36**, 643–669.

Gilman, A. & Cowgill, G. R. (1933). Osmotic relations between blood and body fluids. IV. Pancreatic juice, bile, and lymph. *American Journal of Physiology*, **104**, 467–479.

Ginetzinsky, A. G. (1958). Role of hyaluronidase in the reabsorption of water in renal tubules; the mechanism of action of the antidiuretic hormone. *Nature*, **182**, 1218–1219.

Ginzburg, B. Z. & Katchalsky, A. (1963). The frictional coefficients of the flows of non-electrolytes through artificial membranes. *Journal of General Physiology*, **47**, 403–418.

Ginzburg, H. (1971). Model for iso-osmotic water flow in plant roots. *Journal of Theoretical Biology*, **32**, 147–158.

Ginzburg, H. & Ginzburg, B. Z. (1970a). Radial water and solute flows in roots of *Zea mays*. I. Water flow. *Journal of Experimental Botany*, **21**, 580–592.

Ginzburg, H. & Ginzburg, B. Z. (1970b). Radial water and solute flows in roots of *Zea mays*. II. Ion fluxes across root cortex. *Journal of Experimental Botany*, **21**, 593–604.

Ginzburg, H. & Ginzburg, B. Z. (1971). Evidence for active water transport in a corn root preparation. *Journal of Membrane Biology*, **4**, 29–41.

Glasstone, S., Laidler, K. J. & Eyring, H. (1941). *The Theory of Rate Processes*, 1st edn., New York: McGraw-Hill.

Goldstein, D. A. & Solomon, A. K. (1960). Determination of equivalent pore radius for human red cells by osmotic pressure measurement. *Journal of General Physiology*, **44**, 1–17.

Gordon, J. D. M. (1969). The forces involved in regulating the uptake of water into the blastocoel and archenteron of *Xenopus laevis* embryos. Ph.D. Thesis, University of Edinburgh.

Gortner, R. A. (1932). The role of water in the structure and properties of protoplasm. *Annual Review of Biochemistry*, **1**, 21–54.

Gosselin, R. E. (1967). Kinetics of pinocytosis. *Federation Proceedings*, **26**, 987–993.

Gränicher, H., Jaccard, C., Scherrer, P. & Steinemann, A. (1957). Dielectric relaxation and the electrical conductivity of ice crystals. *Discussions of the Faraday Society*, **23**, 50–62.

Grant, E. H. (1957). Relationships between relaxation time and viscosity of water. *Journal of Chemical Physics.*, **26**, 1575–1577.

Grantham, J. J. (1971. Mode of water transport in mammalian renal collecting tubules. *Federation Proceedings*, **30**, 14–21.

Grantham, J. J. & Burg, M. B. (1966). Effect of vasopressin and cyclic AMP on permeability of isolated collecting tubules. *American Journal of Physiology*, **211**, 255–259.

Grantham, J. J., Cuppage, F. E. & Fanestil, D. (1971). Direct observation of toad bladder response to vasopressin. *Journal of Cell Biology*, **48**, 695–699.

Grantham, J. J., Ganote, C. E., Burg, M. B. & Orloff, J. (1969). Paths of transtubular water flow in isolated renal collecting tubules. *Journal of Cell Biology*, **41**, 562–576.

Grantham, J. J. & Orloff, J. (1968). Effect of prostaglandin E1 on the permeability response of the isolated collecting tubule to vasopressin, adenosine 3′,5′-monophosphate, and theophylline. *Journal of Clinical Investigation*, **47**, 1154–1161.

Gray, J. (1932). The osmotic properties of the eggs of the trout (*Salmo fario*). *Journal of Experimental Biology*, **9**, 277–299.

Green, K. & Green, M. A. (1969). Permeability to water of rabbit corneal membranes. *American Journal of Physiology*, **217**, 635–641.

Green, K. & Otori, T. (1970). Direct measurements of membrane unstirred layers. *Journal of Physiology*, **207**, 93–102.

Green, K. & Pederson, J. E. (1972*a*). Contribution of secretion and filtration to aqueous humor formation. *American Journal of Physiology*, **222**, 1218–1226.

Green, K. & Pederson, J. E. (1972*b*). Effect of temperature on water flow through the ciliary epithelium. *American Journal of Physiology*, **222**, 1227–1229.

Green, K., Seshadri, B. & Matty, A. J. (1962). Independence of transfer of solute and solvent across rat ileum. *Nature*, **196**, 1322–1323.

Green, P. B. & Stanton, F. W. (1967). Turgor pressure; direct manometric measurement in single cells of *Nitella*. *Science*, **155**, 1675–1676.

Grieve, D. W. (1963). The mechanical limitation of swelling in frog sartorious muscle. *Journal of Physiology*, **165**, 71–72P.

Grigera, J. R. & Cereijido, M. (1971). The state of water in the outer barrier of the isolated frog skin. *Journal of Membrane Biology*, **4**, 148–155.

Grim, E. (1953). Relation between pressure and concentration differences across membranes permeable to solute and solvent. *Proceedings of the Society for Experimental Biology and Medicine*, **83**, 195–200.

Grim, E. (1962). Water and electrolyte flux rates in the duodenum, jejunum, ileum and colon and effects of osmolarity. *American Journal of Digestive Diseases*, **7**, 17–27.

Grim, E. (1963). A mechanism for absorption of sodium chloride solutions from canine gall bladder. *American Journal of Physiology*, **205**, 247–254.

Grim, E. & Smith, G. A. (1957). Water flux rates across dog gallbladder wall. *American Journal of Physiology*, **191**, 555–560.

Gross, E. L. & Packer, L. (1967). Ion transport and conformational changes in spinach chloroplast grana. 1. Osmotic properties and divalent cation-induced volume changes. *Archives of Biochemistry and Biophysics*, **121**, 779–789.

Grotte, G. (1956). Passage of dextran molecules across the blood-lymph barrier. *Acta Chirurgica Scandinavica*, Supplement 211, 1–83.

Grotthuss, C. J. T. (1806). Sur la décomposition de l'eau et des corps qu'elle tient en dissolution à l'aide de l'électricité galvanique. *Annales de Chimie*, **58**, 54–74.

Grundfest, H., Kao, C. Y. & Altamirano, M. (1954). Bioelectric effects of ions microinjected into the giant axons of *Loligo*. *Journal of General Physiology*, **38**, 245–282.

Guest, G. M. (1948). Osmometric behaviour of normal and abnormal human erythrocytes. *Blood*, **3**, 541–555.

Guest, G. M. & Wing, M. (1942). Osmometric behavior of normal human erythrocytes. *Journal of Clinical Investigation*, **21**, 257–262.

Gutknecht, J. (1967). Membranes of *Valonia ventricosa*: Apparent absence of water-filled pores. *Science*, **158**, 787–788.

Guyton, A. C., Granger, H. J. & Taylor, A. E. (1971). Interstitial fluid pressure. *Physiological Reviews*, **51**, 527–563.

Haas, C. (1962). On diffusion, relaxation and defects in ice. *Physical Review Letters*, **3**, 126–128.

Haggis, G. H., Hasted, J. B. & Buchanan, T. J. (1952). The dielectric properties of water in solutions. *Journal of Chemical Physics*, **20**, 1452–1465.

Hakim, A., Lester, R. G. & Lifson, N. (1963). Absorption by an *in vitro* preparation of dog intestinal mucosa. *Journal of Applied Physiology*, **18**, 409–413.

Hakim, A. A. & Lifson, N. (1964). Urea transport across dog intestinal mucosa *in vitro*. *American Journal of Physiology*, **206**, 1315–1320.

REFERENCES

Hakim, A. A. & Lifson, N. (1969). Effects of pressure on water and solute transport by dog intestinal mucosa *in vitro*. *American Journal of Physiology*, **216**, 276–284.

Hamburger, J. (1904). *Osmotischer Druck und Ionenlehre in dem medicinischen Wissenschaften, Band III*. Wiesbaden; Bergmann.

Hanai, T., Haydon, D. A. & Taylor, J. (1965). Some further experiments on bimolecular lipid membranes. *Journal of General Physiology*, **48** (No. 5, part 2), 59–63.

Harding, C. V. & Feldherr, C. (1959). Semipermeability of the nuclear membrane in the intact cell. *Journal of General Physiology*, **42**, 1155–1165.

Harris, E. J. (1960). *Transport and Accumulation in Biological Systems*, 2nd edn., pp. 39–42. London; Butterworth.

Harris, E. J. (1963). Distribution and movement of muscle chloride. *Journal of Physiology*, **166**, 87–109.

Harris, E. J. & Van Dam, K. (1968). Changes of total water and sucrose space accompanying induced ion uptake or phosphate swelling of rat liver mitochondria. *Biochemical Journal*, **106**, 759–766.

Harris, F. E. & O'Konski, C. T. (1957). Dielectric properties of aqueous ionic solutions at microwave frequencies. *Journal of Physical Chemistry*, **61**, 310–319.

Harrison, R. G. & Malpas, P. (1953). The volume of human amniotic fluid. *Journal of Obstetrics and Gynaecology of the British Commonwealth*, **60**, 632–639.

Harsch, M. & Green, J. W. (1963). Electrolyte analyses of chick embryonic fluids and heart tissues. *Journal of Cellular and Comparative Physiology*, **62**, 319–326.

Hartley, G. S. (1948). Contribution to a discussion on the 'asymmetrical behaviour of insect cuticle in relation to water permeability' by H. Hurst. *Discussions of the Faraday Society*, **3**, 223.

Hartley, G. S. & Crank, J. (1949). Some fundamental definitions and concepts in diffusion processes. *Transactions of the Faraday Society*, **45**, 801–818.

Hays, R. M. (1968). A new proposal for the action of vasopressin, based on studies of a complex synthetic membrane. *Journal of General Physiology*, **15**, 385–398.

Hays, R. M. (1971). The movement of water across epithelial cells. *Proceedings of the International Union of Physiological Sciences*, Volume VIII, International Congress, Munich, pp. 271–272.

Hays, R. M. (1972a). Independent pathways for water and solute movement across the cell membrane. *Journal of Membrane Biology*, **10**, 367–371.

Hays, R. M. (1972b). The movement of water across vasopressin-sensitive epithelia. In *Current Topics in Membranes and Transport* (eds Bronner, F. & Kleinzeller, A.) pp. 339–366. New York: Academic Press.

Hays, R. M. & Franki, N. (1970). The role of water diffusion in the action of vasopressin. *Journal of Membrane Biology*, **2**, 263–276.

Hays, R. M., Franki, N. & Soberman, R. (1971). Activation energy for water diffusion across the toad bladder; evidence against the pore enlargement hypothesis. *Journal of Clinical Investigation*, **50**, 1016–1018.

Hays, R. M. & Leaf, A. (1962a). Studies on the movement of water through the isolated toad bladder and its modification by vasopressin. *Journal of General Physiology*, **45**, 905–919.

Hays, R. M. & Leaf, A. (1962b). The state of water in the isolated toad bladder in the presence and absence of vasopressin. *Journal of General Physiology*, **45**, 933–948.

Hayward, A. F. (1962a). Aspects of the fine structure of the gall bladder epithelium of the mouse. *Journal of Anatomy*, **96**, 227–236.

Hayward, A. F. (1962b). Electron microscopic observations on absorption in the epithelium of the guinea pig gall bladder. *Zeitschrift für Zellforschung*, **56**, 197–202.

Hazlewood, C. F., Nichols, B. L. & Chamberlain, N. F. (1969). Evidence for the existence of a minimum of two phases of ordered water in skeletal muscle. *Nature*, **222**, 747–750.

Hearst, J. E. & Vinograd, J. (1961). The net hydration of T-4 bacteriophage deoxyribonucleic acid and the effect of hydration on buoyant behaviour in a density gradient at equilibrium in the ultracentrifuge. *Proceedings of the National Academy of Sciences, U.S.A.*, **47**, 1005–1014.

Heinz, E. & Patlak, C. S. (1960). Energy expenditure by active transport mechanisms. *Biochimica et Biophysica Acta*, **44**, 324–334.

Heisey, S. R., Held, D. & Pappenheimer, J. R. (1962). Bulk flow and diffusion in the cerebrospinal fluid system of the goat. *American Journal of Physiology*, **203**, 775–781.

Hempling, H. G. (1960). Permeability of the Ehrlich ascites tumor cell to water. *Journal of General Physiology*, **44**, 365–379.

Hempling, H. G. (1967). Application of irreversible thermodynamics to a functional description of the tumor cell membrane. *Journal of Cellular Physiology*, **70**, 237–256.

Hendry, E. B. (1954). The osmotic properties of the normal human erythrocyte. *Edinburgh Medical Journal*, **61**, 7–24.

Henniker, J. C. (1949). The depth of the surface zone of a liquid. *Review of Modern Physics*, **21**, 322–341.

Henrikson, R. C. (1970). Ruminal epithelium; transport of solutes in the absence of tight junctions (*zonulae occludentes*). *Journal of Anatomy*, **106**, 199.

Henrikson, R. C. & Stacy, B. D. (1971). The barrier to diffusion across ruminal epithelium; A study by electron microscopy using horseradish peroxidase, lanthanum and ferritin. *Journal of Ultrastructure Research*, **34**, 72–82.

Hevesy, G., Höfer, E. & Krogh, A. (1935). The permeability of the skin of frogs to water as determined by D_2O and H_2O. *Skandinavisches Archiv für Physiologie*, **72**, 199–214.

Hill, A. V. (1930). The state of water in muscle and blood and the osmotic behaviour of muscle. *Proceedings of the Royal Society of London* B, **106**, 477–505.

Hill, D. K. (1950a). The effect of stimulation on the opacity of a crustacean nerve trunk and its relation to fibre diameter. *Journal of Physiology*, **111**, 283–303.

Hill, D. K. (1950b). The volume change resulting from stimulation of a giant nerve fibre. *Journal of Physiology*, **111**, 304–327.

Hingson, D. J. & Diamond, J. M. (1972). Comparison of non-electrolyte permeability patterns in several epithelia. *Journal of Membrane Biology*, **10**, 93–135.

Hirsch, H. R. (1967). Relevance of the single-file model to water flow through porous cell membranes. *Currents in Modern Biology*, **1**, 139–142.

Höber, R. (1945). *Physical Chemistry of Cells and Tissues.* 1st edn. Philadelphia; Blakiston Company.

Höber, R. & Höber, J. (1937). Experiments on the absorption of organic solutes in the small intestine of rats. *Journal of Cellular and Comparative Physiology*, **10**, 401–422.

Höber, R. & Ørskov, S. L. (1933). Untersuchengen über die Permeiergeschwindigkeit von Anelektrolyten bei den roten Blutkörperchen verscheidener Tierarten. *Pflügers Archiv für die gesamte Physiologie des Menschen und der Tiere*, **231**, 599–615.

Hodgkin, A. L. (1951). The ionic basis of electrical activity in nerve and muscle. *Biological Reviews*, **26**, 339–409.

Hodgkin, A. L. & Horowicz, P. (1959). The influence of potassium and chloride ions on the membrane potential of single muscle fibres. *Journal of Physiology*, **148**, 127–160.

Hodgkin, A. L. & Katz, B. (1949). The effect of sodium ions on the electrical activity of the giant axon of the squid. *Journal of Physiology*, **108**, 37–77.

Hodgkin, A. L. & Keynes, R. D. (1955). The potassium permeability of a giant nerve fibre. *Journal of Physiology*, **128**, 61–68.

Hodson, S. (1968). Inadequacy of aldehyde fixatives in preserving the ultrastructure of corneal endothelium in rabbit and monkey. *Experimental Eye Research*, **7**, 221–224.

Hogben, C. A. M. (1960). Movement of material across cell membranes. *Physiologist*, **3**, 56–62.

Hokin, M. R. (1967). The Na^+, K^+ and Cl^- content of goose salt gland slices and the effects of acetycholine and ouabain. *Journal of General Physiology*, **50**, 2197–2209.

Holtfreter, J. (1943). Properties and function of the surface coat in amphibian embryos. *Journal of Experimental Zoology*, **93**, 251–323.

Holtfreter, J. (1954). Observations on the physico-chemical properties of isolated nuclei. *Experimental Cell Research*, **7**, 95–102.

Holm-Jensen, I., Krogh, A. & Wartiovaara, V. (1944). Some experiments on the exchange of potassium and sodium between single cells of Characeae and the bathing fluid. *Acta Botanica Fennica*, **36**, 1–22.

Holz, R. & Finkelstein, A. (1970). The water and nonelectrolyte permeability induced in thin lipid membranes by the polyene antibiotics nystatin and amphotericin B. *Journal of General Physiology*, **56**, 125–145.

Hoover, S. T. & Mellon, E. F. (1950). Application of polarization theory to sorption of water vapour by high polymers. *Journal of the American Chemical Society*, **72**, 2562–2566.

Hope, A. B. & Findlay, G. P. (1964). The action potential in *Chara*. *Plant and Cell Physiology*, **5**, 377–379.

Hopkins, A. L. (1960). A method for estimating changes in bound water associated with myocardial contraction. *Biochimica et Biophysica Acta*, **37**, 148–149.

Hopkins, D. A. (1946). The contractile vacuole and the adjustment to changing concentration in freshwater amoebae. *Biological Bulletin of the Marine Biological Laboratory, Woods Hole, U.S.A.* **90**, 158–176.

Hoshi, T. & Sakai, F. (1967). A comparison of the electrical resistances of the surface cell membrane and cellular wall in the proximal tubule of the newt kidney. *Japanese Journal of Physiology*, **17**, 627–637.

Hoshiko, T. & Lindley, B. D. (1964). The relationship of Ussing's flux-ratio equation to the thermodynamic description of membrane permeability. *Biochimica et Biophysica Acta*, **79**, 301–317.

Hoshiko, T. & Lindley, B. D. (1967). Phenomenological description of active transport of salt and water. *Journal of General Physiology*, **50**, 729–758.

Hoshiko, T., Lindley, B. D. & Edwards, C. (1964). Diffusion delay in frog skin connective tissue; A source of error in tracer investigations. *Nature*, **201**, 932–933.

House, C. R. (1964a). The nature of water transport across frog skin. *Biophysical Journal*, **4**, 401–416.

House, C. R. (1964b). Effect of antidiuretic hormone on non-osmotic flow across frog skin. *Nature*, **202**, 1221–1222.

House, C. R. (1965). Rectification of water flow across frog skin. *Biophysical Journal*, **5**, 987–988.

House, C. R. (1968). A discussion of some factors relevant to the study of water transport across frog skin. *Archivio di Scienze Biologiche*, **52**, 209–215.

House, C. R. & Findlay, N. (1966a). Water transport in isolated maize roots. *Journal of Experimental Botany*, **17**, 344–354.

House, C. R. & Findlay, N. (1966b). Mechanism of fluid exudation from isolated maize roots. *Nature*, **211**, 649–650.

House, C. R. & Findlay, N. (1966c). Analysis of transient changes in fluid exudation from isolated maize roots. *Journal of Experimental Botany*, **17**, 627–640.

House, C. R. & Green, K. (1965). Ion and water transport in isolated intestine of the marine teleost, *Cottus scorpius*. *Journal of Experimental Biology*, **42**, 177–189.

Howard, E. (1957). Ontogenetic changes in the freezing point and sodium and potassium content of the subgerminal fluid and blood plasma of the chick embryo. *Journal of Cellular and Comparative Physiology*, **50**, 451–470.

Huang, C. & Thompson, T. E. (1966). Properties of lipid bilayer membranes separating two aqueous phases; Water permeability. *Journal of Molecular Biology*, **15**, 539–554.

Huang, C., Wheeldon, L. & Thompson, T. E. (1964). The properties of lipid bilayer membranes separating two aqueous phases; Formation of a membrane of simple composition. *Journal of Molecular Biology*, **8**, 148–160.

Huettner, A. F. (1941). *Fundamentals of Comparative Embryology of the Vertebrates*, New York; Macmillan.

Huf, E. G. (1955). Ion transport and ion exchange in frog skin. In *Electrolytes in Biological Systems* (ed. Shanes, A. M.). pp. 205–238. Washington, D.C.; American Physiological Society.

Huf, E. G., Parrish, J. & Weatherford, C. (1951). Active salt and water uptake by isolated frog skin. *American Journal of Physiology*, **164**, 137–142.

Huggert, A. & Odeblad, E. (1959a). Studies on the water of the crystalline lens. III. Proton magnetic resonance studies. *Acta Ophthalmologica*, **37**, 93–102.

Huggert, A. & Odeblad, E. (1959b). Proton magnetic resonance studies of some tissues and fluids of the eye. *Acta Radiologica*, **51**, 385–392.

Hunter, A. S. & Hunter, F. R. (1961). Studies of volume changes in the isolated amphibian germinal vesicle. *Experimental Cell Research*, **22**, 609–618.

Hunter, F. R. & De Luque, O. (1959). Osmotic studies of amphibian eggs. II. Ovarian eggs. *Biological Bulletin of the Marine Biological Laboratory, Woods Hole, U.S.A.*, **117**, 468–481.

Huxley, H. E. (1964). Evidence for continuity between the central elements of the triads and extracellular space in frog sartorius muscle. *Nature*, **202**, 1067–1071.

Hyman, C. (1944). Filtration across the vascular wall as a function of several physical factors. *American Journal of Physiology*, **142**, 671–685.

Ibsen, H. L. (1928). Prenatal growth in guinea-pigs with special reference to environmental factors affecting weight at birth. *Journal of Experimental Zoology*, **51**, 51–94.

Ingraham, R. C., Peters, H. C. & Visscher, M. B. (1938). On the movement of materials across living membranes against concentration gradients. *Journal of Physical Chemistry*, **42**, 141–150.

Itagaki, K. (1964). Self-diffusion in single crystals of ice. *Journal of the Physical Society of Japan*, **19**, 1081.

Itoh, M., Izawa, S. & Shibata, K. (1963). Disintegration of chloroplasts with dodecylbenzone sulfonate as measured by flattening effect and size distribution. *Biochimica et Biophysica Acta*, **69**, 130–142.

Ivanova, L. N. & Natochin, Yu. V. (1968). Ugnetenie kal'tsiem deĭstviia antidiureticheskogo gormona i gialuronidazy na osmoticheskuin pronitsaemost' mochevogo puzyria liagushki, *Doklady Akademie Nauk SSSR*, **178**, 489–491. (In Russian.)

Jaccard, C. (1972). Transport properties of ice. In *Water and Aqueous Solutions* (ed. Horne, R. E.) pp. 25–64. New York; Wiley-Interscience.

Jacobs, M. H. (1932). Osmotic properties of the erythrocyte. III. The applicability of osmotic laws to the rate of hemolysis in hypotonic solutions of non-electrolytes. *Biological Bulletin of the Marine Biological Laboratory, Woods Hole, U.S.A.*, **62**, 178–194.

Jacobs, M. H. (1933). The simultaneous measurement of cell permeability to water and to dissolved substances. *Journal of Cellular and Comparative Physiology*, **2**, 427–444.

Jacobs, M. H. (1935). Diffusion processes. *Ergebnisse der Biologie*, **12**, 1–160.

Jacobs, M. H. (1952). The measurement of cell permeability with particular reference to the erythrocyte. In *Modern Trends in Physiology and Biochemistry* (ed. Barron, E. S. G.), pp. 149–171. New York; Academic Press.

Jacobs, M. H., Glassman, H. N. & Parpart, A. K. (1935). Osmotic properties of the erythrocyte. VII. The temperature coefficient of certain hemolytic processes. *Journal of Cellular and Comparative Physiology*, **7**, 197–225.

Jacobson, B. (1953). Hydration structure of deoxyribonucleic acid and its physico-chemical properties. *Nature*, **172**, 666–667.

Johnson, F. R., McMinn, R. M. H. & Birchenough, R. F. (1962). The ultrastructure of the gall-bladder epithelium of the dog. *Journal of Anatomy*, **96**, 447–487.

Johnson, J. A., Cavert, H. M. & Lifson, N. (1953). Kinetics concerned with the distribution of isotopic water in isolated perfused dog heart and skeletal muscle. *American Journal of Physiology*, **171**, 687–693.

Johnson, J. A., Cavert, H. M., Lifson, N. & Visscher, M. B. (1951). Permeability of the bladder to water studied by means of isotopes. *American Journal of Physiology*, **165**, 87–92.

Johnson, J. A. & Wilson, T. A. (1967). Osmotic volume changes induced by a permeable solute. *Journal of Theoretical Biology*, **17**, 304–311.

Jordan, D. O. (1960). *The Chemistry of Nucleic Acids*, 1st edn., p. 186. London; Butterworth.

Kalman, S. M. (1959). Sodium and water exchange in the trout egg. *Journal of Cellular and Comparative Physiology*, **54**, 155–162.

Kamada, T. (1936). Distribution of water osmotically absorbed into the egg of an annelid, *Ceratrocephale osawai*. *Journal of the Faculty of Science, Tokyo University. Sec. IV*, **4**, 215–219.

Kamb, B. (1965). Structure of ice VI. *Science*, **150**, 205–209.

Kamb, B. (1972). Structure of the ices. In *Water and Aqueous Solutions* (ed. Horne, R. E.), pp. 9–24. New York; Wiley-Interscience.

Kamiya, N. & Tazawa, M. (1956). Studies on water permeability of a single plant cell by means of transcellular osmosis. *Protoplasma*, **46**, 394–422.

Karlsson, J-O. & Sjöstrand, J. (1968). Transport of labelled proteins in the optic nerve and tract of the rabbit. *Brain Research*, **11**, 431–439.

Karnovsky, M. J. (1967). The ultrastructural basis of capillary permeability studied with peroxidase as a tracer. *Journal of Cell Biology*, **35**, 213–236.

Karnovsky, M. J. (1970). Morphology of capillaries with special reference to muscle capillaries. In *Capillary Permeability. The Transfer of Molecules and Ions Between Capillary Blood and Tissue*. Proceedings of the Alfred Benzon Symposium II, pp. 341–350. Copenhagen; Munksgaard.

Karnovsky, M. J. (1971). Ultrastructural basis of capillary permeability. *Proceedings of the International Union of Physiological Sciences*. Volume VIII. International Congress, Munich. pp. 267–268.

Katchalsky, A. (1961). Membrane permeability and the thermodynamics of irreversible process. In *Membrane Transport and Metabolism* (ed. Kleinzeller, A. & Kotyk, A.). pp. 69–86. London: Academic Press.

Katchalsky, A. & Curran, P. F. (1965). *Nonequilibrium Thermodynamics in Biophysics*. 1st edn., Cambridge, Mass.; Harvard University Press.

Katchalsky, A. & Kedem, O. (1962). Thermodynamics of flow processes in biological systems. *Biophysical Journal*, **2**, 53–78.

Kauzmann, W. (1959). Some factors in the interpretation of protein denaturation. *Advances in Protein Chemistry*, **14**, 1–57.

Kavanau, J. L. (1964). *Water and Solute-Water Interactions*. 1st edn. San Francisco; Holden-Day.

Kawabe, H., Jacobson, H., Miller, I. F. & Gregor, H. P. (1966). Functional properties of cation-exchange membranes as related to their structures. *Journal of Colloid and Interface Science*, **21**, 79–93.

Kaye, G. I., Wheeler, H. O., Whitlock, R. T. & Lane, N. (1966). Fluid transport in the rabbit gall bladder. A combined physiological and electron microscopic study. *Journal of Cell Biology*, **30**, 237–268.

Kedem, O. (1961). Criteria of active transport. In *Membrane Transport and Metabolism* (ed. Kleinzeller, A. & Kotyk, A.) pp. 87–93, London; Academic Press.

Kedem, O. & Essig, A. (1965). Isotope flows and flux ratios in biological membranes. *Journal of General Physiology*, **48**, 1047–1070.

Kedem, O. & Katchalsky, A. (1958). Thermodynamic analysis of the permeability of biological membranes to non-electrolytes. *Biochimica et Biophysica Acta*, **27**, 229–246.

Kedem, O. & Katchalsky, A. (1961). A physical interpretation of the phenomenological coefficients of membrane permeability. *Journal of General Physiology*, **45**, 143–179.

Kedem, O. & Katchalsky, A. (1963a). Permeability of composite membranes. Part 1. Electric current, volume flow and flow of solute through membranes. *Transactions of the Faraday Society*, **59**, 1918–1930.

Kedem, O. & Katchalsky, A. (1963b). Permeability of composite membranes. Part 2. Parallel elements. *Transactions of the Faraday Society*, **59**, 1931–1940.

Kedem, O. & Katchalsky, A. (1963c). Permeability of composite membranes. Part 3. Series array of elements. *Transactions of the Faraday Society*, **59**, 1941–1953.

Kelly, R. B., Kohn, P. G. & Dainty, J. (1963). Water relations of *Nitella translucens*. *Transactions of the Botanical Society of Edinburgh*, **39**, 373–391.

Keynes, R. D. & Lewis, P. R. (1951). The sodium and potassium content of cephalopod nerve fibres. *Journal of Physiology*, **114**, 151–182.

King, V. (1969). A study of the mechanism of water transfer across frog skin by a comparison of the permeability of the skin to deuterated and tritiated water. *Journal of Physiology*, **200**, 529–538.

Kirschner, L. B., Maxwell, R. & Fleming, D. (1960). Non-osmotic water movement across isolated frog skin. *Journal of Cellular and Comparative Physiology*, **55**, 267–273.

Kishimoto, U. & Ohkawa, T. (1966). Shortening of *Nitella* internode during excitation. *Plant and Cell Physiology*, **7**, 493–497.

Kitching, J. A. (1934). The physiology of contractile vacuoles. I. Osmotic relations. *Journal of Experimental Biology*, **11**, 364–381.

Kitching, J. A. (1936). The physiology of contractile vacuoles. II. The control of body volume in marine Peritricha. *Journal of Experimental Biology*, **13**, 11–27.

Kitching, J. A. (1938). The physiology of contractile vacuoles. III. The water balance of fresh-water Peritricha. *Journal of Experimental Biology*, **15**, 143–151.

Kitching, J. A. (1952). Contractile vacuoles. *Symposia of the Society for Experimental Biology*, **6**, 145–146.

Kitching, J. A. (1954). Osmoregulation and ionic regulation in animals without kidneys. *Symposia of the Society for Experimental Biology*, **8**, 63–75.

Klotz, I. M. (1970). Water; its fitness as a molecular environment. In *Membranes and Ion Transport*. Vol. 1 (ed. Bittar, E. E.), pp. 93–122. New York; Wiley-Interscience.

Kobatake, Y. & Fujita, H. (1964a). Flows through charged membranes. I. Flip-flop current vs. voltage relation. *Journal of Chemical Physics*, **40**, 2212–2218.

Kobatake, Y. & Fujita, H. (1964b). Flows through charged membranes. II. Oscillation phenomena. *Journal of Chemical Physics*, **40**, 2219–2222.

Koefoed-Johnsen, V. & Ussing, H. H. (1953). The contributions of diffusion and flow to the passage of D_2O through living membranes. *Acta Physiologica Scandinavica*, **28**, 60–76.

Kohn, P. G. (1965). Tables of some physical and chemical properties of water. *Symposia of the Society for Experimental Biology*, **19**, 3–16.

Kokko, J. P. (1970). Sodium chloride and water in the descending limb of Henle. *Journal of Clinical Investigation*, **49**, 1838–1846.

Kregenow, F. M. (1971a). The response of duck erythrocytes to non-hemolytic hypotonic media. Evidence for a volume-controlling mechanism. *Journal of General Physiology*, **58**, 372–395.

Kregenow, F. M. (1971b). The response of duck erythrocytes to hypertonic media. Further evidence for a volume controlling mechanism. *Journal of General Physiology*, **58**, 396–412.

Krogh, A., Schmidt-Nielsen, K. & Zeuthen, E. (1939). Osmotic behaviour of frogs eggs and young tadpoles. *Zeitschrift für vergleichende Physiologie*, **26**, 230–238.

Krogh, A. & Ussing, H. H. (1937). A note on the permeability of trout eggs to D_2O and H_2O. *Journal of Experimental Biology*, **14**, 35–37.

Kuhn, W. (1951). Grenze der Durchlässigkeit von Filtrier- und Löslichkeitsmembranen. *Zeitschrift für Elektrochemie*, **55**, 207–217.

Kuhn, W. & Thürkauf, M. (1958). Isotopentrennung beim Gefrieren von Wasser und Diffusionkonstanten von D und ^{18}O im Eis. *Helvetica Chimica Acta*, **41**, 938–971.

Kushmerick, M. J. & Podolsky, R. J. (1969). Ionic mobility in muscle cells. *Science*, **166**, 1297–1298.

Ladbrooke, B. D., Jenkinson, T. K., Kamat, V. B. & Chapman, D. (1968). Physical studies on myelin. I. Thermal-analysis. *Biochimica et Biophysica Acta*, **164**, 101–109.

Lagos, A. E. & Kitchener, J. A. (1960). Diffusion in polystyrenesulphonic acid ion-exchange resins. *Transactions of the Faraday Society*, **56**, 1245.

Lahlou, B. & Fossat, B. (1971). Mécanisme du transport de l'eau et du sel à travers la vessie urinaire d'un poisson téléostéen en eau douce, la truite arc-en-ciel. *Compte rendu de l'Academie des Sciences, Paris*, **273**, 2108–2110.

Lakshminarayanaiah, N. (1967). Permeation of water through cation exchange membranes. *Biophysical Journal*, **7**, 511–526.

Landis, E. M. (1927). Micro-injection studies of capillary permeability. II. The relation between capillary pressure and the rate at which fluid passes through the walls of single capillaries. *American Journal of Physiology*, **82**, 217–238.

Landis, E. M. (1928). Micro-injection studies of capillary permeability. III. The effect of lack of oxygen on the permeability of the capillary wall to fluid and to the plasma proteins. *American Journal of Physiology*, **83**, 528–542.

Landis, E. M. & Gibbon, J. H. (1933). The effects of temperature and tissue pressure on the movement of fluid through the human capillary wall. *Journal of Clinical Investigation*, **12**, 105–138.

REFERENCES

Landis, E. M. & Pappenheimer, J. R. (1963). Exchange of substances through the capillary walls. *Handbook of Physiology (Circulation),* Chapter 29, pp. 961–1034. Washington D.C.; American Physiological Society.

Landis, E. M. & Sage, L. E. (1971). Fluid movement rates through walls of single capillaries exposed to hypertonic solutions. *American Journal of Physiology,* **221,** 520–534.

Lang, M. A. & Gainer, H. (1969). Volume control by muscle fibers of the blue crab. Volume readjustment in hypotonic salines. *Journal of General Physiology,* **53,** 323–341.

Lasek, R. (1968). Axoplasmic flow in cat dorsal root ganglion cells as studied with [^{3}H]-L-leucine. *Brain Research,* **7,** 360–377.

Lassiter, W. E., Mylle, M. & Gottschalk, C. W. (1964). Net transtubular movement of water and urea in saline diuresis. *American Journal of Physiology,* **206,** 669–673.

Lea, E. J. A. (1963). Permeation through long narrow pores. *Journal of Theoretical Biology,* **5,** 102–107.

Leaf, A. (1956). On the mechanism of fluid exchange of tissues *in vitro. Biochemical Journal,* **62,** 241–248.

Leaf, A. (1960). Some actions of neurohypophyseal hormones on a living membrane. *Journal of General Physiology,* **43,** 175–189.

Leaf, A., Anderson, J. & Page, L. B. (1962). Active sodium transport by the isolated toad bladder. *Journal of General Physiology,* **41,** 657–668.

Leaf, A. & Hays, R. M. (1962). Permeability of the isolated toad bladder and its modification by vasopressin. *Journal of General Physiology,* **45,** 921–932.

Leaf, A., Page, L. B. & Anderson, J. (1959). Respiration and active sodium transport of isolated toad bladder. *Journal of Biological Chemistry,* **234,** 1625–1629.

Lee, J. S. (1963). Role of mesenteric lymphatic system in water absorption from rat intestine *in vitro. American Journal of Physiology,* **204,** 92–96.

Lee, J. S. (1968). Isosmotic absorption of fluid from rat jejunum *in vitro. Gastroenterology,* **54,** 366–374.

LeFevre, P. G. (1964). The osmotically functional water content of the human erythrocyte. *Journal of General Physiology,* **47,** 585–603.

Lehninger, A. L. (1964). *The Mitochondrion.* New York; Benjamin.

Levich, V. G. (1962). *Physicochemical Hydrodynamics,* 1st edn. Englewood Cliffs; Prentice-Hall.

Levitt, J. (1966). Winter hardiness in plants. In *Cryobiology* (ed. Meryman, H. T.). pp. 495–563. New York; Academic Press.

Lewis, W. H. (1931). Pinocytosis. *Bulletin of the Johns Hopkins Hospital,* **49,** 17–29.

Lewis, P. R. & Lutwak-Mann, C. (1954). The content of sodium, potassium and chloride in rabbit blastocysts. *Biochimica et Biophysica Acta,* **14,** 589–590.

Leyssac, P. P. (1963). Dependence of glomerular filtration rate on proximal tubular reabsorption of salt. *Acta Physiologica Scandinavica*, **58**, 236–242.

Leyssac, P. P. (1966). The regulation of proximal tubular reabsorption in the mammalian kidney. *Acta Physiologica Scandinavica*, **70**, Supplement 291.

Lichtenstein, N. S. & Leaf, A. (1965). Effect of amphotericin B on the permeability of the toad bladder. *Journal of Clinical Investigation*, **44**, 1328–1342.

Lifson, N., Gruman, L. M. & Levitt, D. G. (1968). Diffusive-convective models for intestinal absorption of D_2O. *American Journal of Physiology*, **215**, 444–454.

Lifson, N. & Hakim, A. A. (1966). Simple diffusive-convective model for intestinal absorption of a nonelectrolyte (urea). *American Journal of Physiology*, **211**, 1137–1146.

Liley, A. W. (1963). Amniotic fluid. In *Modern Trends in Human Reproductive Physiology*, Vol. 1 (ed. Carey, H. M.), pp. 227–244. London; Butterworth.

Lillie, R. S. (1916). Increase of permeability to water following normal and artificial activation in sea-urchin eggs. *American Journal of Physiology*, **40**, 249–266.

Lindemann, B. & Pring, M. (1969). A model of water absorbing epithelial cells with variable cellular volume and variable width of the lateral intercellular gaps. *Pflügers Archiv. European Journal of Physiology*, **307**, R55–R56.

Lindemann, B. & Solomon, A. K. (1962). Permeability of luminal surface of intestinal mucosal cells. *Journal of General Physiology*, **45**, 801–810.

Ling, G. N. (1962). *A Physical Theory of the Living State*, 1st edn., Waltham, Mass.; Blaisdell Publishing Co.

Ling, G. N. (1965). The physical state of water in living cell and model systems. *Annals of the New York Academy of Sciences*, **125**, 401–417.

Ling, G. N. (1966). Cell membrane and cell permeability. *Annals of the New York Academy of Sciences*, **137**, 837–859.

Ling, G. N. & Ochsenfeld, M. M. (1966). Studies on ion accumulation in muscle cells. *Journal of General Physiology*, **49**, 819–843.

Ling, G. N., Ochsenfeld, M. M. & Karreman, G. (1967). Is the cell membrane a universal rate-limiting barrier to the movement of water between the living cell and its surrounding medium? *Journal of General Physiology*, **50**, 1807–1820.

Litovitz, T. A. & Carnevale, E. H. (1955). Effect of pressure on sound propagation in water. *Journal of Applied Physics*, **26**, 816–820.

Loeb, J. (1870). Physiologische Untersuchungen über Ionwirkungen. I. Mitteilung. Versuche am Muskel. *Pflügers Archiv für die gesamte Physiologie des Menschen und der Tiere.* **69**, 1–27.

Loeb, S. (1966). A composite tubular assembly for reverse osmosis desalination. *Desalination*, **1**, 35–49.

Loehry, C. A., Axon, A. T. R., Hilton, P. J., Hider, R. C. & Creamer, B. (1970). Permeability of the small intestine to substances of different molecular weight. *Gut*, **11**, 466–470.

Loeschke, K., Bentzel, C. J. & Csaky, T. Z. (1970). Asymmetry of osmotic flow in frog intestine; functional and structural correlation. *American Journal of Physiology*, **218**, 1723–1731.

Longuet-Higgins, H. C. & Austin, G. (1966). The kinetics of osmotic transport through pores of molecular dimensions. *Biophysical Journal*, **6**, 217–224.

Lorimer, J. W., Boterenbrood, E. I. & Hermans, J. J. (1956). Transport processes in ion-selective membranes. Conductivities, transport numbers and electromotive forces. *Discussions of the Faraday Society*, **21**, 141–149.

Lovelock, J. E. (1953a). The haemolysis of human red blood cells by freezing and thawing. *Biochimica et Biophysica Acta*, **10**, 414–426.

Lovelock, J. E. (1953b). The mechanism of the protective action of glycerol against haemolysis by freezing and thawing. *Biochimica et Biophysica Acta*, **11**, 28–36.

Lovelock, J. E. (1955). Haemolysis by thermal shock. *British Journal of Haematology*, **1**, 117–129.

Løvtrup, S. (1960). Water permeation in the amphibian embryo. *Journal of Experimental Zoology*, **145**, 139–150.

Løvtrup, S. (1962). On the surface coat of amphibian embryo. *Journal of Experimental Zoology*, **150**, 197–206.

Løvtrup, S. (1963a). On the rate of water exchange across the surface of animal cells. *Journal of Theoretical Biology*, **5**, 341–359.

Løvtrup, S. (1963b). Permeability changes in fertilized and activated amphibian eggs. *Journal of Experimental Zoology*, **151**, 79–84.

Løvtrup, S. (1965a). Morphogenesis in the amphibian embryo; fertilization and blastula formation. *Archiv für Entwicklungsmechanik der Organismen*, **156**, 204–248.

Løvtrup, S. (1965b). Morphogenesis in the amphibian embryo. Gastrulation and neurulation. *Acta Universitatis Gothoburgensis. Zoologica Gothoburgensia* I. 1–139.

Løvtrup, S. & Pigon, A. (1951). Diffusion and active transport of water in the amoeba. *Chaos chaos* L. *Compte rendu des travaux du Laboratoire de Carlsberg, Serie Chimique*, **28**, 1–36.

Lucké, B. (1940). The living cell as an osmotic system and its permeability to water (Experiments with egg cells of marine invertebrates). *Cold Spring Harbor Symposia on Quantitative Biology*, **8**, 123–132.

Lucké, B., Hartline, H. K. & McCutcheon, M. (1931). Further studies on the kinetics of osmosis in living cells. *Journal of General Physiology*, **14**, 405–419.

Lucké, B., Hartline, H. K. & Ricca, R. A. (1939). Comparative permeability to water and to certain solutes of the egg cells of three marine invertebrates, *Arbacia, Cumingia* and *Chaetopterus. Journal of Cellular and Comparative Physiology*, **14**, 237–252.

Lucké, B., Larrabee, M. G. & Hartline, H. K. (1935). Studies on osmotic equilibrium and on the kinetics of osmosis in living cells by a diffraction method. *Journal of General Physiology*, **19**, 1–17.

Lucké, B. & McCutcheon, M. (1927). The effect of salt concentration of the medium on the rate of osmosis of water through the membrane of living cells. *Journal of General Physiology*, **10**, 665–670.

Lucké, B. & McCutcheon, M. (1932). The living cell as an osmotic system and its permeability to water. *Physiological Reviews*, **12**, 68–139.

Lucké, B. & Ricca, R. A. (1941). Osmotic properties of the egg cells of the oyster (*Ostrea virginica*). *Journal of General Physiology*, **25**, 215–227.

Luft, J. H. (1965). The ultrastructural basis of capillary permeability. In *The Inflammatory Process* (ed. Zweifach, B. W., Grant, L. & McCluskey, R. T.), pp. 121–159. New York; Academic Press.

Lundberg, A. (1957). The mechanism of establishment of secretory potentials in sublingual gland cells. *Acta Physiologica Scandinavica*, **40**, 35–58.

Lutwak-Mann, C. (1959). Biochemical approach to the study of ovum implantation in the rabbit. *Memoirs. Society for Endrocrinology*, **6**, 35–49.

Lutwak-Mann, C. (1960). Some properties of the early embryonic fluids of the rabbit. *Journal of Reproduction and Fertility*, **1**, 316–317.

Macey, R. I. & Farmer, R. E. L. (1970). Inhibition of water and solute permeability in human red cells. *Biochimica et Biophysica Acta*, **211**, 104–106.

MacGregor, H. C. (1962). The behaviour of isolated nuclei. *Experimental Cell Research*, **26**, 520–525.

Machen, T. E. & Diamond, J. M. (1969). An estimate of the salt concentration in the lateral intercellular spaces of rabbit gall-bladder during maximal fluid transport. *Journal of Membrane Biology*, **1**, 194–213.

Machin, J. (1969). Passive water movements through skin of the toad *Bufo marinus* in air and in water. *American Journal of Physiology*, **216**, 1562–1568.

Mackay, D. & Meares, P. (1959). The electrical conductivity and the electro-osmotic permeability of a cation-exchange resin. *Transactions of the Faraday Society*, **55**, 1221–1238.

MacRobbie, E. A. C. (1971). Phloem translocation. Facts and mechanisms; a comparative survey. *Biological Reviews*, **46**, 429–481.

MacRobbie, E. A. C. & Fensom, D. S. (1969). Measurements of electro-osmosis in *Nitella translucens*. *Journal of Experimental Botany*, **20**, 466–484.

MacRobbie, E. A. C. & Ussing, H. H. (1961). Osmotic behaviour of the epithelial cells of frog skin. *Acta Physiologica Scandinavica*, **53**, 348–365.

Maddrell, S. H. P. (1969). Secretion by the Malpighian tubules of *Rhodnius*. The movements of ions and water. *Journal of Experimental Biology*, **51**, 71–97.

Maddrell, S. H. P. (1971). Fluid secretion by the Malpighian tubules of insects. *Philosophical Transactions of the Royal Society of London* B, **262**, 197–207.

Madras, S., McIntosh, R. L. & Mason, S. G. (1949). A preliminary study of the permeability of cellophane to liquids. *Canadian Journal of Research* B, **27**, 764–779.

Maetz, J. (1968). Salt and water metabolism. In *Perspectives in Endocrinology* (ed. Barrington, E. J. W. & Jørgensen, C. B), pp. 47–162. London; Academic Press.

Maffly, R. H. & Leaf, A. (1959). The potential of water in mammalian tissues. *Journal of General Physiology*, **42**, 1257–1275.

Majno, G. (1965). Ultrastructure of the vascular membrane. *Handbook of Physiology* (*Circulation*), Chapter 64, pp. 2293–2375. Washington, D.C.; American Physiological Society.

Makhlouf, G. M. (1971). Direct evidence for isosmotic gastric secretion. *Gastroenterology*, **60**, 784.

Malmberg, C. G. & Maryott, A. A. (1956). Dielectric constant of water from 0° to 100°C. *Journal of Research of the National Bureau of Standards*, **56**, 1–8.

Malnic, G., Klose, R. M. & Giebisch, G. (1966). Micropuncture study of distal tubular potassium and sodium transport in rat nephron. *American Journal of Physiology*, **211**, 529–547.

Mangos, J. A. & McSherry, N. R. (1971). Micropuncture study of excretion of water and electrolytes by the pancreas. *American Journal of Physiology*, **221**, 496–503.

Marro, F. & Germagnoli, E. (1966). Letter to Editor. *Journal of General Physiology*, **49**, 1351–1353.

Martin, D. W. & Diamond, J. M. (1966). Energetics of coupled active transport of sodium and chloride. *Journal of General Physiology*, **50**, 295–315.

Martín de Julian, P. & Yudilevich, D. L. (1964). A theory for the quantification of transcapillary exchange by tracer dilution curves. *American Journal of Physiology*, **207**, 162–168.

Mast, S. O. (1938). The contractile vacuole in *Amoeba proteus* (Leidy). *Biological Bulletin of the Marine Biological Laboratory, Woods Hole, U.S.A.*, **74**, 306–313.

Mast, S. O. & Doyle, W. L. (1934). Ingestion of fluid by amoeba. *Protoplasma*, **20**, 555–560.

Mast, S. O. & Fowler, C. (1935). Permeability of *Amoeba proteus* to water. *Journal of Cellular and Comparative Physiology*, **6**, 151–167.

Maude, D. L., Shedadeh, I. & Solomon, A. K. (1966). Sodium and water transport in single perfused distal tubules of *Necturus* kidney. *American Journal of Physiology*, **211**, 1043–1049.

Maurice, D. M. (1972). The location of the fluid pump in the cornea. *Journal of Physiology*, **221**, 43–54.

Mauro, A. (1957). Nature of solvent transfer in osmosis. *Science*, **126**, 252–253.

Mauro, A. (1960). Some properties of ionic and nonionic semi-permeable membranes. *Circulation*, **21**, 845–854.

Mauro, A. (1965). Osmotic flow in a rigid porous membrane. *Science*, **149**, 867–869.
Mayerson, H. S., Wolfram, C. G., Shirley, H. H. & Wasserman, K. (1960). Regional differences in capillary permeability. *American Journal of Physiology*, **198**, 155–160.
Mazur, P. (1963). Kinetics of water loss from cells at subzero temperatures and the likelihood of intracellular freezing. *Journal of General Physiology*, **47**, 347–369.
Mazur, P. (1966). Physical and chemical basis of injury in single-celled micro-organisms subjected to freezing and thawing. In *Cryobiology* (ed. Meryman, H. T.) pp. 213–315. New York; Academic Press.
Mazur, S. K., Holtzman, E., Schwartz, I. L. & Walter, R. (1971). Correlation between pinocytosis and hydroosmosis induced by neurohypophyseal hormones and mediated by adenosine 3′,5′-cyclic monophosphate. *Journal of Cell Biology*, **49**, 582–594.
McCafferty, R. E. (1955). A physiological study of the amniotic fluid of the mouse. I. Volume and weight changes of the amniotic fluid compared with the weights of fetus and placenta during gestation. *Anatomical Record*, **123**, 521–530.
McCance, R. A. & Dickerson, J. W. T. (1957). The composition and origin of the foetal fluids of the pig. *Journal of Embryology and Experimental Morphology*, **5**, 43–50.
McCance, R. A. & Stanier, M. W. (1960). The function of the metanephros of foetal rabbits and pigs. *Journal of Physiology*, **151**, 479–483.
McConaghey, P. D. & Maizels, M. (1961). The osmotic coefficients of haemoglobin in red cells under varying conditions. *Journal of Physiology*, **155**, 28–45.
McCutcheon, M. & Lucké, B. (1932). The effect of temperature on the permeability to water of resting and of activated cells (unfertilized and fertilized eggs of *Arbacia punctulata*). *Journal of Cellular and Comparative Physiology*, **2**, 11–26.
McCutcheon, M., Lucké, B. & Hartline, H. K. (1931). The osmotic properties of living cells (eggs of *Arbacia punctulata*). *Journal of General Physiology*, **14**, 393–403.
McHardy, W. J., Meares, P., Sutton, A. H. & Thain, J. F. (1969). Electrical transport phenomena in a cation-exchange membrane. II. Conductance and electroosmosis. *Journal of Colloid and Interface Science*, **29**, 116–128.
McMaster, P. D. & Elman, R. (1926). On the expulsion of bile by the gall bladder: and a reciprocal relationship with the sphincter activity. *Journal of Experimental Medicine*, **44**, 173–198.
Meigs, E. B. (1912). Contributions to the general physiology of smooth and striated muscle. *Journal of Experimental Zoology*, **13**, 497–571.
Meigs, E. B. & Ryan, L. A. (1912). The chemical analysis of the ash of smooth muscle. *Journal of Biological Chemistry*, **11**, 401–414.
Mendoza, S. A., Handler, J. S. & Orloff, J. (1967). Effect of amphotericin B on permeability and short-circuit current in toad bladder. *American Journal of Physiology*, **213**, 1263–1268.

Mercer, E. H. (1959). An electron microscopic study of *Amoeba proteus*, *Proceedings of the Royal Society of London* B, **150**, 216–232.
Mercer, F. V., Hodge, A. J., Hope, A. B. & McLean, J. D. (1955). The structure and function of *Nitella* chloroplasts. *Australian Journal of Biological Sciences*, **8**, 1–18.
Merriam, R. W. (1959). Permeability and structural characteristics of isolated nuclei from *Chaetopterus* eggs. *Journal of Biophysical and Biochemical Cytology*, **6**, 353–360.
Merriam, R. W. (1966). The role of cytoplasmic membranes in the regulation of water concentration within frog oocytes. *Experimental Cell Research*, **41**, 34–48.
Meryman, H. T. (1968). A modified model for the mechanism of freezing injury in erythrocytes, *Nature*, **218**, 333–336.
Meryman, H. T. (1970). The exceeding of a minimum tolerable cell volume in hypertonic suspension as a cause of freezing injury. *Ciba Foundation Symposium* on *The Frozen Cell* (ed. Wolstenholme, G. E. W. & O'Connor, M.), pp. 51–67. London; Churchill.
Meryman, H. T. (1971). Osmotic stress as a mechanism of freezing injury. *Cryobiology*, **8**, 489–500.
Meschia, G. & Setnikar, I. (1958). Experimental study of osmosis through a collodion membrane. *Journal of General Physiology*, **42**, 429–444.
Meyer, H. H. (1899). Zur Theorie der Alkoholnarkose. Erste Mittheilung. Welche Eigenschaft der Anästhetica bedingt ihre narkotische Wirkung? *Archiv für experimentelle Pathologie und Pharmakologie*, **42**, 109–118.
Mikulecky, D. C. (1967). On the relative contributions of viscous flow *vs.* diffusional (frictional) flow to the stationary state flow of water through a 'tight' membrane. *Biophysical Journal*, **7**, 527–534.
Miller, D. M. (1964). Sugar uptake as a function of cell volume in human erythrocytes. *Journal of Physiology*, **170**, 219–225.
Miller, K. W., Paton, W. D. M. & Smith, E. B. (1965). Site of action of general anaesthetics. *Nature*, **206**, 574–577.
Miller, S. (1961). A theory of gaseous anaesthetics. *Proceedings of the National Academy of Sciences, U.S.A.*, **47**, 1515–1524.
Mishima, S. & Hedbys, B. O. (1967). The permeability of the corneal epithelium and endothelium to water. *Experimental Eye Research*, **6**, 10–32.
Missoten, L. (1964). L'ultrastructure des tissus occulaires. *Bulletin de la Societé belge d'ophtalmologie*, No. **136**, 1–200.
Moelwyn-Hughes, E. A. (1964). *Physical Chemistry*, 2nd edn. New York; Macmillan.
Monné, L. & Hardé, S. (1951). On the formation of the blastocoel and similar embryonic cavities. *Arkiv för Zoologi*, **1**, 463–469.
Moody, F. G. & Durbin, R. P. (1969). Water flow induced by osmotic and hydrostatic pressure in the stomach. *American Journal of Physiology*, **217**, 255–261.
Moore, B. & Roaf, H. (1905). On certain physical and chemical properties of solutions of chloroform and other anaesthetics. A contribution to

the chemistry of anaesthesia. *Proceedings of the Royal Society of London* B, **77**, 86–102.

Morgan, J. & Warren, B. E. (1938). X-ray analysis of the structure of water. *Journal of Chemical Physics*, **6**, 666–673.

Morgan, T. & Berliner, R. (1968). Permeability of the loop of Henle, vasa recta, and collecting duct to water, urea and sodium. *American Journal of Physiology*, **215**, 108–115.

Morgan, T. & Berliner, R. W. (1969). In vivo perfusion of proximal tubules of the rabbit; glomerulotubular balance. *American Journal of Physiology*, **217**, 992–997.

Morgan, T., Sakai, F. & Berliner, R. W. (1968). In vitro permeability of medullary collecting ducts to water and urea. *American Journal of Physiology*, **214**, 574–581.

Moszynski, J. R., Hoshiko, T. & Lindley, B. D. (1963). Note on the Curie principle. *Biochimica et Biophysica Acta*, **75**, 447–449.

Motais, R. & Isaia, J. (1972). Temperature-dependence of permeability to water and to sodium of the gill epithelium of the eel *Anguilla anguilla*. *Journal of Experimental Biology*, **56**, 587–600.

Motais, R., Isaia, J., Rankin, J. C. & Maetz, J. (1969). Adaptive changes of the water permeability of the teleostean gill epithelium in relation to external salinity. *Journal of Experimental Biology*, **51**, 529–546.

Mueller, P., Rudin, D. O., Tien, H. Ti & Wescott, W. C. (1962). Reconstitution of excitable cell membrane structure *in vitro*. *Circulation*, **26**, 1167–1171.

Muir, A. R. & Peters, A. (1962). Quintuple-layered membrane junctions at terminal bars between endothelial cells. *Journal of Cell Biology*, **12**, 443–448.

Murrish, D. E. & Schmidt-Nielsen, K. (1970). Water transport in the cloaca of lizards; Active or passive? *Science*, **170**, 324–326.

Narten, A. H., Danford, M. D. & Levy, H. A. (1967). X-ray diffraction study of liquid water in the temperature range 4–200°C. *Discussions of the Faraday Society*, **43**, 97–107.

Natochin, J. V., Janáček, K. & Rybova, R. (1965). The swelling of frog bladder cells produced by oxytocin. *Journal of Endocrinology*, **33**, 171–177.

Needham, J. (1931). *Chemical Embryology*, Vol. 2, 1st edn., pp. 777–789. London; Cambridge University Press.

Nemethy, G. & Scheraga, H. A. (1962a). Structure of water and hydrophobic bonding in proteins. I. A model for the thermodynamic properties of liquid water. *Journal of Chemical Physics*, **36**, 3382–3400.

Nemethy, G. & Scheraga, H. A. (1962b). Structure of water and hydrophobic bonding in proteins. II. Model for the thermodynamic properties of aqueous solutions of hydrocarbons. *Journal of Chemical Physics*, **36**, 3401–3417.

Nernst, W. (1904). Theorie der Reaktions-geschwindigkeit in heterogenen Systemen. *Zeitschrift für Physikalische Chemie*, **47**, 52–55.

Nevis, A. H. (1958). Water transport in invertebrate peripheral nerve fibers. *Journal of General Physiology*, **41**, 927–958.

New, D. A. T. (1956). The formation of sub-blastodermic fluid in hen's egg. *Journal of Embryology and Experimental Morphology*, **4**, 221–227.

Nims, L. F. (1961). Steady state material transfer through biological barriers, *American Journal of Physiology*, **201**, 987–994.

Nims, L. F. & Thurber, R. E. (1961). Ion distribution patterns in stationary state systems. *American Journal of Physiology*, **201**, 995–998.

Nishida, K. (1963). Osmotic swelling of isolated chloroplasts. *Plant and Cell Physiology*, **4**, 247–256.

Nobel, P. S. (1969). The Boyle-van't Hoff relation. *Journal of Theoretical Biology*, **23**, 375–379.

Nobel, P. S. & Packer, L. (1965). Light-dependent ion translocation in spinach chloroplasts, *Plant Physiology*, **40**, 633–640.

Noble-Nesbitt, J. (1970). Water balance in the firebrat, *Thermobia domestica* (Packard). *Journal of Experimental Biology*, **52**, 193–200.

Northrop, J. & Anson, M. (1929). A method for the determination of diffusion constants and the calculation of the radius and weight of the hemoglobin molecule. *Journal of General Physiology*, **12**, 543–554.

Noyes, A. A. & Whitney, W. R. (1897). Ueber die Anflösungeschwindigkeit von festen Stoffen in ihren eigenen Losungen. *Zeitschrift für Physikalische Chemie*, **23**, 689–692.

Nutbourne, D. M. (1968). The effect of small hydrostatic pressure gradients on the rate of active sodium transport across isolated living frog-skin membranes. *Journal of Physiology*, **195**, 1–18.

Öbrink, K. J. (1956). Water permeability of isolated stomach of the mouse. *Acta Physiologica Scandinavica*, **36**, 229–244.

Odeblad, E. (1959). Studies on vaginal contents and cells with proton magnetic resonance. *Annals of the New York Academy of Sciences*, **83**, 189–206.

Odeblad, E., Bhar, B. N. & Lindström, G. (1956). Proton magnetic resonance of human red blood cells in heavy-water exchange experiments. *Archives of Biochemistry and Biophysics*, **63**, 221–225.

Odeblad, E. & Bryhn, U. (1957). Proton magnetic resonance of human cervical mucus during the menstrual cycle. *Acta Radiologica*, **47**, 315–320.

Ogilvie, J. T., McIntosh, J. R. & Curran, P. F. (1963). Volume flow in a series-membrane system. *Biochimica et Biophysica Acta*, **66**, 441–444.

Olmstead, E. G. (1960). Efflux of red cell water into buffered hypertonic solutions. *Journal of General Physiology*, **43**, 707–712.

Onsager, L. (1931a). Reciprocal relations in irreversible thermodynamics; I. *Physical Review*, **37**, 405–426.

REFERENCES

Onsager, L. (1931b). Reciprocal relations in irreversible thermodynamics; II. *Physical Review*, **38**, 2265–2279.

Onsager, L. (1945). Theories and problems of liquid diffusion. *Annals of the New York Academy of Sciences*, **46**, 241–265.

Onsager, L. & Runnels, L. K. (1963). Mechanism for self-diffusion in ice. *Proceedings of the National Academy of Sciences, U.S.A.*, **50**, 208–210.

Orloff, J. & Burg, M. (1971). Kidney. *Annual Review of Physiology*, **33**, 83–130.

Ørskov, S. L. (1947). Solvent water in the human erythrocytes. *Acta Physiologica Scandinavica*, **12**, 192–201.

Oschman, J. L. & Berridge, M. J. (1970). Structural and functional aspects of salivary fluid secretion in *Calliphora*. *Tissue and Cell*, **2**, 281–310.

Oschman, J. L. & Wall, B. J. (1969). The structure of the rectal pads of *Periplaneta americana* L. with regard to fluid transport. *Journal of Morphology*, **127**, 475–509.

Osterhout, W. J. V. (1933). Permeability in large plant cells and in models. *Ergebnisse der Physiologie und Experimentellen Pharmakologie*, **35**, 967–1021.

Osterhout, W. J. V. (1949a). Movements of water in cells of *Nitella*. *Journal of General Physiology*, **32**, 553–557.

Osterhout, W. J. V. (1949b). Transport of water from concentrated to dilute solutions in cells of *Nitella*. *Journal of General Physiology*, **32**, 559–566.

Overbeek, J. T. G. (1952). Electrokinetic phenomena. In *Colloid Science*, Vol. 1 (ed. Kruyt, H. R.), pp. 194–244. Amsterdam; Elsevier.

Overton, E. (1896). Ueber die osmotischen Eigenschaften der Zelle in ihrer Bedeutung für die Toxikologie und Pharmakologie. *Vierteljahrsschrift der Naturforschenden Gesellschaft in Zurich*, **41**, 388–406.

Overton, E. (1899). Ueber die allgemeinen osmotischen Eigenschaften der Zelle, ihre vermutlichen Ursachen und ihr Bedeutung für die Physiologie. *Vierteljahrsschrift der Naturforschenden Gesellschaft in Zurich*, **44**, 88–135.

Overton, E. (1902). Beiträge zur allgemeinen Muskel und Nervenphysiologie. I. Ueber die osmotischen Eigenschaften der Muskeln. *Pflügers Archiv für die gesamte Physiologie des Menschen und der Tiere*, **92**, 115–280.

Packer, L. (1963). Structural changes correlated with photochemical phosphorylation in chloroplast membranes. *Biochimica et Biophysica Acta*, **75**, 12–22.

Packer, L. (1966). Volume changes and contractility of mitochondrial and chloroplast membranes. *Annals of the New York Academy of Sciences*, **137**, 624–640.

Packer, L., Siegenthaler, P. A. & Nobel, P. S. (1965). Light-induced volume changes in spinach chloroplasts. *Journal of Cell Biology*, **26**, 593–599.

Packer, L., Wrigglesworth, J. M., Fortes, P. A. G. & Pressman, B. C. (1968). Expansion of the inner membrane compartment and its relation to mitochondrial volume and ion transport. *Journal of Cell Biology*, **39**, 382–391.

Paganelli, C. V. & Solomon, A. K. (1957). The rate of exchange of tritiated water across the human red cell membrane. *Journal of General Physiology*, **41**, 259–277.

Pak Poy, R. K. F. & Bentley, P. J. (1960). Fine structure of the epithelial cells of the toad urinary bladder. *Experimental Cell Research*, **20**, 235.

Palade, G. E. (1953). Fine structure of blood capillaries. *Journal of Applied Physics*, **24**, 1424.

Palva, P. (1939). Die Wasserpermeabilität der Zellen von *Tolypellopsis stelligera*, *Protoplasma*, **32**, 265–271.

Pappas, G. D. & Smelser, G. K. (1958). Studies on the ciliary epithelium and the zonule. I. Electron microscope observations on changes induced by alteration of normal aqueous humor formation in the rabbit. *American Journal of Ophthalmology*, **46**, 299–318.

Pappas, G. D. & Tennyson, V. M. (1962). An electron microscopic study of the passage of colloidal particles through the blood vessels of the ciliary processes and choroid plexus of the rabbit. *Journal of Cell Biology*, **15**, 227–239.

Pappenheimer, J. R. (1953). Passage of molecules through capillary walls. *Physiological Reviews*, **33**, 387–422.

Pappenheimer, J. R. (1970). Osmotic reflection coefficients in capillary membranes. In *Capillary Permeability. The Transfer of Molecules and Ions between Capillary Blood and Tissue*. Proceedings of the Alfred Benzon Symposium II (ed. Crone, C. & Lassen, N. A.), pp. 278–286. Copenhagen; Munksgaard.

Pappenheimer, J. R., Renkin, E. M. & Borrero, L. M. (1951). Filtration, diffusion and molecular sieving through peripheral capillary membranes. A contribution to the pore theory of capillary permeability. *American Journal of Physiology*, **167**, 13–46.

Pappenheimer, J. R. & Soto-Rivera, A. (1948). Effective osmotic pressure of the plasma proteins and other quantities associated with the capillary circulation in the hindlimbs of cats and dogs. *American Journal of Physiology*, **152**, 471–491.

Parsons, B. J., Smyth, D. H. & Taylor, C. B. (1958). The action of phlorrhizin on the intestinal transfer of glucose and water *in vitro*. *Journal of Physiology*, **144**, 387–402.

Parsons, D. S. (1963). Quantitative aspects of pinocytosis in relation to intestinal absorption. *Nature*, **199**, 1192–1193.

Parsons, D. S. & Paterson, C. R. (1965). Fluid and solute transport across rat colonic mucosa. *Quarterly Journal of Experimental Physiology*, **50**, 220–231.

Parsons, D. S. & Wingate, D. L. (1961). The effect of osmotic gradients on fluid transfer across rat intestine *in vitro*. *Biochimica et Biophysica Acta*, **46**, 170–183.

REFERENCES

Patlak, C. S. (1961). Energy expenditure by active transport mechanisms. II. Further generalizations. *Biophysical Journal*, 1, 419–427.

Patlak, C. S., Goldstein, D. A. & Hoffman, J. F. (1963). The flow of solute and solvent across a two-membrane system. *Journal of Theoretical Biology*, 5, 426–442.

Patlak, C. S. & Rapoport, S. I. (1971). Theoretical analysis of net tracer flux due to volume circulation in a membrane with pores of different sizes. Relation to solute drag model. *Journal of General Physiology*, 57, 113–124.

Pauling, L. (1935). The structure and entropy of ice and of other crystals with some randomness of atomic arrangement. *Journal of the American Chemical Society*, 57, 2680–2684.

Pauling, L. (1960). *The Nature of the Chemical Bond*, 3rd edn. New York; Cornell University Press.

Pauling, L. (1961). A molecular theory of general anaesthesia. *Science*, 134, 15–21.

Peachey, L. D. (1965). The sarcoplasmic reticulum and transverse tubules of the frog's sartorius. *Journal of Cell Biology*, 25, 209–231.

Peachey, L. D. & Rasmussen, H. (1961). Structure of the toad's urinary bladder as related to its physiology. *Journal of Biophysical and Biochemical Cytology*, 10, 529–553.

Peers, A. M. (1956). In general discussion of a symposium entitled: A general discussion of membrane phenomena. *Discussions of the Faraday Society*, 21, 124–125.

Persson, E. (1969). Chloride and water transport in rat cortical nephrons. *Acta Universitatis Upsaliensis. Abstracts of Uppsala Dissertations in Medicine*, 70, 25–28.

Persson, E. (1970). Water permeability in rat distal tubules. *Acta Physiologica Scandinavica*, 78, 364–375.

Persson, E. & Ulfendahl, H. R. (1970). Water permeability in rat proximal tubules. *Acta Physiologica Scandinavica*, 78, 353–363.

Perutz, M. F. (1946). The composition and swelling properties of haemoglobin crystals. *Transactions of the Faraday Society*, 42, 187–195.

Peterson, M. A. & Gregor, H. P. (1959). Diffusion-exchange of exchange ions and nonexchange electrolyte in ion-exchange membrane systems. *Journal of the Electrochemical Society*, 106, 1051–1061.

Pethica, B. A., Thompson, W. K. & Pike, W. T. (1971). Anomalous water not polywater. *Nature*, 229, 21–22.

Philip, J. R. (1958). The osmotic cell, solute diffusibility, and the plant water economy. *Plant Physiology*, 33, 264–271.

Phillips, J. E. (1961). Rectal absorption of water and salts in the locust and blowfly. *Ph.D. Thesis*, University of Cambridge.

Phillips, J. E. (1964a). Rectal absorption in the desert locust, *Schistocerca gregaria*. Forskål. I. Water. *Journal of Experimental Biology*, 41, 15–38.

Phillips, J. E. (1964b). Rectal absorption in the desert locust, *Schistocerca gregaria* Forskål. II. Sodium, potassium and chloride. *Journal of Experimental Biology*, **41**, 39–67.

Phillips, J. E. (1964c). Rectal absorption in the desert locust, *Schistocerca gregaria* Forksål. III. The nature of the excretory process. *Journal of Experimental Biology*, **41**, 69–80.

Phillips, J. E. (1969). Osmotic regulation and rectal absorption in the blowfly, *Calliphora erythrocephala*. *Canadian Journal of Zoology*, **47**, 851–863.

Phillips, J. E. (1970). Apparent transport of water by insect excretory systems. *American Zoologist*, **10**, 413–436.

Phillips, J. E. (1971). Dependence of hyposmotic fluid transport on ion movement within rectal epithelium of locusts. *Proceedings of the International Union of Physiological Sciences*, Vol. IX, International Congress, Munich, pp. 451.

Phillips, J. E. & Beaumont, C. (1971). Symmetry and non-linearity of osmotic flow across rectal cuticle of the desert locust. *Journal of Experimental Biology*, **54**, 317–328.

Picken, L. E. R. (1936). A note on the mechanism of salt and water balance in the heterotrichous ciliate, *Spirostomum ambiguum*. *Journal of Experimental Biology*, **13**, 387–392.

Pidot, A. L. & Diamond, J. M. (1964). Streaming potentials in a biological membrane. *Nature*, **201**, 701–702.

Pitts, R. F. (1968). *Physiology of the Kidney and Body Fluids*. 2nd edn., p. 116. Chicago; Year Book Medical Publishers.

Ponder, E. (1944). The osmotic behavior of crenated red cells. *Journal of General Physiology*, **27**, 273–285.

Ponder, E. (1948). *Hemolysis and Related Phenomena*. 1st edn. New York; Grune & Stratton.

Ponder, E. (1950). Tonicity-volume relations in partially hemolyzed hypotonic systems. *Journal of General Physiology*, **33**, 177–193.

Ponder, E. & Barreto, D. (1957). The behavior, as regards shape and volume, of human red cell ghosts in fresh and in stored blood. *Blood*, **12**, 1016–1027.

Potter, E. L. (1961). *Pathology of the Fetus and Infant*, 2nd ed. Chicago; Year Book Medical Publishers.

Potts, W. T. W. & Rudy, P. P. (1969). Water balance in the eggs of the Atlantic salmon *Salmo salar*. *Journal of Experimental Biology*, **50**, 223–237.

Prescott, D. M. & Zeuthen, E. (1953). Comparison of water diffusion and water filtration across cell surfaces. *Acta Physiologica Scandinavica*, **28**, 77–94.

Price, H. D. & Thompson, T. E. (1969). Properties of liquid bilayer membranes separating two aqueous phases; temperature dependence of water permeability. *Journal of Molecular Biology*, **41**, 443–457.

Prigogine, I. (1961). *Introduction to Thermodynamics of Irreversible Processes*, 2nd edn., New York; Interscience Publishers Inc.

Prusch, R. D. & Dunham, P. B. (1967). Electrical and contractile properties of the isolated contractile vacuole of *Amoeba proteus*. *Journal of General Physiology*, **50**, 1083.

Purdy, D. M. & Hillemann, H. H. (1950). Volume changes in the amniotic fluid of the golden hamster (*Cricetus auratus*) *Anatomical Record*, **106**, 571–575.

Raaflaub, J. (1953). Die Schwellung isolierter Leberzellmitochondrien und ihre physikalish-chemische Beeinflussbarkeit, *Helvetica Physiologica et Pharmalogica Acta*, **11**, 142–156.

Ramsay, J. A. (1953). Active transport of potassium by the Malpighian tubules of insects. *Journal of Experimental Biology*, **30**, 358–369.

Ramsay, J. A. (1954). Active transport of water by the Malpighian tubules of the stick insect, *Dixippus morosus* (Orthoptera; Phasmidae). *Journal of Experimental Biology*, **31**, 104–113.

Ramsay, J. A. (1958). Excretion by the Malpighian tubules of the stick insect, *Dixippus morosus* (Orthoptera, Phasmidae); amino acids, sugars and urea. *Journal of Experimental Biology*, **35**, 871–891.

Ramsay, J. A. (1971). Insect rectum. *Philosophical Transactions of the Royal Society of London* B, **262**, 251–260.

Rand, R. P. & Burton, A. C. (1964). Mechanical properties of the red cell membrane. I. Membrane stiffness and intracellular pressure. *Biophysical Journal*, **4**, 115–135.

Rapoport, S. I. (1966). Convection, diffusion, and electric current through a membrane. *Acta Physiologica Scandinavica*, **66**, 385–394.

Rasio, E. (1970). The permeability of isolated mesentery. Effect of temperature. In *Capillary Permeability. The Transfer of Molecules and Ions between Capillary Blood and Tissue* (ed. Crone, C. & Lassen, N. A.). Proceedings of the Alfred Benzon Symposium. II, pp. 643–646. Copenhagen; Munksgaard.

Rasmussen, H., Chance, B. & Ogata, E. (1965). A mechanism for the reactions of calcium with mitochondria. *Proceedings of the National Academy of Sciences, U.S.A.* **53**, 1069–1076.

Rastogi, R. P., Blokhra, R. L. & Aggarwala, R. K. (1964). Thermoosmosis through membranes. *Indian Journal of Chemistry*, **2**, 166.

Ravdin, I. S., Johnston, C. G., Austin, J. H. & Riegel, C. (1932). Studies of gall-bladder function. IV. The absorption of chloride from bile-free gall-bladder. *American Journal of Physiology*, **99**, 638–647.

Rawlins, F., Mateu, L., Fragachan, F. & Whittembury, G. (1970). Isolated toad skin epithelium; transport characteristics. *Pflügers Archiv. European Journal of Physiology*, **316**, 64–80.

Ray, P. M. (1960). On the theory of osmotic water movement. *Plant Physiology*, **35**, 783–795.

Redwood, W. R. & Haydon, D. A. (1969). Influence of temperature and membrane composition on the water permeability of lipid bilayers. *Journal of Theoretical Biology*, **22**, 1–8.

Reese, T. S. & Karnovsky, M. J. (1967). Fine structural localization of a blood-brain barrier to exogeneous peroxidase. *Journal of Cell Biology*, **34**, 207–217.

Rehm, W. S., Schlesinger, H. & Dennis, W. H. (1953). Effect of osmotic gradients on water transport, hydrogen ion and chloride ion production in the resting and secreting stomach. *American Journal of Physiology*, **175**, 473–486.

Reid, E. W. (1890). Osmosis experiments with living and dead membranes. *Journal of Physiology*, **11**, 312–351.

Reid, E. W. (1892a). Report on experiments upon 'absorption without osmosis'. *British Medical Journal*, **1**, 323–326.

Reid, E. W. (1892b). Preliminary report on experiments upon intestinal absorption without osmosis, *British Medical Journal*, **1**, 1133–1134.

Reid, E. W. (1901). Transport of fluid by certain epithelia. *Journal of Physiology*, **26**, 436–444.

Renkin, E. M. (1954). Filtration, diffusion and molecular sieving through porous cellulose membranes. *Journal of General Physiology*, **38**, 225–243.

Renkin, E. M. (1959). Transport of potassium-42 from blood to tissue in isolated mammalian skeletal muscles. *American Journal of Physiology*, **197**, 1205–1210.

Renkin, E. M. (1964). Transport of large molecules across capillary walls. *Physiologist*, **7**, 13–28.

Renkin, E. M. (1967). Blood flow and transcapillary exchange in skeletal and cardiac muscle. In *Coronary Circulation and Energetics of the Myocardium* (ed. Marchatti, G. & Taccardi, B.), pp. 18–30. Basel: Karger.

Renkin, E. M. & Zaun, B. D. (1955). Effects of adrenal hormones on capillary permeability in perfused rat tissues. *American Journal of Physiology*, **180**, 498–502.

Reuben, J. P., Girardier, L. & Grundfest, H. (1964). Water transfer and cell structure in isolated crayfish muscle fibres. *Journal of General Physiology*, **47**, 1141–1174.

Reuben, J. P., Lopez, E., Brandt, P. W. & Grundfest, H. (1963). Muscle; Volume changes in isolated single fibers. *Science*, **142**, 246–248.

Reynolds, S. R. M. (1953). A source of amniotic fluid in the lamb; the nasopharyngeal and buccal cavities. *Nature*, **172**, 307–308.

Reynolds, S. R. M. (1964). Reversible movement of fluid across the foetal lung. *Nature*, **202**, 197–198.

Rhee, S. W., Green, K., Martinez, M. & Paton, D. (1971). Water permeability of cat corneal endothelium *in vitro*. *Investigative Ophthalmology*, **10**, 288–293.

Rich, G. T., Sha'afi, R. I., Barton, T. C. & Solomon, A. K. (1967). Permeability studies on red cell membranes of dog, cat, and beef. *Journal of General Physiology*, **50**, 2391–2405.

Rich, G. T., Sha'afi, R. I., Romualdez, A. & Solomon, A. K. (1968). Effect of osmolality on the hydraulic permeability coefficient of red cells. *Journal of General Physiology*, **52**, 941–954.

Richards, A. N. & Schmidt, C. F. (1924). A description of the glomerular circulation in the frog's kidney and observations concerning the action of adrenalin and various other substances upon it. *American Journal of Physiology*, **71**, 178–208.

Riddick, D. H. (1968). Contractile vacuole in the amoeba, *Pelomyxa carolinesis*. *American Journal of Physiology*, **215**, 736–740.

Riley, R. L., Merten, U. & Gardner, J. O. (1966). Replication electron microscopy of cellulose acetate osmotic membranes. *Desalination*, **1**, 30–34.

Robbins, E. & Mauro, A. (1960). Experimental study of the independence of diffusion and hydrodynamic permeability coefficients in collodion membranes. *Journal of General Physiology*, **43**, 523–532.

Robinson, J. R. (1953). The active transport of water in living systems. *Biological Reviews*, **28**, 158–194.

Robinson, J. R. (1954). Secretion and transport of water. *Symposia of the Society for Experimental Biology*, **8**, 42–62.

Robinson, J. R. (1960). Metabolism of intracellular water. *Physiological Reviews*, **40**, 112–149.

Robinson, J. R. (1965). Water regulation in mammalian cells. *Symposia of the Society for Experimental Biology*, **19**, 237–258.

Robinson, R. A. & Stokes, R. H. (1959). *Electrolyte Solutions*, 2nd edn. London; Butterworth.

Romanoff, A. L. (1943a). Assimilation of avian yolk and albumen under normal and extreme incubating temperatures. *Anatomical Record*, **86**, 143–148.

Romanoff, A. L. (1943b). Distribution of dry constituents of yolk and albumen in the developing avian egg. *Anatomical Record*, **87**, 303–306.

Romanoff, A. L. (1952). Membrane growth and function. *Annals of the New York Academy of Sciences*, **55**, 288–301.

Romanoff, A. L. & Hayward, F. W. (1943). Changes in volume and physical properties of allantoic and amniotic fluids under normal and extreme temperatures. *Biological Bulletin of the Marine Biological Laboratory, Woods Hole, U.S.A.*, **84**, 141–147.

Röntgen, W. C. (1892). Ueber die Constitution des flüssigen Wassers. *Annalen der Physik und Chemie*, **45**, 91–97.

Rowlinson, J. S. (1959). *Liquids and Liquid Mixtures*, 1st edn., London; Butterworth.

Rugh, R. (1951). *The Frog. Its Reproduction and Development.* 1st edn. Philadelphia; Blakiston Company.

Samoilov, O. Ya. (1946). *Structure of Aqueous Electrolyte Solutions and the Hydration of Ions*, translated by Ives, D. J. G. (1965). New York; Consultants Bureau.

Sanders, E. & Ashworth, C. T. (1961). A study of particulate intestinal absorption and hepatocellular uptake. Use of polystyrene latex particles. *Experimental Cell Research*, **22**, 137–145.

Savitz, D., Sidel, V. W. & Solomon, A. K. (1964). Osmotic properties of human red cells. *Journal of General Physiology*, **48**, 79–94.

Schafer, D. E. & Johnson, J. A. (1964). Permeability of mammalian heart capillaries to sucrose and inulin. *American Journal of Physiology*, **206**, 985–991.

Schafer, J. A. & Andreoli, T. E. (1972). Cellular constraints to diffusion. The effect of anti-diuretic hormone on water flows in isolated mammalian collecting tubules. *Journal of Clinical Investigation*, **51**, 1264–1278.

Scheuplein, R. J. (1966). Analysis of permeability data for the case of parallel diffusion pathways. *Biophysical Journal*, **6**, 1–17.

Schilb, T. P. (1969). Effect of cholinergic agent on sodium transport across isolated turtle bladders. *American Journal of Physiology*, **216**, 514–520.

Schilb, T. P. & Brodsky, W. A. (1970). Transient acceleration of transmural water flow by inhibition of sodium transport in turtle bladders. *American Journal of Physiology*, **219**, 590–596.

Schlögl, R. (1969). Non-linear transport behaviour in very thin membranes. *Quarterly Reviews of Biophysics*, **2**, 305–313.

Schmid, G. (1950). Zur Elektrochemie feinporiger Kapillarsysteme. *Zeitschrift für Elektrochemie*, **54**, 424–430.

Schmid, G. & Schwarz, H. (1952). Zur Elektrochemie feinporiger Kapillarsysteme. V. Strömungspotentiale: Donnan-Behinderung des Elektrolytdurchgangs bein Strömungen. *Zeitschrift für Elektrochemie*, **56**, 35–44.

Schmidt, W. J. (1939). Über die Doppelbrechung des Amöbenplasms. *Protoplasma*, **33**, 44–49.

Schmidt-Nielsen, B. & Schrauger, C. R. (1963). *Amoeba proteus*: Studying the contractile vacuole by micropuncture. *Science*, **139**, 606–607.

Schmidt-Nielsen, K. (1960). The salt-secreting gland of marine birds. *Circulation*, **21**, 955–967.

Schmidt-Nielsen, K. (1965). Physiology of salt glands. In *Sekretion und Exkretion. Funktionelle und morphologische Organisation der Zelle*, pp. 269–282. Berlin; Springer-Verlag.

Schmidt-Nielsen, K. (1969). The neglected interface; the biology of water as a liquid-gas system. *Quarterly Reviews of Biophysics*, **2**, 283–304.

Schneider, L. (1960). Elektronenmikroskopische Untersuchungen über das Nephridialsystem von *Paramaecium*. *Journal of Protozoology*, **7**, 75–90.

Schultz, S. G. & Curran, P. F. (1968). Intestinal absorption of sodium chloride and water. *Handbook of Physiology (Alimentary Canal)*. Chapter 66, pp. 1245–1275. Washington, D.C.: American Physiological Society.

Schumaker, V. N. (1958). Uptake of protein from solution by *Amoeba proteus*. *Experimental Cell Research*, **15**, 314–331.

Sha'afi, R. I., Rich, G. T., Mikulecky, D. C. & Solomon, A. K. (1970). Determination of urea permeability in red cells by minimum method. A test of the phenomenological equations. *Journal of General Physiology*, **55**, 427–450.

Sha'afi, R. I., Rich, G. T., Sidel, V. W., Bossert, W. & Solomon, A. K.

(1967). The effect of the unstirred layer on human red cell water permeability. *Journal of General Physiology*, **50**, 1377–1399.
Shapiro, H. (1966). Osmotic properties of frog nerve. *Comparative Biochemistry and Physiology*, **19**, 225–239.
Shapiro, H. & Parpart, A. K. (1937). The osmotic properties of rabbit and human leucocytes. *Journal of Cellular and Comparative Physiology*, **10**, 147–159.
Shaw, F. H. & Simon, S. E. (1955). The nature of the sodium and potassium balance in nerve and muscle cells. *Australian Journal of Experimental Biology and Medical Science*, **33**, 153–178.
Shaw, J. (1958). Osmoregulation in the muscle fibres of *Carcinus maenas*. *Journal of Experimental Biology*, **35**, 920–929.
Shea, S. M. & Karnovsky, M. J. (1966). Brownian motion; a theoretical explanation for the movement of vesicles across the endothelium. *Nature*, **212**, 353–355.
Shea, S. M., Karnovsky, M. J. & Bossert, W. H. (1969). Vesicular transport across endothelium; simulation of a diffusion model. *Journal of Theoretical Biology*, **24**, 30–42.
Sidel, V. W. & Hoffman, J. F. (1961). Water transport across membrane analogues. *Federation Proceedings*, **20**, 137.
Sidel, V. W. & Solomon, A. K. (1957). Entrance of water into human red cells under an osmotic pressure gradient. *Journal of General Physiology*, **41**, 243–257.
Sigler, K. & Janáček, K. (1969). Water relations of frog oocytes. *Physiologia Bohemoslovaca*, **18**, 147–155.
Sigler, K. & Janáček, K. (1971). The effect of non-electrolyte osmolarity on frog oocytes. *Biochimica et Biophysica Acta*, **241**, 528–538.
Simon, S. E., Shaw, F. H., Bennett, S. & Muller, M. (1957). The relationship between sodium, potassium and chloride in amphibian muscle. *Journal of General Physiology*, **40**, 753–777.
Singwi, K. S. & Sjölander, A. (1960). Diffusive motions in water and cold neutron scattering. *Physical Review*, **119**, 863–871.
Sjölin, S. (1954). The resistance of red cells *in vitro*. A study of the osmotic properties, the mechanical resistance and the storage behaviour of red cells of fetuses, children and adults. *Acta Paediatrica*, Uppsala, **43**, Supplement 98, 1–92.
Skadhauge, E. (1967). *In vivo* perfusion studies of the cloacal water and electrolyte resorption in the fowl (*Gallus domesticus*). *Comparative Biochemistry and Physiology*, **23**, 483–501.
Skadhauge, E. (1969). The mechanism of salt and water absorption in the intestine of the eel (*Anguilla anguilla*) adapted to waters of various salinities. *Journal of Physiology*, **204**, 135–158.
Slack, C. & Warner, A. E. (1973). Intracellular and intercellular potentials in the early amphibian embryo. *Journal of Physiology*, **232**, 313–330.
Smith, H. W. (1951). *The Kidney. Structure and Function in Health and Disease.* p. 485. New York; Oxford Univ. Press.
Smith, M. W. (1970). Active transport in the rabbit blastocyst. *Experientia*, **26**, 736–738.

Smith, P. G. (1969). The ionic relations of *Artemia salina* (L). II. Fluxes of sodium, chloride, and water. *Journal of Experimental Biology*, **51**, 739–757.

Smith, P. G. (1971). The low-frequency electrical impedance of the isolated frog skin. *Acta Physiologica Scandinavica*, **81**, 355–366.

Smulders, A. P., Tormey, J. M. & Wright, E. M. (1972). The effect of osmotically induced water flows on the permeability and ultrastructure of the rabbit gall bladder. *Journal of Membrane Biology*, **7**, 164–197.

Smulders, A. P. & Wright, E. M. (1971). The magnitude of non-electrolyte selectivity in the gallbladder epithelium. *Journal of Membrane Biology*, **5**, 297–318.

Smyth, D. H. & Taylor, C. B. (1957). Transfer of water and solutes by an *in vitro* intestinal preparation. *Journal of Physiology*, **136**, 632–648.

Smyth, D. H. & Wright, E. M. (1966). Streaming potentials in the rat small intestine. *Journal of Physiology*, **182**, 591–602.

Soergel, K. H., Whalen, G. E. & Harris, J. A. (1968). Passive movement of water and sodium across the human small intestinal mucosa. *Journal of Applied Physiology*, **24**, 40–48.

Solinger, R. E., Gonzalez, C. F., Shamoo, Y. E., Wyssbrod, H. R. & Brodsky, W. A. (1968). Effect of ouabain on ion transport mechanisms in the isolated turtle bladder. *American Journal of Physiology*, **215**, 249–261.

Soll, A. H. (1967). A new approach to molecular configuration applied to aqueous pore transport. *Journal of General Physiology*, **50**, 2565–2578.

Sollner, K. (1945). The physical chemistry of membranes with particular reference to the electrical behaviour of membranes of porous character. III. The geometrical and electrical structure of membranes of porous character: some examples of the machine action of membranes. *Journal of Physical Chemistry*, **49**, 265–280.

Solomon, A. K. (1968). Characterization of biological membranes by equivalent pores. *Journal of General Physiology*, **51**, 335s–364s.

Sorenson, A. L. (1971). Water permeability of isolated muscle fibers of a marine crab. *Journal of General Physiology*, **58**, 287–303.

Spanner, D. C. (1954). The active transport of water under temperature gradients. *Symposia of the Society for Experimental Biology*, **8**, 76–93.

Spiegler, K. S. (1958). Transport processes in ionic membranes. *Transactions of the Faraday Society*, **54**, 1408–1428.

Stackelberg, M. V. & Müller, H. R. (1954). Feste gashydrate. II. Struktur und raumchemie. *Zeitschrift für Elektrochemie*, **58**, 25–39.

Stadelmann, E. (1963). Vergleich und Umrechnung von Permeabilitäts-Konstanten für Wasser. *Protoplasma*, **57**, 660–678.

Stallworthy, W. B. (1970). Electro-osmosis in squid axons. *Journal of the Marine Biological Association U.K.*, **50**, 349–363.

Stallworthy, W. B. & Fensom, D. S. (1966). Electro-osmosis in axons of freshly killed squid. *Canadian Journal of Physiology and Pharmacology*, **44**, 866–870.

Starling, E. H. (1896). On the absorption of fluids from the connective tissue spaces. *Journal of Physiology*, **19**, 312–326.
Staverman, A. J. (1951). The theory of measurement of osmotic pressure. *Recueil des travaux chimiques des Pays-Bas et de la Belgique*, **70**, 344–352.
Staverman, A. J. (1952). Non-equilibrium thermodynamics of membrane processes. *Transactions of the Faraday Society*, **48**, 176–185.
Stein, W. D. (1967). *The Movement of Molecules across Cell Membranes*. 1st edn. p. 123. New York; Academic Press.
Steinbach, H. B. (1967). On the ability of isolated frog skin to manufacture Ringer's fluid. *Journal of General Physiology*, **50**, 2377–2389.
Stewart, R. J. & Graydon, W. F. (1957). Ion-exchange membranes. III. Water transfer. *Journal of Physical Chemistry*, **61**, 164–168.
Stewart, D. R. & Jacobs, M. H. (1932). The effect of fertilization on the permeability of the eggs of *Arbacia* and *Asterias* to ethylene glycol. *Journal of Cellular and Comparative Physiology*, **1**, 83–92.
Stobbart, R. H. (1968). Ion movement and water transport in the rectum of the locust *Schistocerca gregaria*. *Journal of Insect Physiology*, **14**, 269–275.
Stokes, R. H. (1950). An improved diaphragm-cell for diffusion studies, and some tests of the method. *Journal of the American Chemical Society*, **72**, 763–767.
Stoner, C. D. & Sirak, H. D. (1969). Osmotically-induced alterations in volume and ultrastructure of mitochondria isolated from rat liver and bovine heart. *Journal of Cell Biology*, **43**, 521–538.
Sussman, M. V. & Chin, L. (1966). Liquid water in frozen tissue; Study by nuclear magnetic resonance. *Science*, **151**, 324–325.
Sutcliffe, J. F. (1962). *Mineral Salts Absorption in Plants*. 1st edn. London; Pergamon.
Swanson, C. H. & Solomon, A. K. (1970). Micropuncture study of the electrolyte secretion of the *in vitro* rabbit pancreas. *Federation Proceedings*, **29**, 845.
Swift, T. J. & Fritz, O. G. (1969). A proton spin-echo study of the state of water in frog nerves. *Biophysical Journal*, **9**, 54–59.
Szent-Györgyi, A. (1957). *Bioenergetics*, 1st edn., New York; Academic Press.
Tait, M. J. & Franks, F. (1971). Water in biological systems. *Nature*, **230**, 91–94.
Talen, J. L. & Staverman, A. J. (1965a). Osmometry with membranes permeable to solvent and solute. *Transactions of the Faraday Society*, **61**, 2794–2799.
Talen, J. L. & Staverman, A. J. (1965b). Negative reflection coefficients. *Transactions of the Faraday Society*, **61**, 2800–2804.
Tanny, G. (1973). Hyperfiltration streaming potential as a probe of water structure in membranes. *Nature*, **242**, 474–475.
Tay, D. K. C. & Findlay, G. P. (1972). Permeability of the duodenum of the toad to non-electrolytes. *Australian Journal of Biological Sciences*, **25**, 931–939.

Taylor, A. C. & Weiss, P. (1965). Demonstration of axonal flow by the movement of tritium labelled protein in mature optic nerve fibers. *Proceedings of the National Academy of Sciences, U.S.A.*, **54**, 1521–1527.

Taylor, A. E. & Gaar, K. A. (1970). Estimation of equivalent pore radii of pulmonary capillary and alveolar membranes. *American Journal of Physiology*, **218**, 1133–1140.

Tazawa, M. & Nishizaki, Y. (1956). Simultaneous measurement of transcellular osmosis and the accompanying potential difference. *Japanese Journal of Botany*, **15**, 227–238.

Tedeschi, H. & Harris, D. L. (1955). The osmotic behaviour and permeability to non-electrolytes of mitochondria. *Archives of Biochemistry and Biophysics*, **58**, 52–67.

Teorell, T. (1936). A method for studying conditions within diffusion layers. *Journal of Biological Chemistry*, **113**, 735–748.

Teorell, T. (1958). Transport processes in membranes in relation to the nerve mechanism. *Experimental Cell Research*, Suppl. 5, The submicroscopic organization and function of nerve cells. 83–100.

Teorell, T. (1959a). Electrokinetic membrane processes in relation to properties of excitable tissues. I. Experiments on oscillatory transport phenomena in artificial membranes. *Journal of General Physiology*, **42**, 831–845.

Teorell, T. (1959b). Electrokinetic membrane processes in relation to properties of excitable tissues. II. Some theoretical considerations. *Journal of General Physiology*, **42**, 847–863.

Teorell, T. (1961). An analysis of the current-voltage relationship in excitable *Nitella* cells. *Acta Physiologica Scandinavica*, **53**, 1–6.

Teorell, T. (1962). Excitability phenomena in artificial membranes. *Biophysical Journal*, **2** (No. 2, part 2), 27–52.

Teorell, T. (1966). Electrokinetic considerations of mechanoelectrical transduction. *Annals of the New York Academy of Sciences*, **137**, 950–966.

Thau, G., Bloch, R. & Kedem, O. (1966). Water transport in porous and non-porous membranes. *Desalination*, **1**, 129–138.

Thull, N. B. & Rehm, W. S. (1956). Composition and osmolarity of gastric juice as a function of plasma osmolarity. *American Journal of Physiology*, **185**, 317–324.

Tolberg, A. B. & Macey, R. I. (1965). Osmotic behavior of spinach chloroplasts. *Biochimica et Biophysica Acta*, **109**, 424–430.

Tomlin, S. G. (1969). Vesicular transport across endothelial cells. *Biochimica et Biophysica Acta*, **183**, 559–564.

Tormey, J. M. (1963). Fine structure of the ciliary epithelium of the rabbit, with particular reference to 'infolded membranes', 'vesicles' and the effects of diamox. *Journal of Cell Biology*, **17**, 641–659.

Tormey, J. M. & Diamond, J. M. (1967). The ultrastructural route of fluid transport in rabbit gall bladder. *Journal of General Physiology*, **50**, 2031–2060.

REFERENCES

Torrey, T. W. (1971). *Morphogenesis of the Vertebrates.* 3rd ed. New York; Wiley.

Trap-Jensen, J. & Lassen, N. A. (1970). Capillary permeability for smaller hydrophilic tracers in exercising skeletal muscle in normal man and in patients with long-term diabetes mellitus. In *Capillary Permeability. The Transfer of Molecules and Ions between Capillary Blood and Tissue* (ed. Crone, C. & Lassen, N. A.). Proceedings of the Alfred Benzon Symposium II. pp. 135–152. Copenhagen; Munksgaard.

Troshin, A. S. (1961). Sorption properties of protoplasm and their role in cell permeability. In *Membrane Transport and Metabolism,* (ed. Kleinzeller, A. & Kotyk, A.), pp. 45–53. London; Academic Press.

Troshin, A. S. (1966). *Problems of Cell Permeability,* 1st edn. London; Pergamon.

Tuft, P. H. (1961). A morphogenetic effect of beta-mercaptoethanol. Distribution of water in the embryo. *Nature,* **191,** 1072–1074.

Tuft, P. H. (1962). The uptake and distribution of water in the embryo of *Xenopus laevis* (Daudin). *Journal of Experimental Biology,* **39,** 1–19.

Tuft, P. H. (1965). The uptake and distribution of water in the developing amphibian embryo. *Symposia of the Society for Experimental Biology,* **19,** 385–402.

Tuft, P. H. & Böving, B. G. (1970). The forces involved in water uptake by the rabbit blastocyst. *Journal of Experimental Zoology,* **174,** 165–172.

Ullrich, K. J., Rumrich, G. & Fuchs, G. (1964). Wasserpermeabilität und transtubulärer Wasserfluss corticaler Nephronabschnitte bei verschiedenen Diuresezuständen. *Pflügers Archiv für die gesamte Physiologie des Menschen und der Tiere,* **280,** 99–119.

Ussing, H. H. (1949). The distinction by means of tracers between active transport and diffusion. *Acta Physiologica Scandinavica,* **19,** 43–56.

Ussing, H. H. (1952). Some aspects of the application of tracers in permeability studies. *Advances in Enzymology,* **13,** 21–65.

Ussing, H. H. (1953). Transport through biological membranes. *Annual Review of Physiology,* **15,** 1–20.

Ussing, H. H. (1965). Transport of electrolytes and water across epithelia. *The Harvey Lectures* 1963–1964. Series 59. pp. 1–30. New York; Academic Press.

Ussing, H. H. (1966). Anomalous transport of electrolytes and sucrose through the isolated frog skin induced by hypertonicity of the outside bathing solution. *Annals of the New York Academy of Sciences,* **137,** 543–555.

Ussing, H. H. (1970). Tracer studies and membrane structure. In *Capillary Permeability. The Transfer of Molecules and Ions Between Capillary Blood and Tissue.* Proceedings of the Alfred Benzon Symposium II (ed. Crone, C. & Lassen, N. A.), pp. 654–656. Copenhagen; Munksgaard.

Ussing, H. H. (1971). Introductory remarks. *Philosophical Transactions of the Royal Society of London* B, **262**, 85–90.
Ussing, H. H. & Andersen, B. (1956). The relation between solvent drag and active transport of ions. *Proceedings of the Third International Congress of Biochemistry*. Brussels 1955 (ed. Liebecq, C.), pp. 434–440. New York; Academic Press.
Ussing, H. H. & Johansen, B. (1969). Anomalous transport of sucrose and urea in toad skin. *Nephron*, **6**, 317–328.
Ussing, H. H. & Windhager, E. E. (1964). Nature of shunt path and active sodium transport path through frog skin epithelium. *Acta Physiologica Scandinavica*, **61**, 484–504.
Ussing, H. H. & Zerahn, K. (1951). Active transport of sodium as the source of electric current in the short-circuited isolated frog skin. *Acta Physiologica Scandinavica*, **23**, 110–127.
Van Os, C. H. & Slegers, J. F. G. (1973). Path of osmotic water flow through rabbit gall bladder epithelium. *Biochimica et Biophysica Acta*, **291**, 197–207.
Vargas, F. F. (1965). Determination of the permeability to water of the membrane of axons from *Dosidicus gigas*. *Biochimica et Biophysica Acta*, **109**, 309–311.
Vargas, F. F. (1968*a*). Filtration coefficient of the axon membrane as measured with hydrostatic and osmotic methods. *Journal of General Physiology*, **51**, 13–27.
Vargas, F. F. (1968*b*). Water flux and electrokinetic phenomena in the squid axon. *Journal of General Physiology*, **51**, No. 5, Part 2, 123s–130s.
Vargas, F. & Johnson, J. A. (1964). An estimate of reflection coefficients for rabbit heart capillaries. *Journal of General Physiology*, **47**, 667–677.
Vargas, F. & Johnson, J. A. (1967). Permeability of rabbit heart capillaries to non-electrolytes. *American Journal of Physiology*, **213**, 87–93.
Vaughan, B. E. (1960). Intestinal electrolyte absorption by parallel determination of unidirectional sodium and water transfers. *American Journal of Physiology*, **198**, 1235–1244.
Vecli, A. & Bianchi, A. (1966). Water transport across isolated frog skin: a critical review of previous techniques and some new experimental results. *Archivio di Scienze Biologische*, **50**, 242–250.
Verney, E. G. (1950). Excretion of water. *Proceedings of XVIII International Physiological Congress*, 60–62.
Vieira, F. L., Sha'afi, R. I. & Solomon, A. K. (1970). The state of water in human and dog red cell membranes. *Journal of General Physiology*, **55**, 451–466.
Villegas, R. & Barnola, F. V. (1961). Characterization of the resting axolemma in the giant axon of the squid. *Journal of General Physiology*, **44**, 963–977.
Villegas, R., Barton, T. C. & Solomon, A. K. (1958). The entrance of water into beef and dog red cells. *Journal of General Physiology*, **42**, 355–369.

REFERENCES

Villegas, L. & Sananes, L. (1968). Independence between ionic transport and net water flux in frog gastric mucosa. *American Journal of Physiology*, **214**, 997–1000.

Villegas, R. & Villegas, G. M. (1960). Characterization of the membranes in the giant nerve fiber of the squid. *Journal of General Physiology*, **43**, No. 5, part 2, 73–103.

Visscher, M. B., Fetcher, E. S., Carr, C. W., Gregor, H. P., Bushey, M. S. & Barker, D. E. (1944). Isotopic tracer studies on the movement of water and ions between the intestinal lumen and blood. *American Journal of Physiology*, **142**, 550–575.

Voute, C. L. & Ussing, H. H. (1970). Quantitative relation between hydrostatic pressure gradient, extracellular volume and active sodium transport in the epithelium of the frog skin (*R. temporaria*). *Experimental Cell Research*, **62**, 375–383.

Wade, J. B. & Discala, V. A. (1971). The effect of osmotic flow on the distribution of horseradish peroxidase within the intercellular spaces of toad bladder epithelium. *Journal of Cell Biology*, **51**, 553–558.

Walker, A. M., Bott, P. A., Oliver, J. & MacDowell, M. C. (1941). The collection and analysis of fluid from single nephrons of the mammalian kidney. *American Journal of Physiology*, **134**, 580–595.

Wall, B. J. (1967). Evidence for antidiuretic control of rectal water absorption in the cockroach *Periplaneta americana* L. *Journal of Insect Physiology*, **13**, 565–578.

Wall, B. J. (1971). Local osmotic gradients in the rectal pads of an insect. *Federation Proceedings*, **30**, 42–48.

Wall, B. J. & Oschman, J. L. (1970). Water and solute uptake by rectal pads of *Periplanta americana*. *American Journal of Physiology*, **218**, 1208–1215.

Wall, B. J., Oschman, J. L. & Schmidt-Nielsen, B. (1970). Fluid transport; concentration of the intercellular compartment. *Science*, **167**, 1497–1498.

Wallin, B. G. (1969). Water permeability in resting and stimulated crayfish nerve. *Journal of General Physiology*, **54**, 462–478.

Walter, J. A. & Hope, A. B. (1971). Nuclear magnetic resonance and the state of water in cells. *Progress in Biophysics and Molecular Biology*, **23**, 1–20.

Wang, J. H. (1954). Effect of ions on the self-diffusion and structure of water in aqueous electrolytic solutions. *Journal of Physical Chemistry*, **58**, 686–691.

Wang, J. H. (1965). Self-diffusion coefficients of water. *Journal of Physical Chemistry*, **69**, 4412.

Wang, J. H., Anfinsen, C. B. & Polestra, F. M. (1954). The self-diffusion coefficient of water and ovalbumin in aqueous ovalbumin solutions at 10°. *Journal of the American Chemical Society*, **76**, 4763–4765.

Wang, J. H., Robinson, C. V. & Edelman, I. S. (1953). Self-diffusion and structure of liquid water. III. Measurement of the self-diffusion of liquid water with H^2, H^3 and O^{18}, as tracers. *Journal of the American Chemical Society*, **75**, 466–470.

Warner, A. C. I. & Stacy, B. D. (1972). Water, sodium and potassium movements across the rumen wall of sheep. *Quarterly Journal of Experimental Physiology*, **57**, 103–119.
Warner, D. T. (1961). Proposed molecular models of gramicidin S and other polypeptides. *Nature*, **190**, 120–128.
Warner, D. T. (1964). Molecular models; IV. A suggested conformation for the protein subunit of tobacco mosaic virus. *Journal of Theoretical Biology*, **6**, 118–136.
Wartiovaara, V. (1944). The permeability of tolypellopsis cells for heavy water and methyl alcohol. *Acta Botanica Fennica*, **34**, 1–22.
Watson, M. L. (1959). Further observations on the nuclear envelope of the animal cell. *Journal of Biophysical and Biochemical Cytology*, **6**, 147–156.
Wedner, H. J. & Diamond, J. M. (1969). Contributions of unstirred-layer effects to apparent electrokinetic phenomena in the gall-bladder *Journal of Membrane Biology*, **1**, 92–108.
Welch, K. (1967). The secretion of cerebrospinal fluid by *Lamina epithelialis*. *Monographs in the Surgical Sciences*, **4**, 155–192.
Welch, K., Sadler, K. & Gold, G. (1966). Volume flow across choroidal ependyma of the rabbit. *American Journal of Physiology*, **210**, 232–236.
Wells, H. S. (1931). The passage of materials through the intestinal wall. I. The relation between intra-intestinal pressure and the rate of absorption of water. *American Journal of Physiology*, **99**, 209–220.
Werth, W. (1961). Vergleichende Untersuchungen über die relative Permeabilität des Protoplasmas für Alkohol und Wasser. *Protoplasma*, **53**, 457–503.
Wheeler, H. O. (1963). Transport of electrolytes and water across wall of rabbit gall-bladder. *American Journal of Physiology*, **205**, 427–438.
White, H. L. (1924). On glomerular filtration. *American Journal of Physiology*, **68**, 523–529.
White, H. L. & Rolf, D. (1962). Osmometric behaviour of blood cells and of whole body cells. *American Journal of Physiology*, **202**, 1195–1199.
Whitlock, R. T. & Wheeler, H. O. (1964). Coupled transport of solute and water across rabbit gallbladder epithelium. *Journal of Clinical Investigation*, **43**, 2249–2265.
Whittembury, G. (1962). Action of antidiuretic hormone on the equivalent pore radius at both surfaces of the epithelium of the isolated toad skin. *Journal of General Physiology*, **46**, 117–130.
Wicke, E. (1966). Structure formation and molecular mobility in water and in aqueous solutions. *Angewandte Chemie*. International Edition in English, **5**, 106–122.
Widdas, W. F. (1951). Changing osmotic properties of foetal sheep erythrocytes and their comparison with those of maternal sheep erythrocytes. *Journal of Physiology*, **113**, 399–411.

REFERENCES

Wigglesworth, V. B. (1932). On the function of the so-called 'rectal glands' of insects. *Quarterly Journal of Microscopical Science*, **75**, 131–150.

Williams, R. J. (1970). Freezing tolerance in *Mytilus edulis*. *Comparative Biochemistry and Physiology*, **35**, 145–161.

Williams, R. J. & Meryman, H. T. (1965). A calorimetric method for measuring ice in frozen solutions. *Cryobiology*, **1**, 317–323.

Willis, E., Rennie, G. K., Smart, C. & Pethica, B. A. (1969). 'Anomalous' water. *Nature*, **222**, 159–161.

Wilson, T. H. (1956). A modified method for study of intestinal absorption *in vitro*. *Journal of Applied Physiology*, **9**, 137–140.

Wind, F. (1937). Versuche zur unmittelbaren Bestimmung des Flüssigkeitsaustritts aus den Blutkapillaren des Mesenterium und des Nierenglomerulus beim Kaltblüter. I. Mitteilung. *Archiv für experimentelle Pathologie und Pharmakologie*, **186**, 161–184.

Wislocki, G. B. (1935). On the volume of the fetal fluids in sow and cat. *Anatomical Record*, **63**, 183–192.

Wolf, A. V., Remp, P. G., Kiley, J. E. & Currie, G. D. (1961). Artificial kidney function; kinetic of hemodialysis. *Journal of Clinical Investigation*, **30**, 1062–1070.

Worthington, C. R. & Blaurock, A. E. (1968). Electron density model for nerve myelin. *Nature*, **218**, 87–88.

Wright, E. M. & Diamond, J. M. (1969a). An electrical method of measuring non-electrolyte permeability. *Proceedings of the Royal Society of London* B, **172**, 203–225.

Wright, E. M. & Diamond, J. M. (1969b). Patterns of non-electrolyte permeability. *Proceedings of the Royal Society of London* B, **172**, 227–271.

Wright, E. M. & Prather, J. W. (1970). The permeability of the frog choroid plexus to non-electrolytes. *Journal of Membrane Biology*, **2**, 127–149.

Wright, E. M., Smulders, A. P. & Tormey, J. M. (1972). The role of the lateral intercellular spaces and solute polarization effects in the passive flow of water across the rabbit gallbladder. *Journal of Membrane Biology*, **7**, 198–219.

Yamada, E. (1955). The fine structure of the gall bladder epithelium of the mouse. *Journal of Biophysical and Biochemical Cytology*, **1** 455–458.

Yamada, K. (1933). Über die Verteilung von Chlor in sich entwickelnder Hühnereiern. *Japanese Journal of Medical Science*. II. *Biochemistry*, **2**, 71–79.

Young, J. A. & Schögel, E. (1966). Micropuncture investigation of sodium and potassium excretion in rat submaxillary saliva. *Pflügers Archiv für die gesamte Physiologie des Menschen und der Tiere*, **291**, 85–98.

Yudilevich, D. L., Renkin, E. M., Alvarez, O. A. & Bravo, J. (1968). Fractional extraction and transcapillary exchange during continuous and instantaneous tracer administration. *Circulation Research*, **23**, 325–336.

Zaduniasky, J. A., Parisi, M. N. & Montoreano, R. (1963). Effect of antidiuretic hormone on permeability of single muscle fibres. *Nature*, **200**, 365–366.

Zerahn, K. (1956). Oxygen consumption and active sodium transport in the isolated and short-circuited frog skin. *Acta Physiologica Scandinavica*, **36**, 300–318.

Zotin, A. I. (1965). The uptake and movement of water in embryos. *Symposia of the Society for Experimental Biology*, **19**, 365–384.

Zweifach, B. W. & Intaglietta, M. (1968). Mechanics of fluid movement across single capillaries in the rabbit. *Microvascular Research*, **1**, 83–101.

AUTHOR INDEX

Abbott, W. O. 402
Abelson, H. T. 213
Abramow, M. 369
Aceves, J. 322
Adair, G. S. 198
Adolph, E. F. 274, 282, 283
Afzelius, B. A. 213
Aggarwala, R. K. 60
Agutter, P. S. 214
Aikman, D. P. 396, 461
Airth, R. L. 244, 245
Altamirano, M. 230, 320, 324, 336–338, 340, 347, 348, 351, 361, 400, 401, 428, 429
Alvarez, O. A. 304, 307–309, 314–317
Andersen, B. 344, 358–360, 366
Anderson, J. 416, 417
Anderson, N. G. 213
Anderson, W. P. 396, 461
Andersson-Cedergren, E. 206
Andreoli, T. E. 108, 109, 115, 123, 125, 127, 131, 132, 141, 142, 322, 360
Anfinsen, C. B. 174
Anson, M. 376
Applebloom, J. W. T. 194
Arisz, W. H. 393
Ashworth, C. T. 411
Austin, G. 91, 93, 118, 165
Austin, J. H. 429
Auty, R. P. 16
Axon, B. T. R. 351

Backmann, E. L. 265
Baker, P. F. 230
Baldwin, H. H. 24

Bangham, A. D. 14
Bangham, D. R. 14
Barker, D. E. 324, 334, 340, 358, 418, 428
Barker, J. N. 282, 283
Barlow, W. 3
Bärlund, H. 354
Barnes, P. 15
Barnola, F. V. 166, 253, 254
Barreto, D. 198
Barrnett, R. J. 411
Barry, B. A. 419
Barry, P. H. 113, 179, 242–247, 249–252, 325, 352, 356, 408
Barton, T. C. 156, 161, 165, 185, 240–242, 253, 255, 256
Bascom, W. D. 15
Battin, W. T. 213
Beament, J. W. L. 420, 421
Beaumont, C. 378
Beck, L. V. 212
Beck, R. E. 131
Bennet-Clark, T. A. 408
Bennett, H. S. 288
Bennett, S. 207
Bentley, P. J. 324, 363, 378, 381
Bentzel, C. J. 214, 215, 323, 378, 381, 385
Berendsen, H. J. C. 27, 28
Bergmann, R. 283
Berliner, R. W. 38, 296, 300, 301, 304, 319, 323, 340, 344, 405, 406
Bernal, J. D. 2, 6, 10, 28
Berntsson, K. 199, 200, 227
Berridge, M. J. 428, 429, 467
Bhar, B. N. 30
Bianchi, A. 408, 409
Birchenough, R. F. 367, 452

Birks, R. I. 205–207
Bjerrum, N. 16
Blaurock, A. E. 35
Blinks, J. R. 202–206
Blinks, L. R. 244
Bloch, R. 92, 115, 117, 123, 125, 126, 136, 137, 184, 341
Blokhra, R. L. 60
Blum, R. M. 165, 238, 241, 379
Bogucki, M. 201
Borrero, L. M. 38, 78, 80, 294–296, 301, 302, 305–307, 313, 315
Bossert, W. H. 162, 165, 185, 281, 310, 396, 452–461, 463, 466–468, 470
Boterenbrood, E. I. 54
Bott, P. A. 404
Boulpaep, E. 405
Bourguet, J. 363
Böving, B. G. 280, 281, 422, 429, 469
Boyle, P. J. 202, 230
Bozler, E. 207–209
Bradbury, S. 175–177
Brading, A. F. 207, 208
Bradley, R. S. 169
Brambell, F. W. R. 279
Brandt, P. W. 175, 202–204, 209–211, 218, 219, 230–232, 313
Bratton, C. B. 31, 32
Braun, G. 405
Bravo, J. 316, 317
Briggs, G. E. 245
Brightman, M. W. 291, 325
Brinster, R. L. 279
Brodsky, W. A. 194, 324, 327, 378, 385, 392, 414, 415, 428, 435, 436, 440–442
Brooks, E. J. 15
Brown, A. L. 367, 452
Brown, D. A. 156
Brown, E. 297
Bruns, R. R. 288, 290, 292, 309, 310, 313, 314
Brusilow, S. W. 428
Bryhn, U. 30
Buchanan, T. J. 23

Buckley, K. A. 194
Buijs, K. 13
Bunch, W. 156, 161, 174, 175, 188
Burg, M. B. 319, 323, 332, 340, 345, 369–372, 406, 460
Burger, M. B. 429
Burton, A. C. 200
Burton, P. R. 221
Bushey, M. S. 324, 334, 340, 358, 418, 428

Campbell, E. S. 6
Capraro, V. 442
Carasso, N. 363
Carnevale, E. H. 11
Carr, C. W. 324, 334, 340, 358, 418, 428
Case, R. M. 429, 450, 465
Cass, A. 115, 123, 125, 127
Cavert, H. M. 300, 320, 324
Cereijido, M. 374–376
Chamberlain, N. F. 32
Chance, B. 217
Chapman, D. 34
Chapman, G. 31
Chapman-Andresen, C. 218, 219
Chappell, J. B. 215
Cherry, I. 15
Chin, L. 30
Chinard, F. P. 309
Choi, J. K. 363
Choppin, G. R. 13
Churaev, N. V. 14
Churney, L. 212
Cirksena, W. J. 405
Civan, M. M. 343, 363, 364
Clarkson, T. W. 356, 419, 420
Claussen, W. F. 25
Clementi, W. F. 316, 317
Cole, D. F. 400, 429
Cole, K. S. 200, 231
Cole, R. H. 16
Coleman, R. 34
Collander, R. 178, 195, 230, 232, 354
Collie, G. H. 18

AUTHOR INDEX

Connick, R. E. 24
Conway, E. J. 194, 202, 230
Cook, J. S. 196–198
Cope, F. W. 31–34, 169, 170
Cortney, M. A. 405
Coulter, N. A. 296
Courant, R. A. 141
Cowgill, G. R. 429
Crank, J. 159, 174
Crawford, J. D. 284
Creamer, B. 351
Crofts, A. R. 215
Croghan, P. C. 98–101
Crone, C. 304, 307, 308, 315
Cross, M. H. 279
Csaky, T. Z. 378, 381, 385
Cuppage, F. E. 363
Curran, P. F. 40, 41, 55, 58, 60, 61, 63, 97, 143, 148–151, 361, 389, 409, 418, 419, 429, 431–434, 443
Currie, G. D. 308
Curtis, H. J. 231

Dainty, J. 38, 39, 84–86, 98–102, 105, 110, 118–120, 122, 161, 162, 178, 181–183, 188, 229, 237, 240, 244–246, 253, 254, 256, 303, 320–322, 327, 339–341, 346, 355, 379, 382, 383, 408, 450, 451
Danford, M. D. 11
Daniel, J. C. 277, 278, 281, 283, 429
Danielli, J. F. 187
Davenport, H. W. 401
Davey, D. F. 205–207
Davies, H. G. 213
Davies, J. 282–284
Davies, M. 323
Davies, R. E. 401
Davis, L. E. 467
Davson, H. 187, 399, 425, 467
Dawson, I. M. 213
De Groot, S. R. 40
De Luque, O. 199
De Maeyer, L. 18, 19
Denbigh, K. G. 60

Dengel, O. 19, 190
Dennis, V. W. 127, 131, 132, 141, 142
Dennis, W. H. 337, 414, 415
Derjaguin, B. V. 14, 90
Despic, A. 140
DeVries, A. L. 262
Diamond, I. 194
Diamond, J. M. 39, 134, 167, 183, 253, 256, 281, 313, 314, 319, 322, 323, 325–329, 335, 340, 348, 351–359, 367, 378–381, 383, 393, 396, 399, 407, 411, 412, 416, 417, 429, 430, 431, 438, 443–450, 452–461, 463, 466–468, 470
Dianzani, M. U. 214
DiBona, D. R. 343, 363, 364, 385
Dick, D. A. T. 33, 91–92, 165, 169–173, 175–177, 179, 183, 184, 189, 197–199, 205, 218, 221, 222, 295, 296
Dick, E. G. 175–177
Dicker, S. E. 377
Dickerson, J. W. T. 282–284
Dietschy, J. M. 335, 352, 356, 399, 417
DiPolo, R. 115, 117, 123, 125, 145–148
Dirks, J. H. 405
Discala, V. A. 364
Dobson, A. 300, 323, 324, 397, 398
Donahue, S. 292
Doyle, W. L. 218
Dunham, P. B. 425
Durbin, R. P. 37, 83, 115, 123, 124, 134–138, 320, 323, 325, 333, 336–341, 347, 348, 351, 399, 401, 402, 428, 429
Dydyńska, M. 202–206

Earley, L. E. 379
Edelman, I. S. 20, 21, 155, 175, 375, 377
Edney, E. G. 420
Edwards, C. 156, 188, 451

Eigen, E. B. 18, 19, 24
Eisenberg, D. 18, 19
Elias, H-G. 53
Elliott, A. B. 377
Elman, R. 399
Enders, A. C. 469
Engelhardt, W. v. 319, 323–325, 335, 337, 365
Enns, T. 309
Erlij, D. 322
Ernst, J. 208
Ershova, I. G. 14
Essig, A. 92–94, 96, 140
Eucken, A. 9
Evans, M. W. 25, 26
Everitt, C. T. 107, 108, 110, 112, 115, 122, 123, 125
Evett, R. D. 367, 452
Eyring, H. 187

Falk, M. 9
Fanestil, D. 363
Farmer, R. E. L. 165, 188, 239–241, 253, 379
Farquhar, M. G. 290, 292, 325
Favard, P. 363
Faxen, H. 79, 131
Fedyakin, N. N. 14, 89
Feeney, R. E. 262
Feidherr, C. M. 213
Fensom, D. S. 98–101, 244, 245, 247, 250, 252
Fenstermacher, J. D. 296
Fernandez, H. L. 221
Fernandez Moran, H. 29
Ferry, J. D. 81
Fetcher, E. S. 324, 334, 340, 358, 418, 428
Fettiplace, R. 127
Fiat, D. 24
Findlay, G. P. 249, 355
Findlay, N. 393–396, 426
Finean, J. B. 29, 34, 35
Finkelstein, A. 108, 115, 123, 125–127, 131, 132, 134, 138
Finney, J. L. 15

Fisher, R. B. 401–403, 418, 419
Fleming, D. 468
Ford, T. A. 9
Forslind, E. 6, 11
Fortes, P. A. G. 214
Fossat, B. 428
Foster, R. E. 165, 238, 241, 379
Fowler, C. 165
Fowler, R. H. 2, 6, 10
Fragachan, F. 322
Franck, J. 414
Frank, H. 83, 320, 323, 340, 347, 351, 401
Frank, H. S. 10, 11, 13, 18, 22, 24–26
Franki, N. 320, 322, 340–342, 346, 373, 416
Franks, F. 14, 30
Franz, T. J. 142, 143, 324, 366, 367, 378
Frazini-Armstrong, C. 206
Frederikson, O. 411
Freeman, A. R. 175, 209–211, 230–232
Frey-Wyssling, A. 40
Fritz, O. G. 30
Frömter, E. 314, 325–327
Fuchs, G. 38, 323
Fugelli, K. 229
Fujita, H. 249
Fullerton, P. M. 418

Gaar, K. A. 299
Gaffey, C. T. 183
Gainer, H. 202, 229, 232
Galey, W. R. 142–144, 168, 367
Gall, J. G. 213
Ganote, C. E. 332, 333, 345, 369–372, 460
Gardner, J. O. 145
Garlick, D. G. 304
Gary-Bobo, C. M. 115, 117, 123, 125, 127–129, 145–148, 189, 198, 377
Gelernter, G. 6
Gelfan, S. 219

Geoghegan, H. 194
George, J. H. B. 141
Germagnoli, E. 448, 449
Gertz, K. H. 405, 406
Gibbon, J. H. 296
Giebisch, G. 391, 405, 429
Gilman, A. 429
Ginetzinsky, A. G. 344–346
Ginzburg, B. Z. 39, 85–86, 102, 106–108, 114–116, 123, 124, 134, 136–139, 181–183, 188, 237, 240, 253, 254, 256, 303, 355, 379, 426, 427
Ginzburg, H. 426, 427
Girardier, L. 165, 202, 230, 232
Glassman, H. N. 195, 258
Glasstone, S. 187
Gold, G. 324, 428, 429
Goldstein, D. A. 39, 83, 151, 253, 254, 386, 389, 433, 436, 443
Gonzalez, C. F. 441
Good, W. 14
Gordes, E. H. 428
Gordon, J. D. M. 270
Goresky, C. A. 309
Gortner, R. A. 30
Gosselin, R. E. 219
Gottschalk, C. 405
Granger, H. J. 299
Gränicher, H. 17
Grant, E. H. 10
Grantham, J. J. 319, 323, 332, 333, 340, 345, 363, 369–372, 386, 460
Gray, J. 201, 202
Graydon, W. F. 106, 108, 356
Green, J. W. 276
Green, K. 108, 109, 296, 324, 325, 377, 400, 418, 428
Green, M. A. 296, 324
Green, P. B. 200
Green, W. A. 34
Gregor, H. P. 106, 108, 131, 324, 334, 340, 358, 418, 428
Grieve, D. W. 203
Grigera, J. R. 374–376
Grim, E. 305, 319, 320, 323, 334, 340, 411, 429

Gross, E. L. 216
Grotte, G. 309, 310, 317
Grotthuss, C. J. T. 19
Gruman, L. M. 360
Grundfest, H. 165, 175, 202–204, 209–211, 230–232
Guest, G. M. 198
Gutknecht, J. 178, 183, 185
Guyton, A. C. 299

Haas, C. 19, 20
Haggis, G. H. 23
Haglund, B. 199, 200, 227
Hakim, A. 334–336, 360, 402–404, 418, 419, 432
Hamburger, J. 212
Hampton, J. C. 288
Hanai, T. 115, 123, 125
Handler, J. S. 344
Hardé, S. 267
Harding, C. V. 213
Hare, D. K. 378, 385
Harper, A. A. 429, 450, 465
Harris, D. L. 214
Harris, E. J. 91, 207, 214
Harris. F. E. 23
Harris, J. A. 320, 323, 340, 428
Harrison, R. G. 283
Harrison, S. C. 356, 407
Harrop, C. J. F. 398
Harsch, M. 276
Hartley, G. S. 174, 381
Hartline, H. K. 165, 199, 402
Hasted, J. B. 18, 23
Haydon, D. A. 107, 108, 110, 112, 115, 122, 123, 125, 127, 129, 130
Hays, R. M. 115, 117, 123, 125, 145, 146, 320, 322, 324, 340–344, 346, 355, 360, 363, 372–374, 416
Hayward, A. F. 367, 452
Hayward, F. W. 274, 275
Hazlewood, C. F. 32
Hearst, J. E. 30
Hedbys, B. O. 296, 324
Heinen, H. 6
Heinz, E. 415

Heisey, S. R. 319, 323, 340, 361, 398
Held, D. 319, 323, 340, 361, 398
Helder, R. J. 393
Helman, S. E. 345, 369, 370
Hempling, H. G. 165, 188, 189, 236
Hendry, E. G. 198
Henniker, J. C. 14
Henrikson, R. C. 325
Hermans, J. J. 54
Hervey, J. P. 402
Hess, W. N. 429
Hevesy, G. 340, 341
Hider, R. C. 351
Higgins, J. A. 367, 452
Hill, A. V. 202
Hill, D. K. 175, 193, 209, 250
Hillemann, H. H. 283
Hills, G. J. 140
Hilton, P. J. 351
Hingson, D. J. 355
Hirsch, H. R. 92, 93
Höber, J. 347, 361
Höber, R. 30, 347, 361
Hodge, A. J. 216
Hodgkin, A. L. 91, 165, 230
Hodson, S. 429, 468
Höfer, E. 340, 341
Hoffman, J. F. 92, 151, 386, 389, 433, 436, 443
Hogben, C. A. M. 411
Hokin, M. R. 428
Holm-Jensen, I. 183
Holter, H. 219
Holtfreter, J. 200, 213
Holtzman, E. 364
Holz, R. 108, 115, 123, 125, 127, 131, 132, 134, 138
Hoover, S. T. 169
Hope, A. B. 27, 113, 114, 178, 179, 181–183, 216, 237, 242–247, 249, 352, 356, 408
Hopkins, A. L., 31, 32
Hopkins, D. A. 423
Horowicz, P. 165
Hoshi, T. 325
Hoshiko, T. 64, 65, 93–95, 256, 451
Hossack, J. 213
House, C. R. 38, 105, 320–322, 324, 331, 340, 341, 346, 356, 378, 382, 383, 393–396, 409, 426, 428, 435, 440, 442, 450, 451
Howard, E. 275
Huang, C. 115, 118, 123, 125, 126
Huettner, A. F. 264
Huf, E. G. 381, 428, 469
Huggert, A. 30
Hunter, A. S. 213
Hunter, F. R. 199, 213
Huxley, H. E. 207
Hyman, C. 294

Ibsen, H. L. 283
Ingelfinger, F. J. 402
Ingraham, R. C. 418
Intaglietta, M. 296, 299
Isaia, J. 320, 324, 339, 340, 377
Itagaki, K. 190
Itoh, M. 216
Ivanova, L. N. 345
Izaguirre, E. 428
Izawa, S. 216

Jaccard, C. 17
Jacobs, M. H. 161, 163, 195, 223–227, 233, 258
Jacobson, B. 28, 131
Janáček, K. 194, 199–201, 208, 226–230, 363
Jard, S. 363
Jenkinson, T. K. 34
Johansen, B. 143, 367
Johnson, F. R. 367, 452
Johnson, J. A. 39, 223–227, 230, 232, 296, 300, 304, 306, 309, 315, 320, 324
Johnston, C. G., 429
Jordan, D. O. 174

AUTHOR INDEX

Kallsen, G. 161, 174, 175
Kalman, S. M. 202
Kamada, T. 212
Kamat, V. B. 34
Kamb, B. 6, 8
Kamiya, N. 181–183, 237
Kao, C. Y. 230
Karlsson, J-O. 221
Karnovsky, M. J. 292, 310, 311, 314–316, 325
Karreman, G. 157–160, 174
Katchalsky, A. 38, 40, 41, 46, 52–55, 58–61, 63, 66, 67, 69–72, 76, 84–86, 89, 97, 106–108, 114–116, 123, 124, 134, 136–139, 143, 151, 235, 305, 306, 386, 389, 409, 433
Katz, B. 230
Kauzmann, W. 18, 19, 26
Kavanau, J. L. 18, 23
Kawabe, H. 131
Kaye, G. I. 367, 437–439, 442, 452
Kedem, O. 38, 40, 46, 52–55, 58, 59, 63, 65–67, 69–72, 76, 84–86, 89, 92–94, 115, 117, 123, 125, 126, 136–138, 143, 151, 184, 235, 305, 306, 341, 386, 389, 433
Kelly, R. B. 181
Keynes, R. D. 91, 250
Kiley, J. E. 308
King, V. 155, 341
Kirschner, L. B. 468
Kishimoto, U. 252
Kitchener, J. A. 131
Kitching, J. A. 219
Klose, R. M. 391, 405, 429
Klotz, I. M. 22
Knowles, C. D. 127
Kobatake, Y. 248, 249
Koefoed-Johnsen, V. 93, 96, 320, 322, 324, 332, 340, 344, 358, 361, 418
Kohn, P. G. 20, 155, 174, 181
Kokko, J. P. 323
Komatsu, S. K. 262
Kregenow, F. M. 229
Krogh, A. 183, 201, 265, 340, 341

Kuhn, W. 19, 110
Kushmerick, M. K. 175

Ladbrooke, B. D. 34
Lagos, A. E. 131
Lahlou, B. 428
Laidler, K. J. 187
Lakshminarayanaiah, N. 93, 115, 117, 122, 123, 125, 126, 137
Landis, E. M. 37, 38, 294–299, 301, 304, 305
Lane, N. 367, 412, 437–439, 442, 452
Lang, M. A. 202, 229
Larrabee, M. G. 199
Lasek, R. 221
Lassen, N. A. 304
Lassiter, W. E. 405
Lea, E. J. A. 33, 91
Leaf, A. 194, 320, 324, 340, 341, 343–346, 360, 363, 372–374, 416, 417
Lee, J. S. 402, 403, 450
LeFevre, P. G. 33, 198
Lehninger, A. L. 215
Lester, R. G. 403, 418, 419
Levich, V. G. 104
Levitt, D. G. 360
Levitt, J. 257, 259
Levy, H. A. 11
Lewis, P. R. 218, 250, 277, 279
Leyssac, P. P. 406, 411
Lichtenstein, N. S. 344
Lifson, N. 300, 320, 324, 334–336, 360, 403, 404, 418, 419, 432
Liley, A. W. 283
Lillie, R. S. 199
Limbrick, A. R. 34
Lindemann, B. 332, 347, 352, 353, 361, 459
Lindley, B. D. 64, 65, 93–95, 256, 451
Lindström, G. 30
Ling, G. N. 29, 33, 157–160, 169, 174, 208
Litovitz, T. A. 11

Loeb, J. 202
Loeb, S. 117, 145
Loehry, C. A. 351
Loeschke, K. 378, 381, 385
Longuet-Higgins, H. C. 91, 93, 118, 165
Lopez, E. 202–204, 230, 232
Lorimer, J. W. 54
Lovelock, J. E. 259, 260
Løvtrup, S. 160, 174, 199, 200, 227, 266, 267
Lowenstein, L. M. 198
Lucké, B. 38, 163–165, 188, 193, 199, 224, 226, 236
Luft, J. H. 288, 290, 313, 314
Lundberg, A. 325
Lutwak-Mann, C. 277, 279, 280

MacDowell, M. C. 404
Macey, R. I. 165, 188, 216, 239–241, 253, 379
MacGregor, H. C. 213
Machen, T. E. 461
Machin, J. 381
Mackay, D. 106, 108, 140, 245
MacRobbie, E. A. C. 221, 244, 330–332, 345, 365
Maddrell, S. H. P. 428, 450, 467, 468
Madras, S. 90, 190
Maetz, J. 320, 324, 340
Maffly, R. H. 194
Maizels, M. 198
Majno, G. 288, 289, 293, 311
Makhlouf, G. M. 450
Malmberg, C. G. 18
Malnic, G. 391, 405, 429
Malpas, P. 283
Mangos, J. A. 405, 466
Marro, F. 442, 448, 449
Martin, D. W. 416, 417
Martin de Julian, P. 307
Martinez, M. 296
Martinoya, C. 320, 336, 338, 340, 347, 348, 351, 361
Maryott, A. A. 18

Mason, S. G. 90, 190
Mast, S. O. 165, 218, 219
Mateu, L. 322
Matthews, J. 419
Matty, A. J. 418
Maude, D. L. 323, 428
Maurice, D. M. 429, 468
Mauro, A. 47, 86, 93, 115, 117, 118, 120–124, 126, 257, 333
Maxwell, R. 468
Mayer, J. E. 414
Mayerson, H. S. 310, 316, 317
Mazur, P. 257–259
Mazur, S. K. 364
McCafferty, R. E. 283
McCance, R. A. 282–284
McConaghey, P. D. 198
McCormack, J. I. 194
McCutcheon, M. 165, 188, 193, 199, 236
McHardy, W. J. 141
McIntosh, J. R. 148–151, 389, 433
McIntosh, R. L. 90, 190
McLauchlan, K. A. 31
McLean, J. D. 216
McMaster, P. D. 399
McMinn, R. M. H. 367, 452
McSherry, N. R. 466
Meares, P. 106, 108, 140, 141, 245
Meigs, E. B. 207
Meiri, A. 396, 461
Mellon, E. F. 169
Mendoza, S. A. 344
Mercer, E. H. 413
Mercer, F. V. 216
Merriam, R. W. 199, 213
Merten, U. 145
Meryman, H. T. 257, 259–261
Meschia, G. 134
Meves, H. 230
Meyer, H. H. 24
Migchelsen, C. 27, 28
Mikulecky, D. C. 93, 233–236, 239–242, 303
Milgram, E. 428
Miller, D. G. 414, 415
Miller, D. M. 33

Miller, I. F. 131
Miller, K. W. 24
Miller, S. 24, 25
Millington, P. F. 35
Mishima, S. 296, 324
Missoten, L. 413
Moelwyn-Hughes, E. A. 4
Monné, L. 267
Montoreano, R. 165, 196, 253, 254
Moody, F. G. 37, 323, 325, 336–339, 348, 399, 401, 402, 428, 429
Moore, B. 24
Moorti, V. R. G. 6
Morgan, J. 3, 10
Morgan, T. 38, 296, 300, 301, 304, 319, 323, 340, 344, 406
Moses, H. L. 332, 345, 369
Moszynski, J. R. 64
Motais, R. 320, 324, 339, 340, 377
Mueller, P. 118
Muir, A. R. 290
Müller, H. R. 26
Muller, M. 207
Mullins, L. J. 183
Murrish, D. E. 396
Mylle, M. 405

Narten, A. H. 11
Natochin, J. V. 345, 363
Needham, J. 283, 284
Nelson, C. D. 244
Nemethy, G. 12, 13, 26
Nernst, W. 104
Nevis, A. H. 89, 156, 185, 188
New, D. A. T. 275
Nichols, B. L. 32
Nickel, W. 319
Nims, L. F. 387
Nishida, K. 216
Nishizaki, Y. 352
Nobel, P. S. 197, 217
Noble-Nesbitt, J. 420, 430
Nolan, M. F. 309
Northrop, J. 376
Noyes, A. A. 104
Nutbourne, D. M. 404

Öbrink, K. J. 320
Ochsenfeld, M. M. 33, 157–160, 174
Odeblad, E. 30
Ogata, E. 217
Ogilvie, J. T. 149–151, 389, 433
Ohkawa, T. 252
O'Konski, C. T. 23
Oliver, J. 404
Olmstead, E. G. 198
Onsager, L. 20, 40, 42, 93
Orloff, J. 332, 333, 344, 345, 369–372, 379, 406, 460
Ørskov, S. L. 195, 198
Oschman, J. L. 428, 463, 465, 467, 469
Osterhout, W. J. V. 161, 181
Otori, T. 108, 109
Overbeek, J. T. G. 99, 100
Overton, E. 202, 354

Packer, L. 214, 216, 217
Paganelli, C. V. 38, 90–91, 153–156, 157, 161, 162, 190
Page, L. B. 416, 417
Pagel, H. D. 405
Pak, Poy, R. K. F. 363
Palade, G. E. 288, 290, 292, 309, 310, 312–314, 316, 317, 325
Palay, S. L. 291
Palva, P. 183
Pappas, G. D. 218, 219, 292, 399, 412
Pappenheimer, J. R. 38, 40, 78, 80, 294, 296, 297, 301–307, 309, 313–315, 319, 323, 340, 361, 398
Parisi, M. N. 165, 196, 253, 254
Parpart, A. K. 165, 195, 258
Parrish, J. 428, 469
Parsa, B. 378, 385
Parsons, B. J. 419
Parsons, D. S. 412, 418
Paterson, C. R. 418
Patlak, C. S. 143, 144, 151, 168, 367, 386, 389, 415, 433, 436, 443

Paton, D. 296
Paton, W. D. M. 24
Pauling, L. 6, 7, 11, 24, 25
Peachey, L. D. 206, 363
Pederson, J. E. 325, 377, 400
Peers, A. M. 107, 108
Persson, E. 319, 340, 342
Perutz, M. F. 30
Peters, A. 290
Peters, H. C. 418
Petersen, S. 15
Peterson, M. A. 106, 108, 131
Pethica, B. A. 15
Philip, J. R. 179, 180, 229
Phillips, J. E. 378, 421, 422, 429, 430, 463, 469
Phillipson, A. T. 398
Picken, L. E. R. 219
Pidot, A. L. 352, 356
Pike, W. T. 15
Pitts, R. F. 414
Podolsky, R. J. 175
Polestra, F. M. 174
Ponder, E. 32, 197, 198
Porter, K. R. 206
Potter, E. L. 283
Potts, W. T. W. 202
Poulson, R. E. 24
Prather, J. W. 355, 356
Prescott, D. M. 156, 162, 165, 185, 186
Pressman, B. C. 214
Price, H. D. 129, 130, 187, 189, 376
Prigonine, I. 40, 64
Pring, M. 459
Prusch, R. D. 425
Purdy, D. M. 283

Quist, A. S. 10

Raaflaub, J. 214, 215
Ramsay, J. A. 413, 429, 430, 467
Rand, R. P. 200

Rankin, J. C. 320, 324, 340
Rapoport, S. I. 143, 144, 168, 367
Rasio, E. 304
Rasmussen, H. 217, 363
Rastogi, R. P. 60
Raumann, G. 60
Ravdin, I. S. 429
Rawlins, F. 322
Rawson, A. J. 402
Ray, P. M. 118, 120, 122
Redwood, W. R. 107, 108, 115, 129, 130
Reese, T. S. 316, 325
Rehm, W. S. 337, 414, 415, 428
Reid, E. W. 365, 418, 451
Remp, P. G. 308
Renkin, E. M. 38, 78–82, 115, 123, 124, 130–133, 136–138, 141, 294, 296, 301–303, 305–308, 310, 313, 315–317, 347
Rennie, G. K. 15
Reuben, J. P. 165, 175, 202–204, 209–211, 230–232
Reynolds, S. R. M. 283
Rhee, S. W. 296
Ricca, R. A. 199
Rich, G. T. 156, 162, 165, 167, 185, 233–242, 253, 255, 256, 303, 379
Richards, A. N. 404
Riddick, D. H. 220, 413, 422–425, 429
Riegel, C. 429
Riehl, N. 19, 190
Riley, R. L. 145
Ritson, D. M. 18
Roaf, H. 24
Robbins, E. 115, 117, 118, 123, 124, 126
Robinson, C. V. 20, 21, 155, 375, 377
Robinson, J. R. 194, 196, 413
Robinson, R. A. 245
Rolf, D. 198
Romanoff, A. L. 274, 275
Romualdez, A. 167, 233, 236–239, 241, 242, 379

AUTHOR INDEX

Röntgen, W. C. 8
Rothstein, A. 419, 420
Routh, J. I. 282, 284
Rowlinson, J. S. 3
Rudin, D. O. 118
Rudy, P. P. 202
Rugh, R. 264
Rumrich, G. 38, 323
Runnels, L. K. 20
Runstrom, J. 265
Rushton, W. A. H. 296
Ryan, H. C. 194
Ryan, L. A. 207
Rybova, R. 363

Sadler, K. 324, 428, 429
Sage, L. E. 38, 297–299
Sakai, F. 319, 325
Samoilov, O. Ya. 10
Samson, F. E. 221
Sananes, L. 401, 428, 429, 469
Sanders, E. 411
Sato, M. 165
Savitz, D. 33, 198
Schafer, D. E. 304
Schafer, J. A. 141, 142, 322, 360
Scheraga, H. A. 12, 13, 26
Scherrer, P. 17
Scheuplein, R. J. 376
Schilb, T. P. 324, 327, 378, 385, 392, 428, 434, 436, 440–442
Schlögl, R. 44
Schlesinger, H. 337
Schmid, G. 352, 358
Schmidt, C. F. 404
Schmidt, W. J. 219, 220
Schmidt-Nielsen, B. 219, 220, 413, 463, 467, 469
Schmidt-Nielsen, K. 265, 396, 397, 410, 422–425, 428, 467
Schneider, K. 220
Schögel, E. 429
Schrauger, C. R. 219, 220, 413, 422–425
Schultz, J. S. 131
Schultz, S. G. 419

Swt

Schumaker, V. N. 218, 219
Schwartz, I. L. 364
Schwartz, R. 365
Schwarz, H. 352, 358
Scott, W. N. 323
Scratcherd, T. 429, 450, 465
Sellers, A. F. 300, 323, 397, 398
Seshadri, B. 418
Setekleiv, J. 207
Setnikar, I. 134
Sha'afi, R. I. 115, 117, 123, 125, 145–148, 156, 162, 165, 167, 185, 188, 190, 233–242, 253, 255, 256, 303, 379
Shamoo, Y. E. 441
Shapiro, H. 165, 209, 212
Shaw, F. H. 33, 207
Shaw, G. T. 323, 397, 398
Shaw, J. 202, 229
Shea, S. M. 310
Shehadeh, I. 323, 428
Shibata, K. 216
Shirley, H. H. 310, 316, 317
Sidel, V. W. 33, 92, 162, 165, 185, 190, 193, 195, 198, 241, 379
Siegenthaler, P. A. 216, 217
Sigler, K. 194, 199–201, 208, 226–230
Simon, S. E. 33, 207
Singwi, K. S. 21
Sirak, H. D. 214
Sjölander, A. 21
Sjölin, S. 165
Sjöstrand, J. 221
Skadhauge, E. 319, 323, 340, 378, 397, 428
Slack, C. 270
Slegers, J. F. G. 319, 340, 376, 377
Smart, C. 15
Smelser, G. K. 412
Smith, E. B. 24
Smith, G. A. 319, 323, 334, 340, 429
Smith, G. H. 213
Smith, H. W. 404, 405
Smith, M. W. 277, 279
Smith, P. G. 320, 324, 332, 340

Smulders, A. P. 38, 329, 341, 342, 347–351, 352, 358, 368, 369, 374, 376, 378, 380, 385–389, 399, 402, 451, 452
Smyth, D. H. 323, 332, 352, 353, 356, 402, 403, 419, 432, 444
Soberman, R. 373, 374
Soergel, K. H. 320, 323, 340, 428
Solinger, R. E. 441
Soll, A. H. 137
Sollner, K. 143
Solomon, A. K. 33, 38, 39, 83, 86, 90, 91, 115, 117, 123, 125, 127–129, 133–135, 143, 145–148, 153–157, 161, 162, 165, 167, 183, 185, 188, 190, 193, 195, 198, 214, 215, 233–242, 253–256, 303, 320, 323, 332, 340, 347, 351–353, 361, 377, 379, 401, 418, 428, 429, 431, 466
Sorenson, A. L. 156, 161, 165, 175, 185, 212
Soto-Rivera, A. 294, 296, 297
Spanner, D. C. 62, 63, 409
Spiegler, K. S. 85, 140
Stackelberg, M. V. 26
Stacy, B. D. 323, 325, 378
Stadelmann, E. 181
Stallworthy, W. B. 247
Stanier, M. W. 284
Stanton, F. W. 200
Starling, E. H. 293, 294, 297
Staverman, A. J. 52–54, 58, 242
Stein, W. D. 189
Steinbach, H. B. 469
Steinemann, A. 17
Stewart, D. R. 224–226
Stewart, R. J. 106, 108, 356
Stobbart, R. H. 421
Stokes, R. H. 245, 376
Stoner, C. D. 214
Sullivan, W. J. 391, 429
Sussman, M. V. 30
Sutcliffe, J. F. 427
Sutton, A. H. 141
Swanson, C. H. 466
Swift, T. J. 30
Szent-Györgyi, A. 29, 30

Tait, M. J. 30
Talaev, M. V. 14
Talen, J. L. 53
Tanny, G. 141
Taube, H. 24
Tay, D. K. C. 355
Taylor, A. C. 221
Taylor, A. E. 143, 299
Taylor, C. B. 402, 403, 419, 432, 444
Taylor, J. 115, 123, 125
Tazawa, M. 181–183, 237, 352
Tedeschi, H. 214
Tennyson, V. M. 399
Teorell, T. 161, 248, 249
Terner, C. 401
Thain, J. F. 141
Thau, G. 92, 115, 117, 123, 125, 126, 136, 137, 184, 341
Thompson, T. E. 115, 118, 123, 125, 126, 129, 130, 187, 189, 376
Thompson, W. K. 15
Thorlacius, S. O. 300
Thull, N. B. 428
Thurber, R. E. 387
Thürkauf, M. 19
Tien, H. Ti. 118
Tolberg, A. B. 216
Tomlin, S. G. 310
Tormey, J. M. 329, 335, 341, 342, 352, 358, 367–369, 374, 378, 380, 385–389, 399, 402, 412, 413, 438, 449–452
Torrey, T. W. 273
Trap-Jensen, J. 304
Troshin, A. S. 29, 33, 169
Troutman, S. L. 108, 109, 115, 123, 125, 141, 142, 360
Tuft, P. H. 265, 266, 268–270, 280, 281, 422, 429, 469
Tuttle, W. S. 194

Ulfendahl, H. R. 319, 340, 342
Ullrich, K. J. 38, 323
Ursino, D. J. 244

AUTHOR INDEX

Ussing, H. H. 40, 93, 94, 96, 143, 201, 279, 305, 320, 322, 324, 325, 327, 330–332, 340, 344, 345, 358–361, 365–367, 418, 451

Valérien, J. 363
Van Bruggen, J. T. 142–144, 168, 324, 366, 367, 378
Van Dam, K. 214
Van Nie, R. 393
Van Os, C. H. 319, 340, 376, 377
Vargas, F. F. 37, 39, 163, 165–168, 183, 184, 201, 248, 296, 304, 306, 309, 315, 336, 348, 352
Vaughan, B. E. 418, 419
Veatch, W. 240
Vecli, A. 408, 409
Verney, E. B. 296
Vieira, F. L. 188, 190
Villegas, G. M. 156, 165, 185, 211
Villegas, L. 183, 401, 428, 429, 469
Villegas, R. 156, 161, 165, 166, 185, 211, 240, 242, 253, 254
Vinograd, J. 30
Visscher, M. B. 320, 324, 334, 340, 358, 418, 428
Voute, C. L. 331, 365

Wade, J. B. 364
Walker, A. M. 404
Wall, B. J. 421, 463, 465, 469
Wallin, B. G. 163, 165
Walter, J. A. 27
Walter, R. 364
Wang, J. H. 20, 21, 23, 155, 174, 375, 377
Wanless, I. R. 244
Warner, A. C. I. 323, 378
Warner, A. E. 270
Warner, D. T. 28
Warren, B. E. 3, 10
Wartiovaara, V. 178, 183
Wasserman, K. 310, 316, 317
Watson, J. A. L. 421

Watson, M. L. 212
Weatherford, C. 428, 469
Wedner, H. J. 329, 352, 356–358, 407
Weigl, A. M. 127, 131, 132, 141, 142
Weinberg, J. W. 31, 32
Weiss, P. 221
Welch, K. 324, 428, 429
Wells, H. S. 402
Wen, W. Y. 11, 13, 18, 22
Werth, W. 229
Wescott, W. C. 118
Whalen, G. E. 320, 323, 340, 428
Wheeldon, L. 118
Wheeler, H. O. 367, 412, 417, 436–439, 442, 445–449, 452
White, H. L. 198, 295
Whitlock, R. T. 367, 412, 417, 436–439, 442, 445, 449, 452
Whitney, W. R. 104
Whittembury, G. 322, 332, 361
Wicke, E. 24
Widdas, W. F. 188
Wigglesworth, V. B. 462
Wilkie, D. R. 202–206
Williams, R. J. 259, 261, 262
Willis, E. 15
Wilson, T. A. 223–227, 230, 232
Wilson, T. H. 335, 403
Wind, F. 295, 296
Windhager, E. E. 325, 327, 331, 366, 391, 405
Wing, M. 198
Wingate, D. L. 418
Wislocki, G. B. 283, 284
Wohlschlag, D. E. 262
Wolf, A. V. 308
Wolfram, C. G. 310, 316, 317
Worthington, B. N. 15
Worthington, C. R. 35
Wrigglesworth, J. M. 214
Wright, E. M. 38, 39, 134, 253, 256, 313, 323, 325, 326, 329, 332, 341, 342, 347–356, 358, 359, 368, 369, 374, 376, 378, 380, 385–389, 399, 402, 451, 452

Wyburn, G. M. 213
Wyssbrod, H. R. 441

Yamada, E. 367, 452
Yamada, K. 275
Young, J. A. 429
Yudilevich, D. L. 304, 307–309, 314–317

Zadunaisky, J. A. 165, 196, 253, 254, 365
Zatzman, M. 323
Zaun, B. D. 294, 296
Zerahn, K. 279, 415, 451
Zetzel, L. 402
Zeuthen, E. 156, 162, 165, 185, 186, 265
Zotin, A. I. 265, 267, 268
Zweifach, B. W. 296, 299

SUBJECT INDEX

Absorption, see Water absorption and secretion
Action potential
　associated water flow 248–252
　electro-kinetic models of action potential 248, 249
　volume efflux from Chara 250–252
Activation energy
　dielectric relaxation of water 12
　self-diffusion in ordinary ice 19
　self-diffusion in water 12, 21
　solute permeability of gall bladder 349
　viscosity of water 12
　water transport across cell membranes 188
　water transport across cellulose acetate membranes 127–129
　water transport across ciliary epithelium 377
　water transport across gall bladder 376
　water transport across gills 377
　water transport across lipid membranes 129, 130
　water transport across skin 376, 377
　water transport across urinary bladder 372–374
Active water transport
　contractile vacuole 422–425
　energy requirements 414–417
　insect rectum 419–422
　insects 430, 431
　mammalian intestine 418, 419
　plant roots 425–427
　rabbit blastocyst 281, 422

　role in water absorption and secretion 413–431
　Xenopus embryo 268–270
Actomyosin
　'complexing' of sodium ions 34
ADH
　effect on L_p of epithelia 323, 324
　effect on P_d of epithelia 319, 320
　mechanism of action 345, 346
ADP
　volume changes of mitochondria 215
Allantoic cavity
　chick embryo 276
　mammalian embryos 283–285
Amniotic cavity
　chick embryo 275, 276
　mammalian embryos 282, 283
Amoeba
　comparison of L_p and P_d values 185
　contractile vacuole 219, 220
　pinocytosis 218, 219
　values of L_p 165
　values of P_d 156
Amphibian embryos
　active water transport 268–270
　archenteron formation 264–270
　blastocoel formation 263–270
Amphotericin B
　action on toad bladder permeability 344
　effect on lipid membranes 115, 123, 125, 127, 131, 138, 141
Anaesthetics 24, 25
Anomalous solvent drag 143, 367
'Anomalous water' 14, 15, 90

SUBJECT INDEX

Antidiuretic hormone
  effect on L_p of epithelia 323, 324
  effect on P_d of epithelia 319, 320
  mechanism of action 345, 346
Aplysia neuron 165
Aqueous humour
  formation 399, 400
  rabbit ciliary epithelium 429
Arbacia egg 163, 165, 188, 199, 200, 224–226, 258 (*see also* Egg)
Archenteron
  formation in amphibian embryos 264–270
Argon (liquid) 3
Arrhenius equation 186, 376
Artemia gills 320, 324, 340
Artificial membranes
  activation energy for water transport 127–130
  comparison of L_p and P_d values 124–127
  reflexion coefficients 134–138
  restricted diffusion of solutes 130–134
  'solute drag' 143
  solvent drag 141–144
  ultrafiltration 132–134
  values of L_p 123
  values of P_d 114
Asymmetrical double-membrane model *see* Composite membranes *and* Curran's double-membrane model
ATP
  volume changes of chloroplasts 216, 217
  volume changes of mitochondria 215
Axon
  activation energy for water transport 188
  anomalous osmotic behaviour 230–232
  comparison of filtration and osmosis 166–169
  comparison of L_p and P_d values 185

cytoplasmic streaming 221
electro-osmosis 247
electro-osmotic permeability 247
equivalent pore radius 254
'osmotic dead space' 211
osmotic relations 209–211
possible effect of osmolarity on L_p 167
possible heteroporous nature 168
reflexion coefficients 254
streaming potentials 248
'transport number effect' 247, 248
values of L_p 165
values of P_d 156
Axoplasmic flow 221

'Backwards channels' in epithelia, 466–468
Barnacle muscle 156, 188
Basement membrane
  capillary walls 288
  continuous capillaries 292
  discontinuous capillaries and sinusoids 293
  fenestrated capillaries 292
  possible hydraulic resistance in epithelia 386
Bird, *see* Chicken, Cormorant, Duck, Goose *and* Petrel
Bjerrum defects
  ordinary ice 16, 17, 19
Blastocoel
  amphibian embryo 263–270
  mammalian embryo 277–282
Blastocyst
  active water transport 281, 422
  osmolarity of absorbate 429
  rate of water uptake 278, 429
  water and salt transport 277–282
Blood
  colloid osmotic pressure 295, 396, 397
Blood-brain barrier
  solute permeabilities 304
  values of L_p 296

SUBJECT INDEX

Boyle-van't Hoff law
  osmotic relations of cells 193, 194
Brain
  NMR study of sodium ions 33
  NMR study of water 31, 32

Capillary
  basement membrane 288, 292
  endothelial cells, 290, 292, 293
  filtration 293-300
  gap junctions 292
  large-pore system 309
  osmosis 297-299
  reflexion coefficients 306
  restricted solute diffusion 303, 315, 316
  route of solute and water transport 311-316
  small-pore system 309
  solute permeabilities 304
  structure 288-293
  values of L_p 296
  values of P_d 300, 301
  vesicles in endothelial cells 290
  vesicular transport 309-311
Cat
  blood-brain barrier 296
  corneal endothelium 296
  erythrocyte 156, 165
  muscle capillaries 296, 304
  pancreas 429, 450, 465
  sublingual gland 325
  sweat gland 428
Cellulose acetate membranes
  activation energies for water transport 127-129
  comparison of L_p and P_d values 125
  reflexion coefficients 147, 148
  solute permeabilities 146, 147
  structure 145
  values of L_p 123, 145, 146
  values of P_d 115, 117, 145, 146
Cerebrospinal fluid, *see* Choroid plexus

Chara
  action potential and water efflux 250-252
  electro-osmosis 243-247
  electro-osmotic permeability 244
  rectification of osmotic flow 237
  reflexion coefficients 253, 254
  thickness of cytoplasmic layer 183
  'transport number effect' in electro-osmosis 244
  values of L_p 182, 183
  values of P_d 178
Chick embryo
  allantoic cavity 276
  amniotic cavity 275, 276
  extra-embryonic coelom 276
  fluid dynamics 271-276
  sub-blastodermic cavity 274, 275
Chicken
  cloaca, 319, 323, 340, 378, 428
  embryo 271-276
  erythrocyte 165, 188, 241
Chick heart fibroblast,
  value of L_p 165
Chloroplasts,
  water relations 216-218
Choroid plexus
  failure of pressure to alter secretory rate 398
  osmolarity of secretion 428
  rate of secretion 428
  value of L_p 324, 326
Ciliary epithelium
  activation energy for water transport 377
  osmolarity of secretion 429
  rate of secretion 429
  role of ultrafiltration in secretion 399, 400
  value of L_p 325
Clathrate
  'anaesthetics' 24, 25
  structure of water 11
  Xenon 24

SUBJECT INDEX

Cloaca
  comparison of L_p and P_d values 340
  mechanism of water absorption 396, 397
  osmolarity of absorbate 428
  rate of water absorption 428
  value of L_p 323
  value of P_d 319
Collagen
  hydration 27, 28
Collecting ducts
  comparison of L_p and P_d values 340
  route of water transport 369–372
  value of L_p 323
  value of P_d 319
Colloid osmotic pressure
  water transport across capillary wall 295
  water transport across cloacal wall 396, 397
Colon, see Intestine
Composite membranes
  artificial membranes 144–151
  cellulose acetate membranes 145–148
  Curran's double-membrane model, experimental examination 148–151
  membranes in parallel 66–69
  membranes in series 70–76
  see Curran's double-membrane model
Conductance
  epithelia 325–327
  phenomenological description 57
Continuous capillaries
  endothelial cells 290
  gap junctions 292
  restricted solute diffusion 303, 315, 316
  route of solute and water transport 311–316
  structure 290–292
  tight junctions 290–292, 313, 314
  values of L_p 296
  vesicles 290
  vesicular transport 309–311
Contractile vacuole
  active water transport 422–425
  associated vesicles 413
  osmolarity of secretion 429
  rate of fluid secretion 429
  significance in cell water relations, 219–221
Cormorant nasal salt gland, 428
Corneal endothelium
  osmolarity of secretion 429
  rate of fluid secretion 429
  structure of endothelial cells 468
  value of L_p 296
Corneal epithelium
  value of L_p 324
Cow
  erythrocyte, 156, 165, 185, 188, 241, 256
  rumen 319, 323, 397, 398
Crab
  muscle 156, 161, 165, 185, 229
  nerve 250
Crayfish
  axon 165
  muscle 165
Crocodile distal tubule 467
Cryoprotection 259, 260
Cubic ice 7, 8
Curran's double-membrane model
  experimental test 148–151
  possible mechanism for water transport in rabbit gall bladder 436–440
  possible mechanism for water transport in turtle urinary bladder 440–442
  rectification of osmotic flow in epithelia 386–389
Cuttlefish axon 193, 209, 250
Cytoplasmic streaming
  axons 221
  Nitella 221

SUBJECT INDEX

Dielectric relaxation time
  ordinary ice 16
  water 12
Diffusion
  mutual diffusion coefficient 174
  mutual diffusion of water and macromolecules inside cells 170–174
  restricted diffusion coefficient 79
  restricted diffusion of solutes in membranes 78–83, 130–134, 303, 315
  self-diffusion coefficients for water 21
  self diffusion of water in cells 157–161, 174
  water diffusion in unstirred layers 104–106, 161, 162, 300, 301, 321, 322
Diffusional permeability to water,
  activation energy, 128, 188, 373–377
  animal cells 152–156
  artificial membranes 114–118
  compared to L_p values for animal and plant cells 184–186
  compared to L_p values for artificial membranes 124, 125
  compared to L_p values for epithelia 339–343
  compared to L_p values for porous membranes 87–89
  compared to L_p values for vasa recta 300–302
  epithelia 319–322
  influence of unstirred layers 110, 116, 161, 162, 300, 321, 322
  method of measurement 38, 152–155
  plant cells 178, 179
  vasa recta 300
Diffusion coefficient, see Mutual diffusion coefficient, Restricted diffusion coefficient, Self-diffusion coefficient

Discontinuous capillaries
  basement membrane 293
  route of solute and water transport 316, 317
  structure 293
Dissipation function
  active transport systems 64
  definition 41
  transport of heat and water 60
  transport of ions and water 55
  transport of uncharged solutes and water 46, 47
Distal tubule
  comparison of L_p and P_d values 340
  osmolarity of reabsorbate 428
  rate of fluid reabsorption 428
  value of L_p 323
  value of P_d 319
Dog
  blood-brain barrier 304
  erythrocyte 156, 165, 185, 188, 190, 191, 237–239, 241, 255, 256
  gall bladder 319, 323, 340, 429
  gastric mucosa 320, 323, 336–339, 340, 347, 361, 400–402, 428
  glomerulus 296
  intestine 319, 320, 323, 334–336, 340, 360, 402, 403, 404, 418, 419, 428
  muscle capillaries 296, 304, 315
  urinary bladder 320, 324
Double-membrane model, see Curran's double-membrane model.
Duck erythrocyte 229
Duodenum, see Intestine

Egg
  activation energy for water transport 188
  comparison of L_p and P_d values 185
  internal pressure of *Arbacia* 200
  osmotic relations 199–202

Egg—*cont.*
  predicted internal pressures of frog eggs 200
  values of L_p 165
  values of P_d 156
  'water-hardening' in fish eggs 201, 202
Ehrlich ascites tumour cell
  activation energy for water transport 188
  value of L_p 165
Electrical conductance
  definition 57
  epithelia 325–327
Electro-osmosis
  axons 247–248
  epithelia 356
  gall bladder 356–358
  influence of unstirred layers 113, 114
  influence on turgor pressure of plant cells 408
  mechanism of water transport in epithelia 407, 408
  particular models applied to porous membranes 98–102
  phenomenological description 54, 97–102
  plant cells 242–247
  plant cell walls 244
  'transport number effect' 114, 244, 247, 248, 356–358
Electro-osmotic permeability
  axons 247
  definition 57, 58
  derived from frictional model 101
  ion-exchange membranes 141
  plant cell, *Chara* 244
Embryo
  active water transport in rabbit blastocyst 281
  active water transport in *Xenopus* 268–270
  allantoic cavity 276, 283–285
  amniotic cavity 275, 276, 282, 283
  archenteron 264–270
  blastocoel 263–270, 277–282
  blastocyst 277–282
  chick 271–276
  extra-embryonic coelom 276, 283
  mammal 276–285
  sub-blastodermic cavity 274, 275
  Xenopus 265–270
Endothelial cells
  continuous capillaries 290
  cornea 468
  discontinuous capillaries 293
  fenestrated capillaries 292, 293
  vesicles 290
  vesicular transport 309–311
Ependyma
  comparison of L_p and P_d values 340
  value of L_p 325
  value of P_d 319
Epithelia, *see* Proximal tubule, Gall bladder, etc.
Equivalent pore radius
  capillary walls 305, 309
  comparison of estimates for artificial membranes 136
  comparison of estimates for epithelia 361
  comparison of estimates for erythrocytes 256
  gall bladder 350, 368
  gastric mucosa 347, 348
  intestine 347
  reflexion coefficients of cellulose membrane 135
  restricted solute diffusion in cellophane membrane 130, 131
  restricted solute diffusion in lipid membrane 132
  theory of determination from comparisons of L_p and P_d 89, 91
  ultrafiltration in artificial membranes 134

SUBJECT INDEX

Erythrocyte
 activation energy for water transport 188
 comparison of L_p and P_d values 185
 equivalent pore-radius 254–256
 internal pressure 200
 NMR studies 30
 osmometric method for determining solute permeabilities 232–234
 osmotic relations 196–199
 Ponder's 'R', 197, 198
 possible effects of osmolarity on L_p 237–240
 ratios of L_p to P_s 195
 rectification of osmotic flow 237–240
 reflexion coefficients 253–256
 values of L_p 165
 values of P_d 156
Extra-embryonic coelom
 chick embryo 276
 mammalian embryo 283
Eye, see Ciliary epithelium, etc.

Fenestrated capillaries
 basement membrane 292
 route of solute and water transport 316, 317
 size of fenestrae 293
 structure 292, 293
 values of L_p 296
Fibroblast
 value of L_p 165
Filtration
 axon 166
 capillary wall 293–300
 compared with osmosis across axon 166–169
 compared with osmosis across capillary 298, 299
 compared with osmosis across epithelia 333–339
 role in fluid secretion by epithelia 399, 400, 400–402

Filtration permeability, P_f
 relation to hydraulic conductivity L_p 88
Fish
 egg 156, 165, 185, 201, 202
 erythrocyte 165, 229
 gall bladder 319, 323, 326, 340, 355, 378, 399
 gills 320, 324, 339, 340, 377
 intestine 323, 324, 428
 muscle 30
 rectal salt gland 429
 skin 324
 urinary bladder 428
'Flickering cluster' model of water 11–13, 18
Flux-ratio test
 for labelled water transport across porous membranes 95–97
 for solute transport across porous membranes 93–95
'Forward channels' in epithelia 452–461, 462–465, 468
Freezing injury to cells 259–262
Frictional coefficients
 artificial membranes 139
 ion-exchange membranes 140, 141
 relation to L_p, ω_s and σ_s 138–140
Frictional model of transport
 applied to comparisons of water permeabilities 92, 93
 applied to electro-osmosis 101, 102
 applied to solute and water transport across cellulose membranes 138–140
Frog
 choroid plexus 326, 355, 356
 egg 156, 160, 165, 185, 199–201, 226–229
 embryo 264
 gastric mucosa 320, 323, 326, 340, 347, 401, 428, 429, 450, 469

Frog—cont.
  glomerulus 296
  intestine 355, 378, 385
  mesentery 296
  muscle 165, 203–207, 207–209, 254
  nerve 30, 209
  skin 320, 321, 322, 324–326, 330–332, 340, 356, 365–367, 374–376, 378, 381–383, 408, 409, 428, 435, 442, 468, 469
  urinary bladder 320, 324, 340

Gall bladder
  activation energy for solute permeation 349
  comparison of L_p and P_d values 340
  effect of temperature on water permeabilities 376
  electro-osmosis 356–358
  equivalent pore radius 350
  estimated salt concentration in intercellular spaces 461
  intercellular spaces 452–461
  local osmosis 443–450
  osmolarity of reabsorbate 429
  possible effect of osmolarity on L_p 379–381
  rates of fluid reabsorption 429
  rectification of osmotic flow 378, 387, 388
  reflexion coefficients 351–355
  route of solute and water transport 367–369
  solute permeabilities 348–351
  'standing-gradient osmotic flow' 452–461
  streaming potentials 352, 353, 356–358
  values of L_p 323
  values of P_d 319
  water reabsorption and Curran's double-membrane model 436–440

Gap junctions
  continuous capillaries 292, 314
  ruminal epithelium 325
Gastric mucosa
  comparison of L_p and P_d values 340
  effect of pressure of passive water flow 336–339
  equivalent pore radius 347, 348
  evidence for local osmosis 450
  osmolarity of secretion 428, 429, 450
  rates of secretion 428, 429
  role of ultrafiltration in 'alkaline secretion' 400–402
  values of L_p 323
  values of P_d 320
Gills
  comparison of L_p and P_d values 339, 340
  effect of temperature on water permeabilities 377
  value of L_p 324
  value of P_d 320
Gland, see Salivary gland, etc.
Glomerulo-tubular balance 404–406
Glycoproteins, role in cryoprotection 262
Goat
  choroid plexus 398
  ependyma 319, 323, 340, 361
  rumen 323, 337, 365, 366
Goose nasal salt gland 428
Gramicidin S, hydration 28
Grotthuss mechanism 19
Guinea-pig
  gall bladder 355, 429
  intestine 355

Haemoglobin,
  osmotic behaviour of erythrocytes 33, 198, 199
  osmotic coefficient 198
Hamster intestine 335

SUBJECT INDEX

Heat of transfer
  definition 61, 62
  relation to Q_{10}, 62
Heavy water
  diffusion coefficient 21
  transport across frog skin 341
  use in NMR studies 31
Helmholtz-Onsager relation 358
Helmholtz-Smoluchowski model of electro-osmosis in porous membrane 98–100
Henle's loop
  comparison of L_p and P_d values 340
  role in glomeruro-tubular balance 405
  values of L_p 323
  values of P_d 319
Horse erythrocyte 165
Human
  erythrocyte 156, 165, 185, 188, 190, 191, 193, 197, 223–236, 237–239, 241, 254, 255, 256, 258, 259–261
  intestine 320, 323, 340, 402, 428
  leucocyte 165
  muscle capillaries 296, 304
  skin 376
  submandibular salivary duct 326
Hydration
  anaesthetics 24, 25
  collagen 27, 28
  DNA 30
  gramicidin S 28
  hydrocarbons 25, 26
  ions 22–24
  lifetimes of ionic hydration 24
  macromolecules 26–28
  myelin 34, 35
  proteins 30, 32
  silk fibroin 28
  tobacco mosaic virus protein 28
Hydraulic conductivity
  activation energy 128, 188, 373–377
  animal cells 162–169
  artificial membranes 123

cellulose acetate membranes 123, 145, 146
compared to P_d values for animal and plant cells 184–186
compared to P_d values for artificial membranes 124
compared to P_d values for capillaries 301
compared to P_d values for cellulose acetate membranes 146
compared to P_d values for epithelia 339–343
compared to P_d values for lipid membranes 125
compared to P_d value for porous membrane 87–89
definition 50
endothelial walls 296
epithelia 322–329
ion-exchange membranes 123
lipid membranes 123
membranes in parallel 67
membranes in series 70–73
methods of measurement 37, 38
plant cells 179–184
Hydride molecules 1, 2
Hydrogen bond
  energy 5, 6
  'flickering cluster' model of water 11
  role in water structure 5, 6
  solute transport across cellulose acetate membranes 147
  water transport across cellulose acetate membranes 128, 129
Hydrophobic bond 26

Ice (*see also* Ordinary ice)
  cubic ice 7
  densities 8
  different forms 7, 8
  high pressure polymorphs 7, 8
  hydrogen bonds 8
  latent heat of fusion 9
  latent heat of vaporization 9

Ice—cont.
  ordinary ice 6, 7
  self-diffusion coefficient 19
  viscosity 16
  vitreous ice 7
Ileum, see Intestine
'Indicator-diffusion' method for determining solute permeabilities of capillaries 307, 308
'Influx-profile' analysis of water exchange in cells, 157–159
Insect
  active water transport 419–422, 430
  (see also Insect rectum, etc)
Insect Malpighian tubule
  evidence for local osmosis 450
  osmolarity of secretion 428, 429
  rates of fluid secretion 428, 429
Insect rectum
  active water transport 419–422
  evidence for 'standing gradient osmotic flow', 462–465
  osmolarity of reabsorbate 429
  rate of fluid reabsorption 429
  rectification of osmotic flow 378
  solute recycling 463–465
Insect salivary gland
  osmolarity of secretion 428
  rate of fluid secretion 428
Intercellular junctions, see Tight junctions, Gap junctions, etc.
Intercellular spaces
  changes in width during osmotic rectification in gall bladder 385–389
  changes in width during reabsorption 452
  estimated salt concentration in gall bladder 461
  estimated salt concentration in insect rectum 463
  'forward channels' in gall bladder 452–461
  role in Curran's double-membrane model applied to gall bladder 438–440
  role in passive water transport across gall bladder 367–369
  role in passive water transport across renal collecting ducts 369–372
  role in passive water transport across skin 365–367
  role in passive water transport across urinary bladder 363, 364
  significance in proximal tubular reabsorption during saline loading 405
Intercellular transport route, see Ultrastructural transport route and Shunt conductance
Interstitial model of water 10, 11
Intestine
  active water transport 418, 419
  comparison of L_p and P_d values, 340
  effect of pressure on passive water and solute movement 334–336
  effect of pressure on power of fluid reabsorption 402–404
  evidence for local osmosis 450
  osmolarity of absorbate 428
  pinocytosis, 412
  rates of fluid absorption 428
  rectification of osmotic flow 378
  relation between active salt flux and passive water flow 419, 420
  values of L_p 323, 324
  values of P_d 319, 320
Ion-exchange membranes
  comparison of L_p and P_d values, 124–126
  electro-osmotic permeabilities 141
  frictional coefficients 140, 141
  values of L_p 123
  values of P_d 115
Ionic hydration
  action of ions on water structure, 22–24

ionic hydration lifetimes 24
primary hydration numbers 245
Ions
    active transport coupled to passive water flow 419, 431, 443
    interaction with water 22–24
    primary hydration numbers 245
    transport path across epithelia 325–327

Jejunum *see* Intestine

Keatite
    relation to water structure 10
Kidney
    action of ADH on renal collecting ducts 369–370
    glomerulo-tubular balance 404–406
    mechanism of action of ADH on epithelia 345, 346
    see also Henle's loop, Renal collecting ducts, etc.

Lattice energy of ice 5
Leucocyte
    value of L_p 165
Lifetime
    'flickering cluster' 12
    ionic hydration 24
    period of molecular vibration 24
    proton-water complex 19
Lipid membranes
    activation energy for water transport 129, 130
    comparison of L_p and P_d values 125–127
    equivalent pore radius 132, 138
    reflexion coefficients 138
    restricted solute diffusion 131, 132
    solvent drag 141, 142
    treatment with Amphotericin B 115, 123, 125, 127, 131, 138, 141
    treatment with nystatin 115, 123, 125, 127, 131, 138
    values of L_p 123
    values of P_d 115
Lizard cloaca 396, 397
Lobster axon, 185, 209–211, 230–232
Local osmosis
    evidence in gall bladder 443–450
    evidence in pancreas 465
    mechanism for water transport 443
    significance of unstirred layers 450–452
    'standing-gradient osmotic flow' 452–461

L_p
    activation energy 128, 188, 373–377
    animal cells 162–169
    artificial membranes 123
    cellulose acetate membranes 123, 145, 146
    compared to P_d values for animal and plant cells 184–186
    compared to P_d values for artificial membranes 124
    compared to P_d values for capillaries 301
    compared to P_d values for cellulose acetate membranes 146
    compared to P_d values for epithelia 339–343
    compared to P_d values for lipid membranes 125
    compared to P_d value for porous membrane 87–89
    definition 50
    endothelial walls 296
    epithelia 322–329
    ion-exchange membranes 123
    lipid membranes 123
    membranes in parallel 67
    membranes in series 70–73
    methods of measurement 37, 38
    plant cells 179–184

Macula adhaerens of epithelia 290–292

Macula occludens
  continuous capillaries 292, 314
  ruminal epithelium 325

Malpighian tubule
  evidence for local osmosis 450
  osmolarity of secretion 428, 429
  rates of fluid secretion 428, 429

Mammalian embryos
  active water transport 281
  allantoic cavity 283–285
  amniotic cavity 282, 283
  blastocoel 277–282
  blastocyst 277–282
  extra-embryonic coelom 283
  fluid dynamics 276–285

Mesentery
  values of L_p 296
  values of solute permeability 304

'Minimum cell volume' hypothesis, ability of cells to withstand low temperatures 259–262

Mitochondria
  ADP and volume changes 215
  ATP and volume changes 215
  'osmotic dead space' 214
  Ponder's 'R' 214, 215

Mouse
  gastric mucosa 320

Muscle (cardiac)
  NMR studies 30
  osmotic relations 208

Muscle (skeletal)
  activation energy for water transport 188
  comparison of L_p and P_d values 185
  equivalent pore-radius 254
  NMR studies 30–32
  osmotic relations 202–207
  reflexion coefficients 254
  sarcoplasmic reticulum 206, 207
  transverse tubules 206, 207
  values of L_p 165
  values of P_d 156

Muscle (smooth)
  osmotic relations 207–209

Mutual diffusion
  role in osmosis in animal cells 170–175
  role in osmosis in capillaries 295, 296
  role in osmosis in plant cells 184

Mutual diffusion coefficient 174

Nasal salt gland
  osmolarity of secretion 428

Necturus
  distal tubule 323, 428
  proximal tubule 323, 326, 378
  gall bladder 325

Nerve
  activation energy for water transport 188
  anomalous osmotic behaviour 230–232
  comparison of filtration and osmosis 166–169
  comparison of L_p and P_d values 185
  cytoplasmic streaming 221
  electro-osmosis 247
  electro-osmotic permeability 247
  equivalent pore radius 254
  NMR studies 30, 31
  'osmotic dead space' 211
  osmotic relations 209–211
  possible effect of osmolarity on L_p 167
  possible heteroporous nature 168
  reflexion coefficients 254
  streaming potentials 248
  'transport number effect' 247, 248
  values of L_p 165
  values of P_d 156

Newt proximal tubule 325

Nitella
  activation energy for water transport 188

comparison of L_p and P_s 195
cytoplasmic streaming 221
electro-osmosis 244, 245
rectification of osmotic flow 237
reflexion coefficients 253, 254
thickness of cytoplasmic layer 183
transcellular osmosis 181, 182
unstirred layer effect on P_d 178
values of L_p 181, 183
value of P_d 178
Nitellopsis
thickness of cytoplasmic layer 183
unstirred layer effect on P_d 178
value of L_p 183
value of P_d 178
NMR studies
hydrated collagen 27, 28
hydrated silk fibroin 28
state of sodium in cells 33, 34
state of water in cells 30–32
Non-linear osmosis
Curran's double-membrane model applied to epithelia 386–389
epithelia 378–389
erythrocytes 237–240
flow-induced deformations of permeability barrier 385, 386
influence of unstirred layers 383, 384
plant cells 237
role of intercellular spaces 385, 386, 387, 388
theoretical treatment for membranes in series 70–74
'Non-solvent' volume *b*
axon 211
definition 193
eggs 199
erythrocytes 196
mitochondria 214
muscle 205–206
'Non-solvent' water
cells 32–34, 193
mitochondria 214
(*see also* 'Non-solvent volume').

Nuclear magnetic resonance studies
hydrated collagen 27, 28
hydrated silk fibroin 28
state of sodium in cells 33–34
state of water in cells 30–32
Nucleus
osmotic relations 212–214
'pore-complex' of nuclear membrane 213, 214

Onsager reciprocal relations 43
Oocyte 175–177, 199–202, 226–229
Ordinary ice
activation energy for self-diffusion 19
Bjerrum defects 16, 17
d.c. conductivity 17
dielectric constant 16
dielectric relaxation time 16
proton mobility 19
self-diffusion 19, 20
structure 6, 7
Osmolarity
effect on L_p of axon 167
effect on L_p of epithelia 379–383
effect on L_p of erythrocytes 237–240
effect on L_p of plant cells 237
fluid secretions and absorbates in epithelia and other tissues 428, 429
Osmosis
across animal cell membranes 162–166
across capillaries 297–299
across epithelia 322–329
across plant cell membranes 179–182
compared with diffusion across animal and plant cell membranes 185
compared with diffusion across capillary walls 301, 302
compared with diffusion across epithelia 339–343

Osmosis—cont.
  compared with filtration across animal and plant cell membranes 166–169
  compared with filtration across capillary walls 293–300
  compared with filtration across epithelia 333–339
  in presence of permeant solutes 223–230
  its nature in porous membranes 118–124
  mechanism for water absorption and secretion in epithelia 391, 398
  mechanism for water absorption in cloaca, 396, 397
  mechanism for water absorption in plant roots 393–396
  mechanism for water absorption in rumen 397, 398
  see also Local osmosis, Non-linear osmosis, Osmotic relations of cells and Transcellular osmosis
'Osmotic dead space' b
  axon 211
  definition 193
  eggs 199
  erythrocyte 196
  mitochondria 214
  muscle 205–206
Osmotic permeability P_{os}
  relation to hydraulic conductivity L_p 88
Osmotic relations of cells
  axons 209–211
  Boyle-van't Hoff law 193, 194
  cardiac muscle 208
  changes in cellular hydration and Bradley isotherm 169, 170
  chloroplasts 216–218
  eggs 199–202
  erythrocytes 196–199
  in presence of permeant solutes 223–230
  mitochondria 214, 215
  nucleus 212–214
  osmometric method for determining solute permeabilities 232–234
  'osmotic dead space' b 193
  role of internal diffusion of water 170–174
  skeletal muscle 202–207
  smooth muscle 207–209
'Osmotic-transient' method
  determination of solute permeabilities of capillary walls 302, 303

Pancreas
  evidence for local osmosis 465, 466
  osmolarity of secretion 429, 450, 465
Paramaecium 220
P_d
  activation energy 128, 188, 373–377
  animal cells 152–156
  artificial membranes 114–118
  compared to L_p values for animal and plant cells 184–186
  compared to L_p values for artificial membranes 124, 125
  compared to L_p values for epithelia 339–343
  compared to L_p values for porous membranes 87–89
  compared to L_p values for vasa recta 300–302
  epithelia 319–322
  influence of unstirred layers 110, 116, 161, 162, 330, 321, 322
  method of measurement 38, 152–155
  plant cells 178, 179
  vasa recta 300
Petrel nasal salt gland 428
P_f (Filtration permeability),
  relation to hydraulic conductivity L_p 88

Phenomenological equations
  active transport systems 63
  general form 42–44
  membranes in parallel 66
  transport of heat and water 60
  transport of ions and water 54
  transport of uncharged solutes and water 48
Pig chorio allantoic membrane 284
Pinocytosis
  capillary wall 309–311
  epithelia 410–413
  significance for cellular water relations 218, 219
Plant cell wall
  elastic properties 180, 181
  electro-osmosis 244
  estimate of L_p 179
Plant cells, see Nitella, Chara, etc.
Plant roots
  active water transport 425–427
  'standing-gradient osmotic flow' 396, 461
  water absorption 393–396
Ponder's 'R'
  erythrocyte 197, 198
  index of 'non-solvent water' in cells 32, 33, 197
  mitochondria 214, 215
  muscle 204
Pore-radius
  capillary walls 305, 309
  comparison of estimates for artificial membranes 136
  comparison of estimates for epithelia 361
  comparison of estimates for erythrocytes 256
  gall bladder 350, 368
  gastric mucosa 347, 348
  intestine 347
  reflexion coefficients of cellulose membrane 135
  restricted solute diffusion in cellophane membrane 130, 131
  restricted solute diffusion in lipid membrane 132
  theory of determination from comparisons of L_p and P_d 89, 91
  ultrafiltration in artificial membranes 134
Porous membranes
  diffusional and viscous flows of water 86–93
  electro-osmosis 97–102
  reflexion coefficient 83–86
  restricted diffusion coefficient 79
  restricted solute diffusion 78–83
  solvent drag 93–97
  unidirectional flux ratio for solutes and labelled water 93–97
P_{os} (Osmotic permeability)
  relation to hydraulic conductivity L_p 88
Potassium ions, primary hydration number 245
Pressure
  effect on gastric mucosa 336–339
  effect on intestine 334–336
  effect on rumen 337, 365, 366
  effect on skin 365
  effect on water absorbing power of intestine 402–404
  internal pressure of Arbacia egg 200
  internal pressure of erythrocyte 200
  predicted internal pressures of frog oocytes 200
  role in water absorption and secretion 220, 398–407
  turgor pressure of plant cells 177, 178, 408
Proton
  mobility 19
  proton-water complex 19
Proximal tubule
  comparison of L_p and P_d values 340
  glomerulo-tubular balance 404–406

Proximal tubule—*cont.*
  osmolarity of reabsorbate 429
  rate of fluid reabsorption 429
  rectification of osmotic flow 378
  values of L_p 323
  values of P_d 319
P_s
  capillary walls 302–309
  compared to L_p values for erythrocytes and *Chara* 195
  definition 50, 51
  epithelia 346–351
  estimated from osmotic behaviour of cells 230–234
  gall bladder 348–351
  membranes in parallel 68, 69
  membranes in series 73–75
  methods of measurement 38, 51, 230–234, 348

Q_{10}
  relation to activation energy 187
  relation to heat of transfer 62
Quartz, relation to water structure, 10

Rabbit
  blastocyst 277–282, 422, 429, 469
  blood-brain barrier 296
  choroid plexus 324, 428
  ciliary epithelium 325, 377, 399, 400, 429
  corneal endothelium 296, 429, 468
  corneal epithelium 324
  gall bladder 319, 323, 325, 326, 328, 329, 335, 340, 348–351, 352–355, 356–358, 367–369, 376, 378, 379–381, 385, 388, 407, 408, 411, 417, 429, 436–440, 444–450, 452, 461
  Henle's loop 323
  intestine 418
  leucocyte 165
  mesentery 296
  muscle capillaries 296, 304, 306
  nerve 31
  pancreas 466
  proximal tubule 323, 406
  renal collecting ducts 319, 323, 332, 333, 340, 369–372, 460
Random net model of water 10
Rat
  brain 31, 32
  distal tubule 319, 323, 340
  erythrocyte 34
  Henle's loop 319, 323, 340
  intestine 323, 326, 332, 347, 353, 356, 361, 402, 403, 418, 419, 429, 432, 450
  mesentery 304
  muscle 31, 32
  muscle capillaries 296
  pancreas 466
  proximal tubule 319, 323, 326, 340, 391, 405, 406, 429
  renal collecting ducts 319, 323, 340, 344
  submandibular gland 326
  submaxillary gland 429
  vasa recta 296, 300, 301, 304
Reabsorption, *see* Water absorption and secretion
Rectal salt gland
  osmolarity of secretion 429
Rectification of volume flow
  Curran's double-membrane model applied to epithelia 386–389
  epithelia 378–389
  erythrocytes 237–240
  flow induced deformations of permeability barrier 385, 386
  influence of unstirred layers 383, 384
  plant cells 237
  role of intercellular spaces 385, 386, 387, 388
  theoretical treatment for membranes in series 70–74

SUBJECT INDEX

Rectum
  active water transport 419–422
  evidence for 'standing gradient osmotic flow' 462–465
  osmolarity of reabsorbate 429
  rate of fluid reabsorption 429
  rectification of osmotic flow 378
  solute recycling 463–465
Red cell
  activation energy for water transport 188
  comparison of L_p and P_d values 185
  equivalent pore-radius 254–256
  internal pressure 200
  NMR studies 30
  osmometric method for determining solute permeabilities 232–234
  osmotic relations 196–199
  Ponder's 'R', 197, 198
  possible effects of osmolarity on L_p 237–240
  ratios of L_p to P_s 195
  rectification of osmotic flow 237–240
  reflexion coefficients 253–256
  values of L_p 165
  values of P_d 156
Reflexion coefficient
  artificial membranes 134–138
  axon 253, 254
  capillaries 306
  cellulose acetate membranes 147, 148
  Chara 253, 254
  definition 52, 53
  erythrocytes 253–255
  lipid membranes 138
  membranes in parallel 68
  membranes in series 75, 76
  methods of measurement 38, 39
  muscle 253, 254
  Nitella 253
  porous membranes 83–86

Renal collecting ducts
  comparison of L_p and P_d values 340
  route of water transport 369–372
  value of L_p 323
  value of P_d 319
Restricted diffusion coefficient
  definition 79
  for water in cellulose acetate membranes 146
  see also Restricted solute diffusion
Restricted solute diffusion
  artificial membranes 130–134
  capillary walls 303, 315, 316
  diffusion coefficient definition 79
  epithelia 346, 347
  lipid membranes 131
  porous membranes 78–83
Rumen
  effect of pressure on passive water transport 337
  value of L_p 323
  value of P_d 319
  water absorption 397, 398

Salivary gland
  changes in osmolarity of saliva 465
  osmolarity of secretion in insect 428
  osmolarity of secretion in rat 429
  rate of secretion in insect 428
Salt glands, *see* Nasal salt gland and Rectal salt gland
Salt permeability; definition 57
Sarcoplasmic reticulum, behaviour in osmotic experiments on muscle 206, 207
Schmid model for elctro-osmosis in porous membranes 100, 101
Sea urchin egg 163, 165, 188, 199, 224–226, 258 (*see also* Egg)
Secretion, *see* Water absorption and secretion

Self-diffusion coefficient for water
 activation energy 21
 inside cells 157–161, 174
 values 21
Self-diffusion of water 20, 21
Sepia axon 193, 209, 250
Sheep
 erythrocyte 188
 rumen 319, 323, 378
Short-circuit current
 pig chorio-allantoic membrane 284
 rabbit blastocyst 279
Shunt conductance of epithelia 327
Sieve coefficient
 cellulose acetate membranes 145
 cellulose membranes 133
Silk fibroin, hydration 28
'Single-file' model for water transport 91–93
Sinusoids
 basement membrane 293
 route of solute and water transport 316, 317
 structure 293
Skin
 activation energy for water transport 374–377
 comparison of L_p and P_d values 340
 electro-osmosis 356
 osmolarity of absorbate 428, 469
 rate of fluid absorption 428
 rectification of osmotic flow 378, 381–383
 route of passive water transport 365–367
 solvent drag 358–360
 unstirred layer effect on P_d 321, 322
 values of L_p 324
 values of P_d 320
Sodium ions
 'complexing' in tissues 33, 34
 primary hydration number 245
'Solute drag' 143, 367

Solute permeability
 capillary walls 302–309
 compared to L_p values for erythrocytes and *Chara* 195
 definition 50, 51
 epithelia 346–351
 estimated from osmotic behaviour of cells 230–234
 gall bladder 348–351
 membranes in parallel 68, 69
 membranes in series 73–75
 methods of measurement 38, 51, 230–234, 348
Solvent drag
 artificial membranes 141–144
 epithelia 358–360
 heteroporous membrane 143, 144
 influence of unstirred layers 113, 141–143, 360
 lipid membranes 141, 142
 porous membranes 93–97
 renal collecting ducts 345
Squid axon 156, 165, 166–169, 185, 188, 193, 209–211, 230–232, 247, 248, 254
'Standing-gradient osmotic flow'
 'backwards channels' in epithelia 466–467
 'forward channels' in gall bladder 452–461
 'forward channels' in insect rectal pads 462–465
 pancreas 465, 466
 plant roots 396, 461
Starling's hypothesis 293
Stomach
 comparison of L_p and P_d values 340
 effect of pressure of passive water flow 336–339
 equivalent pore radius 347, 348
 evidence for local osmosis 450
 osmolarity of secretion 428, 429, 450
 rates of secretion 428, 429

SUBJECT INDEX

role of ultrafiltration in 'alkaline secretion' 400–402
values of L_p 323
values of P_d 320
Streaming potential
axons 248
definition 98
gall bladder 352, 356–358
predicted size in *Nitella* 247
unstirred layer effect 114, 356
Sturgeon embyro 265
Sub-blastodermic cavity in chick embryo 274, 275
Sublingual gland 325
Submandibular gland 326
Submaxillary gland 429
Sweat gland, osmolarity of secretion 428

Temperature
solute transport across gall bladder 349
survival at low temperatures 257–262
water transport across animal and plant cells 186–191
water transport across artificial membranes 127–130
water transport across capillaries 297
water transport across cellulose acetate membranes 128
water transport across epithelia 372–378
water transport across lipid membranes 129, 130
Thermo-osmosis
phenomenological description 60
possible mechanism for water transport 285, 408–410, 430
Tight junctions
continuous capillaries 290–292, 313, 314
epithelia 290–292, 325, 326, 351, 452
role in passive solute and water transport across gall bladder 368
role in passive solute and water transport across renal collecting ducts 369–372
role in passive solute and water transport across skin 366
role in passive solute and water transport across urinary bladder 364
Toad
intestine 355
mesentery 296
oocyte 175–177
skin 320, 324, 332, 340, 359, 360, 361, 377
urinary bladder 320, 324, 326, 340, 344, 346, 360, 362–365, 372–374, 378, 379, 385, 391, 416, 417
Tobacco mosaic virus protein, hydration 28
Tolypellopsis, see *Nitellopsis*
Transcellular osmosis 181, 182
Transport number 58
'Transport number effect'
axons 247, 248
description 114
gall bladder 356–358
plant cells 244
Transverse tubular system of muscle, osmotic experiments 206, 207
Tridymite, relation to water structure 6, 10
Trophoblast cells 277
Tumour cell
activation energy for water transport 188
value of L_p 165
Turgor pressure of plant cells
measurement 200
possible influence of electroosmosis 408
significance in osmotic relations 177, 178

Turtle urinary bladder 324, 326, 378, 385, 392, 393, 428, 436, 440–442

Ultrafiltration
  artificial membranes 132–134
  ciliary epithelium 399, 400
  gastric mucosa 400–402
Ultrastructural transport route
  coupled active salt and water transport across epithelia 325–327, 438, 439, 452, 453
  passive water transport across gall bladder 367–369
  passive water transport across renal collecting ducts 369–372
  passive water transport across skin, 365–367
  passive water transport across urinary bladder 362–365
  solute and water transport across continuous capillaries 311–316
  solute and water transport across fenestrated and discontinuous capillaries 316, 317
Unstirred layer
  definition of thickness 104
  equivalent permeability coefficient 105
  estimates of thickness 106–109
  influence on electro-kinetic experiments 113, 114, 243, 244, 247, 248, 356–358
  influence on L_p measurements 110–112, 327–329
  influence on P_d measurements 110, 161, 162, 300, 321, 322
  influence on P_s measurements 110, 348, 349
  influence on rectification of osmotic flow in epithelia 383, 384
  influence on σ measurements 112, 113

  influence on solvent drag experiments 113, 141–143, 360
  significance as a mechanism of local osmosis 450–452
Urinary bladder
  activation energy for water transport 372–374
  comparison of L_p and P_d values 340
  effect of ADH 344–346
  effect of amphotericin B 344
  osmolarity of reabsorbate 428
  rate of fluid reabsorption 428
  rectification of osmotic flow 378
  route of passive water transport 362–365
  solvent drag 360
  values of L_p 324
  values of P_d 320
  water reabsorption and Curran's double-membrane model 440–442

Valonia
  absence of aqueous pores 185
  comparison of filtration and osmosis 183
  comparison of L_p and P_d values 185
  thickness of cytoplasmic layer 185
  value of L_p 183
  value of P_d 178
Vasa recta
  comparison of L_p and P_d values 301, 302
  solute permeabilities 304
  value of L_p 296
  value of P_d 300
Vesicles
  amoeba 218, 219
  associated with contractile vacuole 413
  ciliary epithelium 413
  endothelial cells 290

gall bladder 411, 412
intestine 412
solute transport across capillary wall 309–311
Viscosity
activation energy for water 12
value for ice 16
value for water 16
Vitreous ice 7, 8

Water absorption and secretion
active water transport 418–431
coupling to active ion transport 419, 420, 431, 443
Curran's double-membrane model 431–436
effect of pressure on intestine 402–404
effect of pressure on proximal tubule 404–406
electrolytes actively transported with water 428, 429
osmolarity of absorbates and secretions 428, 429
rates of fluid transport 428, 429
significance of electro-osmosis 407, 408
significance of local osmosis 442–450
significance of osmosis 391–398
significance of pinocytosis 410–413
significance of thermo-osmosis, 408–410
significance of ultrafiltration 398–402
'standing gradient osmotic flow' 452–469
Water
activation energy for self-diffusion 21
activation energy for viscosity 12
'anomalous water' 14, 15, 90
d.c. conductivity 18, 19
dielectric relaxation time 18
hydrogen bond 5, 6
proton mobility 19
self-diffusion 20, 21
self-diffusion coefficients 21
static dielectric constant 18
structure 8–14
Water diffusion
activation energy for self-diffusion 12
activation energy for water diffusion across cell membranes 188
activation energy for water diffusion across cellulose acetate membranes 127–129
activation energy for water diffusion across epithelia 372–377
mutual diffusion coefficient 79
mutual diffusion of water and macromolecules inside cells 170–174
restricted diffusion coefficient 79
restricted water diffusion in cellulose acetate membranes 146
self-diffusion coefficient 21
self-diffusion of water inside cells 157–161, 174
water diffusion in unstirred layers 104–106, 161, 162, 300 301, 321, 322
see also P_d
'Water-hardening' of fish eggs 201, 202
Water molecule
dipole moment 4
hydrogen bonds 5, 6
'jump and wait' mechanism of self-diffusion 21
period of molecular vibration 24
proton-water complex 19
structure 4–6
Water permeabilities, see Hydraulic conductivity and Diffusional permeability to water
Water structure
clathrate model 11
'flickering clusters' 11–13

Water structure—*cont.*
  hydrogen bonds 5, 6, 9, 10
  interstitial model 10, 11
  random net model 10

Xenon as an anaesthetic 24
Xenopus
  active water transport 268–270
  egg 156, 165, 185
  embryo 265–270
  skin 320, 324, 340
X-ray diffraction studies of isolated membranes 34, 35
Zonula adhaerens
  epithelia 290–292

Zonula occludens
  continuous capillaries 290–292, 313, 314
  epithelia 290–292, 325, 326, 351, 452
  role in passive solute and water transport across gall bladder 368
  role in passive solute and water transport across renal collecting ducts 369–372
  role in passive solute and water transport across skin 366
  role in passive solute and water transport across urinary bladder 364